PLEASE STAMP DATE DUE, BOTH BELOW AND ON CARD

DATE DUE	DATE DUE	DATE DUE	DATE DUE
NOV 1986			
AUG 1986			
Reserved 10-8-6 AY218			
Res AY 218 9-88			
JAN 3 0 1988			
SEP 2 3 1989			
NOV 1 3 1989			
MAY 2 0 1990			
DEC 2 8 1990			

GL-15(3/71)

ACTIVE GALACTIC NUCLEI

edited by
J. E. Dyson

Manchester University Press

Published by Manchester University Press
Oxford Road, Manchester M13 9PL, UK
and 51 Washington Street, Dover,
New Hampshire 03820, USA

British Library cataloguing in publication data

Active galactic nuclei.
  1. Nuclear astrophysics
  I. Dyson, J.E.
  523.1'12   QB857

Library of Congress cataloging in publication data

Active galactic nuclei.

   1. Galactic nuclei--Congresses. I. Dyson, J. E.
(John Edward), 1941-   .
QB857.A36   1985   523.1'12   84-27277

ISBN 0 7190 1097 7

Printed in Great Britain
by Unwin Brothers Limited
The Gresham Press, Old Woking, Surrey
A member of the Martins Printing Group

CONTENTS

PREFACE

The observational investigation of activity in the nuclear regions of
galaxies and the elucidation of the physical phenomena underlying this
activity together form a major area of present day astronomical research
effort. It is clear that this area will remain exceptionally important
and active in the forseeable future. Bernard Pagel's apt paraphrasing
of George Orwell - 'all galactic nuclei are active but some are more
active than others' - is indicative of the enormous scope of the subject.
Perhaps equally indicative of its present state is that one of the most
important problems is the classification of the various types of activity
and the association of the various types with specific classes of galaxy.
What is very evident is that this can only be done by observing galactic
nuclei over the entire accessible range of the electromagnetic spectrum.
The exciting potential of γ-ray astronomy in this regard is extremely
significant.

It was particularly appropriate to hold a conference on active galactic
nuclei in Manchester in 1984 for two reasons discussed in detail in Cyril
Hazard's fascinating introductory article to these proceedings. Firstly,
quasars came of age in 1984, at least in the sense of their reaching the
twenty-first birthday of the announcement of their discovery. Secondly,
much of the pioneering research leading to their discovery took place at
Jodrell Bank then, as now, part of the University of Manchester. It is
only fair to admit that these two important considerations were pointed
out to the organizer who had decided on the subject matter of the conference through the application of self-indulgence.

The very gratifying response to the early announcement of the Manchester
Conference - plus the equally gratifying submission of manuscripts - has
led to the present volume in which practically every paper presented at
the conference is reproduced. The conference was organized around invited
review papers associated with more specialized presentations. The order
of presentation at the conference was more-or-less as reproduced here.
It corresponded roughly to increasing photon energy with theory papers
in their (perhaps rightful) position at the rear. The discussions following the papers have not been reproduced for two main reasons. Firstly,
they often are immediate reactions to presentations which on reflection
speakers wish to amend. Secondly, at least in the organizer's experience,
the most interesting and useful comments tend to be made under circumstances where accurate recording is impossible on technical and/or social
grounds. However, the very wide and thorough range of the contributions
presented here should provide considerable food for thought.

I would like to thank publically all the speakers both for their
attendance at the conference and for submitting their manuscripts with
the very minimum application of carrots or sticks. I am particularly
grateful to the writers of review articles who committed themselves in
advance to an onerous task. These proceedings could not have been
published without the enthusiastic support of Alec McAuley of the
Manchester University Press. Most of the organisational work devolved
onto the staff of the Astronomy Department and as usual, Mrs Ellen
Carling and Miss Joanne Suthers were towers of help and efficiency.
Finally, I would like to thank Franz Kahn who offered every encouragement
for this conference.

J E Dyson

R Antonucci	National Radio Astronomy Observatory, Charlottes-ville, USA
P N Appleton	University of Manchester, UK
D J Axon	Nuffield Radio Astronomy Laboratory, Jodrell Bank, UK
L Baars	Nuffield Radio Astronomy Laboratory, Jodrell Bank, UK
M E Bailey	University of Manchester, UK
P Barr	EXOSAT Observatory, ESA, Darmstadt, W Germany
L Bassani	Istituto Tesre CNR, Bologna, Italy
R D Blandford	California Institute of Technology, Pasadena, USA
P W J L Brand	Edinburgh University, UK
G Branduardi-Raymont	Mullard Space Science Laboratory, Dorking, UK
C Brindle	Hatfield Polytechnic, UK
L Brown	Preston Polytechnic, UK
P F Browne	University of Manchester Institute of Science and Technology, UK
G Burbidge	Kitt Peak National Observatory, Tucson, USA
E Capriotti	Ohio State University, Columbus, USA
T J Carroll	Oxford University, UK
R F Carswell	Institute of Astronomy, Cambridge, UK
S K Chakrabarti	Enrico Fermi Institute, Chicago, USA
C J Clarke	Oxford University, UK
M J Coe	Southampton University, UK
C S Coleman	Oxford University, UK
L Colina	University Observatory, Gottingen, W Germany
R Conway	Nuffield Radio Astronomy Laboratory, Jodrell Bank, UK
A A da Costa	Instituto Superior Tecnico, Lisbon, Portugal
R D Davies	Nuffield Radio Astronomy Laboratory, Jodrell Bank, UK
R Davis	Nuffield Radio Astronomy Laboratory, Jodrell Bank, UK
A J Dean	Southampton University, UK
A I Díaz	Royal Greenwich Observatory, Herstmonceux, UK
M J Disney	University College, Cardiff, UK
W Durodie	University of Manchester, UK
J E Dyson	University of Manchester, UK
I Evans	Mount Stromlo Observatory, ACT, Australia
A C Fabian	Institute of Astronomy, Cambridge, UK
S A E G Falle	Leeds University, UK
I Fejes	Max Planck Institute for Radio Astronomy, Bonn, W Germany
W K Gear	Preston Polytechnic, UK
R Gilmozzi	Royal Greenwich Observatory, Herstmonceux, UK
G Giovannini	Istituto di Radio Astronomia, Bologna, Italy
A Gower	University of Victoria, Canada
P Grimley	University College, Cardiff, UK
J James	University of Manchester, UK
A W Jones	University of Manchester, UK
B Harrison	Nuffield Radio Astronomy Laboratory, Jodrell Bank, UK
C Hazard	University of Pittsburgh, USA
R A James	University of Manchester, UK
M Jones	Institute of Astronomy, Cambridge, UK
W Junor	Nuffield Radio Astronomy Laboratory, Jodrell Bank, UK

F D Kahn	University of Manchester, UK
C Keable	University of Edinburgh, UK
P Kronberg	University of Toronto, Canada
S Kumar	Institute of Astronomy, Cambridge, UK
A Lawrence	Royal Greenwich Observatory, Herstmonceux, UK
J P de S E Lemos	Institute of Astronomy, Cambridge, UK
J A Lopez	University of Manchester, UK
M Maisack	University of Tübingen, W Germany
F Mantovani	Istituto di Radio Astronomia, Bologna, Italy
J M Marcaide	Max Planck Institute for Radio Astronomy, Bonn, W Germany
J Mardaljevic	University of Leicester, UK
P G Martin	University of Toronto, Canada
J Meaburn	University of Manchester, UK
L Mestel	University of Sussex, Falmer, UK
D L Moss	University of Manchester, UK
P K Moore	University of Manchester, UK
S L Morris	Institute of Astronomy, Cambridge, UK
D Murphy	Nuffield Radio Astronomy Laboratory, Jodrell Bank, UK
T Muxlow	Nuffield Radio Astronomy Laboratory, Jodrell Bank, UK
C O'Dea	National Radio Astronomy Observatory, Charlottesville, USA
L Padrielli	Istituto di Radio Astronomia, Bologna, Italy
B E J Pagel	Royal Greenwich Observatory, Herstmonceux, UK
A Pedlar	Nuffield Radio Astronomy Laboratory, Jodrell Bank, UK
M V Penston	Royal Greenwich Observatory, Herstmonceux, UK
E Perez	Royal Greenwich Observatory, Herstmonceux, UK
I Pérez-Fournon	Max Planck Institute for Radio Astronomy, Bonn, W Germany
J J Perry	Institute of Astronomy, Cambridge, UK
M A C Perryman	ESTEC, Noordwijk, Netherlands
R W Porcas	Max Planck Institute for Radio Astronomy, Bonn, W Germany
D Radcliffe	Leeds University, UK
T P Ray	University College, Dublin, Eire
A Reid	Nuffield Radio Astronomy Laboratory, Jodrell Bank, UK
A Robinson	University of Manchester, UK
E I Robson	Preston Polytechnic, UK
H-J Roeser	Max Planck Institute for Astronomy, Heidelberg, W Germany
A Sansom	Royal Greenwich Observatory, Herstmonceux, UK
R Schlickheiser	Max Planck Institute for Radio Astronomy, Bonn, W Germany
P Schneider	Max Planck Institute for Astrophysics, Munich, W Germany
F G Smith	Nuffield Radio Astronomy Laboratory, Jodrell Bank, UK
M G Smith	Royal Observatory, Edinburgh, UK
W Sparks	University of Sussex, UK
L Spinoglio	University of Southampton, UK
D Stannard	Nuffield Radio Astronomy Laboratory, Jodrell Bank, UK
P Stewart	University of Manchester, UK

P Stephens   Nuffield Radio Astronomy Laboratory, Jodrell
                         Bank, UK
R Swinney   Nuffield Radio Astronomy Laboratory, Jodrell
                         Bank, UK
A F Tennant   Institute of Astronomy, Cambridge, UK
P Tomasi   Nuffield Radio Astronomy Laboratory, Jodrell
                         Bank, UK
S Unger   Nuffield Radio Astronomy Laboratory, Jodrell
                         Bank, UK
M Ward   Institute of Astronomy, Cambridge, UK
R Warwick   University of Leicester, UK
K A West   University of Manchester, UK
D A Whitehouse   Mullard Space Science Laboratory, Dorking, UK
P A Whitelock   South African Astronomical Observatory, South
                         Africa
N Whyborn   Nuffield Radio Astronomy Laboratory, Jodrell
                         Bank, UK
M J Wilson   Leeds University, UK
I Wilson   Royal Greenwich Observatory, Herstmonceux, UK
G Zamorani   Istituto di Radio Astronomia, Bologna, Italy
J A Zensus   Max Planck Institute for Radio Astronomy, Bonn,
                         W Germany

THE COMING OF AGE OF QSOs

C. Hazard

Department of Physics and Astronomy,
University of Pittsburgh, Pittsburgh, PA.

Visiting Fellow, Institute of Astronomy,
Madingley Road, Cambridge, UK.

SUMMARY

An account is given of the background and motivation behind the occultation observations which led to the identification of the radio source 3C273 and the recognition of a new class of extra-galactic object the QSOs. Some preliminary results are then presented of a survey carried out compiling in selected regions complete samples of objects with active galactic nuclei. These results suggest that even at magnitudes brighter that B = 19 the generally accepted values for QSO densities may be underestimates or alternatively that there is significant clustering on a scale of at least several degrees. The results further imply that the low densities of QSOs with $z > 3.5$ are not the result of a decrease at this redshift but of a decline which sets in around $z = 2$. Attention is drawn to the wide variety of QSO spectra and the possible important role of BAL QSOs at early epochs. It is emphasized that QSOs cannot be considered as passive objects but that their role as modifiers of their environment must be taken into account.

THE COMING OF AGE OF QSO'S

It is particularly appropriate that this conference on Active Galactic Nuclei should be held in Manchester at this time because 1984 marks the coming of age of this subject which has dominated astrophysics for the past twenty one years and it was here that it really all began. However, since the first papers describing the identification of 3C273 and its recognition as what we now know as a quasi-stellar object, or QSO, appeared in Nature in March 1963, the key role played by the occultation technique developed at Jodrell Bank appears to have been largely forgotten and so a little of the history and background of this work would seem to be appropriate. A reconstruction of scientific history from published papers alone is often difficult because they so rarely contain much, if any, of the relevant background. Furthermore, a significant discovery usually allows earlier work to be reinterpreted so that in later attempts to reconstruct a logical story, this earlier work assumes a significance which it did not at that time possess.For instance it would appear from some accounts that the determination of the position of 3C273 was merely a lucky accident, the moon one night quite fortuitously passing over this particularly interesting object. Nothing could be further from the truth. The occultation studies of 3C273 were the culmination of some five years of work aimed at identifying the so called Class II radio sources but the motivation for which went back to around 1953 when it had become clear, at least at Jodrell Bank, that these sources must be extra-galactic.

When the radio sources were first discovered and catalogued in the late 1940's they appeared to provide the solution to a problem which had plagued Radio Astronomy since its earliest beginnings, namely how to explain the background emission from the Galaxy which had been discovered by Jansky in 1931. The effective black body temperature of this emission of several hundred thousands of degrees at low frequencies ruled out an origin in thermal emission in the interstellar gas which only in the neighbourhood of the hot early type stars reaches a temperature as high as 10,000 K. The discovery of the radio sources led to the suggestion that the background radiation from the Galaxy represented the unresolved radiation from a large number of such sources widely distributed throughout the Galaxy and with a density in the Solar neighbourhood similar to that of the common stars (Smith 1950, Bolton and Westfold 1951). So intractable was the problem of the background radiation that this interpretation was widely adopted in spite of any direct evidence in its support. In particular the sources could not be identified with any object or class of visual objects in the Galaxy with the sole exception of a source in Taurus which Bolton and Stanley (1949) identified with the Crab Nebula. However, their isotropic distribution showed that it was unlikely that the sources as a whole could be identified with such rare bodies as the remnants of supernova explosions and it was generally concluded that they must represent the radiation from some hitherto unknown type of object, probably stellar and common throughout the Galaxy but exceedingly faint in the visible region.

An alternative explanation of the radio sources allowed by their isotropic distribution, namely that they lay not in the galaxy but were extra-galactic objects, received little favour in spite of the identification of two of the stronger sources with the peculiar galaxies NGC 4486 (M87) and NGC 5128 by Bolton, Stanley and Slee (1949) and even after a few of the nearer galaxies had been detected as radio sources. The very strong arguments still supporting the Galactic hypothesis were very ably presented by Ryle at a conference on the Dynamics of Ionized Media held at University College London in 1951. However, at this conference dissenting voices were raised, notably those of Gold (1951) and Hoyle (1951) who argued that, while it was admittedly difficult to explain the sources as extra-galactic, it was at least equally difficult to explain them as galactic and that both possibilities should be borne in mind. In particular they pointed out that if the sources were galactic they must radiate 10^8 times more powerfully than the Sun, the only star detected at radio frequencies, whereas if the radiation from the Galaxy were assumed to arise in some unexplained mechanism in the interstellar gas it was necessary to postulate an extra-galactic nebula radiating only 10^3 times more powerfully than the Galaxy to explain the most intense sources.

A partial clarification of the problem came with the recognition in 1952 that there were in fact two classes of radio sources (a) Class I sources which comprised the more intense sources and showed a strong concentration to the galactic plane and clearly represented a rare type of extra-galactic object, probably supernova remnants and (b) the Class II sources comprising the weaker sources and were isotropically distributed. (Mills 1952) The galactic versus extra-galactic argument now centred on the class II sources but with the difference that if the Class I sources could explain by themselves a large fraction of the emission from the Galaxy, then the most compelling reason for assuming the Class II sources to be also galactic would no longer be valid. In fact a model of the radio emission from the Galaxy by R. Hanbury-Brown and myself (1953) suggested that this was indeed the case. The identification of NGC 4486 and NGC 5128 had already established the existence of "peculiar" galaxies which are much more powerful radio emitters than normal galaxies. The identification of the second strongest source in the sky, Cygnus A, with a 17.5 galaxy (Baade and Minkowski 1954) showed that the poor correlation of the Class II sources with the brightest galaxies could no longer be taken to rule out their extra-galactic nature. Even at ten times its distance, Cygnus A would still be an intense radio source but so faint visually as to be at the then limits of the Palomar 5m telescope. However, as far as I was concerned the most compelling evidence that the Class II sources were extra-galactic was provided by considering their space density and absolute intensity on this assumption and the further assumption that their integrated emission was responsible for the radio emission observed towards the galactic poles. It was easy to show that this emission could be explained by assuming that the space density of the radio sources was similar to that of the extra-galactic nebulae and their mean absolute intensity similar to that for the late type spirals but with a dispersion about the mean of between 2 and 3 magnitudes, or of course a lower density and higher mean luminosity. I am not arguing that this was the commonly accepted view but the above are a summary of the arguments I employed in my thesis in 1953 and presented at a Radio Astronomy Conference at Jodrell Bank in July 1953 and which were reported in Observatory in October of that year. However, the extra-galactic hypothesis seemed so preferable to the hypothesis requiring a common type of galactic object radiating 10^8 times more powerfully than Sun that when I left Jodrell Bank to do my National Service in 1954, I had the feeling that it was widely accepted. It was only on my return in 1957 that I discovered to my surprise that not only had this not been the case (see the recent history by Smith and

Lovell 1984) but also that little further progress had been made in identifying the Class II sources. It was against this background and with the realization that positional accuracies of around 1 arc second might be required to identify the possibly faint optical counterparts of the sources that the occultation technique was developed. When a year or so later I requested a large amount of telescope time on the newly constructed 250 ft Mk 1 telescope one of my arguments to Professor Sir Bernard Lovell was that, although the yield in numbers of sources might be small, even the identification of two or three of them would be of great significance.

The early occultation studies were aimed at studying both predicted occultations and attempting to detect any sources which happened to pass behind the Moon during the period of observation. The technique was at first difficult to apply because digital recording techniques which were later to simplify it were then in their infancy and because the available positions of the strong radio sources were so inaccurate that predicted occultations failed to materialize. However, the observations did result in the measurement of the position of 3C212 (an object which later turned out to be a QSO) to an accuracy of about 3 arc sec (Hazard 1962) However, shortly after this observation which demonstrated the feasibility of the technique I left Jodrell Bank to work on the Hanbury Brown – Twiss interferometer as a member of the Chatterton Astrophysics Department of the University of Sydney. Shortly after taking up this position I gave a lecture on the occultation work at Sydney University, following which I was invited by Dr E. Bowen the director of the CSIRO Radiophysics group and Dr J. Pawsey the head of the radio astronomy group to continue the occultation programme on the new 210 ft steerable telescope at Parkes, NSW. It thus occured that I came to drive regularly the 250 miles from Narrahi to Parkes to use the technique developed at Jodrell Bank. As a guest observer I was fortunate to have Brian Mackey as the Radiophysics collaborator who ensured the smooth working of all the equipment.

The first major success of the Parkes occultation studies occurred in April 1962 with the observation of the immersion of the radio source 3C273, which occurred just as the source was passing outside the field of view of the telescope where the telescope tracking had to be stopped. However, enough of the occultation curve was observed to show the presence of a strong lobe pattern. Up to this time 3C273 had no particular interest but the detection of the lobe pattern showed it to be of small angular size (< 6 arc secs) and that if a second occultation predicted for August 1962 could be observed then both a very accurate position and detailed structural information could be obtained. When I pointed out to Dr J. Bolton (the new head at Parkes after the untimely death of Dr Pawsey) the importance of these upcoming observations, and that the emersion was likely to occur near the limits of the telescope, he took great precautions to ensure their success, including modifications to the telescope to extend the tracking limits. In the event there was time to spare and the whole of the emersion was observed. There was great excitement among all in the control room as the time of occultation neared and the pen on the chart recorder began to trace out the oscillating diffraction pattern as the source neared the Moon's limb. There was even further excitement at emission when the 410 MHz occultation curve showed clear evidence of a double radio source. This double structure was shown even more distinctly in a third occultation observed in October of that year when the higher resolution 1420 MHz observations and a more favourable alignment of the source relative to the Moon's limits showed the two components to have different angular sizes and spectra. The problem now was to derive from these records (see Figure 1) by a process of model fitting the best possible structure for the source and the most accurate positions for its components which, among other things, required corrections for irregularities in the Moon's limb. The calculations of the positions of the Moon's limits at each of my estimated times of immersion and emission wre carried out by Dr W. Nicholson of H.M. Nautical Almanac office and from these I finally estimated the positions of each component to about 1 arc sec. A first preliminary position had just been obtained when I met Dr R. Minkowski who was in Sydney at the time and who immediately produced a copy of a 200 inch plate on which 3C273 had been identified as a galaxy some minutes of arc from my new position. An offset from this galaxy showed that the new position lay close to a bright 13 mag. star and a faint patch of nebulosity which Minkowski likened to an edge-on galaxy. Over the next few months I carried out a much more detailed analysis which finally yielded the very accurate component positions, their angular sizes and spectral indices.

While I was still in the process of refining these measurements, Dr Bolton asked me if he could send the new accurate position to Dr M. Schmidt. Not at that time having any experience of optical work and knowing personally no optical observers, I agreed and it was thus that M. Schmidt became the first person to receive the information on probably the one object which at that time could have led to the interpretation of the spectra of what later came to be called QSOs. It should be noted that this occultation should have turned up a QSO identification is not surprising. 3C212 the source observed previously at Jodrell Bank is now also known to be a QSO as were most of the sources observed in the subsequent occultation studies: some 30% of all the brightest sources are QSOs. Rather the good fortune was that 3C273 was such a nearby object that it was bright and the Balmer series lay in the visible region of the spectrum. In any case some time after this communication to M. Schmidt I received a letter from him in which he informed me of the determination of the redshift of the 13 mag. stellar object and suggesting that we publish our papers simultaneously in Nature (Hazard, Mackey and Shimmins 1963, Schmidt 1963).

It will be recognised from the above account the key role played by the occultation technique developed at the Jodrell Bank station of the University of Manchester in the recognition of this new class of object. As it happened there were at least two completely unrelated programmes in progress which might have led to the discovery of QSOs. One was a programme by H. Palmer and his colleagues on small angular size radio sources and the second some work on three objects 3C481, 3C196 and 3C286 which had been also suggested as identification with stellar objects. However, as Palmer and Kahn (1967) in their book on QSOs written close to the event and while the events were still fresh point out "but before they were explored at all deeply, the problems were completely changed by the successful identification of the radio source 3C273". Recent attributions as to the discovery of QSOs are usually unclear as to the sequence of events. Partly this is due to the fact that 3C273 provided the key by which the spectra of QSOs could be interpreted and that an interpretation of the spectrum of 3C48 based on the 3C273 data appeared in the same issue of Nature as the two discovery papers on 3C273. Up to that time, as far as I am aware, the only mention of 3C48 was an announcement of its identification at a meeting of the American Astronomical Society in December 1960 and it

FIG. 1.   Reprinted from Nature 197, 1037, 1963.

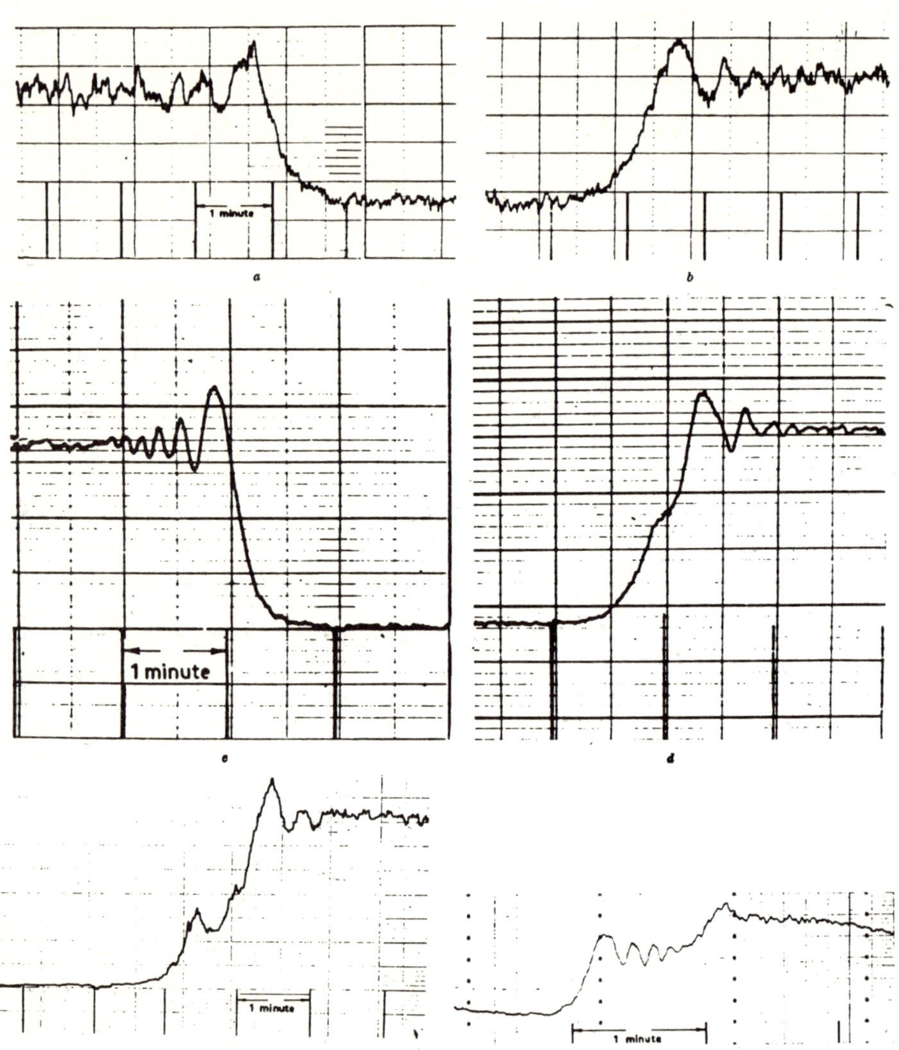

FIG. 1.—Facsimiles of records showing occultations on August 5 and October 26, 1962, at different frequencies. (a) Emersion of August 5, 1962, at 136 Mc/s; (b) immersion of August 5, 1962, at 136 Mc/s; (c) emersion of August 5, 1962, at 410 Mc/s; (d) immersion of August 5, 1962, at 410 Mc/s; (e) immersion of October 26, 1962, at 410 Mc/s; (f) immersion of October 26, 1962, at 1420 Mc/s. Abscissae, U.T.; ordinates, flux density.

was reported in Sky and Telescope that

"there is a remote possibility that it (i.e. 3C48) may be a very distant galaxy of stars; but there is general agreement among the astronomers concerned that it is a relatively nearby star with most peculiar properties"

It is true that there was a paper in press describing the properties of 3C48, 3C196 and 3C286 (Mathews and Sandage 1963) but in this paper they are interpreted also as galactic stars. However, following the discovery of the nature of 3C273 a note was added in press concerning their extra-galactic nature. It is therefore clear that although this paper has been cited as a discovery paper on QSOs the facts do not warrant this claim. A paper discussing the identification of stellar objects and interpreting them as galactic stars and moreover at the time of the 3C273 papers as yet unpublished cannot seriously be claimed as the recongition of QSOs anymore than the identification by Baade and Minkowski of the stellar remnant of the Crab Nebula can be claimed in retrospect as the discovery of pulsars. It should be clear from the above account that the work on 3C273 was not inspired in any way by these stellar identifications, of which I was unaware, but a completely independent investigation Of course if the 3C273 occultation had not been communicated in advance of publication to M. Schmidt the whole story might have been different which is why the two papers in Nature, the occultation paper by Hazard, Mackey and Shimmins and the optical paper by Schmidt must be considered jointly as representing the first publication of the discovery of QSOs.

Active Galactic Nuclei

At the time of their discovery QSOs were considered a unique class of objects whose main characteristics apart from their stellar appearance were (1) their high redshifts which if interpreted as cosmological gave absolute magnitudes for the QSOs significantly brighter than the brightest cluster galaxies (ii) an emission line spectrum in which the large widths of the permitted lines indicated velocities in the emitting gas of up to about 10,000 km/sec (iii) an optical continuum with a strong ultra-violet excess and usually considered to be non-thermal and due to synchrotron emission and (iv) variable optical emission. Over the years, however, it has become accepted by most astronomers that the classical QSO is but one example of activity in the nucleus of a galaxy and that the lineless BL Lac objects, the intrinsically fainter Seyfert galaxies, the N- galaxies etc. are all manifestations, to a greater or lesser degree, of the same basic phenomenon. The driving force for the activity is largely accepted as basically gravitational, arising from accretion onto a compact massive object, possibly a black hole, centred in the galaxy nucleus, as was indeed suggested in one form or another in several contributions at the 1963 Texas Symposium (Robinson et al. 1965). The term Active Galactic Nuclei was coined to cover these various forms of activity. However, it should be emphasized that a single all embracing term for all evidence of activity in a galaxy has its dangers and may tend to obscure genuine and fundamental differences. For instance just as initial different stellar masses imply different evolutionary paths and different fates for galactic stars so different accreting masses may imply different evolutionary paths and different fates for the parent galaxy in different types of Active Nuclei. Furthermore, not all high redshift extra-galactic objects are necessarily evidence of activity due to accretion onto a single massive object nor are all objects which have been classified as AGN's necessarily excited by a non-thermal continuum source. In this respect the very compact so callled "extra-galactic HII regions" but which I will simply refer to as "HII galaxies" are a case in point. Although the low excitation spectra of these objects and their narrow lines generally allow them to be distinguished from the higher excitation and broader lined QSOs their stellar appearance has led several of the higher redshift examples (z > 0.05) to be misclassified as low z QSOs. Typically their luminosities are relatively low and they would be excluded from QSO samples with a lower luminosity limit of M = -23 but examples are known (Hazard, Terlevich, McMahon in preparation) with luminosities comparable to low luminosity QSOs and with line strengths which would make them visible at z > 2. In fact, W. L. W. Sargent and I have recently found narrow emission line objects at z > 2 with velocity widths of $\lesssim$ 600 km/sec which could be high luminosity versions of these "HII galaxies". It is an intriguing question whether or not there may be a relationship between the very luminous "H II galaxies" and the more normal QSOs since the formation of a massive nucleus is presumably preceded by or associated with massive bursts of star formation and whether or not the narrow emission lines seen in some QSOs in addition to the broad lines are also evidence of a thermal contribution to the emission line spectrum.

I doubt if it is generally realized how wide is the variety in the spectra of high redshift objects. Partly this is because many of the selection techniques are biassed to particular types of objects i.e. radio QSOs, which do seem to be more of a uniform group than QSOs in general, or slitless spectroscopy techniques which have been biassed to objects with lines of large equivalent width. In the remainder of this talk I therefore discuss some recent work in which I have been involved which illustrates the diversity of the AGN's and perhaps some modification of existing ideas on QSO density and redshift distributions and of the notion that QSOs can be considered as passive probes of the Universe rather than modifying significantly their environment.

The surface density of QSOs

Among the major changes in AGN research over the past twenty years one of the most striking is in the improvement in the discovery techniques. Improvements in the sensitivity and positional accuracy of radio surveys, the development of new optical search techniques, notably the low dispersion prism Schmidt telescope combination, and the coming into operation of the high speed automatic photometer machines, which make colour selection so much easier to apply and make the discovery of at least certain types of AGN's in particular redshift intervals a relatively simple matter. Coupled with the new and more efficient solid state detectors which have largely replaced the optical plate for spectroscopic work, they have revolutionized the subject. On a single UK Schmidt III aJ objective prism plate it is possible to discover in a few hours more QSOs than were discovered in the past decade of QSO research and to spectroscopically study a large fraction of them in a reasonable period of time.

In spite of these improvements in the search techniques there are still relatively few claimed "complete" samples available and below 18 mag. only in relatively small regions of sky. Most such surveys rely on a single selection criterion, for example colour selection or the selection of strong line objects from slitless spectroscopy studies. There is, however, no single technique which can detect all the various types of AGN now known. As a result I believe that there is still considerable uncertainty regarding the surface density of AGN and in particular there is no conclusive evidence one way or the other as to the existence of significant clustering. Thus in a recent detailed study on quasar evolution Schmidt and Green (1983) supplemented their bright QSO survey data with information from 4 surveys which between them contained a total of only 54 confirmed QSOs with a limiting magnitude fainter than B = 18.

The models of Schmidt and Green predict the numbers of QSOs per sq. degree in different magnitude ranges consistent with the surveys used in their construction. The model predicting the highest QSO densities gives per sq. degree about 0.5 QSO for B < 18, 3 QSO for B < 19 and 18.5 QSO for B < 20. These predictions are in reasonable agreement with a recent QSO sample of Marshall et al. (1983) selected on the basis of UBV colours and which represents the highest density of confirmed QSOs so far published, namely 20 QSO per sq. degree for B < 19.8 over an area of 1.72 sq. degrees. However, some recent work in which I am engaged suggest that at least in some areas of the sky the QSO density is significantly higher than the model predictions. The work is part of a programme aimed at compiling in selected regions of sky complete samples of AGN and emission line galaxies and uses a wide variety of selection techniques to minimise selection effects (Hazard 1980). The most complete data are available for a 4.3 sq. degree area on a plate centred on RA 15H, Decl. +11 which is being studied in collaboration with W. L. W. Sargent. Although the spectroscopic studies so far have been confined largely to objects selected from UK Schmidt IIIaJ objective prism plates with a resultant bias to higher z QSOs and even though this sample is still not complete we have already compared a total of 61 AGN (14 per sq. degree) for z < 3.3. The high success rate we have achieved appears to be partly due to the very high quality of the plate material, partly to the fact that the dispersion of the 44 arc min. prism is well matched to the average widths of the QSO emission lines and partly to our use of selection criteria that are not based solely on the presence of strong emission-features, but which also make use of the form of the continuum spectra and the color information in the 2000Å spectral window. The long uniform density objective prism spectra, indicative of a UV excess is typical of QSOs in the redshift range $1.6 \lesssim z \lesssim 2.1$ while the drop in the continuum level to the blue of Lyα/NV produced by absorption in the "Ly-α forest" is particularly useful for the higher z QSOs. As a result of the different selection procedures, our samples of QSOs appear to be less redshift dependent and exhibit a wider range of spectral parameters than in most prism surveys. The latter property is particularly important at high redshifts where not only the adsorption in "the Ly-α forest", but possibly also the incidence of the BAL phenomenon, increases with increasing redshift and a restriction to strong emission line objects could obscure significant evidence of evolution in young QSOs.

The magnitude and redshift distributions for the 15+11 sample are presented in Figures 2(a) and 2(b) while those of Marshall et al. are shown in 2(c) and 2(d), the B-magnitudes of the latter having been converted to m(J) to facilitate a comparison of the two samples. The

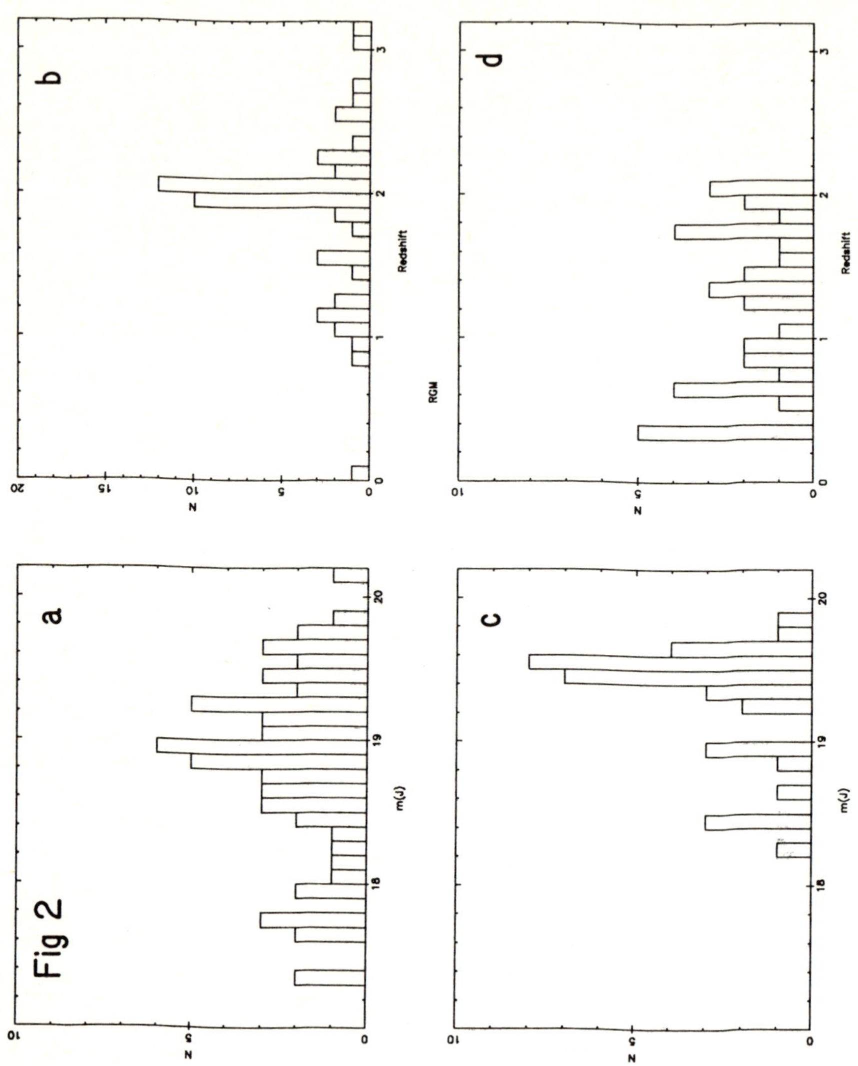

FIG. 2. Comparison of the magnitude and redshift distribution of two samples of QSOs. Figures 2(a) and 2(b) are for the sample of Hazard and Sargent and 2(c) and 2(d) of Marshall et al. (1983) whose B-magnitudes have been converted to m(J). Figure 1(a) omits a Seyfert galaxy with m(J) 14.5 while 1(b) includes only those QSOs for which accurate redshifts are available. The majority of the omitted objects are low redshift QSOs (z < 1.8).   The 15+11 survey should be essentially complete for 2.2 < z < 3.3.

limiting magnitudes of both surveys are approximately the same but it
is clear that if both are drawn from populations with the same
magnitude and redshift distributions then the 15+11 sample is seriously
incomplete fainter than m(J) = 19 and, even if all the objects with as
yet unknown z's are assumed to be low z objects, seriously incomplete
for z < 1.8. A meaningful comparison must therefore be restricted to
the brighter objects although it is probable that further low z QSOs
with z < 1.8 will be discovered in the UBVRI colour search now under
way. As of now, we have in the 15+11 sample 9 AGN (8 QSO + 1 Sy 1
galaxy) for m(J) < 18 and 36 for m(J) < 19, that is 2.1 and 8.4 per sq.
degree respectively. The corresponding values for the Marshall et al.
sample are 0 and 4.6 per sq. degree. Part of this discrepancy is due to
a higher z limit because although the Marshall et al. sample is
nominally complete to z = 2.5 it in fact contains no QSOs with z > 2.1.
It seems unlikely that the colour selection could produce such a severe
cut-off at z = 2.1 so this probably does imply a low density of QSOs in
their survey region with 2.1 < z < 2.5. In any case we have only 3 QSOs
in the 15+11 sample with z > 2.5 and m(J) < 19 and even restricting our
sample to z < 2.2 would still give a density of 7 sq. degree for m(J) <
19, still significantly higher than the Marshall et al. sample. The
discrepancy between the numbers in the 15+11 sample and that of the
highest density Schmidt-Green models are even more striking, their
highest density model predicting 0.5 per sq. degree for B < 18 and 3
for B < 19 for z < 3.5 far too large a discrepancy to be explained by
the small difference (∼ 0.1) between the B and m(J) magnitude systems.
Furthermore, the number of AGN in our sample (36) is too large to allow
it to be easily explained away as a statistical fluctuation in our
region. Either QSO densities are still seriously underestimated even
for objects brighter that 19 mag. or there is evidence here of
significant clustering in scales of several degrees. It should be noted
that the 4.3 sq. degree is not apparently a region of unusually high
density on the 15+11 plate but that it appears typical of the plate as
a whole.
    A notable feature of the redshift distribution shown in Figure 2(b)
is the high density around z = 2 and the steep decline to z = 3, for
contrary to popular opinion the low dispersion prism surveys are not
strongly biassed to QSOs with z ≈ 2. There is, it is true, a bias
against objects with z < 1.5 but for z > 2 QSOs of comparable magnitude
and with lines of similar equivalent width become if anything easier to
detect with increasing z up to the limit of the IIIaJ plates at z = 3.3
(Hazard et al. 1984). The steep fall in the numbers per magnitude bin
going from z = 2 to z = 3 indicated in Figure 2(b) is already
inconsistent with the predicitions of the Schmidt-Green models but even
so probably underestimates the steepness of the decline as we are
certainly not complete at z = 2 for m(J) < 19 and both the objects with
z = 3 are fainter than this limit. A better estimate of the behaviour
in this redshift range is obtained by considering the density of high z
QSOs on the whole plate. A particularly careful search has been made on
the plate for high z QSOs to the plate limits and a total of 9 with 3 <
z < 3.3 have so far been confirmed with m(J) < 20. There are two or
three more possible candidates but even assuming that we are only 50
per cent complete would still give around z = 3 no more than 0.3 QSO
per sq. degree per 0.2 redshift bin. On the other hand there are 11
QSOs with 1.8 < z < 2.2 and m(J) < 19 in the 4.3 sq. degree area or 1.3
per sq. degree. For a q_o = 0.5 universe the magnitude change going from
z = 2 to 3 is approximately 1 magnitude and the volume elements at

these two redshifts approximately equal. Therefore for a constant co-moving density we would expect roughly the same number of QSOs per redshift bin for m(J) < 19 at z = 2 as for m(J) < 20 at z = 3 as compared to the observed ratio of ~ 4:1. There is at least a strong suggestion here for a real fall in the co-moving density of QSOs for z > 2.

The above results have obvious implications regarding the search for QSOs with z > 3.5 and the significance of the difficulty in finding such objects in appreciable numbers. In particular it casts doubt on what appears to be a widely held notion that there is an abrupt fall in QSO densities at a redshift of around 3.5. Rather they suggest that while QSOs with z > 3.5 may be rare this is but a manifestation of a decline which sets in around z = 2 as indeed was implied by early radio surveys.

The spectra of high redshift QSOs

While our main aim in the 15 + 11 region is to compile a complete sample of AGN over a limited area we have also been studying bright QSOs, QSOs with z > 3 and objects with unusual prism spectra found in a search of the whole 36 sq. degrees. We present in Figure 3 a small sample of spectra for objects with z > 2 to illustrate the wide diversity of the spectra of high redshift objects. Among the notable features are the wide range in emission line widths, varying from a FWHM of around 15,000 km/sec to less than 600 km/sec. and the markedly different behaviour in the continuum to the blue of Lyα/NV from object to object and the extent of the fall in the continuum level passing from the red to the blue or of the Lyα/NV emission. The variety of spectra is such that doubts must arise as to whether all these high z objects really are manifestations of the same basic phenomenon. In this connection it may be significant that there appears to be a strong anti-correlation of the relative strength of NV emission with line width suggesting a decrease in the hardness of the ionizing radiation. It may be that the narrow lines are not excited by non-thermal radiation but by intense bursts of star formation possibly connected with the formation of the central massive object and that these narrow line objects are more akin to young galaxies than a normal QSO. However, probably the most interesting spectra are those of the BAL QSOs of which we have already found eight on this one plate, a significant discovery rate as the latest list compiled by Weymann and Foltz (1983) contains only 34 such objects.

The BAL QSOs which have been studied extensively by Weymann and his colleagues are among the most intriguing of the AGN. They are characterized not only by the width of the absorption features which imply velocities up to 0.2c or 60,000 km/sec. but also by the bulk of the absorbing material being in a state of high ionization. While the CIV, SiIV, NV and OVI is usually strong, Lyα absorption is usually weak or absent as also in most cases is the Ly emission. Their relationship to normal QSOs is not clear. Scott, Christiansen and Weymann (1983) consider that the BAL phenomenon may be a property of most QSOs but orientation dependent, thus accounting for the relatively small fraction they form of the whole QSO population. Hazard et al. (1984) on the other hand have suggested that the incidence of the phenomenon may be redshift dependent, increasing with increasing redshift, and have suggested that the BAL QSOs may represent an early phase in the evolution of a QSO. The latter proposal is of interest in connection

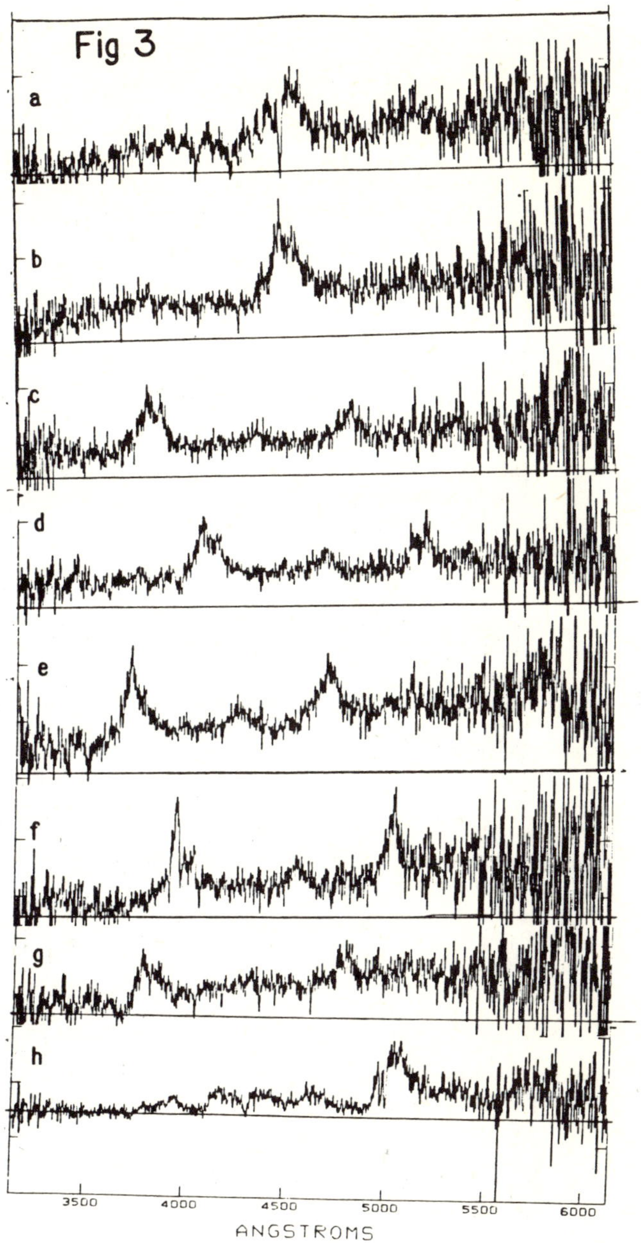

FIG. 3. A sample of spectra of AGN with z >   2. They show the wide range in line widths and the difference in the behaviour to and over Lyα/NV from object to object. The last four spectra and also spectrum 3(h) are BAL QSOs. Note that in 3(o) the peak around 3900Å is NV emission not Lyα.

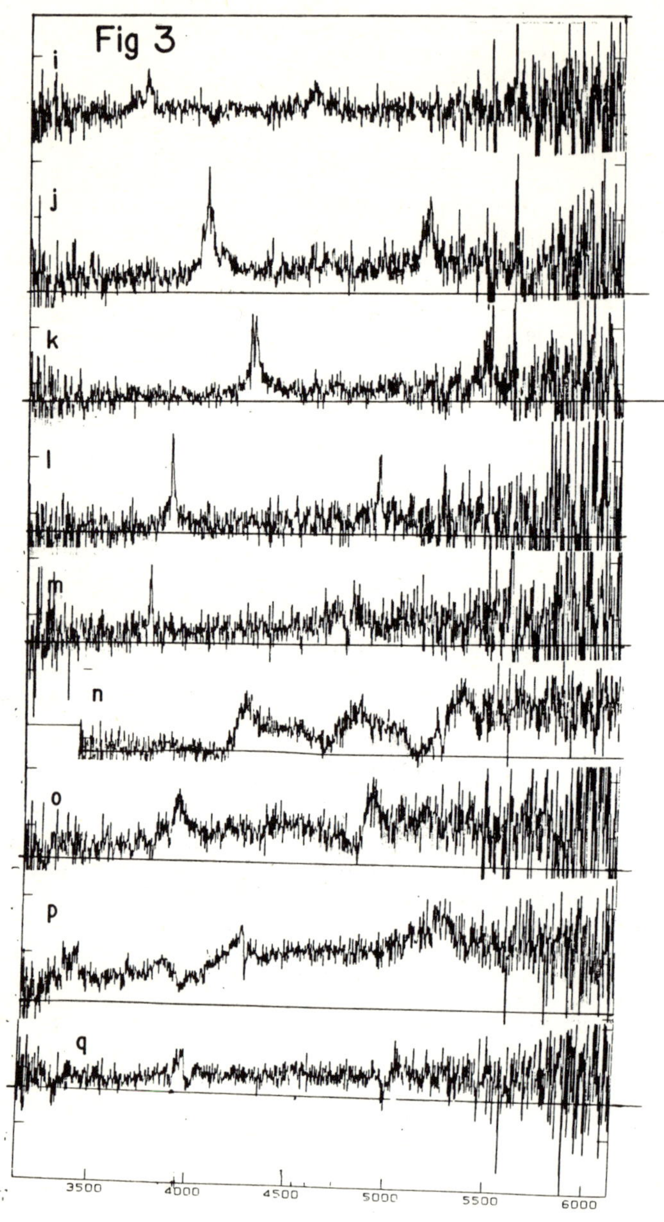
Fig 3
i
j
k
l
m
n
o
p
q
3500
4000
4500
5000
5500
6000

with the suggestion made earlier that there may be a real decrease in
the co-moving density of QSOs for z = 2. Such a decrease would imply
the switching-on of large numbers of QSOs at these redshifts and an
economy of hypotheses could then suggest that the BAL phenomenon is
associated with the switching-on process for while we expect to
reasonably complete for emission line QSOs at 3 < z < 3.3 the more
extreme forms of the BAL phenomenon could certainly not be selected.

QSOs as modifiers of the inter-galactic medium

Irrespective of how they fit in with normal QSOs the BAL objects
show that it is probably a gross over simplification to consider QSOs
as simply passive objects at large redshifts and that consideration
must be given to their interaction with the surrounding inter-galactic
medium just as we cannot ignore the impact of supernova explosions in
the Galaxy on the surrounding interstellar medium. The similarity of
the BAL spectra to supernova spectra has already been remarked on by
Hazard et al. (1984) who pointed out the possible role they may play in
dispersing highly processed material into intergalactic space. Could
they perhaps perform the role for which the Pop III stars have been
proposed or even, just as supernovae initiate star formation in their
neighbourhood could it be that the BAL QSOs perhaps indicate galaxy
formation in the surrounding region? It may be that the variety of BAL
QSOs represents not so much an evolutionary sequence (Hazard et al.
1984) but different initial masses, and that it may be the consequent
differences in the remnants or of the central collapsed object which
gives rise to the wide variety of what are considered as more normal
AGN.
A less speculative implication of the ejection of vast amounts of
material at high velocity is that it is clearly unrealistic to consider
the region surrounding BAL QSOs, and if Scott et al. are correct all
QSOs, as typical of the intergalactic medium as a whole. The
modification of the surrounding medium may provide an explanation of
the radically different behaviour of the continuum over and to the blue
of Lyα/NV from object to object. Figure 4 shows the spectrum of
1500+091 a particularly interesting variant of the BAL phenomenon found
in our 15 + 11 survey where the heavy absorption to, over and to the
blue of Lyα/NV shows a strong resemblance to "Lyα forest" absorption
but which from both its distribution and association with absorption in
SiIV and CIV must be intrinsic to the QSO. It could be that in many
QSOs the absorption lines commonly attributed to absorbers distributed
randomly along the line of sight may at least be partially intrinsic.
Furthermore, if the absorbing material around a QSO is sufficiently
extensive it could be that heavier than usual absorption in a QSO to
the blue of Lyα may be due to its line of sight intercepting such a
concentration of absorbers around an intervening object which may have
passed out of its active optical phase. Some recent observations of a
pair of QSOs in the 15+11 plate separated by 30 arc sec. lend some
credence to this idea. The nearer of the pair 1510+117a is a z = 2.3
BAL QSO while the second 1510+117b is a narrow line z = 2.9
object. High resolution spectra of these two QSOs have recently been
obtained by Sargent and Bokensberg and facsimiles of their unnormalized
spectra are shown in Figure 4. The spectrum of 1510+117(a) is that of a
BAL QSO showing a deep NV absorption trough and a heavily absorbed Lyα
emission feature at a wavelength close to that of OVI in the high

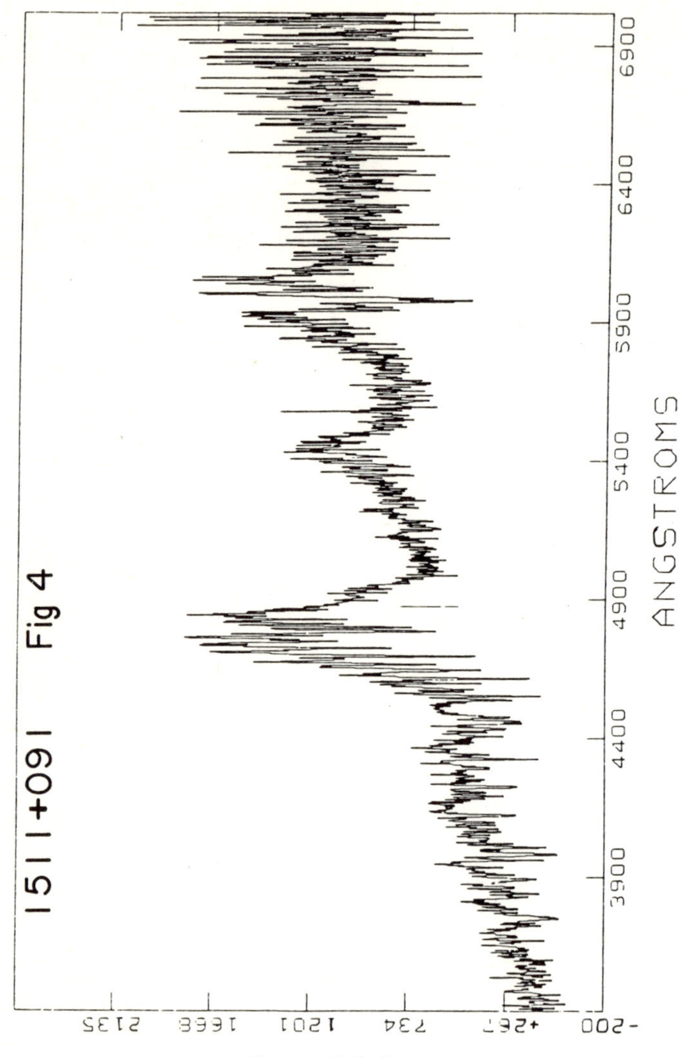

FIG. 4. The spectrum of the 1511+091 a BAL QSO where the absorption over and to the blue of Lyα is intrinsic to the QSO but has the general appearance of "Lyα forest" absorption.

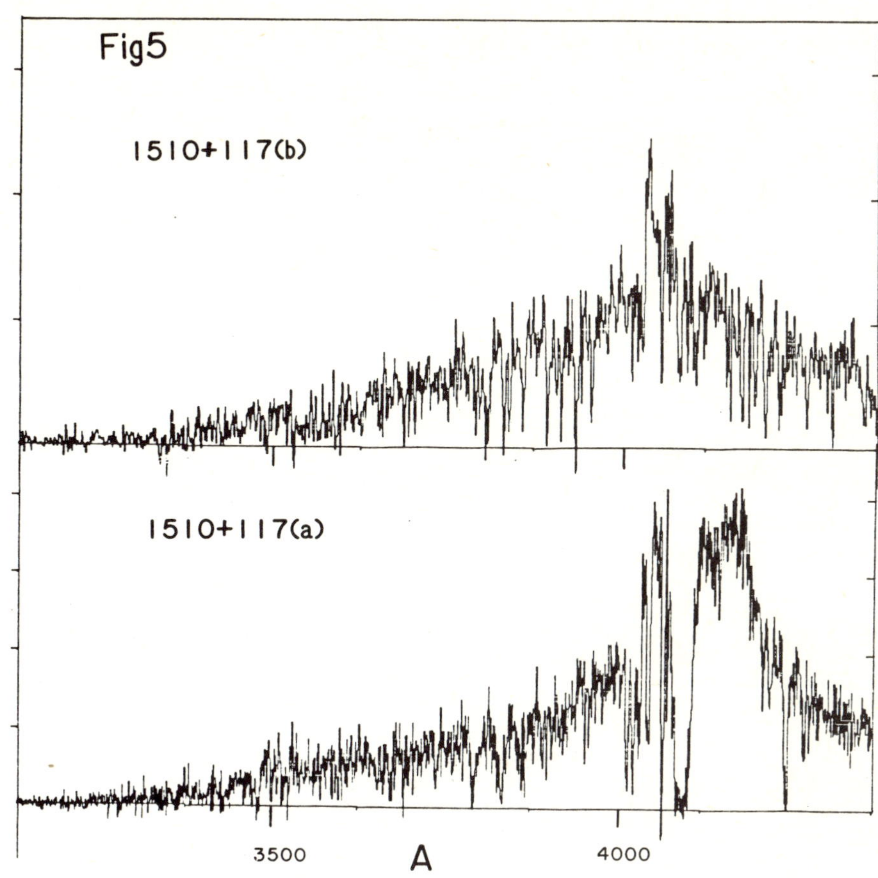

FIG. 5.   The spectra of two QSOs separated by 30 arc sec. The QSO
1510+117(a) is a BAL QSO with z = 2.3 while 1510+117(b) is a narrow
line QSO with z = 2.9. There is evidence of an absorption edge in
the higher z QSO corresponding to a Lyα edge in the lower z QSO and
indications of an enhancement in the "Lyα forest" in 1510+117(b) on
either side of Lyα in the 1510+117(a).
Wavelengths are approximate only.

redshift QSO 1510+117(b). There is clear evidence of absorption over this OVI emission with what appears to be a sharp edge coinciding with the blue edge of the Lyα absorption in 1510+117(a). Furthermore, there appears to me to be a definite enhancement of the "Lyα forest" absorption in the higher redshift object on either side of Lyα in 1510+117(a) extending over a wavelength corresponding to a velocity range of ±20,000 km/sec, consistent with the extent of the intrinsic narrow absorptions seen in 1511+091. However, if this interpretation should prove correct, the separation of the QSOs (30 arc sec.) would imply an extend of the disturbed region around 1510+117(a) of at least a few hundred kpc. This is clearly a pair of QSOs which needs careful study for a confirmation of these admittedly speculative conclusions would imply that the influence of QSO activity is not confined to a small region around the central active object.

Concluding remarks

I hope that my account of the historical background to the subject will in the light of the subsequent papers in this volume give some idea of just how far we have come in this fascinating subject over the past twenty one years. The progress made in understanding these one mysterious objects is one of the great success stories of modern science. However, there are many problems yet to be solved and no doubt even more mysterious objects out there waiting to be discovered. The recent development in search procedures, spectroscopic techniques and high speed scanning machines have now placed at our disposal the means by which we can study and classify for the first time the vast number of faint objects accessible to the large Schmidt survey instruments while the new facilities in the UV, infra-red and X-ray regions of the operation enables us to investigate them over a wider spectral range. It is my belief that the discovery of the QSOs and all that has flowed from it is but a foretaste of the developments we can expect in the next twenty one years.

Acknowledgements

I am grateful to R. McMahon for his assistance in the preparation of this paper and particularly for providing an accurate magnitude calibration for the 15+11 plate. I also wish to thank the UK Schmidt group for providing the objective prism material and the APM group at the Institute of Astronomy whose machine is playing such a large role in analysing the Schmidt telescope material.

References

Baade, W. & Minkowski, R., Ap.J., 206, 14, 1954.
Bolton, J.G. & Stanley, G.J., Aust.J.Sci.Res.A., 2, 139, 1949.
Bolton, J.G., Stanley, G.J. & Slee, O.B., Nature, 164, 101, 1949.
Bolton, J.G. & Westfold, K.G., Aust.J.Sci.Res.A., 4, 476, 1951.
Gold, M., Proceedings of Conference on Dynamics of Ionized Medium,
    University College London. R. Boyd (ed.), 1951.
Hanbury-Brown, R. & Hazard, C., Phil.Mag., 44, 939, 1953.
Hazard, C. as reported in Observatory by R. Hanbury-Brown, 73, 185, 1953.
Hazard, C., M.N.R.A.S., 124, 343, 1962.

Hazard, C., In "Variability in stars and galaxies" (Liege; Institut
     d'Astrophysique), 1980.
Hazard. C., Mackey, M.B. & Shimmins, A.J., Nature, 197, 1037, 1963.
Mazard, C., Morton, D.M., Terlevich, R. & McMahon, R., Ap.J.,
     1984 (in press)
Hoyle, F., Proceedings of Conference on Dynamics of Ionized Medium,
     University College London. R. Boyd (ed.), 1951
Jansky, K.G., Proc.I.R.E., 90, 1920, 1932.
Marshall, H.L., Avni, Y., Braccesi, A., Huchra, J.P., Tananbaum, H.,
     Zamorani, G. & Zitelli, V., J.Center for Astrophysics preprint No.
     1912, 1983.
Mathews, T.A. & Sandage, A.R., Ap.J., 138, 30, 1963.
Mills, B.Y., Aust.J.Sci.Res.A., 5, 268, 1952.
Palmer, H. & Kahn, F.D., "Quasars", Manchester University Press, p. 13,
     1967.
Robinson, I., Schild, A. & Schucking, E.L. (eds.) Quasi-stellar sources
     and gravitational collapse. University of Chicago Press, 1965.
Ryle, M., Proceedings of Conference in Dynamics of Ionized Medium,
     University College London. R. Boyd (ed.), 1951.
Schmidt, M. & Greene, R.F., Ap.J., 269, 352, 1983.
Scott, J.S., Christiansen, W.A. & Weymann, R.J., Ap.J. in press.
Smith, F.G., (Communicated by M. Ryle), Rep.Prog.Phys.XIII, 940, 1950.
Smith, F.G. & Lovell, B., J.H.A., 14, Pt. 3, 155, 1983.
Weymann, R.J. & Foltz, C.B. "Quasars and Gravitational Lenses".
     Proceedings of the 24th. Liege International Astrophysical
     Colloquium, 1983.

Radio observations of active galactic nuclei

R.W. Porcas

Max-Planck-Institut für Radioastronomie
Auf dem Hügel 69, D-5300 Bonn 1, West Germany

SUMMARY

The properties of radio emission from active galactic nuclei are
reviewed. The first two sections give a brief overview of the subject,
and later sections concentrate on observations of the morphology and
variability of the innermost core components of the sources. Emphasis is
given throughout to topics judged to be of most current interest.

INTRODUCTION

Radio techniques

Radio observations of celestial objects are routinely made over a
range of four decades in frequency, the observing window being limited
by the ionosphere below $\sim$ 30 MHz and by atmospheric absorption above
$\sim$ 300 GHz. Radio telescopes of a variety of different designs are employed, and, unlike their optical counterparts, their resolution is
essentially always diffraction limited. The simplest antennas, consisting of a single parabolic reflecting dish, have resolutions of at best
about one arcminute at cm wavelengths. Observations of most extragalactic objects with such instruments are thus limited to the measurement
of integral source properties such as total flux density, broad band
spectrum, polarisation and variability.

Extensive use is made of the technique of aperture synthesis, whereby a spaced array of antennas is used to form a network of Michelson
interferometers whose outputs are combined to form a radio map by
Fourier transformation. The Very Large Array (VLA) in New Mexico
(Thompson et al. 1980; Napier et al. 1983) and Multi-Element Radio Link
Interferometer (MERLIN) operated at Jodrell Bank (Davies et al. 1980)
use this technique and achieve sub-arcsecond resolution – rather better
than the best optical seeing – corresponding to a few kiloparsecs for
the most distant extragalactic objects.

The highest resolution is obtained by the technique of Very Long
Baseline Interferometry (VLBI), whereby antennas in different countries
and continents are used to form interferometers with baselines many
thousands of kilometers in length. The radio signals received at each
antenna are recorded on magnetic tape, together with accurate timing
information and the interferometers are formed "off-line" by cross-
correlation of the signals from the various tapes (Clark 1973; Rogers
et al. 1983). Here a resolution of a few tenths of a milliarcsecond
(mas) can be achieved, providing direct structural information unavail-
able at any other wavelength band. Even for the most distant active
galaxies and quasars the morphology of compact regions on the scale of
a few parsecs can be investigated.

Radio spectra

Unlike at optical wavelengths, the radio sky away from the Galactic plane is dominated by emission from extragalactic objects. One of the simplest properties which can be measured for radio sources is the broad band, integrated source spectrum. A sample of such spectra is shown in Fig. 1 (from Kühr et al. 1979). Three basic shapes can be recognized.

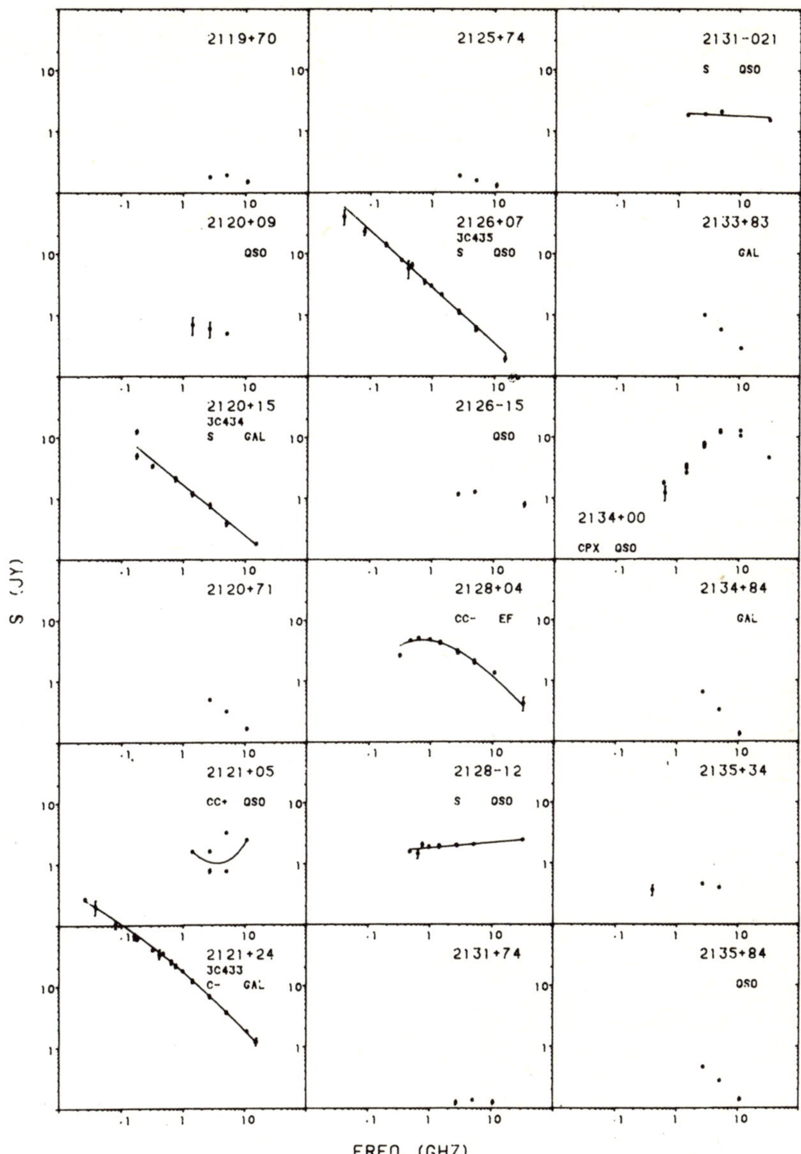

FIG. 1. Broad band spectra of radio sources, from Kühr et al. 1979.

a) <u>Steep spectra</u> obeying a power law of the form flux density (S) $\propto$ frequency $(\nu)^{\alpha}$, with spectral index α typically having a value of −0.8.
b) <u>Peaked spectra</u> characterised by convex shape and a distinct turn-

over frequency

c) <u>Flat spectra</u> with spectral index $\simeq 0$ and often exhibiting signs of variability.

It is widely accepted that the radiation received from extragalactic sources is incoherent synchrotron emission produced by a relativistic plasma and magnetic field present in the source. The spectrum expected from a uniform synchrotron source (e.g. Moffet 1975) is characterized by a turnover frequency, below which a power law slope of +2.5 is expected in the optically thick region if the distribution of particle energies itself obeys a power law, $N(E) \propto E^{-\gamma}$. Above the turnover frequency, the optically thin part of the spectrum also exhibits a power law behaviour, with a spectral index reflecting the particle energy spectrum, $\alpha = (1-\gamma)/2$.

We can readily identify the steep spectra and peaked spectra seen in extragalactic radio sources with the optically thin and turnover regions of the synchrotron spectrum. One of the major reasons for accepting the synchrotron emission process is the agreement between the deduced particle energy spectrum and that directly measured for cosmic rays. Another is the linear polarisation observed in the optically thin regions of many sources, the measured degree of polarisation never exceeding the maximum theoretical value for synchrotron radiation. A third reason is that the measured brightness temperature rarely exceeds, but often approaches, the theoretical maximum value of $\sim 10^{12}$ K for synchrotron radiation (Kellermann and Pauliny-Toth 1969). The flat spectra seen in some sources may be explained either as:

a) the superposition of multiple, homogeneous synchrotron components with different turnover frequencies (Cotton et al. 1980) or

b) radiation from a single, inhomogeneous component with, perhaps, a radial dependence of the magnetic field and/or particle energy density (Condon and Dressel 1973).

Morphology of active galaxies

Radio maps of 4 of the strongest radio sources, Cygnus A, Hercules A, Virgo A and Centaurus A, are shown in Fig. 2, which serves to emphasize

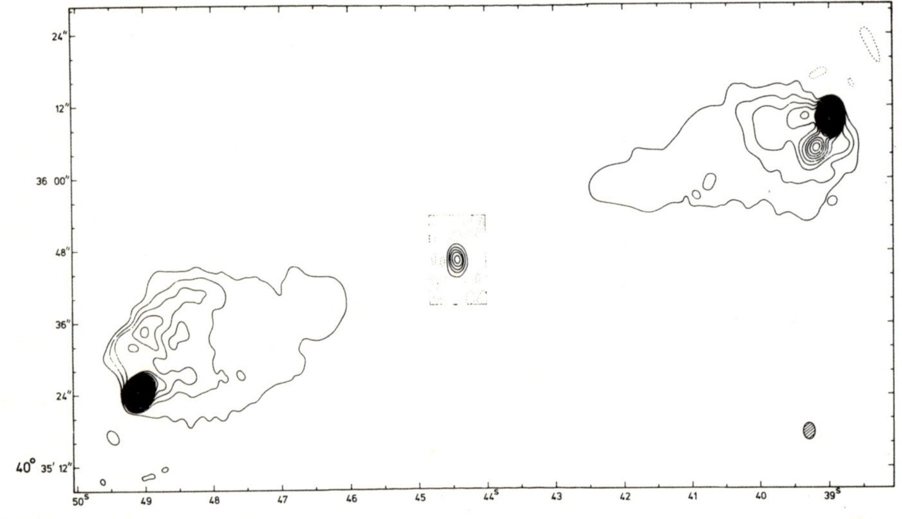

FIG. 2.   Arcsecond scale maps of four strong radio sources:
   2a) Cygnus A, from Hargrave and Ryle 1974

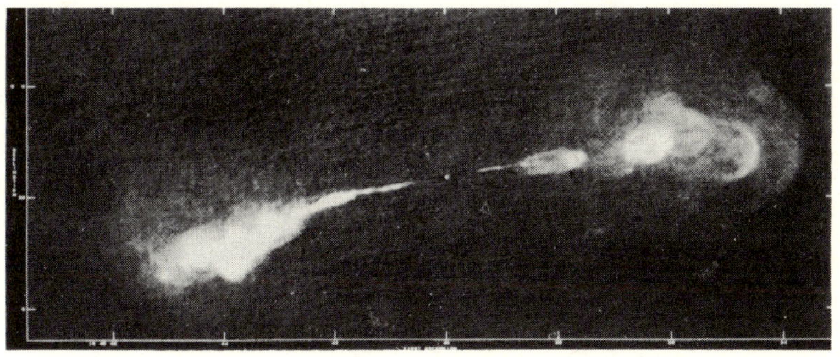

2b) Hercules A, from Dreher and Feigelson 1984

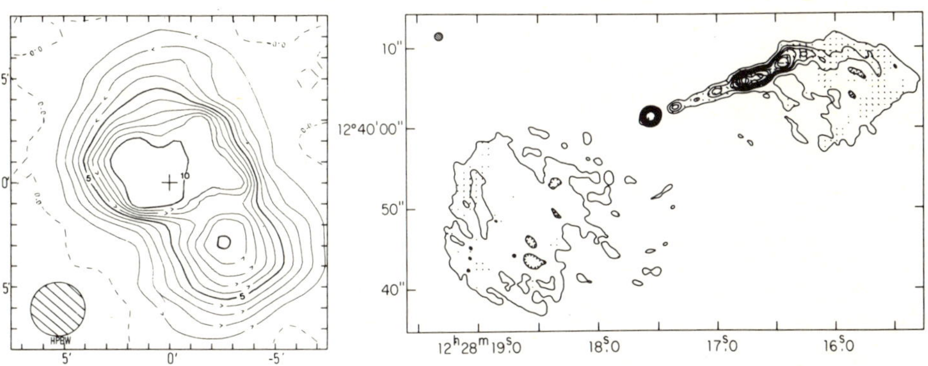

2c) Virgo A, from Andernach et al. 1979 and Owen et al. 1980

the dominance of active galaxies in the radio source population. Of
these, the closest, Centaurus A, has been termed the "Nearest Active
Galaxy" (Burns et al. 1983). A number of features can be identified in
the maps which characterize the typical radio morphology of active
galaxies and quasars:

   a) lobes, extended regions of relatively low brightness, and steep
spectrum, often displaced symmetrically on either side of the optical
galaxy, and extending out to distances as large as a few Mpc.

   b) hot spots, regions of enhanced brightness within the lobe structure

   c) jets, long thin traces of emission connecting the lobes and hot
spots with the source centre. Recent VLA observations of Cygnus A
(Perley, in preparation) also reveal such a structure.

   d) nucleus, the compact, generally flat spectrum, core of the galaxy.

   These features are usually described and discussed in the framework
of a model in which energy generated around a massive compact central
object (black hole?) emerges from the nucleus in narrow beams of plasma,
which interact with the interstellar and intergalactic medium to form
the most extended emission regions. Thus the morphology of active galaxies depends both on the nature of the energy flow and also the surrounding medium. To illustrate this point, Fig. 3 shows a VLA map of the
Seyfert galaxy NGC1068 (Wilson and Ulvestad 1983). The radio morphology
clearly exhibits features analogous to the more luminous radio galaxies
but the overall extent of the emission is only 2 kpc. Presumably the
much smaller size reflects the higher gas density found in such galaxies.

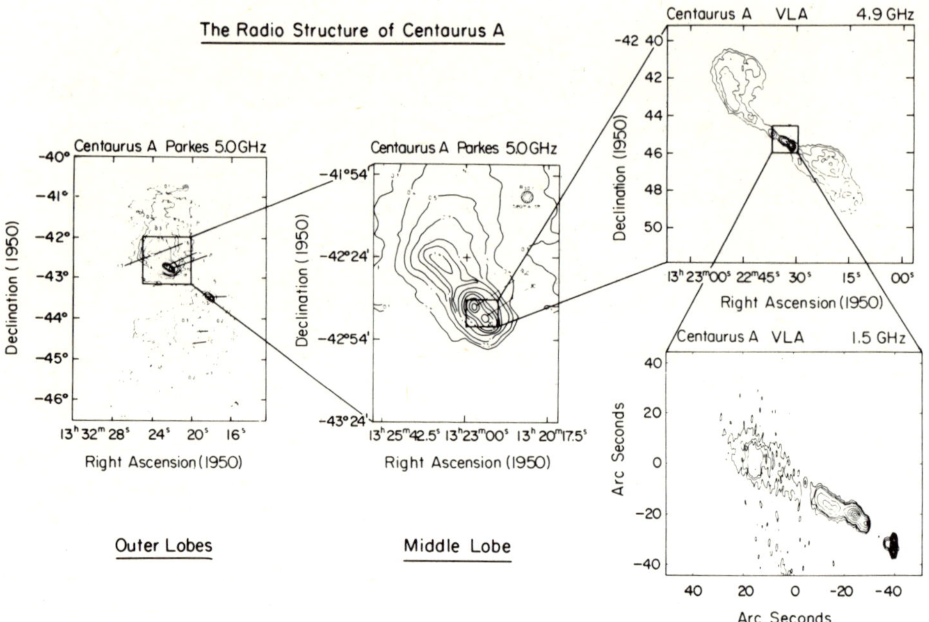

2d) Centaurus A, from Burns et al. 1983

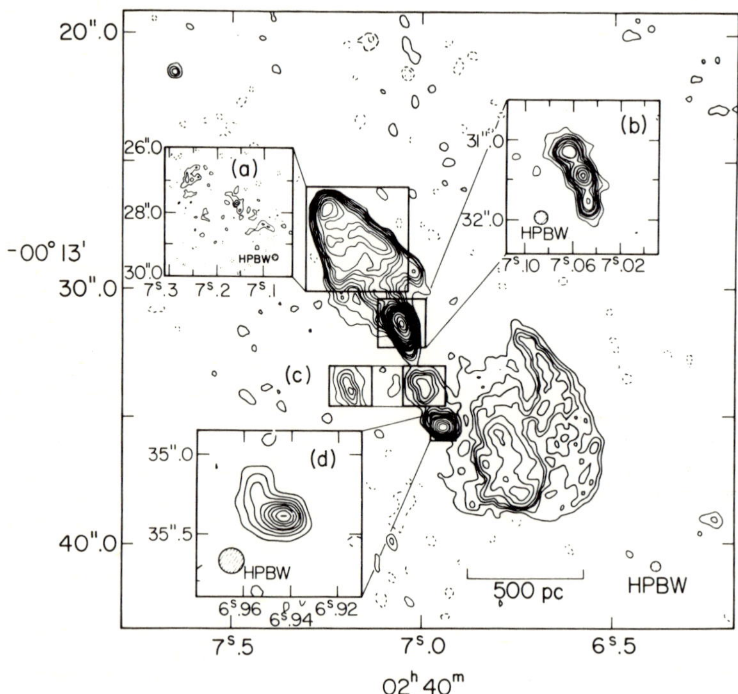

FIG. 3.   VLA map of NGC1068, from Wilson and Ulvestad 1983

STATISTICS OF ACTIVE GALAXIES

Radio surveys

As mentioned in the previous section, flux density limited radio surveys conducted away from the Galactic plane yield optical identifications essentially exclusively with extragalactic objects: quasars, BL Lac objects, radio galaxies, Seyferts and "empty fields" (= blank on the Palomar Sky Survey but generally turning out to be faint galaxies when observed to fainter limiting magnitudes, e.g. Lebofsky et al. 1983; Riley et al. 1980; Lilly and Longair 1982). The distribution of spectral types found in such a survey naturally depends on the observing frequencies. For the strong sources in the NRAO/MPIfR 5 GHz survey (Kühr et al. 1981; Witzel 1981) there are approximately 50% steep spectrum sources and 50% with flat or peaked spectra. Most of the steep spectrum sources are objects dominated by emission from lobes and jets, but a small number are steep spectrum compact (SSC) sources, having angular sizes $\lesssim$ 2 arcsec. The flat spectrum sources are all dominated by emission from a compact core less than 1 arcsecond in size. The important conclusion is that essentially _all_ strong extragalactic radio sources are associated with active galactic nuclei, either exhibiting flat spectrum emission from the active nucleus itself or emission from jets and lobes clearly associated with past or present activity in the nucleus.

Surveys of quasars

Radio observations of optically selected samples of quasars give some idea of the frequency of occurrence of strong radio emission from active galaxies. Typically some 10% of quasars in such samples show detectable radio emission (e.g. Strittmatter et al. 1980). Observations by Condon et al. (1980) of the brightest quasars from the Palomar-Green sample with the VLA have yielded a higher detection rate (8 out of 16) and suggest that there is a correlation between the strengths of the optical and radio luminosities. Kellermann et al. (1983) have made very deep observations of the Palomar-Green sample using the VLA and have detected 66% of the quasars above a level of 250 μJy. An extrapolation suggests that essentially all would be detected above a level of 100 μJy.

Although radio maps for large samples of optically selected quasars have not yet been obtained, maps of radio selected samples are available. Owen et al. (1978) and Owen and Puschell (1984) observed quasars from the Jodrell Bank 966 MHz Survey (Cohen et al. 1977; Porcas et al. 1980) and found that even for those (steep spectrum) quasars dominated by emission from the outer lobes, essentially all show radio emission from a compact, nuclear core component above a level of 6 mJy. This contrasts with similar observations of empty field radio sources of comparable lobe flux density, whose core components, if present, must be at least a factor of 30 weaker (Owen et al. 1982; Longair 1975).

Sizes of compact sources

An important question is that of the typical size of the compact components seen in maps of $\sim$ 1 arcsecond resolution. Zensus et al. (1984) have made short VLBI observations of an unbiased sample of 57 flat spectrum sources, chosen from the NRAO/MPIfR 5 GHz survey, with flux densities greater than 1 Jy and spectral index flatter than −0.5. These observations utilized transatlantic baselines at 5 GHz, giving resolutions of $\sim$ 1 mas (equivalent to 5 - 10 pc for typical quasars). Histograms of the degree of compactness (flux in VLBI structure/total source flux) for the various identification categories are shown in Fig. 4. These reveal that essentially all flat spectrum components have a significant fraction of their emission coming from regions at least as small

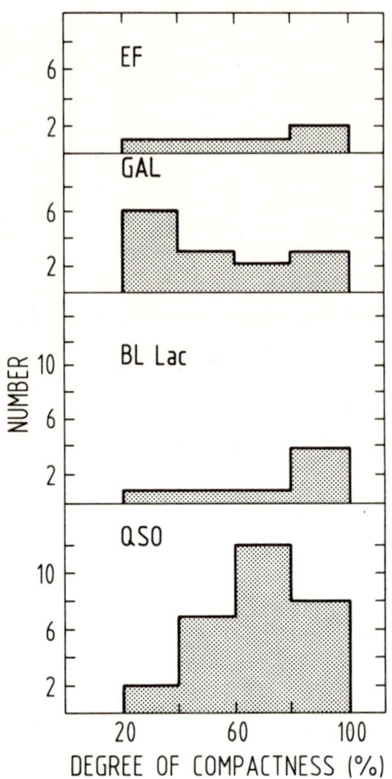

FIG. 4. Percentage of flux density in compact structure for strong,
flat spectrum sources, from Zensus et al. 1984

as a few pc.

The steep spectrum compact sources have been studied with the VLA by
van Breugel et al. (1984) and with VLBI by Fanti et al. (in preparation).
The VLA observations show that the SSC sources have complex morphology
and typical sizes of a few kpc. The sources have high surface brightness
and low degree of polarisation. Their complex structure, and small size
compared to other steep spectrum sources may result from the interaction
of the energy flow with a dense gas environment. Another source in this
class, 3C380, has been observed by Wilkinson et al. (1984) using MERLIN.
The map, shown in Fig. 5, shows a very complex morphology and also a
compact, central component.

VARIABILITY

<u>Introduction</u>
Flux density variability at cm wavelengths was reported by Dent
(1965) for a number of flat spectrum, extragalactic objects and since
then the study of the frequency and time behaviour of the variability
has provided one way of probing the details of the physical processes
in the nuclei of active galaxies and quasars. The observed time scales
of the variations ($\sim$ months to years) indicate that the flat spectrum
radio emission comes from very small regions. Discussion of variability

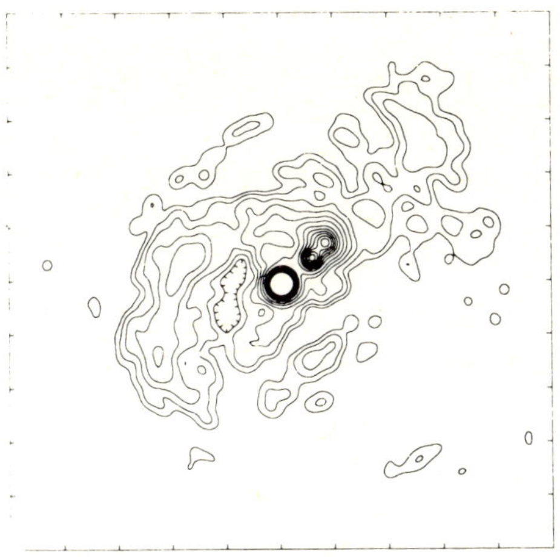

FIG. 5.   MERLIN map of 3C380, from Wilkinson et al. 1984. Tick interval
   = 1 arcsec

is conventionally divided into high frequency (> 1 GHz) and low frequency properties. An excellent review of the subject is given by Fanti
and Salvati (1980).

High frequency variability
   Most high frequency variability studies are carried out with single
dish antennas and a number of monitoring programs have established the
basic properties of the variations (e.g. Andrew et al. 1978). Some
general results from these studies are:
   a) essentially all compact, flat spectrum sources show flux density
variations on time-scales ranging from a few months to a few years
   b) the percentage of the mean flux density which varies is typically
20 to 50%, but a few sources show > 100% variations
   c) the flux density variations are larger at higher frequencies
   d) the occurrence of single isolated outbursts ("flares") is rare, and
usually one sees a superposition of such events in the spectrum at any
one time
   This latter property complicates the interpretation of the spectrum
of a single flare and its time evolution, but high frequency variability
is normally understood in terms of the "standard" model of an adiabatically expanding synchrotron emitting cloud (Pauliny-Toth and Kellermann
1966; van der Laan 1966). The simplest model assumes a cloud expanding
linearly with magnetic flux conserved and an instantaneous injection of
relativistic electrons. There is qualitative agreement between theoretical prediction and observation; at frequencies where the flux density is
rising the spectral index is positive (optically thick) although rarely
reaches its theoretical slope of +2.5. There is disagreement in some
other details, for example, the flux variations are generally too large
at lower frequencies and occur after too short a time delay. At most
frequencies the onset of a flare is too slow and its decay too rapid. It
is generally thought, however, that these discrepancies are not too
serious, and can be accommodated by small adjustments to the simple
expanding cloud assumptions.
   Monitoring observations at 90 GHz have recently been reported by
Epstein et al. (1982), and these have shown that some sources show

variations on time-scales as short as a few days to weeks. Heeschen (1982)
has recently reported radio flux density "flickering", with an amplitude
of ∿ 2%, and a time-scale of a few days for a sample of flat spectrum
sources observed at 3.3 and 5.0 GHz.

Variations in the polarisation properties of compact radio sources
have also been monitored, notably by Altschuler and Wardle (1976, 1977).
These variations in general are very complicated and are not always
associated with outbursts in flux density. Indeed, there are some sources
which show large changes in percentage polarisation and/or position
angle while the total flux density remains constant. The BL Lac object
0300+47 is a source which exhibits such behaviour (Fig. 6). More recently there have been a number of reports of the phenomenon on "polarisation rotators" (Altschuler 1980; Aller et al. 1981; O'Dea et al. 1983).

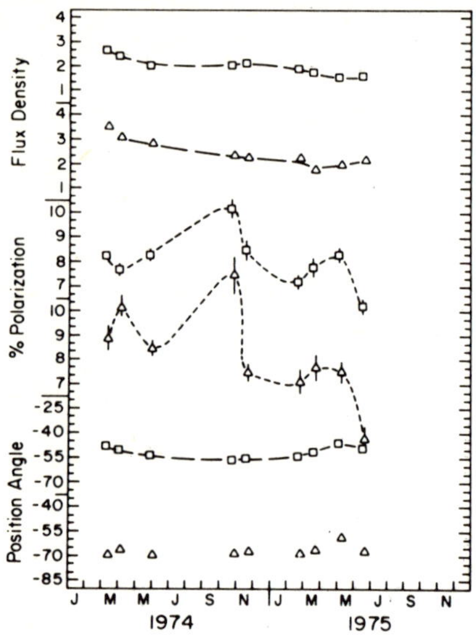

FIG. 6. Variation of flux density and polarisation of 0300+47, from
   Altschuler and Wardle 1976. 2.7 GHz points are squares, 8.1 GHz triangles

These are sources in which a flux density outburst is accompanied by a
linear variation of the polarisation position angle with time. In some
cases this change is the same at two or more frequencies and so cannot
be attributed to varying Faraday rotation. The simplest explanation is
that there is a rotating or revolving structure in the radio emitting
region (Aller et al. 1981).

Low frequency variability

Radio flux density variations at 920 MHz were reported by Sholomitzkii
(1965) for the quasar CTA102, but were largely disbelieved until variations were again reported at 408 MHz in 4 sources (including CTA102) by
Hunstead (1972). The phenomenon is now well established as the result
of a number of systematic monitoring programs (see Proceedings of the
Green Bank Workshop on Low Frequency Variability, Cotton and Spangler
1982). Fig. 7 shows the 408 MHz flux density variations in the sources
BL Lac (2200+420) and 3C454.3 (2251+158). Some general results are:
   a) the variations are much larger than expected from the high fre-

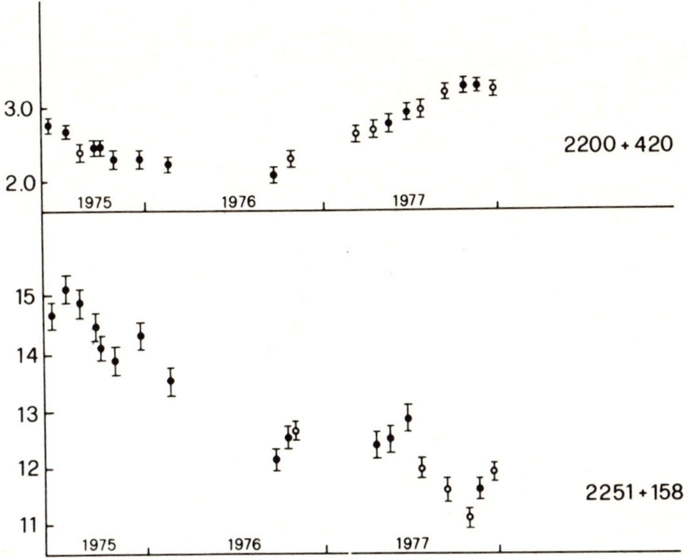

FIG. 7.   408 MHz flux density variations of 2200+420 and 2251+158, from
   Fanti et al. 1979

quency variations
   b) the time-scales are much shorter than expected from the high frequency variations
   c) amongst the compact, flat spectrum sources the phenomenon is as
common as high frequency variability
   d) a few "normal" (steep) spectrum sources show low frequency variability, e.g. 3C159 (see Padrielli et al. these Proceedings)
   e) in a few sources, such as BL Lac, there appears to be a connection
between high and low frequency events, whilst for others, such as
3C454.3, there does not seem to be any specific relationship
   The unexpectedly large and rapid variations at low frequencies create
a serious problem for the synchrotron cloud model because of the very
high brightness temperatures they imply. In particular the maximum
angular size, derived from taking the variation time-scale as the light
crossing time and using the redshift as a distance indicator, is much
smaller than the minimum angular size deduced from the maximum allowed
brightness temperature of $\sim 10^{12}$ K for synchrotron radiation (Kellermann
and Pauliny-Toth 1969). This difficulty can be overcome if the synchrotron cloud expands at relativistic speeds (Rees 1967) or if the emission
comes from a relativistic jet (Blandford and Königl 1979). However, for
some sources such as PKS1524-13 and CTA102, very high Lorentz factors
of $\gamma \sim 20$ to 30 are required to explain the variability.
   An alternative explanation for rapid low frequency variability is
slow scintillation of the interstellar medium. This was suggested by
Shapirovskaya (1978) and has recently been investigated by Rickett et
al. (1984) after the discovery of a correlation between the slow variation time-scale and dispersion measure for pulsars by Sieber (1982).
Such interstellar "focussing" can account for the variations at low
frequencies unaccompanied by high frequency variations, and also,
incidentally, for the "flickering" observed by Heeschen (1982).

SMALL SCALE MORPHOLOGY

<u>General</u>

Observations of active galaxies with the VLBI technique are at present the only direct means in astronomy of obtaining structural information and morphologies of the innermost regions of the nucleus. Observations using transatlantic baselines at frequencies as high as 22 GHz are regularly used to explore structures on scales down to a few times 10^{-4} arcseconds, corresponding to a few parsecs in even the most distant quasars. As an example, VLBI maps of the flat spectrum cores of the active galaxies Cygnus A, Virgo A and Centaurus A are shown in Fig. 8.

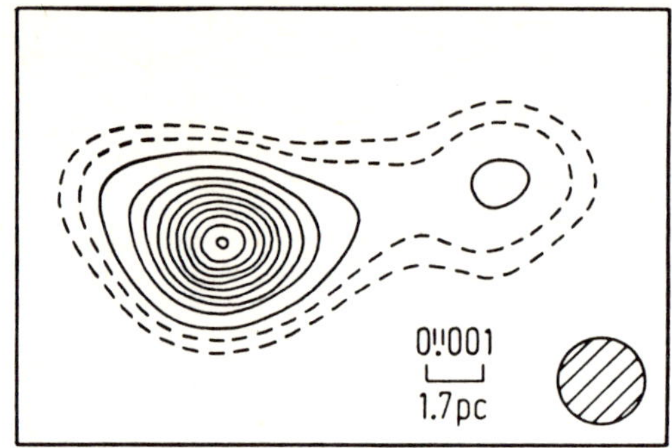

FIG. 8. VLBI maps of cores of active galaxies
   8a) Cygnus A, from Kellermann et al. 1981

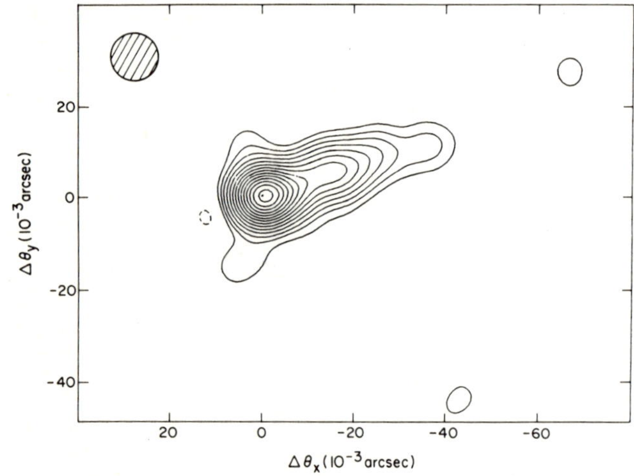

   8b) Virgo A, from Reid et al. 1982

The 10 GHz Cygnus A map (Kellermann et al. 1981) reveals not only a compact core of dimensions significantly smaller than 1 pc but also an extension closely aligned with (and on the same side of the nucleus as) the narrow jet seen in the VLA map of Perley. A very similar morphology is seen in the 1.7 GHz map of the core of Virgo A (Reid et al. 1982). The map of Centaurus A (Preston et al. 1984) was obtained using a

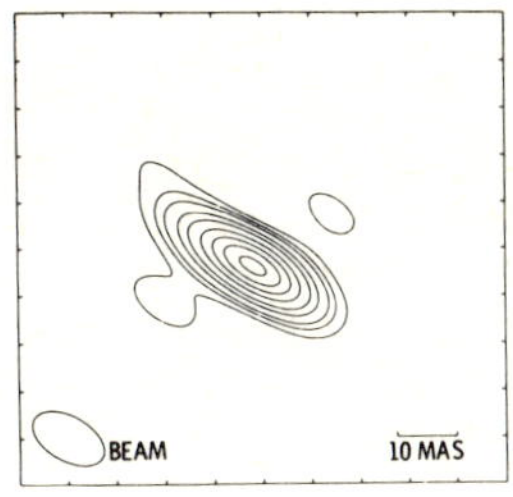

8c) Centaurus A, from Preston et al. 1984

Southern Hemisphere array of telescopes operating at 2.3 GHz; it shows
an extension of the core of about 40 mas ($\sim$ 1 pc) in position angle (pa)
51°, closely aligned with that of the VLA jet (Fig. 2d). Although these
radio cores are not very luminous compared with those of the more closely
studied flat spectrum quasars (nor are they known to be variable), these
observations serve to illustrate some general properties of nuclear
morphology:
    a) the existence of parsec-scale emitting regions
    b) structures closely aligned with the radio lobes on a scale 5 orders
of magnitude larger
    c) asymmetric structure in sources which are relatively symmetric on
the large scale (e.g. Cygnus A)
    A note about VLBI maps is in order here. Because each antenna in a
VLBI array uses an independent local oscillator, and observes through
its own local atmosphere (and weather!) mutual coherence of the recorded
signals only lasts a short time (typically 1 to 20 minutes, depending
on observing frequency). In general it is not possible to calibrate the
relative telescope instrumental phases within this "coherence time" and
so correlation only produces amplitudes and telescope-corrupted phases.
The method of hybrid mapping (e.g. Cornwell and Wilkinson 1981) has been
developed, which uses a Fourier transform of the fringe visibility data
and the "CLEAN" deconvolution algorithm (Högbom 1974), together with
estimates of the visibility phases derived from "closure phase" relations (Rogers et al. 1974). The maps thus produced contain no information about the absolute position of the structure, and so "hybrid" maps
are presented with axes showing only relative right ascension and
declination. This, and other, "self calibration" methods of image formation in radio astronomy have recently been reviewed by Pearson and
Readhead (1984a).
    Hybrid maps are essential for displaying the detailed morphology of
sources exhibiting very complex structure. The first hybrid map, made
by Wilkinson et al. (1977) at 610 MHz, is shown in Fig. 9a, and shows
the quasar 3C147. The structure consists of a compact core, and a long,
well collimated jet extending over a distance of 200 mas. Observations
of the core of the same source by Preuss et al. (1984) at 5 GHz (and
hence much higher resolution) are presented in Fig. 9b. This reveals a
very complex structure. A number of directions of features from larger
scale maps are indicated by broken lines, and show that, like Cygnus A,
Virgo A and Centaurus A, 3C147 too has a long term "memory" of direction. Fig. 10 shows a 22 GHz map of the peculiar galaxy 3C84 (= Perseus
A = NGC1275). Because of the high observing frequency and the (relative)
proximity of this source (z = 0.018) this map has a very high linear
resolution at the source ($\sim$ 0.1 pc) and thus provides a unique view of
the innermost regions of the nucleus of an active galaxy. The source has

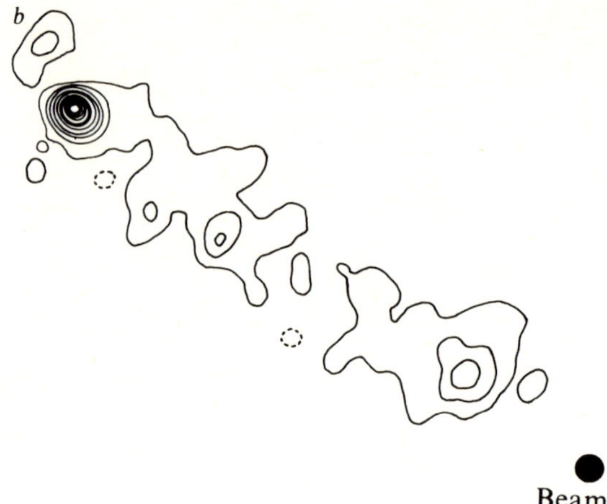

FIG. 9.   VLBI maps of 3C147
  9a) 610 MHz from Wilkinson et al. 1977. Length of jet = 200 mas

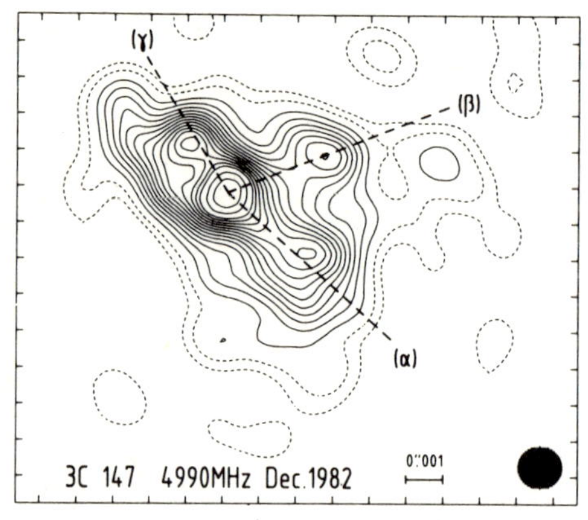

  9b) 5 GHz from Preuss et al. 1984

a northern core component and an extended region emerging to the S.W.,
which leads towards a long southward extension of ∿ 10 mas.

Classification of morphologies

    As in other fields of astronomy, it is highly desirable that schemes
for the classification of objects should address real properties of the
objects themselves, and not those aspects which depend on the instrumental parameters with which they are observed, such as resolution, sensitivity and dynamic range. On the other hand, Professor Burbidge has
emphasized in this conference the importance of keeping to simple,
observed properties, rather than those which reflect a model-dependent
prejudice. To some extent these are competing demands and must be born
in mind whilst classifying milliarcsecond structures.
    The largest survey of milliarcsecond morphologies conducted so far
is that of Pearson and Readhead (1984b), who have observed a sample of

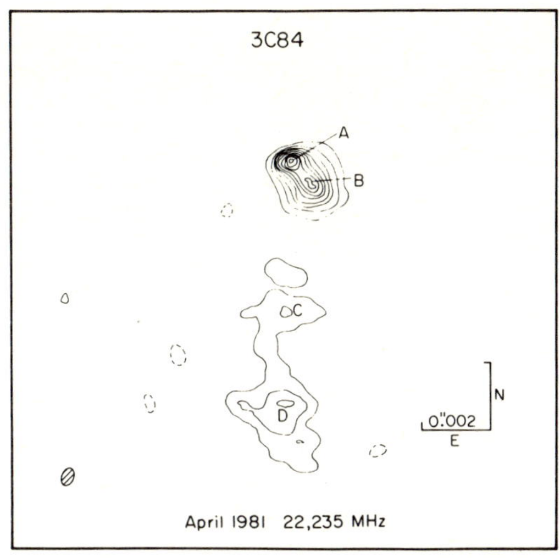

FIG. 10.   VLBI map of 3C84 at 22 GHz, from Readhead et al. 1983

65 sources with flux densities stronger than 1.3 Jy at 5.0 GHz, of which
45 have been mapped with VLBI. They have grouped the sources into 6
categories on the basis of a number of "archetypical" sources.

a) <u>Very compact sources</u>: These are sources which are only barely
resolved, if at all, with the highest resolution VLBI observations,
containing > 95% of their flux density in an unresolved core. Examples
are the sources 0016+731 and 1739+522. A property of this class is that
they are associated exclusively with quasars and BL Lac objects, and are
highly variable.

Most sources which are resolved have structures with a well defined
source axis on the milliarcsecond scale. These may be classified as
symmetric or asymmetric, by analogy with the D1 and D2 classification of
Miley (1971) for arcsecond scale structures.

b) <u>Asymmetric sources</u>: Unlike the D2 class of arcsecond scale double
structure, a very large fraction of sources fall into this class. The
well known sources 3C345 and 3C273 (Fig. 11) are typical examples,

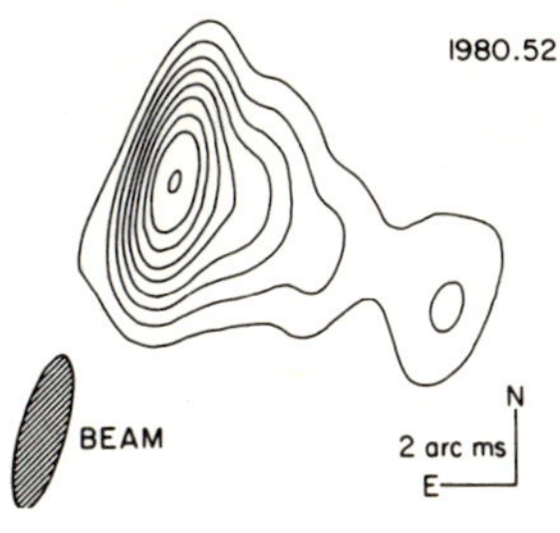

FIG. 11.   VLBI map of the asymmetric structure of 3C273, from Pearson
    et al. 1982

exhibiting a compact, normally inverted spectrum, component at one end
of the structure and a lower brightness feature of steeper spectrum at
the other. This latter component may sometimes be many beamwidths long,
giving rise to the name "core-jet" for this class.

c) <u>Symmetric sources</u>: These are also known as "compact doubles"
(Phillips and Mutel 1982b) by analogy with the Dl class of arcsecond
double sources. They have neither the structural nor spectral one-sided-
ness of class (b) but consist of 2 components of almost equal brightness
and spectrum. The prototype of this class is CTD93 (Fig. 12). The separa-

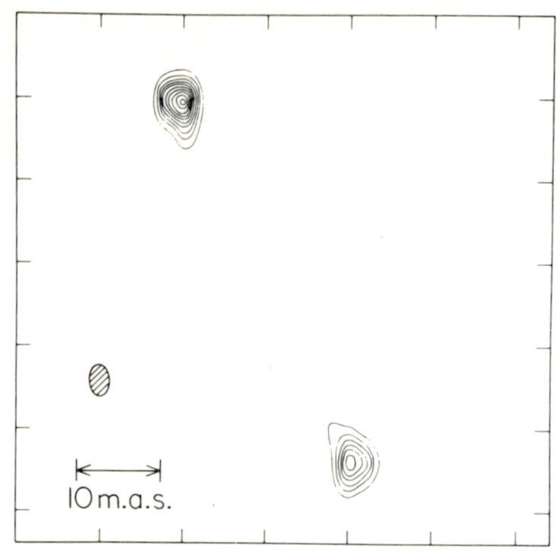

FIG. 12. VLBI map of the symmetric structure of CTD93, from Phillips
    and Mutel 1980

tion of the components is usually much larger than the extension of the
asymmetric sources. Other properties of these sources are the lack of
variability, low polarisation and lack of significant structure on the
arcsecond scale. They also tend to have very faint optical counterparts
(Zensus and Porcas these Proceedings).

Sources which are resolved but do not clearly belong in classes (b)
or (c) can be classified as:

d) <u>Complex sources</u>: This class would include 3C147 (Fig. 9b) and
3C84 (Fig. 13). Here, however, we cannot be sure that this category
represents a distinct type of source morphology; observations of 3C84
at 22 GHz (and consequently higher resolution) reveal an asymmetric
"core-jet" like structure (Fig. 10) similar to the class (b) sources.
We can also include the source 0710+439 in the complex class (Readhead
et al. 1984). Fig. 14 illustrates the structure of this triple source
at 5 and 10 GHz, which consists of a central, flat spectrum "mini core-
jet" component, straddled by 2 outer, steep spectrum "mini-lobe" components. Clearly, if the central component were weaker (or if we observed
at a lower frequency) we would identify this as a compact double source;
if the southern component were weaker (or if the map had less dynamic
range) it would be classified as asymmetric.

Pearson and Readhead present a further two categories of source based
not on the morphology of the milliarcsecond structure but on that of the
arcsecond scale. These are central components of classical doubles and
steep spectrum compact sources. It is not apparent that these form
distinct morphological classes based on their milliarcsecond structure

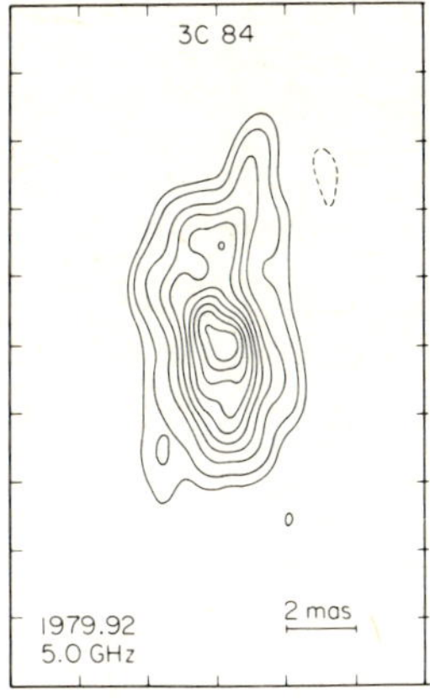

FIG. 13. VLBI map of the complex structure of 3C84 at 5 GHz, from Unwin
et al. 1982

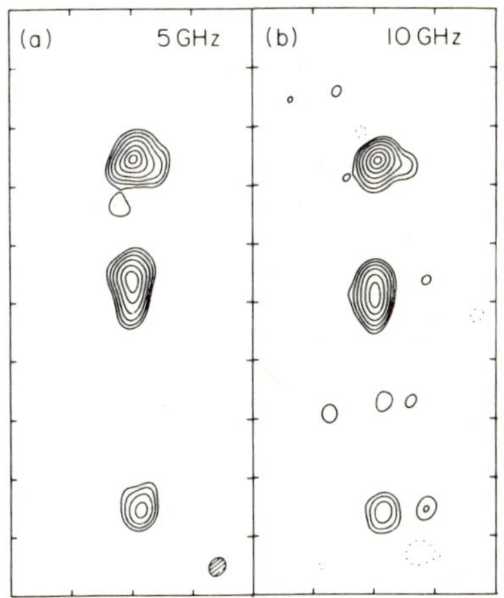

FIG. 14.   VLBI map of 0710+439 at 5 GHz and 10 GHz, from Readhead et al.
1984. Tick interval = 4 mas

alone. The central component of the classical double source Cygnus A
(Fig. 8a) clearly belongs to the asymmetric class (b). Likewise, the
milliarcsecond core of the SSC source 3C216 (Fig. 15a) also belongs to
this class, although the SSC source 3C147 is complex. Where these
categories of source do differ is in the relationship between their
milliarcsecond structures and the outer structure. Thus, central components of classical doubles always show rather good alignment between
the milliarcsecond source axis and that of the arcsecond double structure. SSC sources, by contrast, generally show a marked degree of misalignment and bending between the small scale and larger scale. This is
illustrated in Fig. 15b, c for the case of 3C216. Asymmetric, class (b)
sources which exhibit asymmetric (D2) structure on the large scale
generally show an intermediate degree of misalignment, although with a
clear continuity between the small and large scale asymmetry.

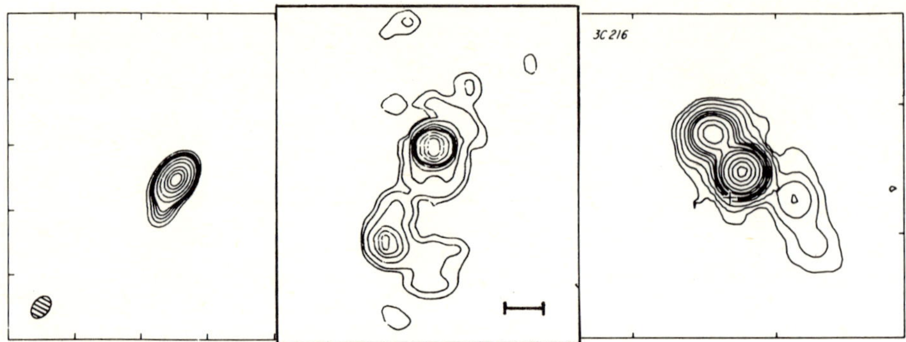

FIG. 15.   Structure of the SSC quasar 3C216
   a) mas scale, from Pearson and Readhead 1984b. Tick interval = 4 mas
   b) Intermediate scale, from Porcas and Fejes (in preparation).
      Bar = 50 mas
   c) arcsecond scale, from Pearson and Readhead 1984b. Tick interval =
      2 arcsec

Spectral and polarisation distributions

   As noted previously, different components of the milliarcsecond
structure of sources frequently have different spectra. It is of interest to find out to what extent sources with very flat total source
spectra can be decomposed into a number of components with different
spectral peaks. The source 0735+178 has a very flat spectrum over at
least three decades in frequency and has been observed with VLBI at
several different frequencies by Cotton et al. (1980). The data for this
source are modelled well by four sub-components which indeed have peaks
at different frequencies (Fig. 16). It has been shown by Cook and

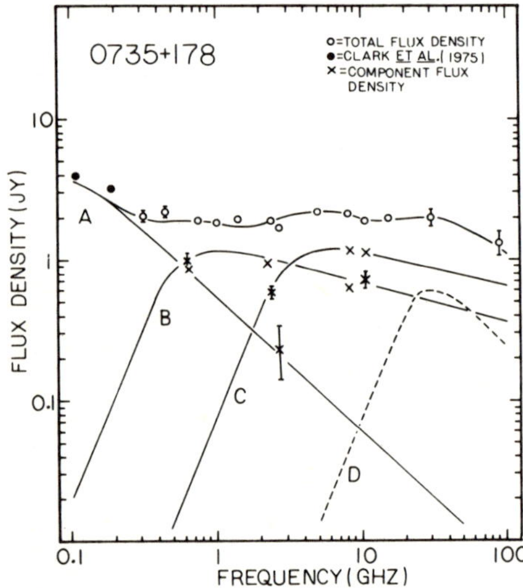

FIG. 16.   Radio Spectrum of the "cosmic conspiracy" source 0735+178,
   from Cotton et al. 1980

Spangler (1980) that such a combination of homogeneous synchrotron components can only rarely result in a flat spectrum if they are truly
independent; the frequent occurrence of this phenomenon - the source
2021+614 is another, less complex case (Wittels et al. 1982) - implies
a "cosmic conspiracy" amongst the components, which cannot, therefore,
emit independently (see also Pacholczyk 1981).

In a few cases, spectral index mapping of milliarcsecond structure
has been attempted. Unwin et al. (1982) and Readhead et al. (1983) have
made spectral index maps of 3C84. There are three main difficulties
encountered in making VLBI spectral index maps:

a) most of the strong compact sources are variable, so that observations must be made quasi-simultaneously

b) the lack of absolute position information makes the registration
of maps at different frequencies somewhat speculative

c) VLBI observations with a given array produce maps whose resolution
scales with frequency

The first two of these problems have been overcome by Marcaide and
Shapiro (1984) who made VLBI observations of the pair of quasars 1038+
528 A, B at 2.3 and 8.4 GHz simultaneously. These two sources are only
33 arcseconds apart on the sky and can thus be observed simultaneously
because they both appear within the individual antenna primary beam
patterns. Furthermore, the measured relative visibility phase on each
baseline is a good estimate of the true visibility phase difference because both quasars suffer the same telescope phase errors, and thus the
relative position of the two quasars can be determined at each frequency.
This in turn provides a very strong constraint on the registration of
the map pairs at different frequencies, particularly because of the
(apparently fortuitous) circumstance that the angle of elongation of the
quasars is in one case almost parallel, and the other perpendicular to
the line joining the two quasars. The spectral index maps of 1038+528 A,
B are shown in Fig. 17. As can be seen, each has an inverted spectrum

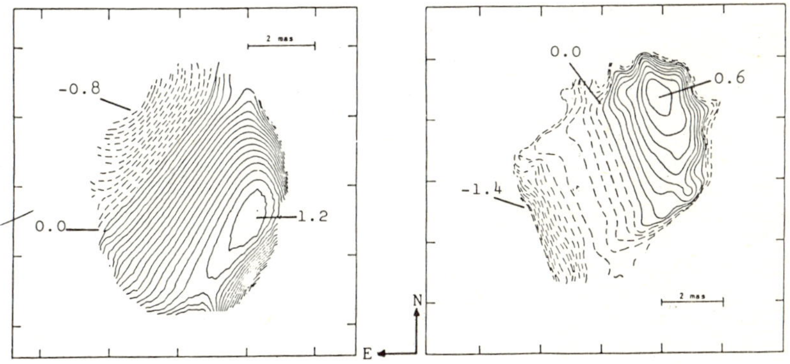

FIG. 17.   Spectral index maps of the quasar pair 1038+528 A, B, from
Marcaide and Shapiro 1984

core component at one end of the structure and an extended component,
with spectral index which steepens with distance from the core. An
important result from this work is the discovery that the position of the
peak of the brightness distribution of 1038+528 A is frequency dependent.

Determination of the milliarcsecond scale polarisation structure in
sources is particularly difficult because of the high sensitivity

required to detect the rather low percentage polarisation, and the calibration problems brought about by the presence of significant telescope
cross-polarisation residuals. Cotton et al. (1984) have made polarisation VLBI observations of the OVV quasar 3C454.3 at 2.3 GHz. This source
consists of an inverted spectrum core and a jet component about 6 mas
away in pa -65° (e.g. Pauliny-Toth et al. 1984). A plot of the polarised
flux and position angle versus distance from the core along this direction is plotted in Fig. 18. The core itself is relatively weakly polar-

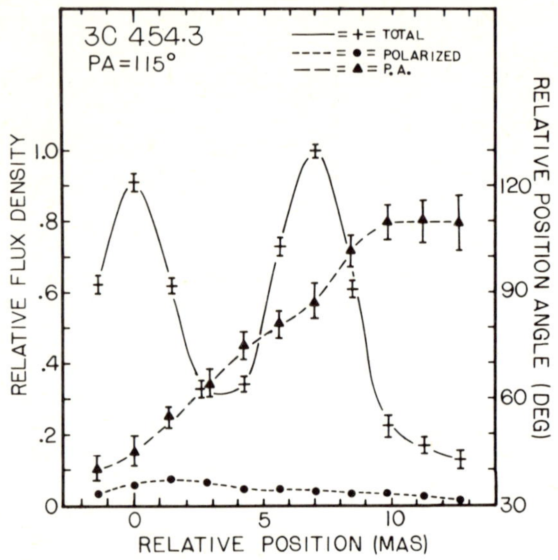

FIG. 18. Plot of polarised flux and position angle along the VLBI
   source axis of 3C454.3, from Cotton et al. 1982

ised; in addition the position angle of the polarisation swings through
an angle of $\sim 90^\circ$ with distance along the source axis. Roberts et al.
(1984) have observed the BL Lac object OJ287 at 5 GHz and also find
weak polarisation in the core, and a highly polarised jet, with significant changes in the polarisation angle along the source.

STRUCTURAL VARIATIONS

Introduction

The capability of VLBI to image regions as small as a few light years
in objects at cosmological distances provides a unique opportunity to
study structural changes in galaxies and quasars over time periods of
just a few years. During the first decade of VLBI observations, the
relatively poor sensitivity arising from the narrow (2 MHz) bandwidth
and poor receivers and coherence times required that effort be concentrated on the very strongest sources. Repeated observations of a number
of these soon uncovered the phenomenon of "superluminal" motion and this
is dealt with in a later section. The strongest source at cm wavelengths,
3C84, however, shows slower, but no less violent, structural changes.
Fig. 19 shows a sequence of maps made by Romney et al. (1984) at 10.7 GHz.
At all epochs the structure consists of a compact component in the north,
which observations at other frequencies (see Fig. 10) reveal to be an

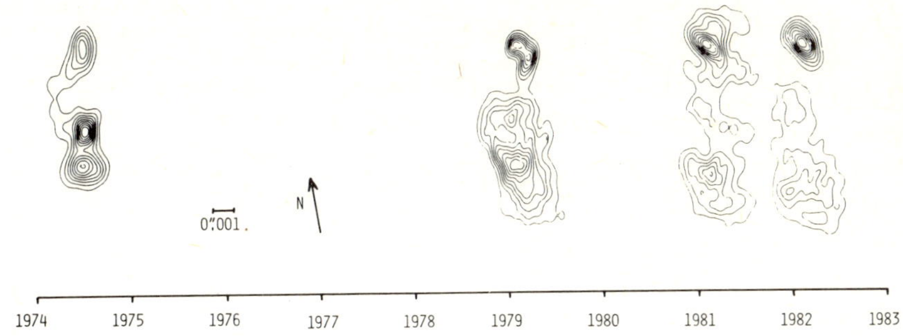

FIG. 19.   VLBI maps at 10.7 GHz of 3C84 at different epochs, from
   Romney et al. 1984

inverted spectrum core, and more diffuse components extending to the
south. The later epochs show a gradual increase in the strength of the
core component (corresponding to a flare seen in the total source flux
density in 1980) and a decrease in the diffuse components. They also
show a marked increase in separation between the core and the southern-
most feature. If one assumes that the core is stationary, then the
southern component is separating from the core at $\sim$ 0.24 mas/yr, corresponding to a velocity of $\sim$ 0.4c (taking H_0 = 55 km s^{-1} Mpc^{-1} and q_0 =
0.05, used throughout this review, and assuming a cosmological origin
for the redshift).

Superluminal sources

   The most dramatic result from VLBI monitoring of strong, compact
variable sources has been the discovery that in a number of sources,
separation velocities appear to exceed the velocity of light. The first
report of such "superluminal" behaviour for the sources 3C279 and 3C273,
(Whitney et al. 1971) was based on data taken with a single baseline
interferometer, and the structural changes were inferred from fitting
models to the amplitude of the visibility function. Ten years later,
instrumental and computing advances had proceeded to the point that a
sequence of hybrid maps of the source 3C273, produced from 10 simultaneous baselines,could be presented (Pearson et al. 1981). The measured
expansion rate in 3C273 corresponds to 10c.
   Although the number of sources exhibiting superluminal motion has now
risen to at least 7, with many others showing similar but (as yet) less
well established behaviour, it is only for a few sources that detailed
observations at a number of frequencies have been made over many years.
The best example is the quasar 3C345. A detailed study of this source
has been presented by Unwin et al. (1983). The source is of the asymmetric morphological type; a sequence of VLBI maps at both 10.7 and 5.0 GHz
is shown in Fig. 20. At each epoch and frequency the maps show a compact,
unresolved core to the east, and a number of "knots" in an elongated
"jet" region to the west. Further out, on the arcsecond scale, a jet is
also seen on this same side, although at a slightly different position
angle (Browne et al. 1982). No VLBI counter-jet is seen above 2% of the
core brightness.
   The VLBI maps of 3C345 during the period 1979 to 1981 show two knots,
whose positions with respect to the core are independent of frequency,
but move outwards with time. The proper motions of the knots are similar
(outer knot C_2 = 0.42 mas/yr, inner knot C_3 = 0.31 mas/yr) and differ by

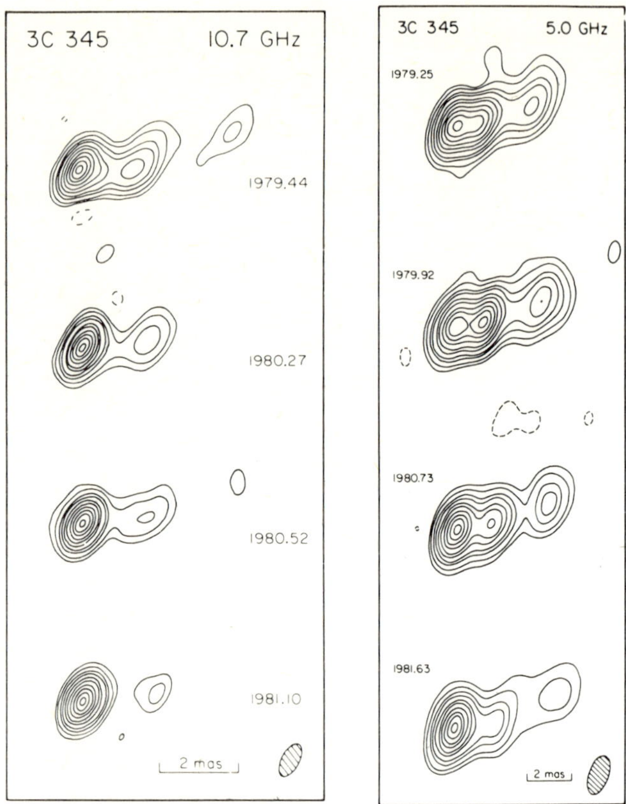

FIG. 20.   VLBI maps of 3C345 at different epochs, from Unwin et al. 1983

less than two standard deviations. The corresponding apparent transverse speeds are $\sim$ 15c and there is no departure from a linear separation rate. The position angles of the components do differ significantly, however (Cohen et al. 1983). In particular, the pa of C_2 is the same as the orientation of the source structure during the observed motions between 1971 and 1975. If the component observed at that time is identified as C_2, then an acceleration of C_2 must have occurred. The knots themselves clearly "decay" as they move out from the core. Fig. 21 shows the spectral index of the core and knots C_2 and C_3 as a function of time. The knots exhibit steep spectra and the slopes steepen with time.

Since 1981, observations at both 10.7 GHz (Biretta et al. 1983) and 22 GHz (Moore et al. 1983) have shown that a new knot, "C_4", has appeared. The position angle of C_4 is different from those of both C_2 and C_3, and is changing as the knot moves outwards; however, the locus of successive positions of C_4 seems to be a straight line which does not pass through the core. The proper motion has also changed, from 0.17 to 0.27 mas/yr. Three maps made at 22 GHz are shown in Fig. 22, which show the motion of the new components.

Until recently, all measured superluminal motions referred only to the relative motion of components within sources, because the lack of

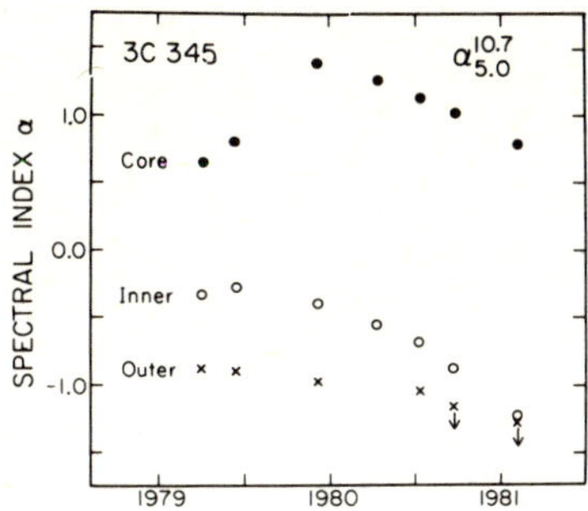

FIG. 21. Plot of spectral indices of core and knots at different epochs for 3C345, from Unwin et al. 1983

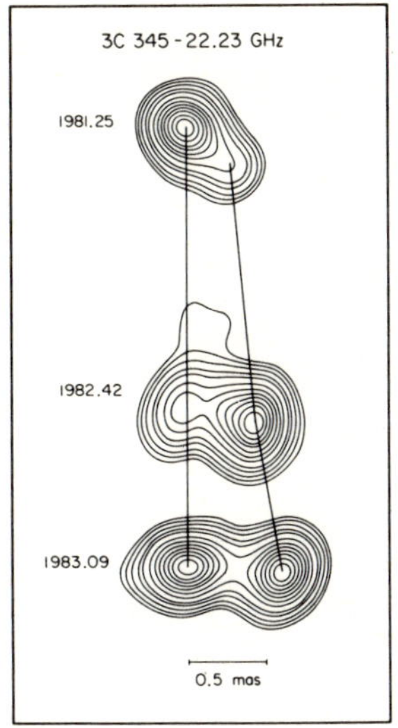

FIG. 22. VLBI maps at 22 GHz of 3C345 at different epochs, from Moore et al. 1984

absolute positions for VLBI structures does not permit an independent registration of source maps from different epochs. However, measurements of the relative visibility phase of the source pair 3C345 and the nearby quasar NRAO512 have recently allowed maps of 3C345 to be aligned with reference to the external position of NRAO512 (presumed fixed). Preliminary results reported by Bartel et al. (1984) have confirmed that the core of 3C345 has remained "fixed" for about a decade (< 0.02 mas/yr) and it is indeed the knots which move at superluminal velocities (see Fig. 23.).

A key kinematic question which is crucial for the interpretation of superluminal behaviour is whether or not successive knots follow the same trajectory, and the associated question of the constancy of motion of any particular knot. At present the different components of 3C345 have only been unambiguously discriminated in the recent observations resulting in maps, and no knot has yet moved far enough to occupy the

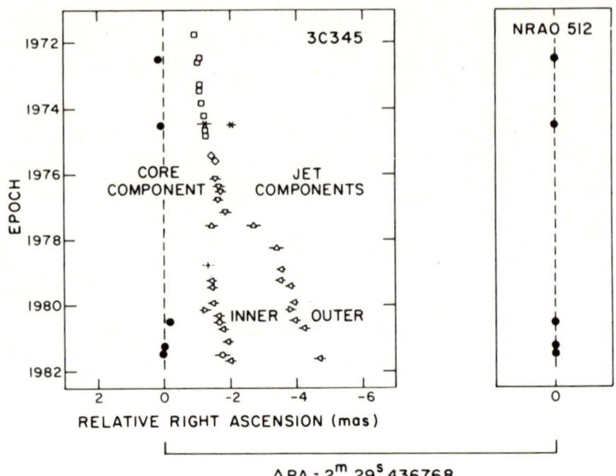

FIG. 23. Proper motion of the mas components of 3C345 with respect to
the position of NRAO512, from Bartel et al. 1984

distance from the core formerly held by its predecessor.

The observational status of superluminal sources has been reviewed
recently by Cohen and Unwin (1984) and Porcas (1983); here the current
situation is presented at the end of this section in the form of two
tables, listing the known superluminal sources, good candidates and
related sources with references (Table 1), and a summary of observational properties (Table 2). A few sources are worthy of special comment
here:

3C179 (Porcas 1981) is the only confirmed superluminal source which
was selected for monitoring on the strength of its outer, arcsecond
scale double-lobed structure and not primarily on the basis of the
strength of its core (its core flux density is 100 times weaker than
that of 3C273). A map of the arcsecond scale structure, made by combining MERLIN and VLA data (Shone et al., in preparation), is shown in
Fig. 24.

4C39.25 (Shaffer 1984) has been monitored for many years and exhibits
a structure consisting of 2 components, whose separation remained fixed
at 2.0 mas from 1972 until 1979. Recent observations suggest that the
separation is now decreasing at a rate of 0.05 to 0.1 mas/yr, corresponding to a velocity of between 1.4c and 2.8c. This is the only known
example of apparent superluminal contraction.

The popular model of superluminal motion

Whilst there have been many attempts to explain the phenomenon of
superluminal motion (Scheuer (1984) has recently reviewed them), the
most widely discussed model is the relativistic plasma jet (Blandford
and Königl, 1979). In this model the core and knot components of the
superluminal sources represent, respectively, radio emission from the
base of the jet, and shocks or plasmons in the jet flow. The critical
parameters of the model are the Lorentz factor, γ, of the bulk motion
of the plasma, and the angle, θ, of the jet with respect to the observer's line of sight. When γ is large and θ small, two effects happen:

a) superluminal motion of the knot with respect to the core as high
as γc can be observed (when $\theta \sim \gamma^{-1}$), because the observer's view of the
early part of the knot motion is delayed with respect to later parts,

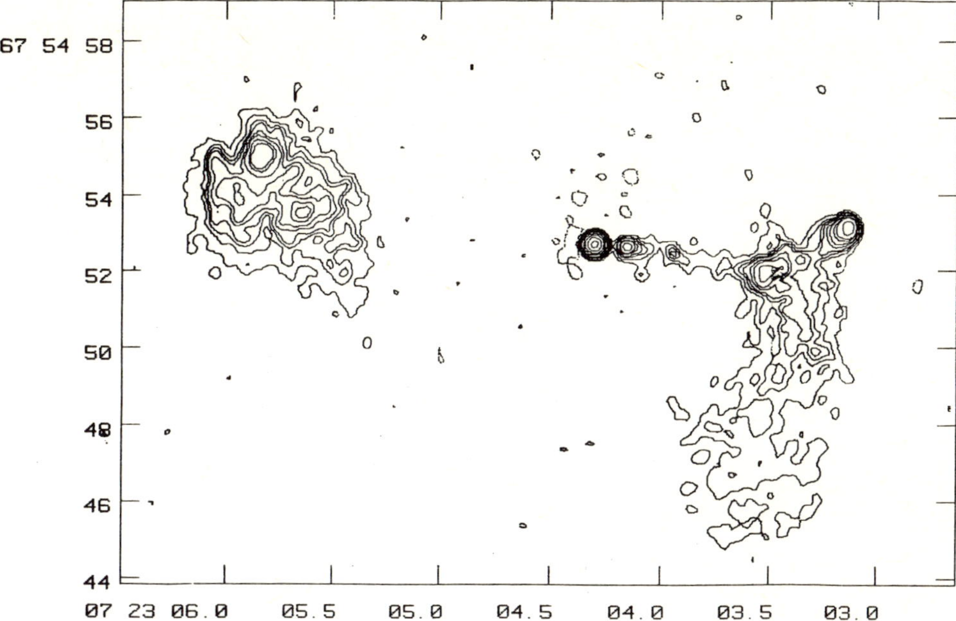

FIG. 24.   Combined MERLIN/VLA 1.6 GHz map of 3C179, from Shone et al.
(in preparation)

and hence the apparent duration of the motion is shortened.

b) <u>Doppler boosting</u> of the flux density of both the knot and core
occurs, the knot emission coming from a region in the jet with motion
towards the observer, the core emission also coming from plasma with
motion directed towards the observer, but from a <u>fixed</u> position at the
base of the jet (where the jet becomes optically <u>thick</u>).

We may expect the jet axis to be orientated sufficiently close to
the line of sight to produce an apparent superluminal motion of γc in
only a fraction $\sim \gamma^{-2}$ cases. An important question raised by the model
is thus the nature of the other $(1-\gamma^{-2})$ objects. Scheuer and Readhead
(1979) suggested that since these objects would have very weak radio
emission from the jet they could be identified with the radio quiet
quasars. However, the realization that some of the superluminal sources
have significant extended (apparently unboosted) structure on the arcsecond scale (e.g. Browne et al. 1982) has led to the suggestion that
the unaligned counterparts of the superluminal sources are themselves
radio sources, the classical double-lobed quasars (Orr and Browne 1982).

Some consequences and tests of the relativistic jet model are as
follows:

a) The conditions of high γ and small θ which give rise to the superluminal effect also produce a high Doppler factor, δ, and this explains
the rather high incidence of the effect amongst sources selected on the
basis of strong core flux density (see Table 1).

b) In samples of sources selected without any bias towards strong
core flux density, the effect should be rare (Porcas 1984). VLBI
observations of such a sample are in progress (Zensus and Porcas 1984).

c) Because of the generally good alignment of milliarcsecond and
arcsecond scale structures in classical double sources with weak cores,

we would expect any large scale structure of superluminal sources to
have a smaller projected size than an average source at the same redshifts. Observations by Schilizzi and de Bruyn (1983) show that this is
not the case!

d) There is redshift-independent evidence for high Doppler factors
in some sources. Calculation of the expected inverse-Compton X-ray flux
from measurements of the radio angular size and turnover frequency
overestimate the observed X-ray flux by very large factors. The conclusion is that the rest-frame of the emission is different from ours, and
Lorentz factors of $\sim$ 5 to 10 are sufficient to explain the discrepancy.
The superluminal motion in the source NRAO140 was "predicted" by
Marscher and Broderick (1981, 1982) on the basis of this argument.

e) Preferential Doppler enhancement of the approaching jet could
make an intrinsically symmetric source appear one-sided. However, since
the large and small-scale jets are always on the same side, the cause
of the asymmetry must be the same for both, either intrinsic or Doppler
favouritism.

f) The recent observation of acceleration of knot C_4 in 3C345, and
the "backwards" motion in 4C39.25 provide new challenges to the
relativistic jet model.

CONCLUSIONS

Inevitably in a brief review such as this it is impossible to do full
justice to such a large area of research. More detailed reviews and
reference lists can be found in the following publications: Begelman et
al. (1984), Bridle and Perley (1984), Kellermann and Pauliny-Toth (1981),
Preuss (1983, 1984). Large collections of relevant articles are published
in IAU Symposium No. 97 "Extragalactic Radio Sources" and IAU Symposium
No. 110 "VLBI and Compact Radio Sources".

I thank my colleagues Ismael Pérez-Fournon, Jon Marcaide, Eugen Preuss
and Anton Zensus for help in preparing this review, and authors and
publishers for permission to use published material.

Figs. 2B, 5 and 9A are reprinted from NATURE, Copyright 1977, 1984
Macmillan Journals Limited.

TABLE 1. <u>Superluminal (SL) and related sources</u>

Source	5 GHz Flux Density (Jy)	v/c*	Comment	Reference
3C279	16	6.5	The first SL source but difficult observationally	Unwin and Biretta 1984
3C273	40	∼10	The brightest SL source; three knots have ∼ same velocity	Unwin and Biretta 1984
3C345	10	∼15(C_2,C_3) 6→10 (C_4)	The best observed case; the core is stationary, new component has new velocity	Unwin et al. 1983 Biretta et al. 1983 Moore et al. 1983 Bartel et al. 1984
3C120	10	4–7	The only "galaxy"; multiple knots, possibly with different velocities	Walker et al. 1984 Walker et al. 1982
3C179	0.3	∼10	The weakest SL source; core of classical double	Porcas 1984 Porcas 1981
BLLac	5	4.4		Mutel and Phillips 1984 Phillips and Mutel 1982a
NRAO140	2	9	"Predicted" SL from lack of X-rays, and <u>is</u>.	Marscher and Broderick 1982
3C454.3	17	20?	Tentative velocity; possibly example of SL brightening	Pauliny-Toth et al. 1984
4C39.25	8	0, –2	Previously constant separation, now contracting	Shaffer 1984
0735+17	2	2.6		Bååth 1984
1928+73	3	6		Eckart et al. in preparation
2134+00	12	0	Fixed separation	Pauliny-Toth et al. 1984
3C446	5	0	SL brightening of fixed components	Brown et al. 1981
3C84	50	∼0.4	"Sub-luminal" motion of diffuse component	Romney et al. 1984
3C147	8	–	"Predicted" SL motion not yet observed	Preuss et al. 1984
3C263	0.15	(2±2)	Early result from weak core sample	Zensus, Hough et al. (in preparation)

*velocities calculated assuming $H_0 = 55$ km s^{-1} Mpc^{-1}, $q_0 = 0.05$

TABLE 2. <u>Summary of observational properties of superluminal sources</u>
   <u>(with exceptions)</u>.

1) <u>Milliarcsecond morphology is of "core-jet" type</u>
        a) core is compact, inverted spectrum
        b) knots in jet are steep spectrum
        c) no counter-jets are seen

2) <u>Arcsecond morphology is of "core-jet" type</u>
   <u>but</u> a) in most cases (except 3C273) there is also two-sided emission
           (Schilizzi and de Bruyn 1983), and in the case of 3C179 there
           is classical double-lobed structure (Porcas 1981, 1984)
        b) the mas and arcsecond jets are <u>always</u> on the same side.
        c) no counter-jets are seen.

3) <u>Knots in the jet have frequency-independent separation</u>

4) <u>Knot spectra steepen, with time scale of $\sim$ a few years</u>
        a) well established only for 3C345 and 3C273

5) <u>Motion is always outwards, never inwards</u>
        a) except for 4C39.25 (Shaffer 1984)

6) <u>Velocity and position angle is constant for each knot</u>
   <u>but</u> a) knot C_2 in 3C345 may have accelerated on the basis of pre-
           1975 data
        b) knot C_4 in 3C345 has changed velocity close to core
        c) knot C_4 in 3C345 has changed direction if it indeed emerged
           from the core

7) <u>Successive knots in same source have same velocity and position angle</u>
   <u>but</u> a) knots of 3C120 may have different velocities
        b) knots C_2, C_3 of 3C345 may have different velocities
        c) knot C_4 of 3C345 has different velocity from knots C_2, C_3
        d) knot C_3 of 3C345 will <u>either</u> change direction (and violate
           point 6) <u>or</u> has different direction from knot C_2

REFERENCES

Aller HD, Hodge PE, Aller MF 1981 Ap J 248:L5-L8
Altschuler DR 1980 A J   85:1559-1564
Altschuler DR, Wardle JFC 1976 Mem RAS 82:1-67
Altschuler DR, Wardle JFC 1977 MNRAS 179:153-178
Andernach H, Baker JR, von Kap-herr A, Wielebinski R 1979 A·A 74:93-99
Andrew BH, MacLeod JM, Harvey GA, Medd WJ 1978 A J 83:863-899
Bååth LB 1984 IAU Symp No 110:127-130
Bartel N, Ratner MI, Shapiro II, Herring TA, Corey BE 1984 IAU Symp No
    110:113-116
Begelman MC, Blandford RD, Rees MJ 1984 Rev Mod Phys 56:255-351
Biretta JA, Cohen MH, Unwin SC, Pauliny-Toth IIK 1983 Nature 306:42-44
Blandford RD, Königl A 1979 Ap J 232:34-48
Bridle AH, Perley RA 1984 Ann Rev A A 22: (in press)
Brown RL, Johnston KJ, Briggs FH, Wolfe AM, Neff SG, Walker RC 1981 Ap
    L 21:105-110
Browne IWA, Clark RR, Moore PK, Muxlow TWB, Wilkinson PN, Cohen MH,
    Porcas RW 1982 Nature 299:788-793
Burns JO, Feigelson ED, Schreier EJ 1983 Ap J 273:128-153
Clark BG 1973 Proc IEEE 61:1242-1248
Cohen AM, Porcas RW, Browne IWA, Daintree EJ, Walsh D 1977 Mem RAS 84:
    1-44
Cohen MH, Unwin SC 1984 IAU Symp No 110:95-103
Cohen MH, Unwin SC, Pearson TJ, Seielstad GA, Simon RS, Linfield RP,
    Walker RC 1983 Ap J 269:L1-L4
Condon JJ, Dressel LL 1973 Ap L 15:203-207
Condon JJ, O'Dell SL, Puschell JJ, Stein WA 1980 Nature 283:357-358
Cook DB, Spangler SR 1980 Ap J 240:751-758
Cornwell TJ, Wilkinson PN 1981 MNRAS 196:1067-1086
Cotton WD, Geldzahler BJ, Marcaide JM, Shapiro II, Sanroma M, Rius A
    1984 Ap J (in press)
Cotton WD, Geldzahler BJ, Shapiro II 1982 IAU Symp No 97:301-303
Cotton WD, Spangler SR (eds) 1982 Proc of the Green Bank Workshop on
    "Low Frequency Variability in Extragalactic Radio Sources" publ NRAO
Cotton WD, Wittels JJ, Shapiro II, Marcaide J, Owen FN, Spangler SR,
    Rius A, Angulo C, Clark TA, Knight CA 1980 Ap J 238:L123-L128
Davies JG, Anderson B, Morison I 1980 Nature 288:64-66
Dent WA 1965 Science 148:1458
Dreher JW, Feigelson ED 1984 Nature 308:43-45
Epstein EE, Fogarty WG, Mottmann J, Schneider E 1982 A J 87:449-461
Fanti R, Ficarra A, Mantovani F, Padrielli L, Weiler K 1979 A A Suppl
    36:359-369
Fanti R, Salvati M 1980 Proc of the 5th European Regional Meeting in
    Astronomy "Variability in Stars and Galaxies" publ Inst. d'Astro-
    physique Liège:C.3.1-C.3.18
Hargrave PJ, Ryle M 1974 MNRAS 166:305-327
Heeschen DS 1982 IAU Symp No 97:327-328
Högbom JA 1974 A A Suppl 15:417-426
Hunstead RW 1972 Ap L 12:193-200
Kellermann KI, Downes AJB, Pauliny-Toth IIK, Preuss E, Shaffer DB,
    Witzel A 1981 A A 97:L1-L4
Kellermann KI, Pauliny-Toth IIK 1969 Ap J 155:L71-L78
Kellermann KI, Pauliny-Toth IIK 1981 Ann Rev A A 19:373-410
Kellermann KI, Sramek R, Shaffer D, Schmidt M, Green R 1983 Proc of the
    24th Liège International Astrophysical Colloquium "Quasars and
    Gravitational Lenses":81-83, University of Liège Institute d'Astro-
    physique

Kühr H, Nauber U, Pauliny-Toth IIK, Witzel A 1979 "A Catalogue of Radio
    Sources" MPIfR Preprint No 55
Kühr H, Pauliny-Toth IIK, Witzel A, Schmidt J 1981 A J 86:854-863
Lebofsky MJ, Rieke GH, Walsh D 1983 MNRAS 203:727-734
Lilly SJ, Longair MS 1982 IAU Symp No 97:413-422
Longair MS 1975 MNRAS 173:309-326
Marcaide JM, Shapiro II 1984 Ap J 276:56-59
Marscher AP, Broderick JJ 1981 Ap J 247:L49-L52
Marscher AP, Broderick JJ 1982 Ap J 255:L11-L15
Miley GK 1971 MNRAS 152:477-490
Moffet AT 1975 in "Galaxies and the Universe" Stars and Stellar Systems
    9:211-281 eds Sandage A, Sandage M and Kristian J, Univ. of Chicago
    Press
Moore RL, Biretta JA, Readhead ACS, Bäåth L 1984 IAU Symp No 110:109-110
Moore RL, Readhead ACS, Bäåth L 1983 Nature 306:44-46
Mutel RL, Phillips RB 1984 IAU Symp No 110:117-118
Napier PJ, Thompson R, Ekers RD 1983 Proc IEEE 71:1295-1320
O'Dea CP, Dent WA, Balonek TJ, Kapitzky JE 1983 A J 88:1616-1625
Orr MJ, Browne IWA 1982 MNRAS 200:1067-1080
Owen FN, Hardee PE, Bignell RC 1980 Ap J 239:L11-L15
Owen FN, Porcas RW, Neff SG 1978 A J 83:1009-1020
Owen FN, Puschell JJ 1984 A J 89:932-957
Owen FN, Puschell JJ, Laing RA 1982 IAU Symp No 97:435-436
Pacholczyk AG 1981 Ap L 21:87-92
Pauliny-Toth IIK, Kellermann KI 1966 Ap J 146:634-645
Pauliny-Toth IIK, Porcas RW, Zensus A, Kellermann KI 1984 IAU Symp No
    110:149-152
Pearson TJ, Readhead ACS 1984a Ann Rev A A 22:(in press)
Pearson TJ, Readhead ACS 1984b IAU Symp No 110:15-24
Pearson TJ, Unwin SC, Cohen MH, Linfield RP. Readhead ACS. Seielstad GA.
    Simon RS. Walker RC 1981 Nature 290:365-368
Pearson TJ, Unwin SC, Cohen MH, Linfield RP, Readhead ACS, Seielstad GA,
    Simon RS, Walker RC 1982 IAU Symp No 97:355-356
Phillips RB, Mutel RL 1980 Ap J 236:89-98
Phillips RB, Mutel RL 1982a Ap J 257:L19-L21
Phillips, RB, Mutel RL 1982b A A 106:21-24
Porcas RW 1981 Nature 294:47-49
Porcas RW 1983 Nature 302:753-754
Porcas RW 1984 IAU Symp No 110:157-161
Porcas RW, Urry CM, Browne IWA, Cohen AM, Daintree EJ, Walsh D 1980
    MNRAS 191:607-614
Preston RA, Jauncey DL, Tsioumis A, Meier DL, Skjerve LJ, Ables J,
    Batchelor R, Faulkner J, Hamilton PM, Haynes R, Johnson B, Louie A,
    McCulloch P, Moorey G, Morabito DD, Nicolson G, Niell AE, Robertson
    JG, Royle GWR, Slade MA, Slee OB, Watkinson A, Wehrle AE, Wright A
    1984 IAU Symp No 97:67-69
Preuss E 1983 in "Astrophysical Jets" Proc of an international workshop
    held in Torino:1-25 eds Ferrari A, Pacholczyk AG, publ. D. Reidel Co.
Preuss E 1984 IAU Symp No 110:251-256
Preuss E , Alef W, Whyborn N, Wilkinson PN, Kellermann KI 1984 IAU Symp
    No 110:29-30
Readhead ACS, Hough DH, Ewing MS, Walker RC, Romney JD 1983 Ap J 265:107-
    131
Readhead ACS, Pearson TJ, Unwin SC 1984 IAU Symp No 110:131-134
Rees MJ 1967 MNRAS 135:345-360
Reid MJ, Schmitt JHMM, Owen FN, Booth RS, Wilkinson PN, Shaffer DB,
    Johnston KJ, Hardee PE 1982 Ap J 263:615-623

Rickett BJ, Coles WA, Bourgois G 1984 A A 134:390-395
Riley JM, Longair MS, Gunn JE 1980 MNRAS 192:233-241
Roberts DH, Potash RI, Wardle JFC, Rogers AEE, Burke BF 1984 IAU Symp
    No 110:35-38
Rogers AEE, Cappallo RJ, Hinteregger HF, Levine JI, Nesman EF, Webber JC,
    Whitney AR, Clark TA, Ma C, Ryan J, Corey BE, Counselman CC, Herring
    TA, Shapiro II, Knight CA, Shaffer DB, Vandenberg NR, Lacasse R,
    Mauzy R, Rayhrer B, Schupler BR, Pigg, JC 1983 Science 219:51-54
Rogers AEE, Hinteregger HF, Whitney AR, Counselman CC, Shapiro II,
    Wittels JJ, Klemperer WK, Warnock WW, Clark TA, Hutton LK, Marandino
    GE, Rönnäng BO, Rydbeck OEH, Niell AE 1974 Ap J 193:293-301
Romney JD, Alef W, Pauliny-Toth IIK, Preuss E, Kellermann KI 1984 IAU
    Symp No 110:137-140
Scheuer PAG 1984 IAU Symp No 110:197-205
Scheuer PAG, Readhead ACS 1979 Nature 277:182-185
Schilizzi RT, de Bruyn AG 1983 Nature 303:26-31
Shaffer DM 1984 IAU Symp No 110:135-136
Shapirovskaya NYa 1978 Sov Ast 22:544-547 (English translation)
Sholomitskii GB 1965 Sov Ast 9:516 (English translation)
Sieber W 1982 A A 113:311-313
Strittmatter PA, Hill P, Pauliny-Toth IIK, Steppe H, Witzel A 1980 A A
    88:L12-L15
Thompson AR, Clarke BG, Wade CM, Napier PJ 1980 Ap J Suppl 44:151-167
Unwin SC, Biretta JA 1984 IAU Symp No 110:105-107
Unwin SC, Cohen MH, Pearson TJ, Seielstad GA, Simon RS, Linfield RP,
    Walker RC 1983 Ap J 271:536-550
Unwin SC, Mutel RL, Phillips RB, Linfield RP 1982 Ap J 256:83-91
van Breugel W, Miley G, Heckman T 1984 A J 89:5-22
van der Laan H 1966 Nature 211:1131-1133
Walker RC, Benson JM, Seielstad GA, Unwin SC 1984 IAU Symp No 110:121-124
Walker RC, Seielstad GA, Simon RS, Unwin SC, Cohen MH, Pearson TJ, Lin-
    field RP 1982 Ap J 257:56-62
Whitney AR, Shapiro II, Rogers AEE, Robertson DS, Knight CA , Clark TA,
    Goldstein RM, Marandino GE, Vandenberg NR 1971 Science 173:225-230
Wilkinson PN, Booth RS, Cornwell TJ, Clark RR 1984 Nature 308:619-621
Wilkinson PN, Readhead ACS, Purcell GH, Anderson B 1977 Nature 269:764-
    768
Wilson AS, Ulvestad JS 1983 Ap J 275:8-14
Wittels JJ, Shapiro II, Cotton WD 1982 Ap J 262:L27-L30
Witzel A 1981 in Proc of the 5th Göttingen-Jerusalem Symposium:99-110
    eds Fricke KJ, Shaham J, Vandenhoek and Ruprecht
Zensus JA, Porcas RW 1984 IAU Symp No 110:163-164
Zensus JA, Porcas RW, Pauliny-Toth IIK 1984 A A 133:27-30

SIMULTANEOUS DUAL-WAVELENGTH VLBI OBSERVATIONS OF THE COMPACT RADIO
SOURCE NEAR THE GALACTIC CENTER

J.M. Marcaide
Max-Planck-Institut für Radioastronomie, Bonn, F.R.G.

N. Bartel, N.L. Cohen, M.V. Gorenstein, I.I. Shapiro
Harvard-Smithsonian Center for Astrophysics, Cambridge, MA, U.S.A.

B.E. Corey, A.E.E. Rogers
Haystack Observatory/NEROC, Westford, MA, U.S.A.

R. Bonometti
Massachusetts Institute of Technology, Cambridge, MA, U.S.A.

R.A. Preston
Jet Propulsion Observatory, Pasadena, CA, U.S.A.

J.D. Romney
National Radio Astronomy Observatory, Charlottesville, VA, U.S.A.

SUMMARY

Preliminary results from simultaneous VLBI observations at 3.6 and 13
cm confirm that, in this wavelength range, the size of the compact radio
source near the Galactic center varies as λ^2, where λ is the wavelength
of observation. These results also establish a flux density upper limit
at 3.6 cm of 5 mJy for any source in Sgr A West smaller than 8 mas
(80 AU).

1. INTRODUCTION

Three milestones in the study of the compact radio source near the
Galactic center were:
1. the discovery of the compact radio object itself in Sgr A West by
   Balick and Brown (1974);
2. the determination by Davies, Walsh and Booth (1976) that the source
   size varies as λ^2, where λ is the wavelength of observation and their
   suggestion that, therefore, the measured sizes are not intrinsic; and
3. the claim by Kellermann et al. (1977) that the source has another,
   even more compact, source embedded in it. From their 3.6 cm data,
   they estimated 10 AU as the size of this core.
   Kellermann et al.'s result was confirmed by Geldzahler et al. (1979),
using data taken in 1977, but without supporting details being given.
   The group led by Lo (see for example Lo et al., 1981) failed in many
attempts, since 1975, to detect this core at observing wavelengths between 1.35 to 6 cm. However, in most cases they did detect the (less)
compact source. They found the source size at 2.8 cm to be consistent
with the λ^2 law, but at 1.35 cm they found the size to be about 50%
larger than expected, which they interpret as evidence for detection of
the intrinsic source. At the three epochs for which they observed the
source at 3.6 cm, the size determined along PA $\sim$ 136 deg were 14, 18 and
13 mas (1 mas = 0''.001), each with standard errors of about 2 mas. They
do not assign any significance to the large difference between the estimate for the second epoch and those from the other two. [The sizes for
the first two epochs are the FWHM of the circular gaussian model used to

fit the data. To analyze the data from the last epoch they used an elliptical gaussian model and found that the source appeared elongated along PA $\sim$ 107 deg with eccentricity 0.8 and major axis 15 mas (Lo,1984).]

The determination of the positions of the radio, optical, and infrared sources in the Galactic center region is difficult and presently there are large discrepancies in the estimates from different groups (see, for example, Willner et al., 1983). For the compact radio source the most reliable is probably Brown et al.'s (1981) position, based on VLA data.

Using data that span 6 years, Backer and Sramek (1982) have attempted to determine the proper motion of the compact radio source with respect to two nearby (assumed extragalactic) compact radio sources. They obtained an upper bound of about 3 mas/year for PA $\simeq$ 46 and a bound about ten times higher for PA $\simeq$ 136 deg.

2. OBSERVATIONS

We observed the Galactic center at 3.6 cm on May 8th, 1983 and simultaneously at 3.6 and 13 cm on May 10th, both days with a seven antenna VLBI array and MkIII instrumentation. Our purpose was threefold:
1. to try to detect the core with the most sensitive equipment available;
2. to check the λ^2 law with dual-wavelength data obtained simultaneously; and
3. to estimate the position of the compact radio source in the reference frame of extragalactic radio sources.

We have preliminary results relevant to our first two goals but none as yet for the third. Our expectations are that we will have a one standard deviation error of about 25 mas in each of right ascension and declination.

3. RESULTS

The radio telescopes at Effelsberg[1] (100 m diameter) and Robledo[2] (64 m diameter) formed our most sensitive interferometer for detecting the core component while being insensitive to the compact source which should be totally resolved for this baseline. We estimate that 5 mJy is the 8σ upper limit for detection, appropriate for the MkIII system and part of the parameter space that we searched for evidence of this core. This limit is about 40 times lower than the flux density of the core estimated by Kellermann et al. (1977). Since all of our calibrators were detected, we are confident that our equipment was performing well.

Our preliminary results are presented in Fig. 1. At 3.6 cm, the source was detected 10 times (over a range of 4 1/2 hours) in the Goldstone[3]-OVRO[4] (DO) interferometer, but only once with the Haystack[5]-Green Bank[6] (KG) interferometer. From the data from the interferometer

---

[1] near Bonn, F.R.G., belongs to the Max-Planck-Institut für Radioastronomie

[2] near Madrid, Spain, belongs to National Aeronautics and Space Administration

[3] in the Mojave Desert, CA, belongs to the National Aeronautics and Space Administration

[4] in the Owens Valley, CA, belongs to the California Institute of Technology

[5] in Westford, MA, belongs to the Northeast Radio Observatory Corporation

[6] in Green Bank, WV, belongs to the National Radio Astronomy Observatory

DO, we estimate the diameter of the source to be 16 mas, with a standard
error of about 3 mas. At 13 cm, the source was detected only three times,
over a range of about one hour, with the DO interferometer. Except for a
missed observation at the beginning, the other failures to detect the
source were due to its being totally resolved. We estimate the size at
this wavelength to be 195 ± 30 mas.

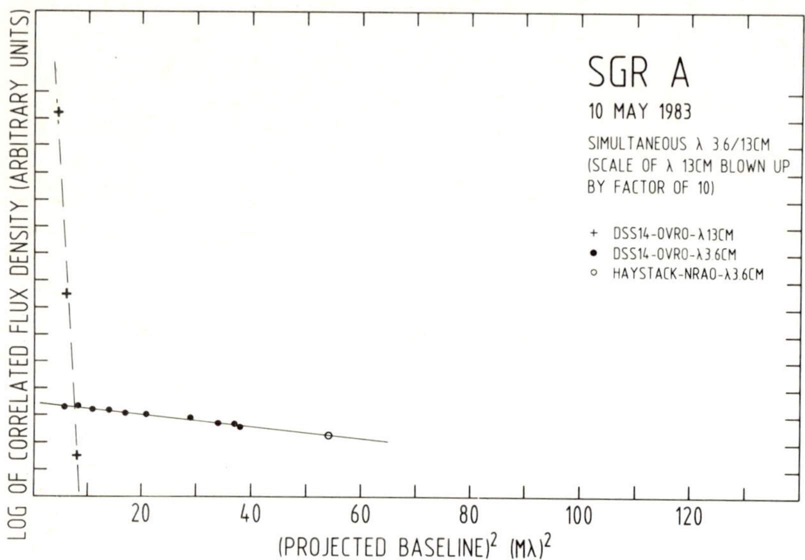

Fig. 1. Logarithm of the fringe amplitudes of the compact radio source
   near the Galactic center at 3.6 and 13 cm vs. the square of the projected baselines. Gaussian models at each wavelength are represented
   by straight lines whose slopes are proportional to the square of the
   scale sizes. (The radiotelescopes indicated by DSS14 and NRAO in this
   figure are those in Goldstone and Green Bank, respectively. See text.)

Our preliminary check on the λ^2 dependence shows that the source size
increases as $\lambda^{1.9 \pm 0.2}$.

Our best estimates of the total flux density of the compact source
are 1.2 and 0.7 Jy at 3.6 and 13 cm, respectively. These preliminary
estimates are uncertain by $\sim$ 20%. The corresponding spectral index is
$\alpha \sim 0.4$ ($S \alpha \nu^{\alpha}$) for this wavelength range.

4. DISCUSSION

We confirm that for 3.6 and 13 cm the compact source in Sgr A West
has a λ^2-dependent size, where λ is the wavelength of observation. The
13 cm wavelength represents the longest at which this source has been
observed with VLBI. Our best estimate of the diameter of the source at
3.6 cm, based on a model of a circular gaussian, is 16 ± 3 mas. Our datum from the KG interferometer which is sensitive to structure along
PA $\sim$ 100 deg, does not indicate any extension in this direction. The rest
of our data were sensitive to structure along PA $\sim$ 136 deg.

Our results from data obtained 2 days apart indicate that there are
no discernible flux density or structural variations on these time
scales, at least for these epochs. We can also rule out any time varia-

tions larger than 5% for time scales from 5 to 800 sec. This lack of
variability, and our inability to detect the core component at a flux
density level about 40 times lower than found twice in earlier observations, may indicate that the Galactic center nucleus has phases of low
activity. In any case, our very low upper limit on the flux density of
any core should constrain the proposed physical conditions for the
"central engine". A series of frequent VLBI observations of the Galactic
center are needed to test the variability conjecture. At the very least,
since 1975 our "active" Galactic nucleus has not been very active in the
radio domain. It remains to be seen whether it ever becomes active in
this domain.

Acknowledgement
    J.M.M.'s research at the Max-Planck-Institut für Radioastronomie was
supported in part by the NASA-JPL Contract 956542.

References
Backer DC, Sramek RA 1982 Ap J 260:512
Balick B, Brown RL 1974 Ap J 194:265
Brown RL, Johnston KJ, Lo KY 1981 Ap J 250:155
Davies RD, Walsh D, Booth RS 1976 M N R A S 177:319
Geldzahler BJ, Kellermann KI, Shaffer DB 1979 A J 84:186
Kellermann KI, Shaffer DB, Clark BG, Geldzahler BJ 1977 Ap J Lett 214:
    L61
Lo KY, Cohen MH, Readhead ACS, Backer DC 1981 Ap J 249:504
Lo KY 1984 in IAU Symposium No 110 "VLBI and Compact Radio Sources", ed.
    R. Fanti, K.I. Kellermann and G. Setti (Dordrecht: Reidel), p. 265
Willner SP, Schild RE, Pipher JL 1983, A J 88:177

High resolution observations of optically weak flat spectrum radio
sources

J. Anton Zensus and Richard W. Porcas

Max-Planck-Institut für Radioastronomie, Bonn, F.R.G.

SUMMARY

We present high resolution radio maps of 4 flat spectrum radio objects which have very faint optical counterparts.These objects show
either relatively compact 'core-jet' structure or are 'compact double'
sources. We suggest a possible link between the absence of optical
activity and the existence of compact radio double structure.

INTRODUCTION

Very Long Baseline Interferometry (VLBI) observations have shown
that essentially all flat or inverted spectrum radio sources exhibit
compact structure on the milliarcsecond (m.a.s.) scale (Zensus et al.
1984). Many such sources show 'core-jet' structures conventionally interpreted by relativistic jet models (Blandford and Königl 1979). Most are
identified with 'Active Galaxies' (e.g. quasars, BL Lac objects), the
activity being indicated by the presence of strong, non-thermal, optical
continuum and broad line emission (see e.g. Kellermann and Pauliny-Toth
1981). However, a small number of flat spectrum radio sources do not
have 'active' optical counterparts (i.e. they are 'empty field' identifications).
In order to investigate whether this apparent lack of nuclear activity in these sources can be linked to the morphology of the compact
radio regions, we have made high resolution, VLBI observations of 4
such objects (see table 1). They have been selected from the MPIfR-NRAO
5 GHz surveys (Kühr et al. 1981; references therein) and fulfil the
following criteria:
a) they are listed as 'empty fields' from Palomar Sky Survey identifications (subsequently, all four sources have been identified with
   faint objects, see table 1).
b) they have radio spectra which are flat over a wide range of frequencies. The radio spectra are shown in Fig. 1.

TABLE 1.   Source List

Name	5 GHz Survey Flux Density (Jy)	Identification
0026+34	1.27	Gal m_V=20.2 (Peacock et al. 1981)
0218+35	1.16	Gal m_V=20.0 (Johnson 1974)
0602+67	1.06 (variable)	SO (Meisenheimer and Röser 1983)
1504+35	1.10 (variable)	BL/OVV (Kühr, priv. communication)

The initial transatlantic VLBI observations at 5 GHz were followed by
other VLBI and also MERLIN observations for 2 of the sources. Details
of the observations will be published elsewhere. Here we present results
from this project and give a short summarizing discussion.

RESULTS

0026+34
The 5 GHz map of this source shows double structure with two
almost unresolved components of $\sim$ 25 m.a.s. separation. This appears
also in a 1.66 GHz map made subsequently (Figure 2a). A comparison of
the two maps suggests that the components have similar, flat spectra.
Both maps contain almost the total flux density of the source.

0218+35
The 5 GHz VLBI data for this source indicate the existence of two
components of several hundred m.a.s. separation and therefore could not
be easily mapped. A map obtained at 1.6 GHz with the European VLBI Network confirms the double structure with an angular separation of
$\sim$ 340 m.a.s. This is shown in the MERLIN map at 5 GHz (Figure 2b).
Again the two components show similar, flat spectra. However, the
higher dynamic range of the MERLIN map reveals low surface brightness
emission around the weaker Northern component.

0602+67
This source is only slightly resolved at 5 GHz (Figure 2c), with
most of the flux density contained in the map.

1504+37
The 5 GHz map (Figure 2d) shows a 'core-jet' structure, with bending
occurring in the extended region. Most of the total flux density is
contained in this map.

DISCUSSION

The morphologies of the 4 sources observed in this project fit into
the classification scheme of Pearson and Readhead (1984) for the structures of flat spectrum radio sources. The 2 objects with variable flux
density, 0602+67 and 1504+37, have simple asymmetric but compact structures, similar to many 'core-jet' type sources. In contrast, 0026+34
and 0218+35 show the characteristic features of the 'compact double'
class proposed by Phillips and Mutel (1982), exhibiting widely spaced,
symmetric double structure and similar component spectra. Phillips and
Mutel (1982) selected such sources on the basis of a peaked radio spectrum with a turnover frequency around 1 GHz. They interpreted the
symmetric structure as emission from non-relativistically moving regions,
possibly energized by relativistic beams with a small angle to the line-
of-sight. Recent studies show that such truly symmetric structures are
rather rare, even amongst those sources with peaked radio spectra
(Phillips and Mutel, 1984). We note that of the six compact doubles
they found at least three are optically very faint ('empty field' or
m_B > 20). This, together with our results, seems to suggest a link between compact double structure and optical faintness.

FIG.1. Radio Continuum
spectra of the four
sources (from Kühr
et al. 1981 updated)

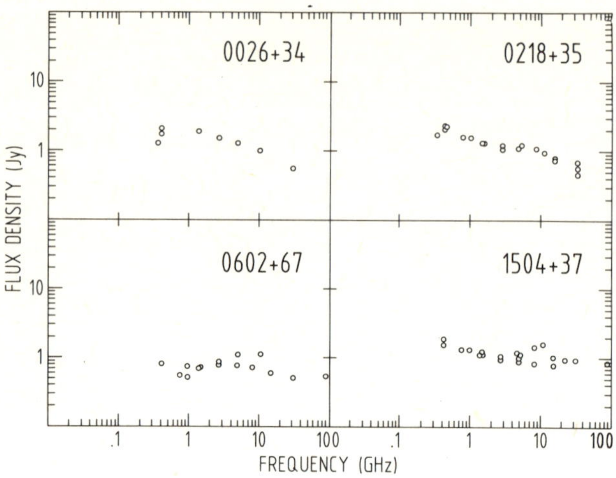

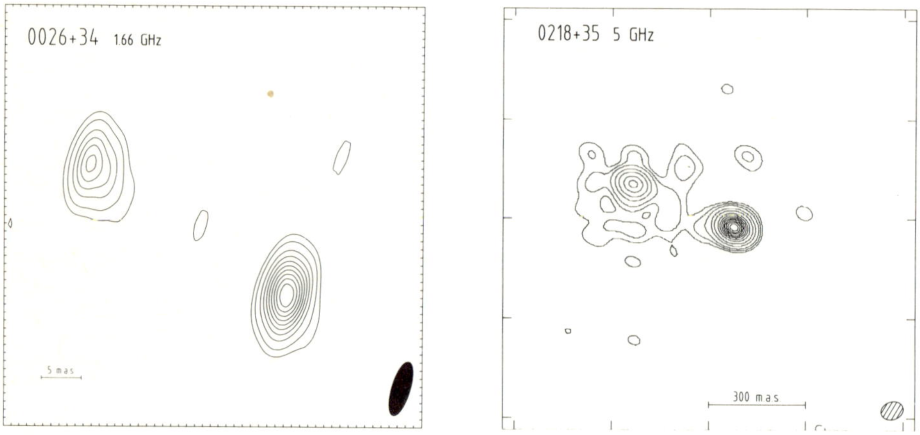

FIG. 2. Maps of the four sources:
  a) 0026+34   b) 0218+35   c) 0602+67   d) 1504+37
  For comments see text.

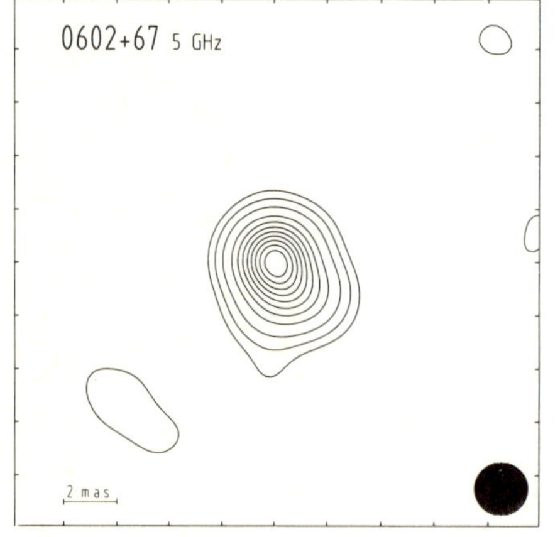

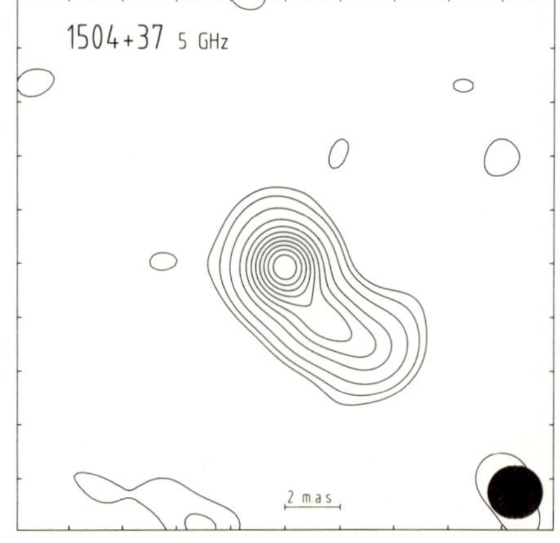

ACKNOWLEDGEMENT

We thank Dr. H. Kühr for providing an identification prior to publication and Dr. Arno Witzel for helpful comments.

REFERENCES

Blandford RD, Königl A 1979, Astrophys J 232:34
Johnson KH 1974, Astron J 79:1006
Kellermann KI, Pauliny-Toth IIK 1981, Annual Rev Astron Astrophys
    eds. G. Burbidge, D. Layzer and J.G. Phillips 19:373
Kühr H, Witzel A, Pauliny-Toth IIK, Nauber U 1981, Astron Astrophys
    Suppl 45:367
Meisenheimer K, Röser HJ 1983, Astron Astrophys Suppl 51:41
Peacock JA, Perryman MAC, Longair MS, Gunn JE, Westphal JA 1981,
    Monthly Notices Roy Astron Soc 194:601
Pearson TJ, Readhead ACS 1984, Proc IAU Symposium No. 110, eds. R.
    Fanti, K.I. Kellermann, G. Setti, publ. Reidel Dordrecht, p. 15
Phillips RB, Hodges MW, Mutel RL 1984, Proc IAU Symposium No. 110, eds.
    R. Fanti, K.I. Kellermann, G. Setti, publ. Reidel, Dordrecht, p. 63
Phillips RB and Mutel RL 1982, Astron Astrophys 106:21
Zensus JA, Porcas RW, Pauliny-Toth IIK 1984, Astron Astrophys 133:27

Very Large Array Observations of 4C 50.55 (BG 2122+50)

P. Tomasi

University of Manchester, Nuffield Radio Astronomy Laboratories,
Jodrell Bank, Macclesfield, Cheshire SK11 9DL

and

Instituto di Radioastronomia C.N.R., Via Irnerio 46, 40126 Bologna,
Italy

SUMMARY

In this paper the author presents the preliminary results of the Very
Large Array observations of the source 4C 50.55 at 1464.9 MHz. The
source, on the Galactic Plane, was already recognized as extragalactic
by previous low resolution observations. In the present observations it
shows a strong one side extended jet and a barely resolved nucleus. Its
classification as powerful radio source with active nucleus is suggested.

INTRODUCTION

The source 4C 50.55 (Gower et al. 1967; R.A. (1950) = $21^h22^m44.2^s$,
Dec(1950) = $50^o51.1'$) was selected from the BG survey, a survey of the
Galactic plane at 408 MHz (Fanti et al. 1974), for observations with the
Westerbork radio telescope at 1415 MHz. At that time, from a statistical
analysis on the BG survey (Fanti et al. 1975), a Galactic origin for
this source, was guessed. It should be remembered, however, that in
the whole selected sample (42 objects) we expected 3 ± 2 extragalactic
object (Fanti et al. 1981).
    In fact, the Westerbork observations at 1415 MHz (Fanti et al. 1981)
revealed what appeared to be a large angular diameter ($\theta \simeq 12'$) extragalactic double source, with a slight asymmetry, and a rather strong
nucleus. The total flux density at 1415 MHz is 2.75 ± 0.12 Jy, and a
research for compact sources discovered an unresolved component (upper
diameter limit 18"x20") at the position of the central core, with a
flux density of 188 ± 13 mJy (Fanti et al. 1981).
    More observations were produced by Mantovani et al. (1982), using
the Effelsberg-100m radio telescope at 1720 and 10,700 MHz, and by Goss
et al. (1984) using the Westerbork radio telescope at 608.5 MHz and the
Effelsberg-100m radio telescope at 4750 MHz, studying the complex
Galactic area surrounding the source 4C 50.55. New measured integrated
flux densities together with previous ones on the Baars et al. scale
(1977) were provided in these papers. The resulting best-fit spectral
index of the integrated flux densities is α = -0.55 ± 0.03 ($S \propto \nu^\alpha$). No
evidence of deviation from a straight line appears between 178 MHz and
10.7 GHz. The spectral index distribution is almost constant along
the source, but the nucleus shows a very flat spectrum (Mantovani et al.
1982).
    A linearly polarized component from the emission on the western lobe
at 4750 MHz (7% of the emission in the area) was discovered by Goss et
al. (1984) suggesting the presence of a jet in the area. This and the
spectral index distribution are very similar to many other giant radio

galaxies.For that the source was observed with higher resolution in order to study in detail its structure.

OBSERVATIONS

The source was inserted in an already scheduled Very Large Array program (VLA of the National Radio Astronomy Observatory) at 1464.90 MHz, and observed with B configuration, that means the spatial frequencies lower than about 1 Kλ were missed, and for that, part of the extended emission from the source is not visible.

In order to reduce the problems connected with the primary beam attenuation and the field of view, due to angular extension of the source, the radio telescope was pointed at four different positions along the source, observed for about 20 minutes each. In the data presented here, only three of these observations were combined together. The relevant parameters of the observations are summarized in Table 1.

TABLE 1: OBSERVING PARAMETERS

Frequency (MHz)	1464.9
Bandwidth (MHz)	50
Integration time (sec)	30
N. antennas	26
Duration (min)	20 x 3 (20)*
Configuration	B
Min. baseline (Kλ)	1
Telescope position (1950)	$21^h22^m44.7^s$ $50^o47'01.0''$ 57.2^s $45'24.0''$ $23^m10.0^s$ $44'02.0''$ ($23^m21.6^s$ $42'30.0''$)*

(*) Not used in the present data reduction

The data were processed using the Astronomical Image Processing System (A.I.P.S.) resident on the VAX of the Astronomy Department in Manchester. In a first step, each data base was processed independently using Clean and self-calibration. In this first step only few clean components on the central source were inserted in the task performing the self-calibration. Then the UV-data were combined together using a new multi-purpose task produced by Baath (1984). After that more Clean and self-calibration iterations were used, until a reasonable convergence was achieved. A "zero-spacing" of 1.2 Jy was inserted to minimize the zero level offset, and improve the final cleaned result.

The cleaned map was restored with a 4"x4" beam, just a bit larger than the VLA resolution at this frequency , in order to improve the signal to noise ratio, and it is shown in Fig. 1.

DISCUSSION

As it is possible to see in Fig.1, the source has a shape in gross as expected from lower resolution observations; a barely resolved central core (diameter: 1.2"x1.2", gaussian fit), with a total flux density of 178 mJy, 14% of the total flux density in the map; a strong

one sided jet, far away from the nucleus (starting 2 arc min away from
the core), about 2 arc min long. It appears well confined for a while,
with ripples and slightly increasing surface brightness, then the surface brightness almost goes down to the r.m.s. noise level (0.5 mJy/beam)
with a sharp bend. An increase of the emission is following in what
seems to be a jet elongation, with the same direction as the previous
section. In this area the jet shows less confinement and perhaps some
evidence of no confinement at all, with a nearly linear perpendicular
expansion. The whole jet is well aligned and confined in a cone,
centred on the nuclear source, about 10^0 wide. No clear evidence of
unresolved hot-spot, but the north-western edge seems sharper than the
opposite side. In fact, toward the central source the jet bends and
disappears in what appears to be the residual of the extended emission
region (partially lost in these observations). This is shown in Fig.2,
where a crosscut along the major axis of the source is displayed. The
total flux density of the north-western jet is 670 mJy, ≈50% of the
total flux density of the map.
   On the opposite side of the core, there is evidence of diffuse
emission, nearer to the core, spread in a wider angle (in fact nearly
a quarter of a circular shell is visible), and perhaps a very short
"remnant" of counterjet, with the highest brightness temperature area
properly collimated with the other side jet. The total flux density of
this region is 242 mJy.
   Far away in the south-western direction, but not exactly collimated

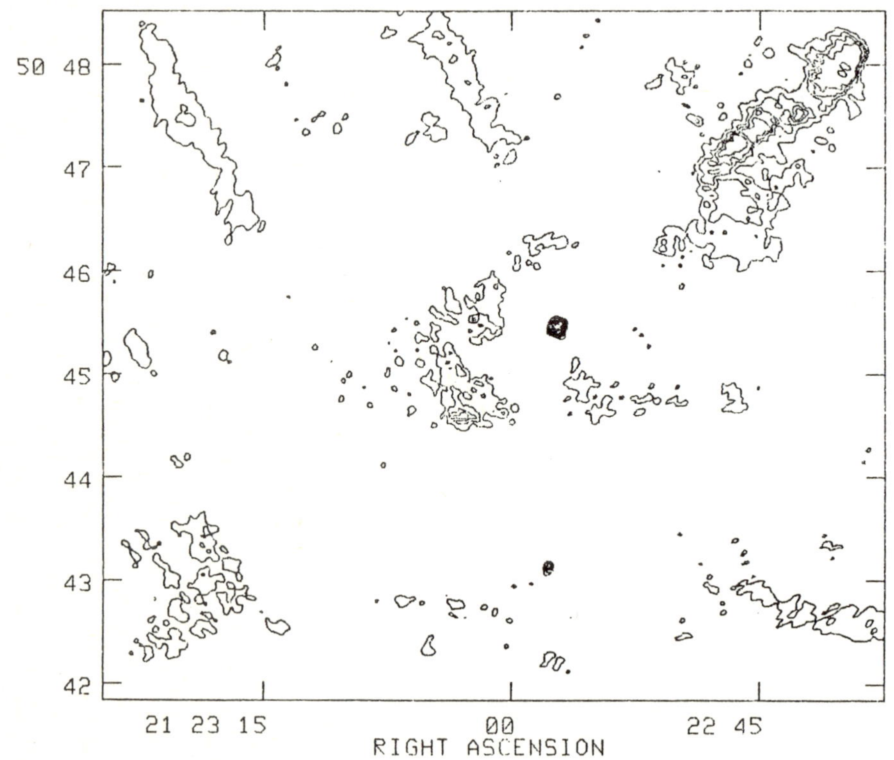

Fig.1 - Contour map of 4C50.55 at 1469.4 MHz. Peak flux 169 mJy. Contour
levels are -1,+0.5,+1,+1.5,+2,+4,+8,+16,+32,+64% of the peak flux.

with the other regions (≈5 degrees clockwise) a diffuse emission is visible with a total flux density of 202 mJy. In this area the surface brightness is very low, but an elongation along the axial symmetry of the source is still visible. No evidence of hot-spot in this area.

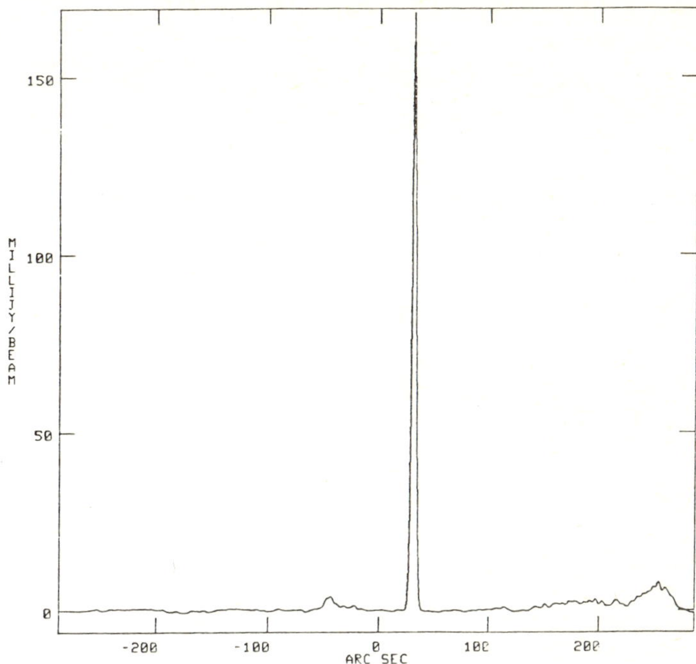

Fig.2 - Crosscut along the major axis of the source 4C50.55. Centre of the cut at $21^h22^m59.57^s$, $50^o45'07.00''$. Rotation angle -47°.

FINAL REMARK

In the lack of a sure optical identification, almost nothing is possible to say on the distance and then on the power of the radio source. In fact, the optical identification provided by Mantovani et al. (1982) is not only a dubious one (a foreground star is highly probably), but even if it is real, its magnitude from the Palomar Sky Survey is drastically affected by the extinction, at least 3^m in this direction (Neckel and Klare, 1980). Moreover, high resolution observations at only one radio frequency don't allow a detailed study of the dynamics of the jet and of the whole source. However, it is possible to suggest some hypotheses on the nature of the source.
1) If the optical identification is real, 4C 50.55 should be a nearby galaxy. The optical identification should be the core of the galaxy, strongly absorbed by our Galaxy. In this case it is possible to guess an intermediate radio power source, with not too large physical dimension. The nearly compact core and othe prominence of the one side jet suggest a rather active central "engine". Different boundary conditions might explain the East-West asymmetry.
2) The optical identification is a spurious one, the source may have larger distance and then higher power. Its similarity with the largest radio galaxy (3C236 and NGC6251) is particularly striking. However, it should be noticed that the lack of emission between the core and the extended emission, the dominance of the jet and the

presence of a rather strong nucleus favour the hypothesis that
4C50.55 is a strong active radio galaxy of intermediate distance.
3) Laying on the Galactic plane it is possible that the source is, at
least partially, a Galactic source. Some probably confusing sources
are visible in Fig.1, and in general confusion has affected the
quality of the final map, but it should be remembered that the shape
and the spectral index distribution at lower resolution, resemble so
closely the extragalactic radio galaxy, that this hypothesis can be
ruled out.

<u>Acknowledgements</u>
    I wish to thank Dr. Hans de Ruiter and Dr. Paola Parma for the
observations and calibrations of present data, and Dr. Laars Baath for
invaluable help in the data reduction.

REFERENCES

Baars, J.W.M., Genzel, R., Pauliny-Toth, I.I.K., Witzel, T., 1977.
    Astron.Astrophys., 61, 99.
Baath, L., 1984. Private communication.
Fanti, C., Felli, M., Ficarra, A., Salter, C.J., Tofani, G., Tomasi, P.,
    1974. Astron.Astrophys.Suppl., 16, 43.
Fanti, C., Felli, M., Panagia, N., Tofani, G., Tomasi, P., 1975.
    Astron.Astrophys. 45, 277.
Fanti, C., Mantovani, F., Tomasi, P., 1981. Astron.Astrophys.Suppl.,
    43, 1.
Goss, W.M., Mantovani, F., Salter, C.J., Tomasi, P., Velusamy, T.,
    1984. Astron.Astrophys., in press.
Gower, J.F.R., Scott, P.F., Wills, D., 1967. Mem.Roy.Astron.Soc., 71,
    49.
Mantovani, F., Nanni, M., Salter, C.J., Tomasi, P., 1982. Astron.
    Astrophys., 105, 176.
Neckel, Th., Klare, G., 1980. Astron.Astrophys.Suppl., 42, 251.

Broadband Spectral Evolution of Outbursts in Extragalactic Radio
Sources

C. P. O'Dea[1], W. A. Dent[2], and T. J. Balonek[3]

(1) NRAO Edgemont Road, Charlottesville, VA 22903
(2) University of Massachusetts, Astronomy Dept., GTWR-B, Amherst, MA
    01003
(3) Williams College, Physics and Astronomy Dept., Williamstown, MA
    01267

SUMMARY

Broadband (2.7-90 GHz) radio observations of the spectral evolution
of outbursts in 12 extragalactic sources are presented. These data
represent a small part of a comprehensive study in progress of the
properties of about 100 variable extragalactic sources. The
preliminary results are qualitative in nature, and there are
counterexamples to all of these general trends. The outbursts tend to
appear first at mm wavelengths (or higher) and decrease in amplitude
as they propagate to lower frequencies (though not as much as
predicted by the simple expanding cloud model). The outbursts broaden
with time and can extend over a decade in frequency at cm and mm
wavelengths. In some sources, an apparent anti-correlation between
activity at cm and mm wavelengths can occur if the timescale for the
outbursts to propagate from mm to cm wavelengths is comparable to the
timescale between outbursts.

INTRODUCTION

The discovery of variability of the cm wavelength radio emission in
compact extragalactic sources (Dent 1965) opened up new possibilities
for the study of the physics of the emitting region. The study of the
evolution of the spectrum of the radio outbursts probes the physical
conditions, velocity, particle energy gain and loss mechanisms, and
geometry on size scales of a few light months to light years in the
environment of the central engine.

However, previous work has been hampered by the limited frequency
coverage available in the data. In order to properly test the various
models for the outbursts, the frequency dependence of the flux density
at the spectral peak must be known over a large range in frequency.
In addition, the value of the electron energy index can only be
obtained from broadband observations which enable the shape of the
outburst spectrum to be determined.

In this paper, broadband (2.7-90 GHz) observations of outbursts in
12 sources are presented. Because of the small number of sources and
the selection effects involved, the conclusions are qualitative and
should be considered preliminary. A study of the systematic
properties of the spectral evolution in a sample of about 100 sources
is currently in progress.

OBSERVATIONS

For over a decade, a team of observers led by Dent have monitored
the flux densities of about 100 variable sources at five radio
frequencies. The 2.7 GHz data are taken at intervals of 3-4 months
using the NRAO 92 m telescope and include measurements of the linear
polarization (Kapitzky 1976). The data at 8 and 15.5 GHz are taken
with the Haystack 37 m antenna at intervals of about 1 month (Dent
et al. 1974; Dent and Kapitzky 1976). The observations at 31.4 and
89.6 GHz are obtained using the NRAO 11 m telescope at intervals of
roughly 3 months (Dent and Hobbs 1973; Hobbs and Dent 1977). The
telescope parameters and observing techniques are described in the
above references.

Outburst spectra of 12 variable extragalactic sources are given
below. Further details and discussion of the observations of
individual sources can be found in O'Dea et al. (1983a): 1308+326;
O'Dea et al. (1983b): 0048-097, 0607-157, 0727-115, and 2200+420
(BL Lac); O'Dea et al. (1984, in preparation): 0133+476, 0235+164,
and 2131-021; and Balonek (1982): 1510-08, 1921-29, 2223-052 (3C446),
and 2251+158 (3C454.3).

CONCLUSIONS

Based on the observations of these 12 sources, the results are the
following:

(1) The outbursts tend to appear first at mm wavelengths (or
higher, cf. Ennis et al. 1982; Epstein et al. 1982; Robson et al.
1983).

(2) The amplitude tends to decrease as the outbursts propagate to
lower frequencies with time, though not as much as predicted by the
simple expanding source model (van der Laan 1966; Pauliny-Toth and
Kellermann 1966); see also Altschuler and Wardle 1977; Andrew et al.
1978; Epstein et al. 1982; Ennis et al. 1982. This suggests that
continuing injection (Peterson and Dent 1973) or reacceleration
(Pacholczyk and Scott 1976) is taking place, or that the source is
expanding primarily in one direction, i.e., along a jet (Pacholczyk
1981).

(3) The outbursts tend to be broadband, extending over a decade in
frequency at cm and mm wavelengths (cf. Epstein et al. 1982; Ennis
et al. 1982). This suggests that the outbursts originate in
inhomogeneous regions, e.g., with a tapered structure as suggested for
the quiescent spectra (Hoyle and Burbidge 1966; Condon and Dressel
1973; de Bruyn 1976; Marscher 1977).

(4) Typically, the outburst spectra broaden with time, e.g.,
0133+476, 0727-115, 1510-089, and BL Lac, (as the emitting region
becomes more inhomogeneous?). However, in 1308+326 the outburst
spectrum narrowed.

(5) In sources where the timescale for the outbursts to propagate
from mm to cm wavelengths is comparable to the timescale between
outbursts, the total flux density spectrum "ripples" as the different
parts of the spectrum respond simultaneously to different events,
e.g., 0235+164, 0607-157, and 2131-021. This produces an apparent
anti-correlation between activity at cm and mm wavelengths. A long
time baseline is needed to distinguish the effects of individual
outbursts.

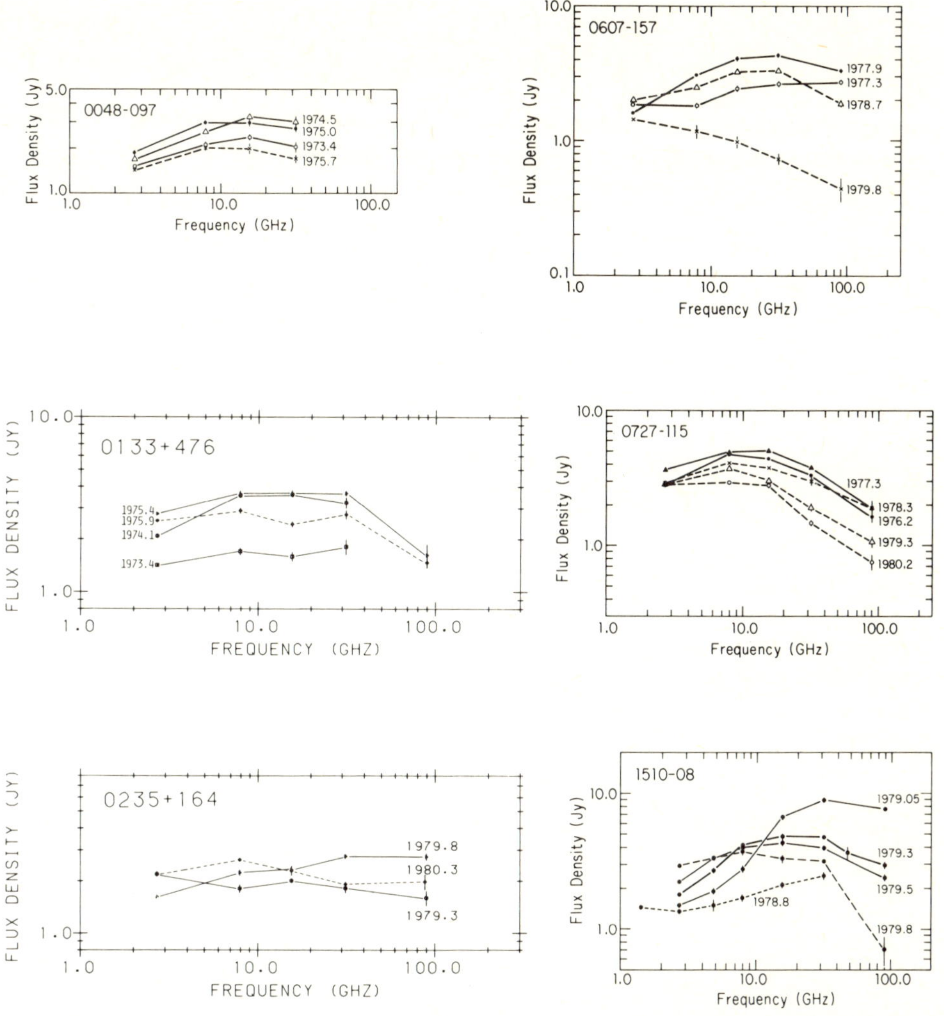

Fig. 1. Broadband spectral evolution of outbursts in the extragalactic sources 0048-097, 0133+476, 0235+164, 0607-157, 0727-115, and 1510-08. See text for details.

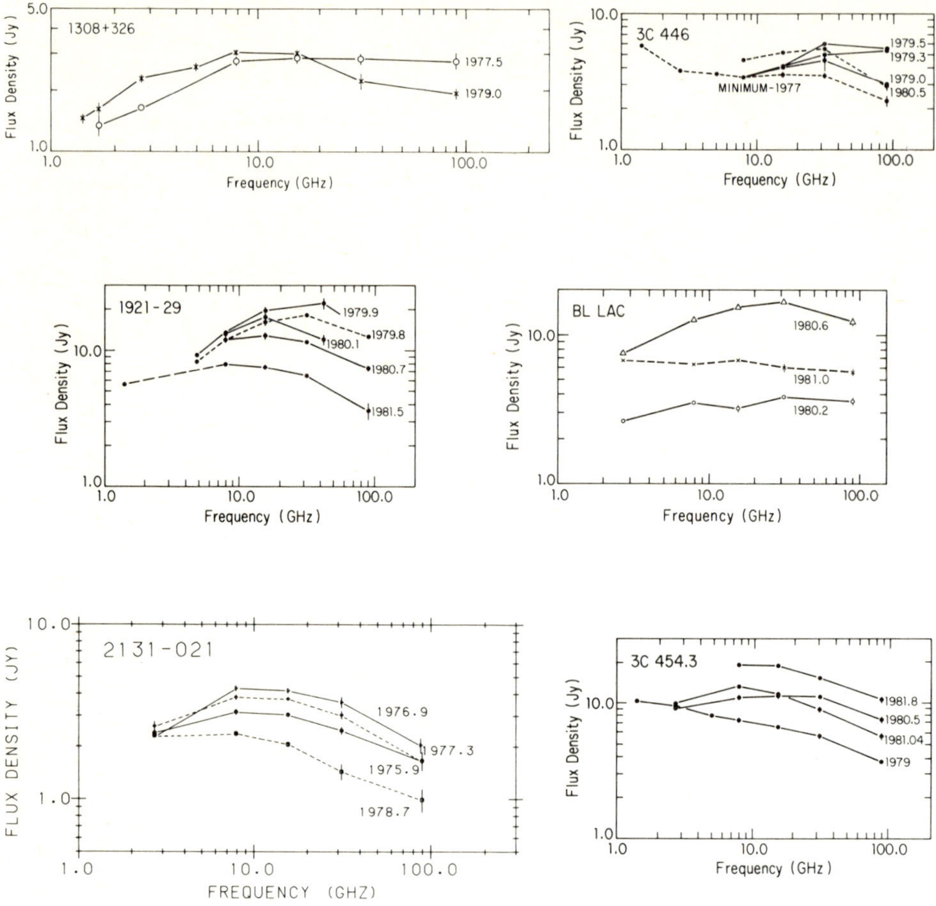

Fig. 2. Broadband spectral evolution of outbursts in the extragalactic sources 1308+326, 1921-29, 2131-021, 2223-052 (3C446), 2200+420 (BL Lac), and 2251+158 (3C454.3). See text for details.

REFERENCES

Altschuler, D. R., and Wardle, J.F.C. 1977 M.N.R.A.S. 179:153.
Andrew, B. H., MacLeod, J. M., Harvey, G. A., and Medd, W. J. 1978
    A. J. 83:863.
Balonek, T. J. 1982 Ph.D. thesis, Univ. of Massachusetts, Amherst.
Condon, J. J., and Dressel, L. L. 1973 Ap. Letters 15:203.
de Bruyn, A. G. 1976 Astr. Ap. 52:439.
Dent, W. A. 1965 Science 148:1458.
Dent, W. A., and Hobbs, R. W. 1973 A. J. 78:163.
Dent, W. A., and Kapitzky, J. E. 1976 A. J. 81:1053.
Dent, W. A., Kapitzky, J. E., and Kojoian, G. 1974 A. J. 79:1232.
Ennis, D. J., Neugebaur, G., and Werner, M. 1982 Ap. J. 262:451.
Epstein, E. E., Fogarty, W. G., Mottmann, J., and Schneider, E. 1982
    A. J. 87:449.
Hobbs, R. W., and Dent, W. A., 1977 A. J. 82:257.
Hoyle, F., and Burbidge, G. R. 1966 Ap. J. 144:534.
Kapitzky, J. E. 1976, Ph.D. thesis, Univ. of Massachusetts, Amherst.
Marscher, A. P. 1977 Ap. J. 216:244.
O'Dea, C. P., Dent, W. A., and Balonek, T. J. 1983a Ap. J. (Letters)
    266:L1.
O'Dea, C. P., Dent, W. A., Balonek, T. J., and Kinzel, W. 1984
    in preparation.
O'Dea, C. P., Dent, W. A., Balonek, T. J., and Kapitzky, J. E. 1983b
    A. J. 88:1616.
Pacholczyk, A. G. 1981 Ap. Letters 21:87.
Pacholczyk, A. G., and Scott, J. S. 1976 Ap. J. 210:311.
Pauliny-Toth, I.I.K., and Kellermann, K. I. 1966 Ap. J. 146:634.
Peterson, F. W., and Dent, W. A. 1973 Ap. J. 186:421.
Robson, E. I., Gear, W. K., Clegg, P. E., Ade, P.A.R., Smith, M. G.,
    Griffin, M. J., Nolt, I. G., Radostitz, J. V., and
    Howard, R. J. 1983 Nature 305:194.
Van der Laan, H. 1966 Nature 211:1131.

THE LOW FREQUENCY VARIABLE SOURCE 3C159

I. Browne,[1] F. Mantovani[2], T. Muxlow[1], L. Padrielli[2], J.D. Romney[3]

1. University of Manchester, Nuffield Radio Astronomy Labs,
   Jodrell Bank, Macclesfield, Cheshire, SK11 9DL, England.
2. Istituto di Radioastronomia, Via Irnerio 46, 40126 Bologna, Italy.
3. N.R.A.O., Edgemont Rd., Charlottesville, VA 22901, U.S.A.

Summary – Observational data on the steep spectrum source 3C159, obtained with Merlin, VLA, European VLBI network, are presented. This source exhibits low frequency variability and shows a radio structure atypical of this class of objects.

1. Introduction

Generally, low frequency variable sources have a flat spectrum at radio frequencies and their structures are dominated by a compact core in which the variability is presumed to take place, (see for example Padrielli and Romney, 1982). However, there is a small fraction of low frequency variable sources which have steep spectra. In the sample of sources monitored at 408 MHz since 1975 with the E-W arm of the Bologna Telescope 4 steep spectrum sources are classified as variables, (Fanti et al., 1982).

A study of these sources could be very important to test the conventional models that explain the low frequency variability in terms of relativistic beaming at small angles to the line of sight.

We are continuing our programme to observe the structure of these sources with Merlin and the European VLBI Network (EVN). The present observations of 3C159 are part of this larger investigation. This source shows radio morphological characteristics atypical of other low frequency variable sources and also of other steep spectrum ones.

2. Observations

3C159 was added to the list of monitored sources as calibrator because it was known as a strong radio source at 408 MHz (S=6.4 Jy) with a straight steep spectrum, ($\alpha = 0.84$, $S \propto \nu^{-\alpha}$).

Statistical analysis of the data after 5 years of monitoring showed it to have a reduced $\chi^2 = 4$ with 52 degrees of freedom. Following Fanti et al (1981), the reduced χ^2 is defined as:

$$\chi^2 = \Sigma \ ((S_i - \bar{S})^2 / \ \sigma_i^{\ 2})/(n-1)$$

where σ_i is the expected error of the measurement. A value of 4 corresponds to a probability < 0.001 that the observed variations are due to random fluctuations. The rms. is 0.22 Jy corresponding to peak to peak fractional variability of 10%.

Fig. 1 shows the light-curve of the source. 3C159 shows two main periods of activity one in 1980 and one in 1981.

The time-scale of the variations is quite short, of the order of 2 or 3 months. For comparison the light-curve of 3C165 is shown. 3C165 has a reduced χ^2 equal to 0.6. This source lies only 17 degrees away in Declination and 20 minutes in R.A. Furthermore, an observation of 3C165 always follows the observation of 3C159.

In conclusion there seems little doubt that the 408 MHz flux density detected by the Bologna telescope in the direction of 3C159 varies by about 0.6 Jy.

It is known however that there is at least one relatively strong confusing source, 4C40.16, which lies within the beam of the Bologna Telescope (4'x 100'). This source contributes ~ 0.7 Jy to the peak of 3C159 and it should vary about 80%, in order to explain the observed variation.

Merlin maps at 18 cm were obtained for both sources. The 3C159 observation, done on Nov. 81, has a resolution of 0.75 arcsec and 4C40.16 has a resolution of 0.25 arcsec (Fig. 2 and Fig. 3 respectively).

Important points to note are:

i) Both objects have a double structure typical of extragalactic sources.
ii) Neither source has a compact central component.
iii) Any compact structure in the lobes of either source is weak.

Inspection of visibility data enables us to say that for 3C159 there is less than 100 mJy in an unresolved component. Furthermore a VLBI observation on 3C159 was carried out during the 18-cm European Network session in June 1983. Again no compact component was detected. The fringe search was done in a field of 150 arcsec around the radio centroid and we can exclude the existence of a compact component stronger than 30 mJy.

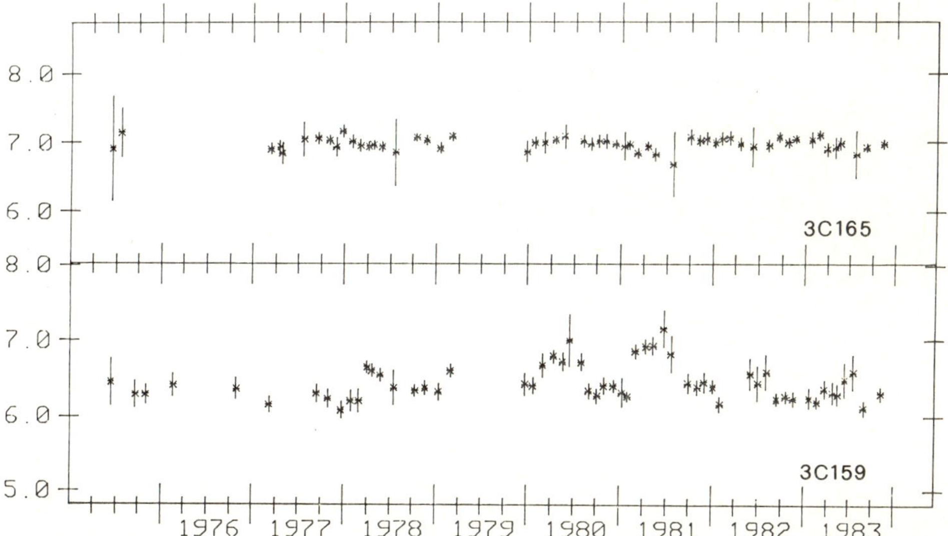

Fig.1 – The light-curve of 3C165 and 3C159 at 408 MHz.

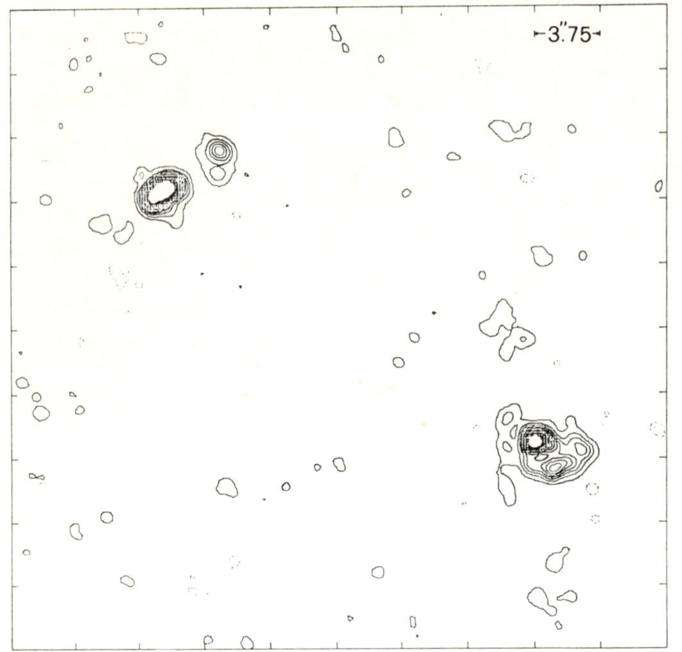

Fig.2 - Merlin map at 18-cm of 3C159. Peak flux is 235.9 mJy/beam.

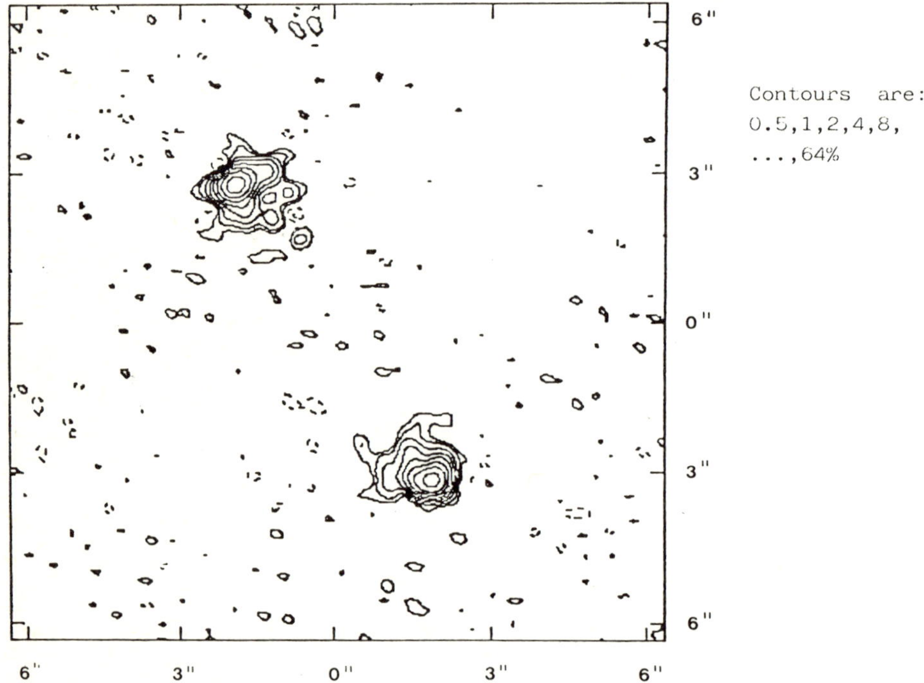

Fig.3 - Merlin map at 18-cm of 4C40.16. Peak flux is 215.4 mJy/beam.

Recently (Feb. 84) a VLA map was obtained at 5 GHz, shown in Fig. 4. The north-east lobe shows a twin hot-spot. Such a multiple hot spot morphology is seen in only 10% of the complete 3CR sample, (Laing, 1980). One such source with a similar morphology is the quasar 3C351, (Kronberg and Clarke, 1980).

The weak core coincides with some optical fuzz visible on the Palomar Sky Survey print. A spectroscopic measurement of this fuzz was made (Tytler and Browne, in preparation) and it reveals the existence of two emission line regions with z=0.48. The velocity of the two regions differs by 600 Km/sec.

3C159 is therefore a powerful radio galaxy with a 408 MHz luminosity of 1.6×10^{28} W Hz^{1} and a linear diameter of 80 Kpc. At a redshift of 0.48 the observed amplitude and time-scale of variability would imply the existence of a component of flux density 1 Jy with a brightness temperature $\sim 6\ 10^{14}$ K. Yet the radio maps show no compact regions with flux density in excess of a few tens of mJy. What then can be going on?

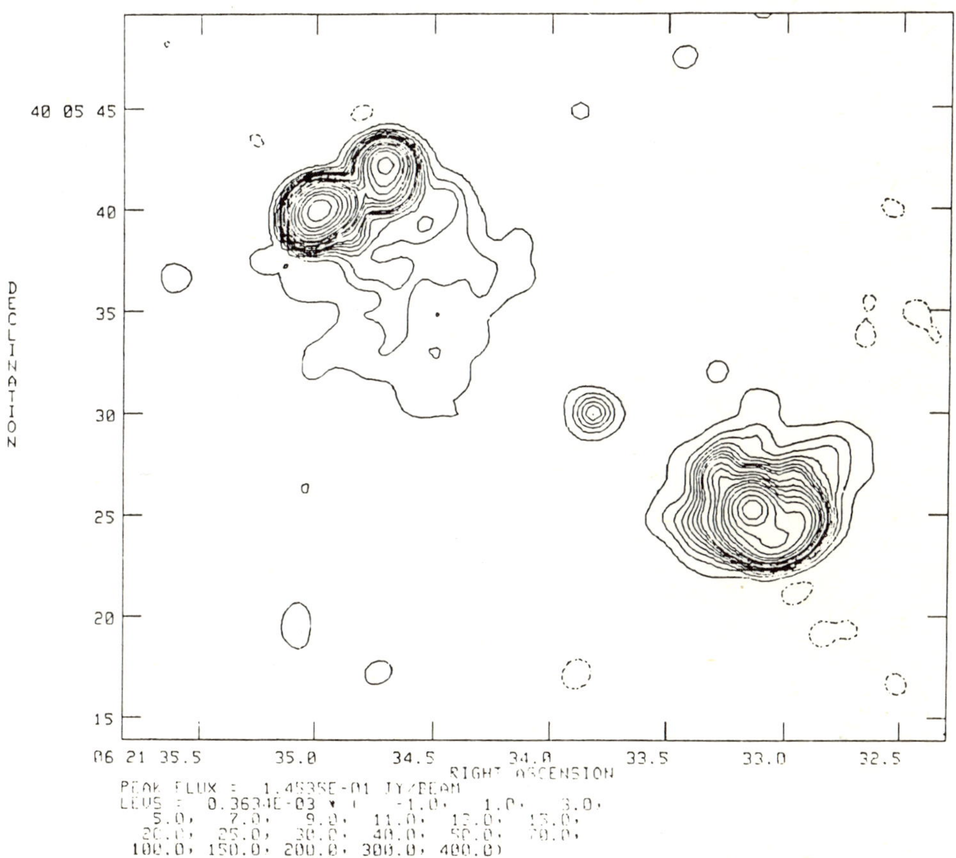

Fig.4 - VLA contour map of 3C159 at 6-cm.

3. Discussion

A few possible explanations of the apparently contradictory observations are listed below:

1) A confusing source may be responsible for the 408 MHz variability. 4C40.16 has a double classical structure as has 3C159 and it would have to vary by about 80% to be responsible for the observed variations, which is very unlikely. We are not aware of the presence of other candidates in the beam. This possibility should be further investigated with, for example, WSRT observations of the field at 50-cm.

2) Day to night changes in ionospheric Faraday rotation might cause the observed variations, if the source is linearly polarized at 408 MHz by say 10-15%. In fact the E-W arm of the Bologna Telescope measures only the linear polarization in a fixed direction. We point out that the high points in the light curve occur at approximately the same time of year.

There are no polarization measurements of 3C159 at 408 MHz and the source is only 6% polarized at 21 cm (Tabara and Inoue, 1980). Therefore unless the degree of polarization shows exceptional behaviour and increases at lower frequencies, this explanation also seems unlikely, even if we cannot exclude it.

3) The variations might have been due to a transient compact component already extinguished when the source was observed with Merlin, EVN and the VLA. This is possible because the source was at its minimum during these observations. However, this component would have to have a very unusual spectral behaviour, because it has not been detected in simultaneous observations at high frequency, (Aller and Aller, 1982).

4) The 408 MHz measurement errors may be greater than expected and the variability not significant. It seems to us very unlikely that a coincidence of errors can produce the observed light-curve. This curve shows alternatively periods of high and low emission, that remain rather constant for several months. However, simultaneous observations with independent telescopes would be necessary to establish the variability beyond any doubt, but such observations do not exist.

At present we cannot choose which, if any, of these possible explanations is correct. Nevertheless 3C159 is a very peculiar object and worthy of further study.

References

H.D. Aller and M.F. Aller: Proceedings of a workshop on "Low Frequency Variability of Extragalactic Radio Sources", N.R.A.O. Green Bank, April 21-22, 1982

C.Fanti, R. Fanti, A. Ficarra, F. Mantovani, L. Padreilli, K.W. Weiler: Astron;Astrophys. Suppl.Ser. 45,61, 1981.

C. Fanti, R. Fanti, A. Ficarra, L. Gregorini, F. Mantovani, L. Padrielli: 1982, Astron.Astrophys. 118,171

P.P. Kronberg and J.N. Clarke: 1980, Astron.J. 85,973

R.A. Laing: 1981, Mon.Not.R.Ast.Soc. 195,261

L. Padrielli and J.D. Romney: 1982, "Astrophysical Jets", Ed. by A. Ferrari and A.G. Pacholczyk, D. Reidel Publishing Company

H. Tabara and M. Inoue: 1980, Astron.Astrophys.Suppl.Ser., 39,379.

Young supernovae in the starburst galaxy M82

S.W. Unger, A. Pedlar, D.J. Axon, P.N. Wilkinson
Nuffield Radio Astronomy Laboratories, Jodrell Bank, Macclesfield
Cheshire SK11 9DL

P.N. Appleton

Department of Astronomy, University of Manchester, Manchester M13 9PL

SUMMARY

   MERLIN observations at 18 cm detect 18 compact (<0.25 arcsec) radio
sources within the central few hundred parsecs of M82. M82 is well
known as a 'starburst' galaxy, and it is probable that these objects
are either radio supernovae or young supernova remnants. The distribution of radio sources appears asymmetric with respect to the centre
of the galaxy, and we suggest that this is due to the current region of
star formation lying within a molecular ring.

INTRODUCTION

   The irregular galaxy M82 has been intensively studied in all wavebands since the discovery by Lynds & Sandage (1963) of a system of Ha
filaments extending along the galactic minor axis out to 3 kpc from the
galactic plane. The central few hundred parsecs of M82 appears optically as a complicated maze of star clusters, giant HII regions, and
dust lanes (e.g. O'Connell & Mangaro 1978) and is also an extended
radio (Kronberg & Wilkinson 1975), infra-red (Rieke et al. 1980) and
X-ray (Watson, Stanger & Griffiths 1984) source. Although the nature
of the filamentary system is still unclear (e.g. Axon & Taylor 1978),
it seems likely that the current activity within the central few hundred parsecs of M82 is powered by a massive burst of star formation,
probably triggered by an interaction with the nearby companion M81
(Rieke et al. 1980). M82 is therefore an example of a 'starburst'
galaxy (Weedman et al. 1981), and its proximity makes it an ideal
prototype for studying this phenomenon in detail; at an assumed
distance of 3 Mpc, one arcsecond corresponds to about 15 parsecs.
   Previous radio observations of M82 have suggested the presence of a
number of discrete compact (typically <2 arcsec) radio sources embedded in the extended radio source (Kronberg & Wilkinson 1975). By far
the strongest of these is the object 41.9+58, which has a steep spectrum
at frequencies above 1 GHz, and whose flux density at these frequencies
is known to be decreasing with an e-folding time of about 15 years
(Kronberg & Biermann 1983). This object is known to possess milliarcsecond scale structure (e.g. Shaffer & Marscher 1979, Jones et al.
1979). The other compact radio sources are weak compared to the diffuse radio emission and so an attempt to study them requires high
resolution radio observations which resolve out the diffuse radio
emission. Recent 6 cm VLA observations by Kronberg & Biermann (1983)
detect about 30 compact radio sources. We have independently observed
M82 using MERLIN (Davies, Anderson & Morison 1980) at a wavelength of

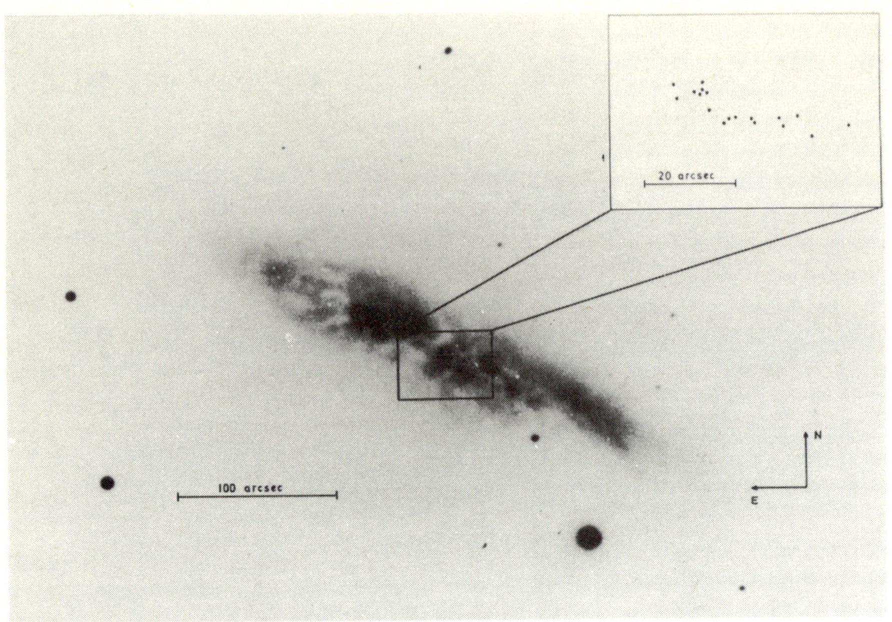

Plate 1   The positions of the 18 compact radio sources discussed in
           the text compared with an R band optical photograph of M82
           (Blackman, Axon & Taylor 1979).

18 cm and detect 18 compact radio sources with flux densities at 18 cm
greater than 2 mJy, all coincident with objects seen by Kronberg &
Biermann (1983). None of the compact radio sources appear to be resolved
by the 18 cm observations and so we can place upper size limits of
about 0.25 arcsec ($\sim$3 pc). In plate 1 we show the location of the compact radio sources on an optical photograph of M82 (Blackman, Axon &
Taylor 1979).

<u>What are the compact radio sources?</u>
     As the compact radio sources lie in a region thought to be undergoing
a massive burst of star formation (Rieke et al. 1980) it seems reasonable to associate them with this starburst. Since the brightness temperatures of the compact radio sources are all greater than about 10^5K,
we can rule out the idea that the compact radio sources are HII regions.
The most plausible explanation is therefore that they are either supernova remnants or young radio supernovae located in the disc of M82.
     Although a high supernova rate is predicted for M82 by Rieke et al.
(1980) on the basis of infra-red observations, the large optical extinctio
($A_V \approx 26$; Rieke et al. 1980) to the centre of the galaxy means that
these supernovae would not be observed optically. Radio emitting
counterparts to optically observed supernovae have however been observed
in a few nearby galaxies (Weiler et al. 1983), and for M82 radio observations may provide the only direct means of observing such supernova
events (c.f. NGC4258; van der Hulst et al. (1983)). The few existing
observations of radio supernovae (Weiler et al. 1983, van der Hulst et
al. 1983) indicate that the radio emission turns on shortly after the
optical supernovae, with a luminosity of up to 200 times that of Cas A.
The radio emission appears initially at high frequencies and then moves

to lower frequencies as the supernova evolves. Lifetimes vary; the
flux density of the radio supernova in NGC4258 halved in 1 year (van der
Hulst et al. 1983), whilst SN1979c in M100 shows no sign of fading 4
years after the original event (Weiler et al. 1983). However, it is
apparent that the time spent by a supernova remnant in the radio supernova phase is short compared with the age of a classical supernova
remnant, and if the compact radio sources are radio supernovae then we
might expect them to show variability in both the total flux density
and the spectral peak on timescales of between a few and a few tens of
years.

The alternative is that some of the compact radio sources are classical supernova remnants similar to Cas A. At the distance of M82, Cas
A would have a flux density at 18 cm of 1.6 mJy, just below our detection
limit of 2 mJy, and an angular size of 0.1 arcsec. If some of the
weaker compact radio sources are classical supernova remnants, then we
expect them not to show variability and to have steep radio spectra.

Distribution of the compact radio sources

Low resolution VLA observations of M82 (Kronberg, Bierman & Schwab
1981) show the diffuse radio source to have a central plateau, away
from which the radio emissivity falls steeply. In Fig.1 we plot the
positions of the compact radio sources detected by MERLIN on the low
resolution VLA map, and it can be seen that the distribution is asymmetric, with many of the compact radio sources (and all of the strongest
ones) lying close to the steep gradient in radio emissivity to the
south of the central plateau, whilst there are none to the north of
the plateau.

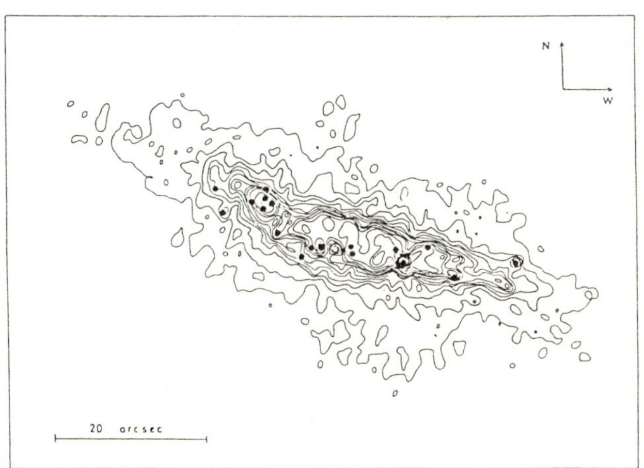

Figure 1   The positions of the 18 compact radio sources discussed in
the text plotted on a low resolution VLA map of the diffuse
radio source (Kronberg, Biermann & Schwab 1981).

A possible reason why the radio sources appear to lie to the south
of the nucleus of M82 is that this side is nearer to us, and that radio
emission from sources to the north is absorbed by ionised gas in the
centre of the galaxy. However, the asymmetry seen in the MERLIN 18 cm
observations is also seen in the VLA 6 cm observations, and an emission
measure of nearly 10^8 pc cm^{-6} would be required to provide significant

free-free absorption at such a high frequency. This is physically
implausible as the optically thick thermal emission from the required
gas would give rise to a flux density at 6 cm an order of magnitude
greater than that observed by Kronberg et al. (1978).

The observed asymmetry must therefore reflect the true distribution
of compact radio sources rather than being due to obscuration effects,
and in this context it is interesting to note that both the Hα and X-
ray emission associated with the filamentary system of M82 are asymmetric in the same sense, being much stronger (by a factor of about 10)
to the south of the galaxy than to the north (Watson et al. 1984, Axon
& Taylor 1978). The asymmetry in the distribution of the radio sources
could be due either to them lying below, or to them being distributed
inhomogeneously within, the tilted galactic disc. A number of observations suggest a model in which the compact radio sources lie within
that part of a nearly edge-on ring which is tilted to the south. Rieke
et al. (1980) noted that the infra-red source in the nucleus of M82 is
substantially more centrally condensed at 2 μm than at 10 μm. The
emission at 2 μm is probably due to a cluster of old red giant stars
situated close to the dynamical centre of M82, whilst the emission at
10 μm is probably due to hot dust associated with a more recent region
of star formation, prompting Rieke et al. to suggest that the region of
star formation is moving outwards from the centre of M82. Further
evidence in favour of this view has been provided by high resolution
VLA observations of HI and OH absorption against the nuclear continuum
radio source, which show the presence of a toroidal ring of cold gas
in regular rotation, centred on the dynamical centre of M82 and extending from an inner radius of 170 pc to an outer radius of 340 pc
(Weliachew et al. 1984). This ring has strong parallels with the
molecular ring in our own galaxy (Scoville 1972), and is likely to
represent the region of current star formation in M82.

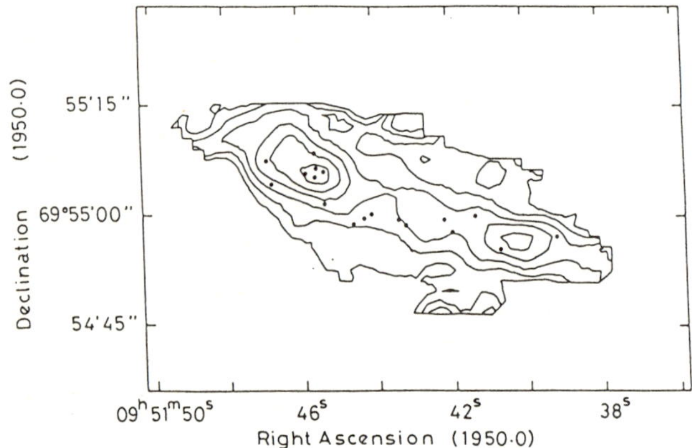

Figure 2   The positions of the 18 compact radio sources discussed in
           the text plotted on a map of HI optical depth (Weliachew,
           Fomalont & Greisen 1984)

Fig. 2 shows the positions of the compact radio sources in relation
to the HI ring and suggests strongly that the two are associated. In
general the compact radio sources lie towards the south of the HI
distribution, and the ring model therefore requires that star formation

is not occurring uniformly throughout the ring. Evidence for patchy
star formation in rings of neutral material in other galaxies has been
provided by the structure of the 'hotspot' galaxies, and in particular
by recent infra-red and radio observations of the face-on galaxy NGC1097
(Telesco & Gatley 1981; Wollstencroft, Tully & Perley 1984).

In Fig.3 we give a schematic view of the nuclear regions of M82
according to our model. Within the errors, the near infra-red star
cluster of Rieke et al. (1980) is coincident with the kinematic centre
of M82. Although the total extent of the far infra-red source appears
similar to that of the HI ring, the twin peaks of the far infra-red
map of Rieke et al. (1980) lie inside the peaks in the HI distribution,
whilst the bulk of the diffuse radio source also lies inside the inner
edge of the HI ring. A comparison with X-ray observations suggests
that the ring laterally confines the X-ray emitting gas, forcing it to
expand along the minor axis of M82. An attractive feature of this model
is that the filamentary system and supernovae are related, in that the
hot gas responsible for the X-ray emission originates in supernova
explosions, and that the Hα filaments are caused by rapid cooling of
this gas as it expands into the cool neutral hydrogen surrounding M82.
This model can also provide a natural explanation for the asymmetry of
the filamentary structure if the external pressure due to the surrounding HI is larger to the north of the galaxy than to the south.

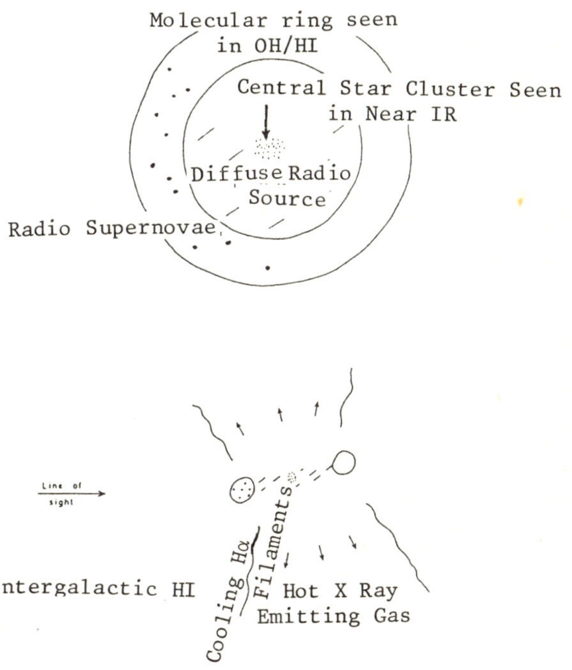

Figure 3 A schematic diagram of M82 according to our model, showing
both a plane view and a cut through the central regions of
the galaxy

CONCLUSION

Our observations of M82 detect 18 compact radio sources within the central few hundred parsecs of this 'starburst' galaxy. These objects are probably either radio supernovae or supernova remnants. The radio sources are distributed asymmetrically with respect to the centre of M82, and we have suggested that they are associated with a molecular ring similar to that seen in our own galaxy.

REFERENCES

Axon, D.J. & Taylor, K., 1978. Nature 274, 37.
Blackman, C.P., Axon, D.J. & Taylor, K., 1979. Mon.Not.R.astr.Soc., 189, 751.
Davies, J.G., Anderson, B. & Morison, I., 1980. Nature 288, 64.
Jones, D.L., Sramek, R.A. & Terzian, Y., 1981. Astrophys.J., 246, 28.
Kronberg, P.P. & Wilkinson, P.N., 1975. Astrophys.J., 200, 430.
Kronberg, P.P., Biermann, P. & Schwab, F., 1981. Astrophys.J., 246, 751.
Kronberg, P.P. & Biermann, P., 1983 IAU Symposium 101, p.583.
Lynds, C.R. & Sandage, A.R., 1963. Astrophys.J., 137, 1005.
O'Connell, R.W. & Mangano, J.J., 1978. Astrophys.J., 221, 62.
Rieke, G.H., Lebofsky, M.J., Thompson, R.I., Low, F.J. & Tokunaga, A.T. 1980. Astrophys.J., 238, 24.
Scoville, N.Z., 1972. Astrophys.J., 175, L127.
Shaffer, D.B. & Marscher, A.P., 1979. Astrophys.J., 233, L105.
Telesco, C.M. & Gatley, I., 1981. Astrophys.J., 247, L11.
van der Hulst, J.M., Hummel, E., Davies, R.D., Pedlar, A. & van Albada, G.D., 1983. Nature 306, 566.
Watson, M.G., Stonger, V. & Griffiths, R.E., 1984. Preprint.
Weedman, D.W., Feldman, F.R., Balzano, V.A., Ramsey, L.W., Sramek, R.A. & Chi-Chao, W., 1981. Astrophys.J., 248, 105.
Weiler, K.W., Sramek, R.A., van der Hulst, J.M. & Panagin, N., 1983. In IAU Symposium 101, p.171.
Weliachew, L., Fomalont, E.B. & Grisen, E.W., 1984. Astron.Astrophys. Submitted.
Wilkinson, P.N. & de Bruyn A.G., 1984. Mon.Not.R.astr.Soc., submitted.
Wolstencroft, R.D., Tully, R.B. & Perley, R.A., 1984. Mon.Not.R.astr. Soc., 207, 889.

DISCOVERY OF AN ENTIRE POPULATION OF VARIABLE RADIO SOURCES IN THE
NUCLEUS OF M82

P. P. Kronberg

Scarborough College and Department of Astronomy,
University of Toronto,
Toronto M5S 1A1, Canada.

SUMMARY

MORE THAN 40 DISCRETE RADIO SOURCES HAVE BEEN DISCOVERED IN THE
INNER 600 pc NUCLEAR REGION OF M82. SUBSEQUENT MONITORING OVER A
PERIOD OF 2.7 YEARS REVEALS THAT MOST OF THE BRIGHTEST SOURCES ARE VARY-
ING ON A REMARKABLY SHORT TIME SCALE OF A FEW MONTHS TO A FEW YEARS.
AT CURRENT DECAY RATES, EIGHT OF THE BRIGHTEST TEN SOURCES (PRESUMED
RADIO SN AND SNR'S) WILL BE SCARCELY VISIBLE AT 0.3" RESOLUTION ABOVE
THE AMORPHOUS RADIO "BACKGROUND" WITHIN CA. 35 YEARS.

Recent observations with the NRAO Very Large Array (VLA)[1] by
Kronberg, Biermann, and Schwab (1984)(KBS) have revealed an entire
population of radio supernova and supernova remnant candidates within
the inner 600 pc of M82. A 6cm. map using the VLA in its extended "A"
-configuration is shown in Figure 1. The HPBW is 0.34 arcseconds.
This map reveals more than 40 discernable discrete radio sources, whose
luminosities range from approximately that of Cassiopeia A (the most
radio-luminous supernova remnant in our Galaxy), to 150 times as bright.

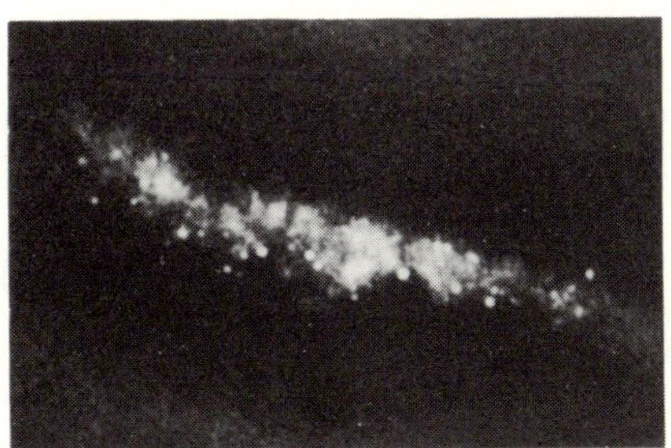

FIG. 1.   5GHZ radio map of the nucleus of M82 at epoch 1981.1 made
    with the VLA. This is a slightly more "exposed" version of that
    published by Kronberg and Biermann (1983). The resolution is 0.34"
    At the adopted distance of 3.2 Mpc, 1"=15 pc. The horizontal
    dimension of the photograph is 51.5".

Since early 1982, R.A. Sramek (NRAO) and I have used the VLA to
repeatedly monitor this population of radio sources in M82's nucleus.
We have made the remarkable discovery that nearly all of the brightest
sources have decreased in intensity over a 2.7 year period ending in

[1] The National Radio Astronomy Observatory is operated by Associated
    Universities Inc., under contract with the National Science
    Foundation.

October 1983.   One source, which was nearly 10 times as luminous as
Cassiopeia A, has "turned off" in only a few months. Most of the
remaining 10 strongest ones are declining so rapidly that they will
have faded into the background in a mere 30 years from now.

If we assume an exponential rate of decay, eight of the 10 brightest
sources have e-folding flux density decay times which are less than 25
years. The shortest is a few months and the median is ca. 10 years.
One of the brightest sources, on the other hand, has barely changed at
all over our preliminary 2.7 year monitoring period. The decay rates
observed lead to the first direct estimate of the supernova production
rate in a starburst-type galaxy nucleus, which is 1 every 4 to 5 years.
This likewise enables us to make the first direct, observation-based
estimate of the rate of energy input into the active nuclear region,
which is $\sim$ 6X10{10} Lo for an assumed energy release per supernova of
10{52} ergs. Interestingly, this is, within reasonable errors, close to
the excess infrared luminosity of M82. (Telesco and Harper, 1980).

From our preliminary analysis outlined above, we have already established the following facts concerning the dynamically evolving
population of supernova/supernova remnants in M82: (i) there is a large
range of luminosity decay times, which suggests that the brightest 20
radio sources consist of either different types of supernovae, or that
the behaviour of the SN/SNR radio emission is strongly influenced by
external factors such as the ambient interstellar density and structure.
Most probably both are true. (ii) The observation of the "rapid
turnoff" source (whose decay time was at most a few months) by KBS in
February 1981 was presumable very close in time to the actual supernova
event. If even a small fraction of the 40 sources in our 6 cm VLA map
are of this type, then their appearance as new sources (presumably
fresh supernovae) will be even more frequent than the number estimated
above. (iii) Statistically, the fainter sources have smaller decay
rates than the more luminous ones. This would clearly be expected for
a power law rate of flux decay, if all initial luminosities were
comparable.

Our discovery thus has revealed the "engine room" of the mysterious
activity in M82, and by implication, all similar active, or "starburst"
galaxies. These are characterized by a disturbed optical appearance,
kpc sized strong radio emission near their nucleus, and "excess" far
infrared emission which greatly exceeds the extrapolated non-thermal
radio flux density.

REFERENCES

Kronberg, P. P., and Biermann, P., 1983, I.A.U. Symposium 101, 583.
   (D. Reidel & Co.).
Kronberg, P. P., Biermann, P. and Schwab, F.R., 1984., Ap.J. (in press).
Telesco, C.M., and Harper, D.A., 1980, Ap.J., 235, 392.

Radio Emission from the Nuclei of Sbc Galaxies

R.D. Davies

University of Manchester, Nuffield Radio Astronomy Laboratories,
Jodrell Bank, Macclesfield, Cheshire

SUMMARY

A study of the radio emission from 105 Sbc (T=4) galaxies is in
progress using several interferometer and aperture synthesis systems.
The aim is to derive the profile of nuclear radio emission from a well-
chosen sample of normal spiral galaxies. Some 5 percent of the galaxies
have discernible optical activity while 40 percent show radio emission
within the central 20 arcsec and at least 10 percent have emission
within the central 1 arcsec. The radio emission is being correlated
with data at optical, infrared and X-ray wavelengths.

1. INTRODUCTION

A number of major surveys have been made of the radio continuum
emission from normal galaxies (e.g. Condon & Dressel 1978, Hummel 1980,
van der Hulst et al. 1981). These observations, as well as studies of
individual galaxies, have shown that synchrotron emission can originate
both in the nuclear regions and in the main body of normal galaxies.
The well-known emission from the spiral arms of our Galaxy and of the
nearby spiral galaxies is the result of relativistic electrons spiral-
ling in the spiral arm magnetic fields. It is still not clear where
these relativistic electrons come from - a substantial fraction will
originate in supernovae, and some may come from a central source. The
emission from the nuclear regions is the major interest of the present
investigation. In some galaxies (e.g. NGC 1275 and 4151) this is the
dominant emission. Such emission is generally associated with nuclear
activity at other wavelengths. Extreme cases of such activity are
quasars and radio galaxies. Seyfert galaxies represent a less violent
form of activity; they comprise several percent of normal galaxies and
are the most active members. I present here a progress report on an
investigation of the profile of radio activity in a whole sample of
normal spiral galaxies, not limiting attention to the most active
members.
The programme galaxies were chosen to be of type Sbc (T=4) which
shows a low scatter in optical properties. A sample of the 105 brightest Sbc galaxies was selected from RC2 (de Vaucouleurs, de Vaucouleurs
& Corwin 1976) at Dec >-33°. They were first observed in the λ21 cm
line of neutral hydrogen to provide systemic velocities, neutral hydrogen (HI) integrals and velocity widths from which HI and total masses
were estimated. Fig. 1 is a plot of the neutral hydrogen mass against
linear optical diameter for the programme galaxies. It shows the well-
known fact that within the one morphological class the galaxies range

from dwarfs to giants with linear diameters between 10 and 70 kpc. Their integral properties (blue luminosity L, total mass M_T and HI mass M_H) may be summarized as follows:

$$\log\,(L/L_\odot) = 9.7 \text{ to } 11.1$$

$$\log\,(M_T/M_\odot) = 10.2 \text{ to } 12.0$$

M_H/L is constant over the luminosity range.

Our study concerns the nuclear activity of a given morphological class which contains a range of $\sim$30:1 in total mass and luminosity. This is a report on the radio continuum projects in progress.

2.   THE RADIO SURVEYS

Radio continuum surveys of the majority of the galaxies in the sample have been made with interferometers of different resolutions to investigate a range of linear scales in the sample. The MkIA-MkII interferometer (baseline 434 m) was used in a preliminary survey at λ18 cm to make a study on a scale of one arcminute which corresponds to several kiloparsecs at the typical redshift (1000–3000 kms^{-1}) of the sample.

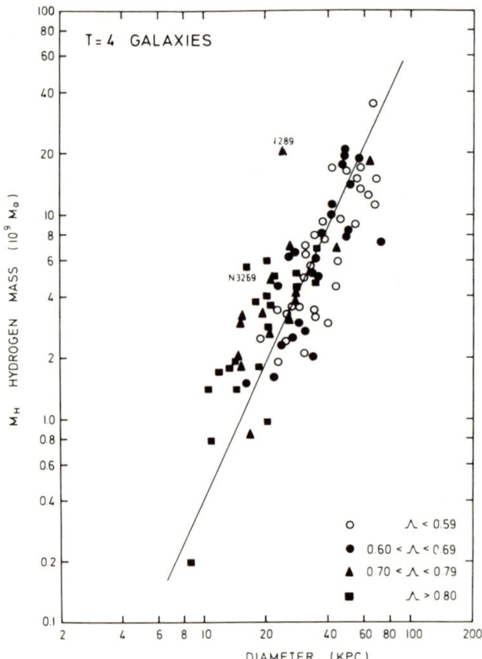

FIG. 1.   Neutral hydrogen (HI) masses of Sbc galaxies plotted against their linear optical diameters. Corrected luminosity classes (Λ) are indicated by different symbols. Note that the HI diameters of NGC 289 and 3259 are about 3 times the plotted optical diameters.

30 of the galaxies showed a flux density greater than 50 mJy. These
galaxies were subsequently observed with MERLIN at λ18 cm on short Hour
Angle tracks which provided resolutions in the range 0.25 to 2.0 arcsec
corresponding to several hundred parsecs. 8 galaxies were detected with
a flux density greater than 15 mJy. These included NGC 613, 2339, 3066,
4303, 4536, 5005, 5248 and 6907.

A more comprehensive survey has recently been undertaken using the
Very Large Array (VLA), New Mexico, at λ20 cm in the C configuration
where the resolution is $\sim$20 arcsec. 88 Sbc galaxies were observed and
of these 72 (82 percent) showed disk or central emission. 37 (42 percent) galaxies showed emission within the central $\sim$20 arcsec with a flux
density greater than $\sim$2 mJy. Maps of 71 Sbc galaxies are being prepared
for publication by E. Hummel, A. Pedlar, J.M. van der Hulst and the
author. Subsequently the luminosity function of central activity in
normal Sbc galaxies will be derived.

3. THE RELATION TO OTHER WAVELENGTHS

3.1 Optical

Our sample of 105 Sbc galaxies includes 3 Seyfert galaxies (NGC 4051,
4258 and 6814), a number consistent with the known fraction of Seyferts
in the general population of spiral galaxies. NGC 5005 has "extremely
broad" emission lines in its nucleus and could therefore be classified
as a Seyfert (Kennicutt & Kent 1984). NGC 3310 shows strong signs of
nuclear activity through its kinematics, morphology and high Hα flux
(see for example Balick & Heckman 1981). NGC 4258 has well-known radio
synchrotron arms which extend from the anomalous optical Hα arms that
traverse the normal optical and HI arms (van Albada & van der Hulst 1982).
Thus we have 6 galaxies which show major abnormalities in our sample. It
will not be surprising to discover a spectrum of activity levels in the
sample as a whole.

Heckman et al. (1983) in their optical spectroscopic study of the
radio-loud nuclei of spiral galaxies find that, on the whole, these
nuclei have stronger emission lines than radio-quiet nuclei of similar
morphological type. They found that this emission-line gas was at least
as spatially extended as the radio continuum emission. Their study includes the Sbc galaxy NGC 4536 which shows Hα emission extending 33 arcsec
along position angle 130$^\mathrm{o}$. We find that this galaxy has extended continuum radio emission as well as a nuclear source with a diameter $\leqslant$1 arcsec.
Heckman et al. find that only one third of the nuclei have HII region
spectra (starbursts) while the majority require ionizing radiation which
is harder than the Lyman continuum of OB stars. It will be of interest
to see if our Sbc galaxies have properties consistent with this result.
Those which have a lower level of nuclear radio emission might then be
expected to have optical emission similar to HII regions.

3.2 Infrared

Studies of the infrared emission from the nuclear regions of a
galaxy can be used to distinguish between the old stars of the nuclear
bulge and the longer wavelength excess which arises from nuclear activity. This nuclear activity can be in the form of starbursts or a central "engine" – in both cases excess 10μ IR emission can be generated.
Condon et al. (1982) examined a statistically complete sample of 33
bright spiral galaxies which were strong radio sources and found that the
radio flux of the cores was well-correlated with their 10μ flux. They
concluded that this correlation indicated that bursts of star formation
and the subsequent supernovae were responsible for the 10μ and radio

emission. This conclusion has been challenged by subsequent workers
including Heckman et al., as discussed above. Nevertheless the correlation between radio and IR emission in the cores of galaxies has held up
in many statistical studies (e.g. Rieke et al. 1980, van der Hulst et al.
1981). The dispute relates to the origin of the 10μ emission – in a
starburst or in some violent event in the galactic nucleus. In both
scenarios the actual radiating medium is supposed to be interstellar dust.

The resolution of this problem will require both a comparison of the
spatial morphology of the emission in the IR and radio continuum and
optical spectroscopic studies. The former approach is being taken by
Dr R.D. Wolstencroft and the author in an examination of the Sbc galaxy
sample using the U.K. Infrared Telescope on Mauna Kea.

3.3 X-rays

A high level of radio emission from a galaxy is accompanied by strong
X-ray emission in the case of quasars and Seyfert galaxies (Fabbianno et
al. 1982). It is not clear to what extent this correlation between radio
and X-ray activity is exhibited by galaxies with less active nuclei. We
have observed 10 Sbc galaxies with EINSTEIN observatory and have detected
4. The data base is being interrogated to add more Sbc candidates.

4. THE SUPERNOVA CONTRIBUTION

If a burst of star formation is responsible for at least a fraction
of the nuclear activity of galaxies, then supernovae will make a contribution to the radio emission. I will present in this section a caution-
ary tale which demonstrates that supernova rates in galaxies may be
higher than is deduced optically.

During our VLA survey of Sbc galaxies a radio supernova was discovered
in a bright spiral arm of NGC 4258 (van der Hulst et al. 1983). This
supernova had not been detected in the various continuing optical
searches, which make routine inspections down to the 15th magnitude.
However, subsequent inspection of the existing optical material revealed
a supernova between 16 and 17th magnitude. The radio supernova was
easily detected (NGC 4258 has a corrected redshift 516 kms^{-1}) and had
similar intrinsic radio flux to supernovae found previously in nearby
galaxies. There are two obvious reasons why the supernova was not
detected optically: the supernova reached maximum light at the time of
year when NGC 4258 was setting in the early evening and, secondly, most
supernova search programmes do not go fainter than 15–16th magnitude.
Further, since the supernova was found in a strongly developed spiral
arm it may have suffered several magnitudes of absorption.

Clearly, present optical estimates of supernova rates are underestimates. In the context of the present discussion, supernova rates
in starburst galaxies are higher than will be measured optically.
Indeed the heavy obscuration implied by the 10μ emission from dust may
preclude their detection at all in the optical range. Searches at radio
wavelengths where dust obscuration has no effect and where the first
phase of radio emission will last for up to 10 years are likely to be
most promising.

5. THE CAPTURE OF MATERIAL FROM AN ADJACENT GALAXY

Statistical investigations of multiple galaxy systems have shown a
greater detection rate for radio emission from galaxies in multiple
systems as compared with isolated galaxies (e.g. Wright 1974, Stocke
1978). In a more definitive aperture synthesis survey of spiral

galaxies Hummel (1981) found that the central radio sources in double galaxies are on average 4 times more powerful than the central sources in isolated galaxies. No significant difference was found however in the disk components. This higher central radio emissivity of multiple systems is widely regarded as being the result of the transfer of mass from the companion to the central gravitational well. Dahari (1984) has drawn a similar conclusion about the occurrence of the central (optical) activity of Seyfert galaxies. He found a definite excess of close companion galaxies for Seyfert galaxies as compared with a control sample of field galaxies. An optical spectroscopic survey of disk galaxies by Stauffer (1982) showed that those with active nuclei preferentially occur in groups while isolated galaxies seldom have strong nuclear emission lines.

Our study allows us to investigate these matters in some detail. The abundant redshift data on nearby galaxies enables group membership to be established with more certainty than previously. Furthermore the HI integrated profile shape is a good indicator of gravitational interaction with a companion (Davies & Johnson, in preparation) whereas proximity in projected distance is not a reliable indicator. Our sample contains some 20 galaxies with evidence of gravitational interaction with a companion. Two galaxies (NGC 876/7 and NGC 2207/IC 2163) are members of tidally distorted systems and have above average radio power.

6. CONCLUSION

A range of radio powers is found in the central regions of normal galaxies. The more active objects have attracted attention in the past. It is now important to investigate the whole activity profile of this central activity in order to understand better the processes at work. This profile of activity presumably reflects the time profile of activity of individual galaxies. For example, normal galaxies spend several percent of their lifetime in the Seyfert phase.

One clear problem which has emerged in the understanding of the nuclear activity of normal galaxies is the question of the nature of the central engine. Is it a starburst phenomenon, or is it the violent outburst resulting from gravitational collapse onto a massive central object? Both scenarios have been suggested and both may apply in different objects. Clearly optical, spectroscopic, infrared and radio studies will be required to disentangle this complex problem. Radio studies of the morphology of central objects may help distinguish between the two scenarios. Active nuclei of the Seyfert type appear to produce more jet-like radio morphologies. Supernova emission in a central starburst is likely to produce a more amorphous radio source.

<u>Acknowledgements</u>
This progress report on radio observations of Sbc galaxies and associated studies at infrared and X-ray wavelengths is the result of a substantial collaborative programme. In the radio, collaborators include Dr. L. Hart, Mr. S.C. Johnson, Drs. A. Pedlar, R. Booler, J.M. van der Hulst and E. Hummel; in the infrared Dr. R.D. Wolstencroft; and at X-rays, Dr. J.P. Pye.

REFERENCES

Balick B, Heckman T 1981 Astr Astrophys 96:271
Condon JJ, Dressel A 1978 Astrophys J 221:456
Condon JJ, Condon MA, Gisler G, Puschell JJ 1982 Astrophys J 252:102
Dahari O 1984 Preprint
de Vaucouleurs G, de Vaucouleurs A, Corwin HG 1976 Second Reference
    Catalogue of Bright Galaxies. University of Texas Press, Texas
Fabbianno G, Feigelson E, Zamorani G 1982 Astrophys J 256:397
Heckman TM, van Breugel W, Miley GK, Butcher HR 1983 Astr J 88:1077
Hummel E 1980 Astr Astrophys Suppl 41:151
Hummel E 1981 Astron Astrophys 96:111
Kennicutt RC, Kent SM 1984 Preprint
Rieke GH, Lebofsky MJ, Thompson RI, Low FJ, Tokunaga AT 1980 Astrophys J
    238:24
Stauffer JR 1982 Astrophys J 262:66
Stocke JT 1978 Astron J 83:348
van Albada GD, van der Hulst JM 1982 Astr Astrophys 115:263
van der Hulst JM, Crane PC, Keel WC 1981 Astr J 86:1175
van der Hulst JM, Hummel E, Davies RD, Pedlar A, van Albada GD 1983
    Nature 306:566
Wright AE 1974 Mon Not R astr Soc 167:251 & 273

MERLIN observations of the radio structure of Seyfert and related
nuclei

A. Pedlar, S.W. Unger & R.V. Booler

University of Manchester, Nuffield Radio Astronomy Laboratories,
Jodrell Bank, Macclesfield, Cheshire SK11 9DL

SUMMARY

We summarise MERLIN observations of Seyfert type nuclei. At 0.4 GHz
a number of these nuclei show low frequency turnovers which can be
interpreted in terms of free-free absorption by ionised gas in the forbidden line region (FLR). The emission measures deduced are of order
10^5 pc cm^{-6} which imply a path length of 0.1 pc if the electron density
is 10^3 cm^{-3}.

Many of these nuclei have structures consistent with collimated
emission – a phenomenon which relates them to more energetic radio
galaxies and quasars.

Although the more extended components are in approximate pressure
balance with the FLR gas, high resolution observations have revealed a
number of small components with high internal pressures and hence these
components must be expanding into the FLR.

INTRODUCTION

Although the nuclei of Seyfert galaxies have been studied optically
for several decades, it is only recently that radio techniques have had
sufficient resolution and sensitivity to investigate the detailed radio
structure of these nuclei. Most radio Seyfert nuclei have an angular
size of a few arcseconds and steep non-thermal spectra, and hence
MERLIN (Davies, Anderson & Morison 1980) with 1", 0.25" and 0.08"
resolution at 0.4 GHz, 1.6 GHz and 5 GHz respectively, is well suited
to study both spectral effects and source structures on subarcsecond
scales. In table 1 we list a selection of Seyfert and other nearby
active nuclei which have been observed by MERLIN and in the following
sections we describe briefly the main results of these studies.

Low frequency turnovers due to free-free absorption

The possibility of detecting high frequency free-free emission from
the forbidden line regions (FLR) of Seyferts has been discussed by
Ulvestad, Wilson & Sramek (1981), who conclude that at 5 GHz such
emission contributes at most only a few percent of the total, mostly
non-thermal, radio emission. Hence the unambiguous detection of such
emission would be extremely difficult, requiring highly accurate
spectral decomposition.

An alternative is to observe the nuclei at low frequencies and
measure the free-free absorption of the non-thermal emission. This is
possible as the free-free optical depth of ionised gas is given by

$$\tau_c \approx 8.2 \times 10^{-2} \, T_e^{-1.35} \nu^{-2.1} \, E \quad \text{(Mezger \& Henderson 1967)}$$

TABLE 1   <u>Nearby active galactic nuclei studied with MERLIN</u>

GALAXY	REFERENCE
NGC 4151	Booler et al. (1982)
NGC 1068	Pedlar et al. (1983a)
NGC 1275 (Perseus A)	Pedlar et al. (1983b)
Markarian 3	Pedlar et al. (1984)
NGC 1218	Unger et al. (1984a)
Markarian 348	Unger et al. (1984b)
M82	Unger et al. (in preparation)
3C120	Browne et al. (1982)
NGC 4486 (Virgo A)	Charlesworth & Spencer (1982)
NGC 5506	
MCG 8-11-11	
Markarian 6	Booler, Ph.D. Thesis (1984)
NGC 4278	Unger, Ph.D. Thesis (in preparation)
NGC 1052	
Markarian 463	
NGC 6500	

Hence at 0.4 GHz an electron temperature of 10^4K results in a measurable (>30%) decrease in the non-thermal emission if the emission measure (E) is $\gtrsim 10^5$ pc cm^{-6}. As the electron densities in the FLR are $\gtrsim 10^3$ cm^{-3} (Koski 1978) a path length greater than 0.1 pc should result in detectable absorption, provided that a significant fraction of the non-thermal radio nucleus is covered by FLR gas.

From our sample it is possible to select seven nuclei which have approximately linear, non-thermal spectra at frequencies higher than 1 GHz. Hence if we assume the non-thermal spectral index to remain constant, we can extrapolate the spectra to estimate the flux density at 0.4 GHz. Of these nuclei three have components which show evidence for low frequency turnovers (NGC 1068, Markarian 3 & NGC 7674), whereas no measurable turnovers were observed in NGC 4151, Markarian 6 & MCG-8-11-11. There is also evidence for a turnover in NGC 5506, although because of the diffuse structure of this source, further work is required to confirm this result.

The cases of NGC 1068 and Markarian 3 have been discussed in some detail (Pedlar et al. 1982, 83a), and synchrotron self-absorption can be ruled out as the component sizes are too large. The low frequency spectra for NGC 1068, Markarian 3, NGC 7674 and NGC 4151 are shown in Figs. 1-3. Upper limits on these spectra at frequencies less than 0.4 GHz represent measurements with large (>1 arcmin) beams.

It is interesting to note that in NGC 1068 only the component coincident with the nucleus shows definite evidence for a turnover – this is not expected as only this component is completely covered by Walkers (1968) FLR clouds (Fig.4). In Mkn 3 only the western, more compact component, appears to be absorbed. The low frequency turnovers detected are all consistent with E = 2-5 x 10^5 pc cm^{-6} and hence must imply a high covering factor and an effective path length of a ˉ

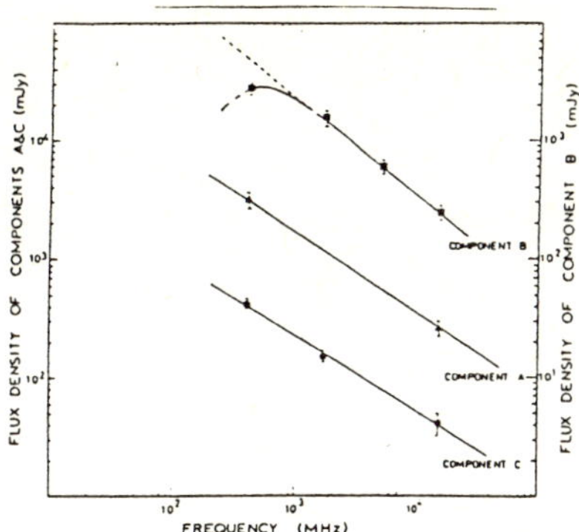

Fig.1   Low frequency spectra of radio components in NGC 1068
        (Adapted from Pedlar et al. 1983a).

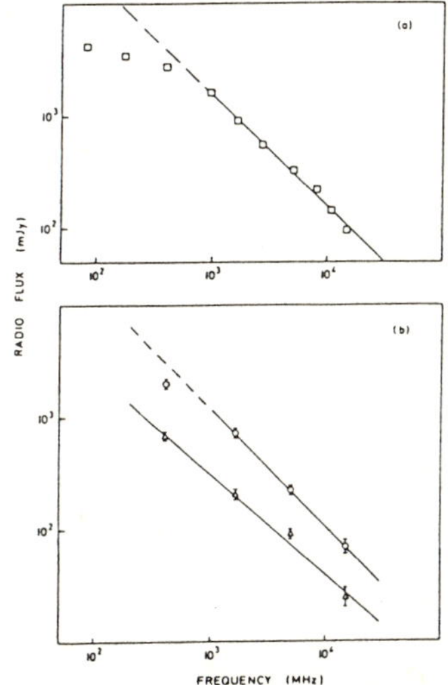

Fig. 2 (a)   Total flux radio spectrum of Mkn 3
       (b)   Radio spectra of eastern (triangles) and western
             (circles) components of Mkn 3

few tenths of parsecs.

   The radio emission associated with NGC 4151, despite the fact that
it appears to be embedded in a dense FLR region (Booler et al. 1982)
does not show a significant decrease from the extrapolated flux at 0.4
GHz (Fig.3). This suggests $E < 10^5$ pc cm^{-6}. The agreement between the
151 MHz flux density with arcminute resolution (Baldwin, private
communication) and the extrapolated value suggests that an emission
measure as low as 10^4 may be present.

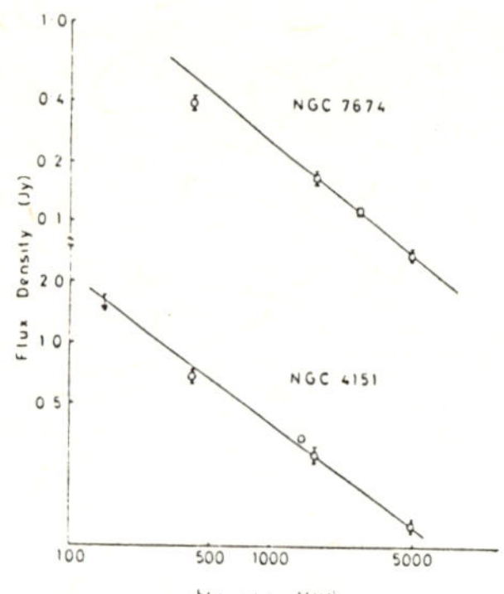

Fig.3 Low frequency spectra of NGC 7674 and NGC 4151

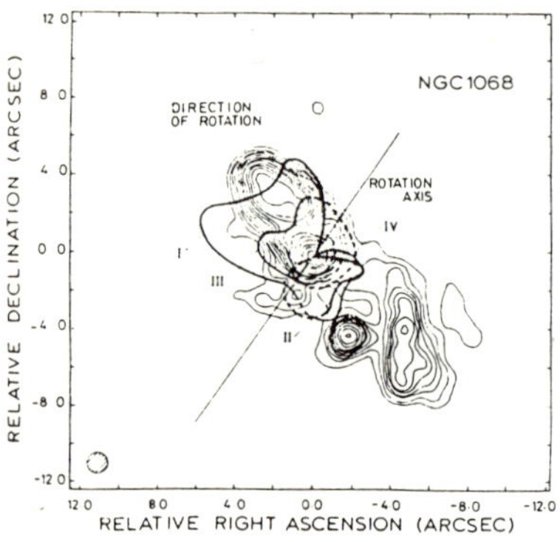

Fig.4 The outline of forbidden-line clouds I, II, III and IV as
       measured by Walker (1968) superposed on the 73 cm radio map

Evidence for collimated emission from Seyfert nuclei

Angular resolutions as high as 0.08" have enabled MERLIN to investigate the radio structure of nearby Seyfert nuclei on scales as small as
10 pc. This has revealed a number of linear or double structures consistent with collimated ejection from the nucleus over scales up to a
few kpc. The best examples of these are NGC 4151 (Booler et al. 1982),
NGC 1068 (Pedlar et al.1983a) Markarian 3 (Pedlar et al. 1984),
Markarian 348 (Unger et al. 1984), NGC 1052, Markarian 6 and NGC 7674.
Most of the other nuclei observed show some elongation which higher
resolution observations may show to be consistent with collimated
emission. A 5 kpc radio jet was discovered in the peculiar Seyfert
NGC 1275 (Pedlar et al. 1983b) and a 1 kpc jet in the nucleus of the
SO galaxy NGC 1218 (Unger et al. 1984a). Typical examples of radio
jets in Seyfert nuclei are shown in Figures 5 and 6.

These results, together with higher frequency studies at the VLA
(e.g. Wilson 1982), have led to the suggestion that collimated beams
of particles are emitted from Seyfert nuclei in a manner similar to
those postulated to exist in more energetic radio galaxies and quasars.
Unlike classical double radio sources the proposed beams in Seyferts
do not appear to escape from the galaxy and possibly dissipate much of

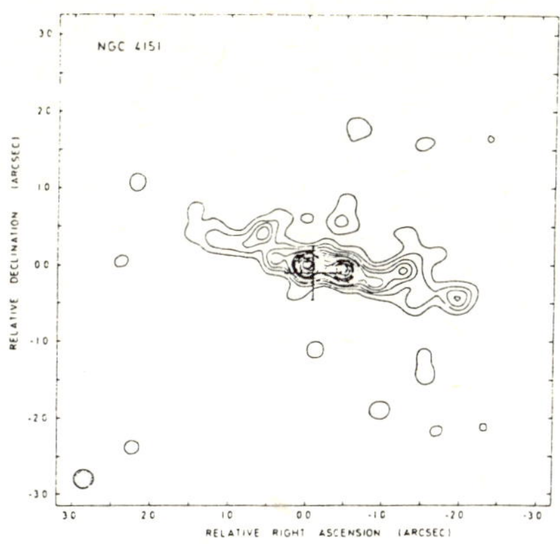

Fig.5   1.6 GHz MERLIN map of the nucleus of NGC 4151

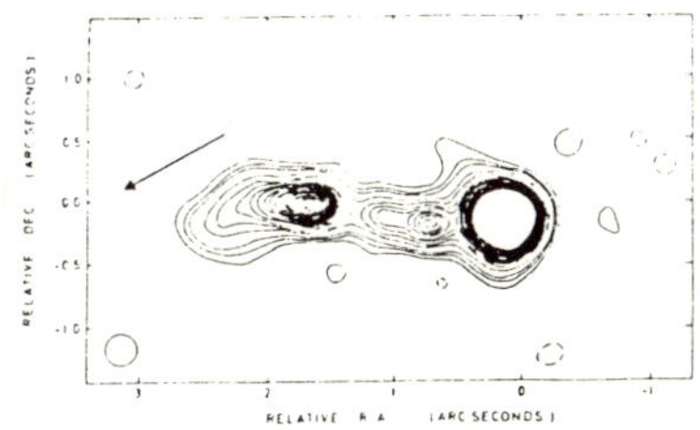

Fig.6   1.6 GHz MERLIN map of the nucleus of Markarian 3

their energy in the FLR.

High pressure radio components in Seyfert nuclei

The thermal pressure of the ionized gas in the FLR can be calculated if the electron density and temperature are known. Koski (1978)
gives estimates of these parameters for a number of Seyferts which
imply typical thermal pressures for the FLR of $\sim 10^{-9}$ dynes cm^{-2}.

The pressure due to the relativistic particles/magnetic fields
responsible for the radio synchrotron emission can be estimated from
the radio flux density and angular size using minimum energy arguments
(e.g. Miley 1980). Most radio Seyferts have extended components
($\gtrsim 100$ pc) which appear to be in approximate pressure balance with the
FLR gas. However high resolution MERLIN observations have revealed a
number of small (~ 10 pc) radio components which have relativistic
particle pressures in excess of 10^{-8} dynes cm^{-2} (see table 2). These
components clearly cannot be confined by the FLR gas and must be
expanding at approximately their internal sound speed into the FLR
gas and may be an important influence on its dynamics.

TABLE 2   High pressure components in Seyfert nuclei

Galaxy	Distance from nucleus pc	Pressure dynes/cm^2	Reference
NGC 4151	30	10^{-8}	Booler et al. 1982
Mkn 348	50	3×10^{-7}	Unger et al. 1984
NGC 1068	25	3×10^{-7}	Unger Ph.D. Thesis (in preparation)

REFERENCES

Booler, R.V., Pedlar, A. & Davies, R.D., (1982). Mon.Not.R.astr.Soc., 199, 229.

Browne, I.W.A. et al. (1982). Nature 299, 788.

Charlesworth, M. & Spencer, R.E., (1982). Mon.Not.R.astr.Soc., 200, 933.

Davies, J.G., Anderson, B. & Morison, I., (1980). Nature 288, 64.

Koski, A.T., (1978). Astrophys.J., 223,56.

Mezger, P.G. & Henderson, A.P., (1967). Astrophys.J., 147, 471.

Miley, G. (1980). Ann.Rev.Astron.Astrophys. 18, 165.

Pedlar, A., Booler, R.V., Spencer, R.E. & Stewart, O.J., (1983a). Mon.Not.R.astr.Soc., 202, 647.

Pedlar, A., Booler, R.V. & Davies, R.D., (1983b). Mon.Not.R.astr.Soc., 203, 667.

Pedlar, A., Unger, S.W. & Booler, R.V., (1984). Mon.Not.R.astr.Soc., 207, 193.

Ulvestad, J.S., Wilson, A.S. & Sramek, R.A. (1981). Ap.J. 247, 419.

Unger, S.W., Booler, R.V. & Pedlar, A., 1984a. Mon.Not.R.astr.Soc., 207, 679.

Unger, S.W., Pedlar, A., Neff, S. & de Bruyn, G., (1984b). Submitted to Mon.Not.R.astr.Soc.

Walker, M.F., 1968. Astrophys.J., 151, 71.

Wilson, A.S., (1982). Highlights of Astronomy, IAU Patras.

PROPERTIES OF THE RADIO CORES IN ELLIPTICAL GALAXIES

G. Giovannini

Istituto di Radioastronomia
via Irnerio 46
I-40126 Bologna   Italy

SUMMARY

A statistical study of the properties of the radio cores in elliptical
galaxies has been performed using a sample of 138 radio galaxies taken
from the 3CR and B2 catalogues.
   A statistically significant correlation is found between the core and
the total radio power. Moreover, a correlation between the core radio
power and the absolute optical magnitude is also found: optically
brighter galaxies tend to have higher core radio powers.

Introduction

   Several authors have carried out studies on the statistical properties
of small diameter ($\leqslant$few arcsec) radio sources (cores) in the central
regions of elliptical radio galaxies to study a possible relation between
the radio power of cores and the power of the outer components. Bridle
and Fomalont (1978) do not find any significant correlation between the
core radio powers (P_c) and the total radio powers (P_t); Fanti and Perola
(1977) conclude that a correlation of the type $P_c \propto P_t^{0.5}$ is equally
possible as a null correlation. Fabbiano et al. (1984) studying a
complete sample of 43 3CR radio galaxies find $P_c \propto P_t^{0.75}$.
   An obvious difficulty in the study of such a correlation is to obtain
consistent measurements of the nuclear radio power. In fact: 1) cores
are often variable, 2) different angular resolution of different
instruments and different distances of radio galaxies give flux
measurements from different linear regions. These difficulties introduce
an additional dispersion to any correlation which involves the core
fluxes. To minimize this problem we have selected three samples of radio
galaxies for a total of 138 objects with the widest range of total radio
power presently available: $23.0 < \log P_t < 27.0$ W/Hz at 408 MHz;
$H_o = 100$ km sec^{-1} Mpc^{-1}, $q_o = 1$.

The samples are:

1) The B2 "bright" sample: 50 elliptical galaxies obtained from the identification of B2 radio sources ($S(408) \geqslant 0.25$Jy) with Zwicky galaxies ($m_{pg} \leqslant 15.7$).

2) The B2 "faint" sample: 44 radio galaxies obtained from the identification of B2 radio sources with galaxies fainter than the CGCG limit and $m_v \leqslant 16.5$.

3) The 3CR sample: 49 radio galaxies from the 3CR catalogue, excluding those already belonging to the previous samples. The optical limit is $m_v < 20$ and the radio limit is 10 Jy at 178 MHz; $\delta > 10°$ and $|b_{II}| > 15°$.

Before investigating possible correlations we define what we consider to be a radio core: we define as "core" any unresolved source coinciding with the optical center of the galaxy. Since VLBI measurements of our radio galaxies are very rarely available, we use data obtained at 6 cm with the WSRT, the Cambridge 5 km radio telescope or the VLA. The typical angular resolution is of few arcsec corresponding (at the average distance of our galaxies) to linear sizes of a few kpc. Therefore there is the possibility that the core fluxes are sometimes overestimated, by a factor which we estimate not to exceed, in the worst cases, $\sim 40\%$ of the core flux. Given the large range of P_c covered by the objects in our samples and the large dispersion of the correlation found, none of our conclusions can be significantly affected by this possible overestimate.

Statistical analysis

Even if the use of a regression analysis between intrinsic luminosities can induce a spurious correlation because of the redshift dependence ($P \propto Sz^2$), it has been shown by Feigelson and Berg (1983) that the determination of an intrinsic correlation is not inhibited provided that upper limits are taken into proper account. On the other hand a flux-flux diagram can hide a correlation between luminosities expecially in samples with a large intrinsic dispersion (Feigelson and Berg, 1983).

Fig.1 shows P_c versus P_t for all the galaxies of the three samples together. In $\sim 70\%$ of our galaxies a core radio emission is detected (96/138). The best fit linear regression of log P_c versus log P_t gives
$$\log P_c = 8.74 + (0.57 \pm 0.06) \log P_t$$
The fit has been obtained with the Maximum Likelihood Method of Avni et al. (1980) which makes full use of all the available data including the upper limits. The main assumption which assures the applicability of this method is that all the objects in the sample (detection and upper limits) are drawn from the same parent population. Our result means that a core of high power is more likely to be found in a strong source, but the ratio P_c/P_t decreases when P_t increases. However, the large dispersion indicates that the core radio power is not a good indicator of the total radio power.

To test if there is the possibility of a different correlation between P_c and P_t for low and high luminosity radio galaxies we have analyzed separately the galaxies with P_t higher and lower than 10^{25} W/Hz at 408 MHz. The best fit slopes are: 0.52 ± 0.16 and 0.63 ± 0.14 respectively;

also the normalizations are perfectly consistent with each other, showing
that the correlation between P_c and P_t can be represented by a single law
for our entire data set covering a range larger than 10^4 in total radio
luminosity.

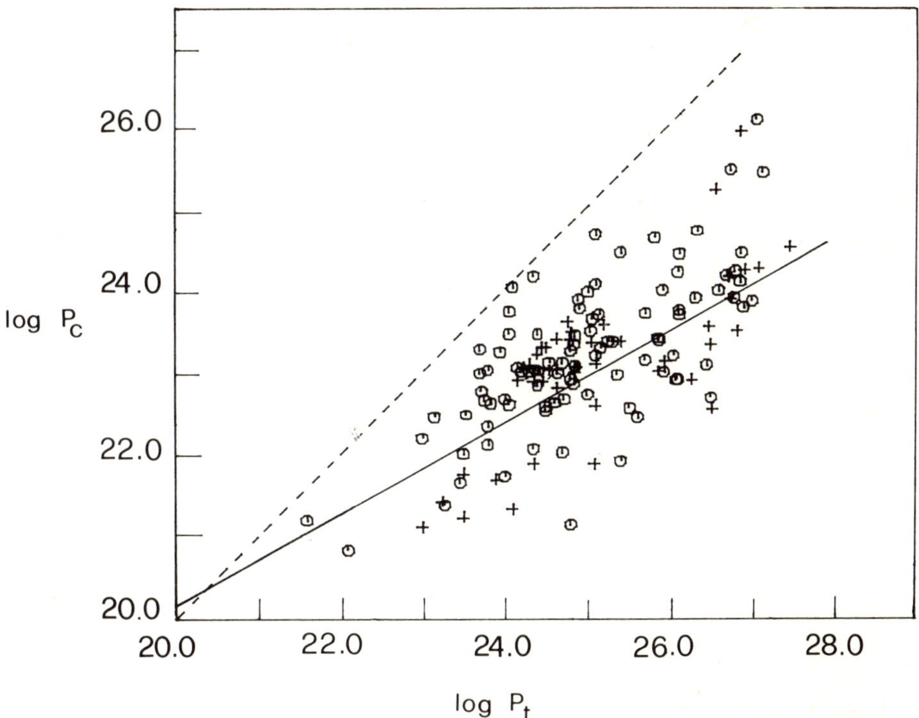

Fig.1. Core radio power at 5.0 GHz versus total radio power at 408 MHz.
The circles represent the core detections, while the crosses represent
the upper limits for the core radio powers. The dashed line corresponds
to $P_c = P_t$; the solid line is the best fit linear regression of log P_c
versus log P_t.

We have also computed the radio luminosity function (RLF, see Avni et
al. 1980 for a description of the method used) of the cores in three
different ranges of P_t (Fig. 2).

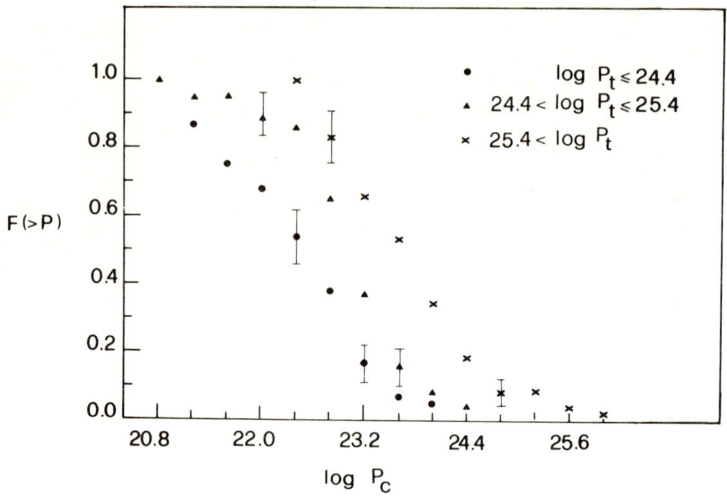

Fig.2. Integral luminosity distribution of the cores in three different
ranges of total radio power.   The bars represent one sigma uncertainty.

The shift among the three curves is quantitatively in agreement with
the correlation found above between P_c and P_t.   Note that the derivation
of the RLF makes no assumption on the functional shape of the
distribution.

In order to test if the core radio luminosity depends also on the
absolute magnitude of the parent galaxy, we have computed a linear
regression of P_c as a function of both P_t and M_v simultaneously, of the
form:

$$\log P_c = A_m |M_v| + A_t \log P_t + A$$

The best estimates of the parameters are $A_m = 0.32$, $A_t = 0.52$.   The value
$A_m = 0$, i.e. no correlation with M_v, is excluded at more than 99% level.
Also in this case we have computed the RLF of cores in three classes of
M_v (Fig. 3).   The shift among the curves is not appreciably influenced
by the distribution of P_t in the three classes of M_v and is in agreement
with the previous result.   This shows the existence of a significant
correlation between the core radio power and the optical magnitude.

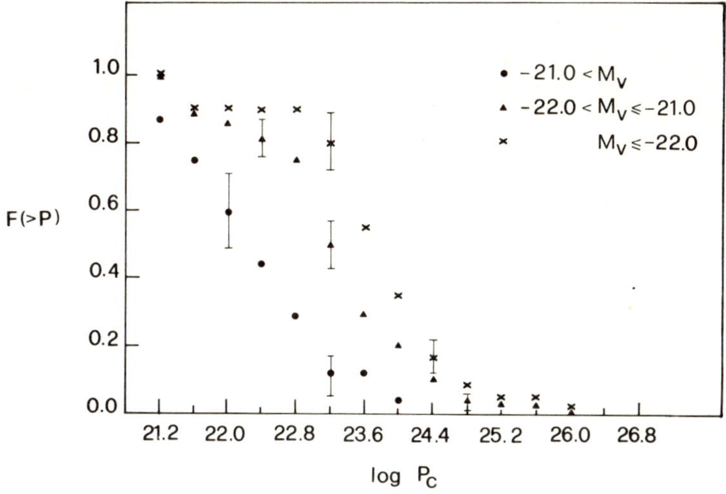

Fig.3. Integral luminosity distribution of the cores in three different ranges of absolute visual magnitude.   The   bars represent one sigma uncertainty.

This correlation can   be   understood   if   we   assume   that M_v is   an indicator   of the mass of the parent galaxy.   A galaxy with a larger mass may have a higher quantity of interstellar gas;   a fraction of it could reach the nucleus and power a compact radio source.

References

Avni, Y., Soltan, A., Tananbaum, H., Zamorani, G.: 1980, Ap. J. **238**,800

Bridle, A.H., Fomalont, E.B.: 1978, Astron. J. **83**,704

Fabbiano, G., Miller, L., Trinchieri, G., Longair, M., Elvis, M.: 1984, Ap. J. **277**,115

Fanti, R., Perola, G.C.: 1977, IAU Symp. **74**, p.171

Feigelson, E.D., Berg, C.J.: 1983, Ap. J. **269**,400

Extended Radio Emission Associated with Blazars

R.R.J. Antonucci

National Radio Astronomy Observatory
Edgemont Road
Charlottesville, Virginia   22903

SUMMARY

Blazars have many extraordinary properties which can be explained by
a radiating beam undergoing relativistic bulk motion directed towards
the observer. This beam may possibly be identified with jets in the
nuclei of normal double radio sources. Under this hypothesis, blazars
should show substantial extended radio emission, and it should appear
distorted and foreshortened from our point of view. J. S. Ulvestad and
I have made deep VLA maps of a large sample of blazars, and we detect
extended radio emission generally consistent with the hypothesis.

Blazars are defined by two very strongly correlated characteristics.
They are highly variable in optical flux, showing changes of a factor
of two in days, and sometimes a factor of 100 in months. They also
exhibit high optical polarization--sometimes over 30%. The
polarization amount and position angle generally vary in hours to
weeks. The word "blazars" is used to denote the union of two
historical classes, the BL Lac objects and the optically violently
variable (OVV) QSOs (Angel and Stockman 1980).

In addition to these defining properties, many blazars show certain
characteristic radio properties. Almost all have strong variable
compact radio sources. Some have radio fluxes and variability
timescales apparently conflicting with the Compton limit on brightness
temperature, and some show apparent superluminal motions.

These properties could be explained by a relativistic radiating beam
directed towards the observer (e.g., Blandford and Königl 1979). We
see jets of material connecting the core sources to extended lobes of
normal double radio sources, so it is natural to speculate that the
bases of these jets are in fact the beams needed to explain blazars
(Blandford and Rees 1978, Browne 1983). According to this "Unified
Scheme," if we could look at a normal double from along the jet
direction we would identify it as a blazar.

This idea has several obvious consequences. The space density of
normal doubles must be right, $\sim 4\pi\gamma^2$ times the blazar space density
(where γ is the relativistic velocity factor of the radiating jet).
Blazars and normal doubles must agree in all aspect-independent
parameters such as extended radio power, emission line luminosity and

host galaxy type. The extended radio emission must have the right
appearance and the right projected linear size. Finally we expect
statistical correlations between the aspect-dependent properties such
as core dominance and shape of the extended radio emission. The only
properties I will present data on here are the morphologies, powers
and projected linear sizes of the blazar extended radio emission. We
will discuss some of the other properties in our forthcoming paper
(R.R.J. Antonucci and J. S. Ulvestad, in preparation).

Some previous mapping of blazars has been reported by Stannard and
McIlwrath 1982 (BL Lac object), Browne et al. 1982 (superluminal
sources), Schilizzi and de Bruyn 1983 (superluminal sources), Ulvestad
et al. 1983 ("UJW"), and Wardle et al. 1984 ("WMA"). The last two
projects provided 20 cm A array VLA maps of 27 out of the 54 definite
blazars known at the time of the Angel and Stockman (1980) review
paper. For this project J. S. Ulvestad and I have (1) mapped the rest
of the blazars at 20 cm with the VLA A array, and (2) reprocessed the
Ulvestad et al. 1983 and Wardle et al. 1984 data sets using additional
editing and self-calibration to obtain substantial increases in
dynamic range.

Cylindrically-symmetric double sources seen end on would appear as
circular halos surrounding point sources, so we were quite surprised
by the maps in Figure 1! No symmetric doubles or head tails had been
reported previously among blazars. But perhaps these maps aren't as
unexpected as they seem. Double radio sources are often bent or
distorted, especially among the low luminosity sources. Figure 2
shows several quasar maps taken from the survey of Hintzen et al.
(1983). Perhaps if viewed from the direction of the radio axes,
these would appear as blazars. In looking at the maps in
Figure 2, consider the morphology and the (modest) degree of
foreshortening expected if they were viewed end-on. The third one
might look like a small classical double. The sources in Figure 1 do
have small projected linear sizes (Antonucci and Ulvestad 1984).
Figure 3 shows some of our more typical maps. They present something
closer to the expected morphology.

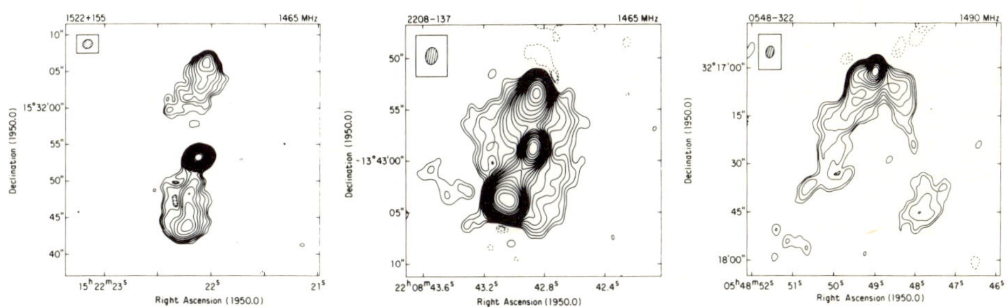

Figure 1   Some blazars with unexpected morphologies.

Less subjective than interpreting the morphologies is consideration
of the power in the extended emission. Figure 4 shows histograms of
the extended powers of the blazars, and also the relatively weak B2
catalog sources and the relatively strong 3C catalog (Laing et al.
1983) sources. The extended powers are very substantial, and this is

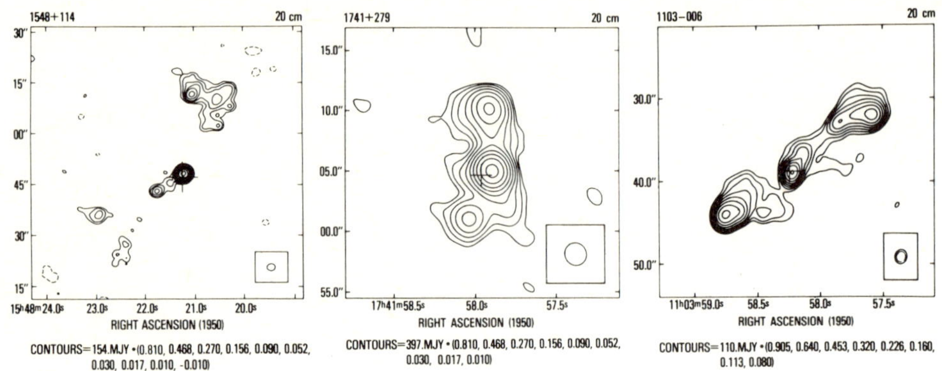

Figure 2   Maps of normal quasars for comparison.

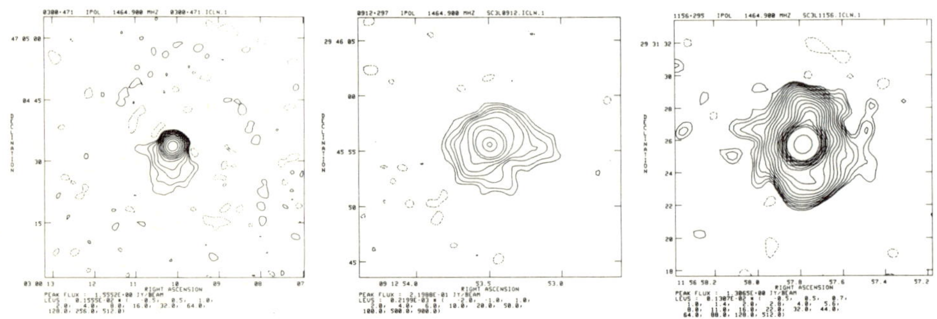

Figure 3   More typical blazar maps.

a strong plus for the hypothesis under discussion. We do run into a serious problem, however, in that we don't know a priori which sample of "normal" doubles to compare the blazars with. The blazars weren't originally selected by extended radio flux, or any single property. If the Unified Scheme is correct, some blazars come from a B2-like parent population and some from a 3C-like parent population. This is not surprising.

Though the blazars have strong extended emission, their projected linear sizes are fairly small (Figure 5). They are typically much smaller than the 3C sources and even somewhat smaller than the B2 radio galaxies. The main worry with this test is that the surface brightness sensitivity isn't constant for all of the objects in Figure 5. We have applied various tapers to the blazar maps to see whether more structure appeared as we increased our surface brightness sensitivity. The objects in which some of the structure appeared to be "resolved out" are marked with lower limits in Figures 3 and 4. We think that the projected linear size histograms are qualitatively correct, and that the degree of foreshortening is reasonable. This is another success for the Unified Scheme.

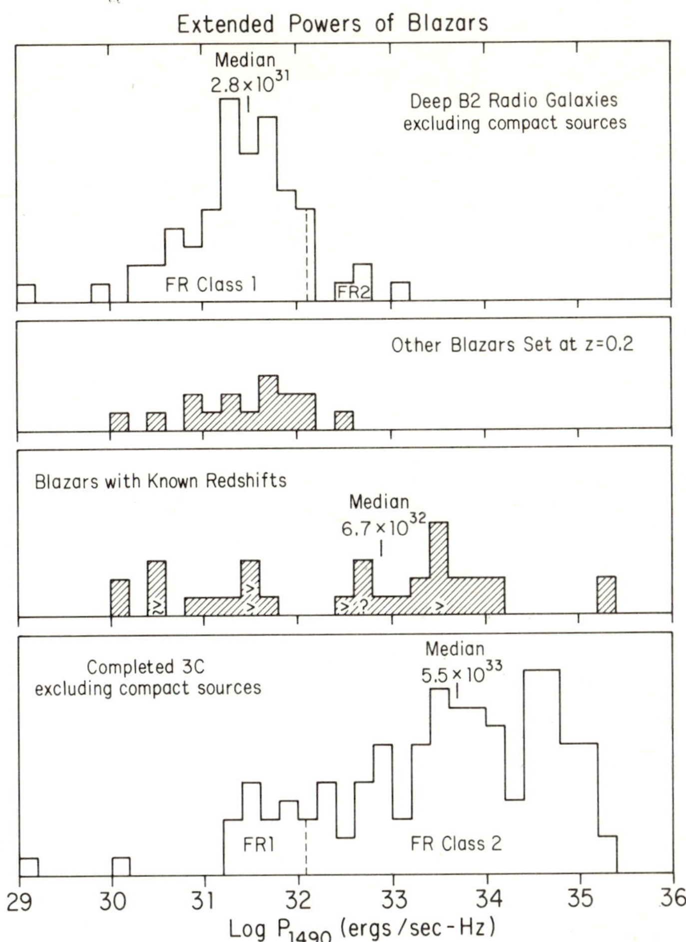

Figure 4   The extended radio power of blazars compared with normal extragalactic sources. The blazar data are from this project; UJW; WMA; Perley et al. 1982 (1150+497, 1253-055, 1641+399, and 2251+158); Noordam and de Bruyn 1982 (0316+413); and Stannard and McIlwrath 1982 (0235+164). Question marks indicate uncertain structure or redshift. Five blazars, all lacking definite redshifts, show no extended emission. They aren't shown as upper limits in here or in Figure 5 because we don't know whether their extended emission is too weak or too small to be detected.

REFERENCES

Angel, J.R.P., and Stockman, H. S. 1980 Ann. Rev. Astr. Ap. 18, 321.
Antonucci, R.R.J., and Ulvestad, J. S. 1984 Nature 308, 617.
Blandford, R. D., and Königl, A. 1979 Ap. J. 232, 34.
Blandford, R. D., and Rees, M. J. 1978 in Pittsburgh Conference on BL lac Objects, ed. A. M. Wolfe (Pittsburgh: Univ. of Pittsburgh Press), p. 328.

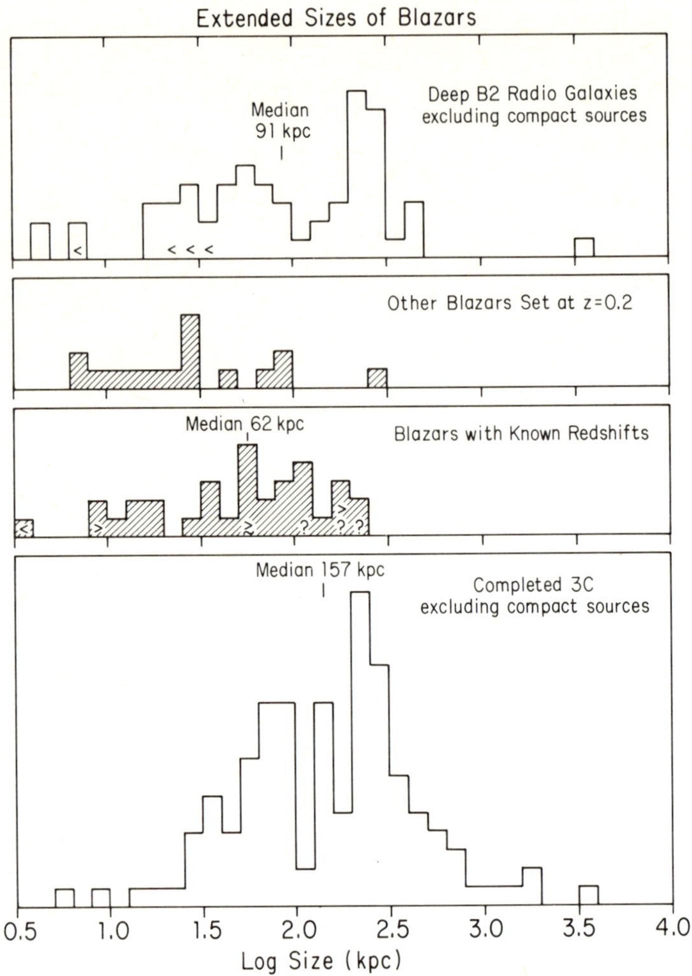

Figure 5   The projected linear sizes of blazars compared with those of
normal extended sources.

Browne, I.W.A. 1983 MNRAS 204, 23p.
Browne, I.W.A., Clark, R. R., Moore, P. K., Muxlow, T.W.B.,
    Wilkinson, P. N., Cohen, M. H., and Porcas, R. W. 1982 Nature 299,
    788.
Hintzen, P., Ulvestad, J., and Owen, F. 1983 A. J., 88, 709.
Laing, R. A., Riley, J. M., and Longair, M. S. 1983 MNRAS 204, 151.
Noordam, J. E., and de Bruyn, A. G. 1982 Nature 299, 597.
Perley, R. A., Fomalont, E. B., and Johnston, K. J. 1982 Ap. J. Lett.
    255, L93.
Schilizzi, R. T., and de Bruyn, A. G. 1983 Nature 303, 26.
Stannard, D., and McIlwrath, B. K. 1982 Nature 298, 140.
Ulvestad, J. S., Johnston, K. J., and Weiler, K. W. 1983 Ap. J.
    266, 18.
Wardle, J.F.C., Moore, R. L., and Angel, J.R.P. 1984 Ap. J., in press.

Infrared Observations of the Activity in Galactic Nuclei

Malcolm G. Smith

Royal Observatory, Blackford Hill, Edinburgh, U.K.

SUMMARY

Although covering a few related topics, this review is largely
confined to observations of extragalactic objects in which non-thermal
and star-forming activity are found together in the central regions. A
descriptive catalogue of a representative series of such objects is
presented, roughly ordered according to the degree of dominance of the
non-thermal activity. Though the occurrence of both forms of activity
may be much more common in galactic nuclei than generally realised, the
existence of a firm physical link between the two mechanisms has yet to
be established. It is hoped that this article will stimulate systematic observations of well-defined radio, X-ray and infrared-selected
samples of galactic nuclei, in order to establish this physical link.

CONTENTS

1. INTRODUCTION

1.1 The two main sources of infrared radiation from active galactic nuclei - starbursts and non-thermal power-law activity.

The sensitivity of infrared equipment on satellites and large ground-based telescopes is now reaching levels where large-scale systematic work on the properties of even quite normal galaxy nuclei can really get underway (e.g. Willner et al. 1984;   de Jong et al. 1984;   Glass 1984). Becklin et al. (1984b) report results of a 10μm survey of activity in the nuclei of the 100 brightest Virgo galaxies; they observed 60 spirals, 34 SO galaxies and 17 ellipticals. "Taken as a sample, the 100 brightest Virgo galaxies, including spirals, SOs and ellipticals, show activity in their nucleus, as evidenced by their average 10μm emission". Scoville et al. (1983) showed that there is a continuum of nuclear activity in all spiral galaxies, rather than discrete on-off phases. Based on earlier detailed studies of 10μm emission in galactic and extragalactic sources - see, e.g., the classic study of the starburst in M82 by Rieke et al. (1980) and the direct VLA observations of the related supernovae reported at this conference by Kronberg - they concluded that star formation may be occurring in the nuclei of virtually all spiral galaxies, independent of Hubble type (see also the review of molecular clouds in galaxies by Morris and Rickard 1982). Joseph et al. (pers. comm.) have, on the other hand, found a marked dependence of the K-L colour on morphological type: later-type galaxies show a greater level of 3.8μm activity in their nuclei. The infrared luminosity in newly-formed stars is typically 2×10^9 $L_\odot$ for spirals in the Virgo sample; unless the initial mass function in the galaxy nuclei is heavily biased towards masses above $3M_\odot$, mass loss from the recently formed stars cannot maintain a steady state, for too much recycled material becomes locked up in low-mass stars (see also Young, Gallagher and Hunter, 1984).

Becklin et al. (1984b) are less certain of the source of the energy in the nuclei of SO and elliptical galaxies. They report an average 5-15μm luminosity of 10^8 $L_\odot$ in these nuclei. This is equivalent to an average bolometric luminosity in newly-formed stars of $\sim 10^9$ $L_\odot$. On the other hand they cannot rule out the possibility that the 10μm luminosity in SO and elliptical nuclei comes predominantly from a non-thermal source at the very centre. Similar uncertainties hold for Seyfert galaxies, for which infrared excess is an almost universal characteristic (Rieke 1978). It is interesting that de Jong et al. (1984) failed to detect any of the $\sim$26 elliptical galaxies that they observed with the IRAS satellite.

The survey in Virgo illustrates quite well the astrophysics coming out of much recent infrared work on galactic nuclei (for reviews of earlier work and more extensive material see, e.g. Rieke and Lebofsky 1979;   Stein and Soifer 1983;   Smith 1984). Most of the energy radiated from these nuclei originates (a) from intense star-formation activity and/or (b) from non-thermal activity (whose original source of energy is presumed to be gravitational, rather than thermonuclear processing in stars). Figures 2-4 show broad-band spectra of objects in which these forms of activity occur. This review will examine these two major energy sources and the state of evidence for a possible link between them.

1.2 On triggering an active nucleus and the relationship between different classes of AGN.

A highly simplified and speculative picture of the relation between these main energy sources can be built up; it starts perhaps with an interaction between galaxies which redistributes material from the outer parts into orbits which plunge towards the potential well in the central regions of the galaxy (see, also the paper by Morris at this conference) and triggers intense star-formation activity (see, e.g., Arp 1966; Toomre and Toomre 1972; Larson and Tinsley 1978; Gunn 1979). Some of the redistributed material, along with mass loss from the hot stars, accretes at the centre onto a massive central object (e.g. Weedman et al. 1981; Bailey 1982; Norman and Silk 1983; Carroll and Kwan 1983; Thuan 1984; but see also van der Hulst, Crane and Keel 1981; Weedman 1983). Bailey has shown how the star formation acts as a trigger to enable viscous accretion from the inner disc to proceed in towards the compact nucleus. The gravitational energy released emerges as non-thermal radiation (Blandford, this conference).

The possibility that galaxy interactions may be the initial triggering mechanism for a relation between starbursts and non-thermal power-law activity, has resulted in a flurry of observational and theoretical papers in recent years. For example, Condon and Dressel (1978) found that one-sixth of the radio sources stronger than 35mJy at 2380MHz, in their complete sample of bright galaxies, are compact, and are all confined to the nuclear region (i.e. within a few hundred parsecs of the centre); they found a strong tendency for these compact radio sources to occur in paired galaxies. Hummel (1980) found that the central sources in interacting galaxies are typically 2 or 3 times stronger at 1.4 GHz than the central sources in isolated spirals. Condon et al. (1982) mapped the radio cores of a subset of radio-loud optically-bright spiral galaxies. They found that the radio sources were usually confined to the central 1 kpc of the galaxies and were co-extensive with regions of intense star formation ($\sim 10^{10}L_{\odot}$ - $10^{11}L_{\odot}$). Condon et al. concluded that the 1413 MHz flux from these sources is consistent with synchrotron radiation from supernova remnants produced at the rate of $\lesssim$ 1yr^{-1}. They also found that the radio and 10µm fluxes of the extended core sources are roughly proportional and concluded that "Nearly all of the strong, extended radio sources are found in galaxies with nearby companions, so most of the episodic bursts of star formation are apparently triggered by galaxy-galaxy interactions".

Direct mergers of galaxies may also supply some of the fuel. Schweizer (1980) concluded that for NGC 1316 (Fornax A) ".... infall of gas into the nucleus is likely to have supplied the fuel for past radio outbursts. The presence of two dust lanes that spiral directly into the nucleus suggests that some fueling from the disk continues and further outbursts may be expected"; models for the merging of galaxies in an expanding universe have been discussed by Roos (1980).

Recent infrared data are in accord with the radio and optical results cited above. De Jong (1984) has reported an overrepresentation of multiple and interacting systems in the IRAS mini-survey of spiral galaxies. He and his collaborators (Soifer et al. 1984; de Jong et al. 1984) also find a wide range of values of the ratio, λ, of infrared to blue luminosity in galaxies. In his talk, de Jong cited the classic example of the triplet M82, M81 and NGC 3077. M81 dominates in the blue, whereas M82 is much brighter than the other two galaxies at infrared wavelengths. De Jong finds that simple interaction models coupled to starburst activity (close encounters producing stronger shorter-lived starbursts) predict a probability P (λ) $\sim\lambda^{-7/3}$ of observing the i.r./blue ratio λ - a result entirely consistent with the recent IRAS observations. Joseph (1984) has found K-L excesses in many spiral-galaxy nuclei which are members of interacting sytems; he has IRAS data for 6 of these pairs and finds a peak in their continuum spectrum near 100μm in all cases as expected from the K-L colours. L measurements may well prove the best ground-based indicator for dust in star-formation regions. Joseph et al. (1984) suggest that "interactions induce starbursts with an efficiency approaching 100%".

Galaxy interactions and mergers appear to have significance in triggering activity even at luminosity levels appropriate to QSOs. Stockton (1982) has made optical studies of a number of objects having small angular separations from QSOs. He concludes that "similar compact objects may well exist in the neighborhoods of a significant fraction ($\sim$20%) of all low-redshift QSOs and.... that the presence of these close companions is related to the QSO activity.... the QSO activity results from fresh material being brought into the nuclei of these galaxies during the interaction...." Hutchings and Campbell (1983) arrive at similar conclusions and find that about a third of the QSOs they studied reside in clusters or small groups of galaxies; Bothun and Schommer (1982) have investigated the effect of the environment (cluster or group) on interactions between galaxies. Stockton and MacKenty (1983) found two luminous QSOs which each appear to have resulted from encounters between two galaxies of roughly equal mass; they argue that the galaxies involved possessed extensive rotating disks before their encounters. "These two examples provide support for the position that luminous QSOs are frequently activated by violent interactions".

Terlevich and Melnick (1984) have recently investigated the effect of 'warmers' on the evolution of a metal-rich giant HII region; 'warmers' are massive mass-loss stars - extreme WC or WO Wolf-Rayet stars - which can reach an effective temperature > 100,000°K. They find that the spectrum of a sufficiently massive starburst HII region evolves into a Seyfert type 2 spectrum and then into a Low-Ionisation Nuclear Emission Region (LINER) spectrum. Theoretical models based on isochrones for mass-loss stars in combination with stellar atmospheres for hydrogen and helium-burning stars - plus an initial mass function (IMF) with $dN/dm \propto m^{-3}$ - give an ionisation source for G.Ferland's ionisation code 'CLOUDY'. The output line ratios fit those observed for the corresponding types of narrow-emission-line galaxy. This is not surprising, for Terlevich and Melnick find that "the ionising continuum corresponding to a 3Myr old cluster with warmers, is very nearly a power law of slope about -1.5 and with cut off at about 20 Rydbergs". Nevertheless the predicted ratio of the blue and ultraviolet NeIII/NeV lines is close to that observed whereas power-law ionisation models give values an order of magnitude larger. The predicted HeII λ4686Å line strengths are also in better agreement with observations. Terlevich and Melnick's models provide the following evolutionary sequences:

| 'Large' starburst | | 'Small' starburst | |
| $M \sim 10^9 M_\odot$ | | $M \sim 10^8 M_\odot$ | |
TIME INTERVAL (Myr)	NUCLEAR PHASE	TIME INTERVAL (Myr)	NUCLEAR PHASE
0-2.5	Starburst	0-2.5	Starburst
3-5	Seyfert 2	3-7	LINER (Red); $W_{H\beta} \sim 200$Å
5-7	LINER (Blue); $W_{H\beta} \sim 40$Å		No Seyfert 2 phase

Let us nevertheless be quite clear from the outset that an unambiguous observational link between starburst nuclei and non-thermal emission from Seyfert nuclei and quasars has yet to be established. As Balzano (1983) points out "Star-burst nuclei have specific observational properties.... "[as discussed by Balzano and Weedman 1981 and by Feldman et al. 1982]" .... and can be readily distinguished from Seyferts.... there is no evidence of a nonthermal radiation source in any of our [starburst] sample objects". However a number of authors have been careful to emphasise that optical surveys (such as the Markarian survey used by Balzano) are biased against discovering objects with even quite small amounts of dust extinction (see, e.g., Rieke 1978; Keel 1980; Rieke and Lebofsky 1980; Condon et al 1982; Rudy 1984; Rudy et al. 1984), whereas the Virgo survey by Becklin and collaborators and the far infrared work by Rickard and Harvey (1984) remind us that substantial dust emission is a common feature at least of spiral-galaxy nuclei. The work of Phillips, Charles and Baldwin (1983) suggests that Seyfert-like activity is also ".... considerably more common in the nuclei of spirals than had been suspected" (see also Keel 1983). Selection effects in the searches for active galactic nuclei (AGN) could easily cause bias against those nuclei in which star-formation and Seyfert-like activity co-exist; this bias could distort our view of the overall properties and evolution of AGN. To illustrate this point, consider quasars from the 3C catalogue; they are generally strong radio sources with extended double radio lobes and steep radio spectra. For some considerable time, it was believed by many that such quasars represent fairly the entire quasar population. We now believe that most quasars are not strong radio sources, and many of the few that are have compact radio structure and flat spectra!

Surveys at a range of different discovery frequencies such as Becklin's survey in Virgo, are important in order to become aware of such selection bias. X-ray surveys have already located a few highly reddened AGN (McAlary et al. 1983; Stocke et al 1983), and results from IRAS are awaited with keen interest (see, e.g., Becklin et al 1984a; de Jong et al 1984; de Jong 1984). Davies has reported at this conference on the status of a survey of radio emission from the 105 brightest Sbc galaxies in the RC 2 catalogue (de Vaucouleurs, de Vaucouleurs and Corwin 1976); Davies and Wolstencroft plan an infrared survey of these same nuclei. Heckman et al. (1983) have found that $\gtrsim$ 85% of a sample of bright ($m_v \lesssim$ 18) galaxies with compact nuclear radio sources have a non-stellar optical-infrared nuclear continuum. Observations of a sample of 30 late-type galaxies at wavelengths from 40μm to 160μm have led Rickard and Harvey (1984) to note a "Strong correlation with non-thermal radio continuum emission and radio-to-optical flux ratios [which] suggests that massive star formation underlies a wide range of galactic activity".

Osterbrock (1984) has emphasised that it is difficult, if not impossible to define an exact lower limit to a [nonthermal] galactic nucleus: he states that "All the observational phenomena we see in Seyfert galaxies [are] also evidently occurring at a lower level in many more galactic nuclei, not yet recognised as Seyferts. Particularly in the lower intensity ones, the observed emission lines result partly from gas photoionised by the featureless continuum source characteristic of [nonthermal] galactic nuclei, extending far into the ultraviolet, and partly from gas photoionised by hot stars".

It is also important, though sometimes difficult, to distinguish between non-thermal activity associated with supernovae (formed as a result of a starburst - Harwit and Pacini 1975; Rieke et al. 1980;

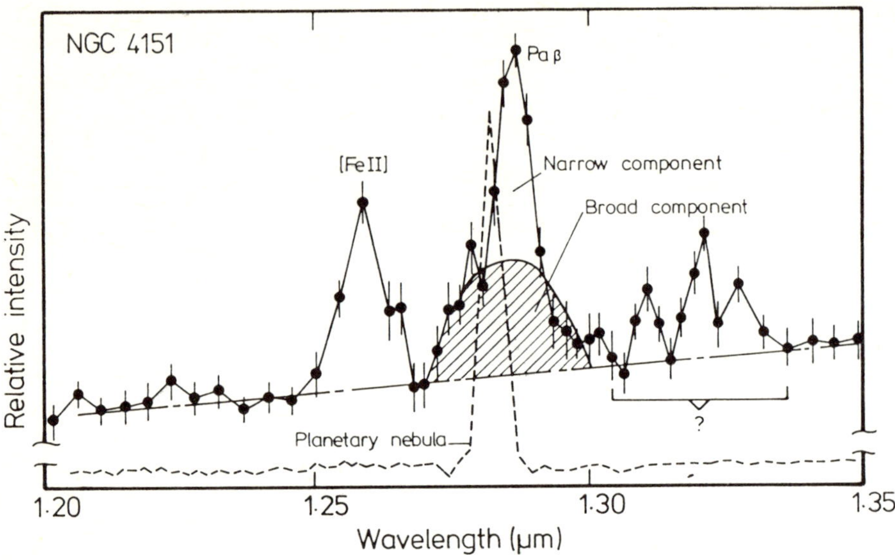

Figure 1:-   Paschen-beta and [FeII] 1.26µm emission as observed in the nucleus of NGC 4151, using the CGS2 spectrometer on UKIRT. The narrow component of Pa β is resolved. The [FeII] 1.26µm line is thought to be a low-temperature (500-5000°K) coolant in post-shock recombination zones (Shull, personal communication).

Condon et al. 1982; Sandqvist, Jörsäter and Lindblad 1982; Klein
1984; Kronberg, this conference; see also section 2.3) and non-
thermal, power-law activity usually assumed to result from gravitational collapse and synchrotron radiation of the energy released.

In section 2 we shall examine a sequence of AGN in which the
presence of a non-thermal component becomes gradually more and more
obvious, until it dominates over the emission from molecules, dust and
even gas.

1.3 Infrared techniques for observing AGN.

I will first give a brief description of the rapidly developing
infrared techniques which are beginning to allow much more detailed
studies of the two main sources of activity in galactic nuclei. Of
particular interest is the development of sensitive infrared imaging.
Figures 5 and 6 in section 2.4 illustrate two sites where one observes
the two forms of activity simultaneously. As discussed in more detail
in that section the blue (visual) images are dominated by radiation
from giant complexes of ionised hydrogen - sites of large-scale star-
formation activity. At longer wavelengths, and especially in the
infrared, the Seyfert nucleus dominates. As two-dimensional arrays are
being brought into service on infrared telescopes on the ground and in
satellites (e.g. Moneti et al. 1983; Arens et al. 1984; Tresch-
Fienberg et al. 1984; Goebel et al. 1984), much more of this kind of
imaging data will become available. The emergence of various kinds of
linear detector arrays is allowing infrared spectroscopy to emerge from
its infancy. One- or two-hour integrations on very bright Seyfert
nuclei are at last proving sufficient to resolve the narrow-line
component at reasonable signal-to-noise ratio (Figure 1). However, for
extragalactic work, we are still very far short of the data rates
provided by optical spectrometers. Multi-frequency broad-band
photometric studies of Seyfert nuclei, quasars and BL Lac objects are
now becoming commonplace throughout the window 1µm to 1mm. Starburst
and Seyfert nuclei are being studied in extensive surveys at JHKL, 10µm
and 20µm (e.g. Rieke 1978; Ward et al. 1982; McAlary et al. 1983;
Elvis et al. 1984; Lawrence et al. 1984 - see also Figures 2 and 4 and
the paper by Lawrence at this conference) and by IRAS at 12,25, 60 and
100µm. Studies of such nuclei at millimetre wavelengths are just
beginning and will accelerate as the large new millimetre-wave
telescopes come on line (e.g. Hildebrand et al. 1977; Longmore et al.
1984; Cunningham et al. 1985). As the level of non-thermal activity
increases, it becomes easier to find large numbers of extragalactic
objects for study at far infrared and millimetre wavelengths (Figures
3, 11 and 12) as well as in the near (1-5µm) and mid (5-35µm)
infrared - where at lower luminosities radiation from stars and dust
had dominated. Multifrequency broad-band polarimetry is being
increasingly used, particularly in studies of BL Lac objects. A
limited amount of infrared spectropolarimetry can be expected on the
very brightest nuclei (see, e.g., Aitken et al. 1984).

2. A CATALOGUE OF NUCLEAR ACTIVITY AS OBSERVED AT INFRARED WAVELENGTHS:

Let us examine observations of a series of galaxies in which the
presence of non-thermal activity becomes more and more obvious relative
to the activity associated with star formation: the sequence is
NGC3690, The Milky Way, M82 and other starburst galaxies, NGC1097 and
NGC1365, NGC1667, NGC5135 and IC5135, NGC1068, 3CR galaxies, NGC7469,
3C273 and BL Lac objects.

2.1 NGC3690

Spectacular star-formation activity can take place outside the nuclei of some galaxies. The significance of NGC3690 (which is in interaction with IC694) is that although the activity is spread through the body of the galaxy, the scale of the activity ($\geq 7 \times 10^{10}$ L$_\odot$)is comparable to the brightest analogous events in galactic nuclei (Gehrz, Sramek and Weedman 1983; IRAS observers - e.g. Rowan Robinson 1984 - have recently found even more extreme examples, such as NGC 6240 and Arp 220). The current starburst in NGC3690 corresponds to the formation of $> 10^9$ M$_\odot$ of massive stars. Observations of strong CO indicate the presence of $\sim 10^{10}$ M$_\odot$ of molecular hydrogen (Solomon, de Zafra and Barret in preparation). Direct detection of distributed line emission from vibrationally excited H_2 has been reported in this interacting galaxy by Fischer et al. (1983); the total observed luminosity in the $v = 1 \to 0$ S(0) line alone is approximately 1.5×10^7 L$_\odot$ In sources in the Milky Way, this line is thought to be excited by shocks. Fischer et al. derive a mass of $> 2 \times 10^4$M$_\odot$ of hot molecular hydrogen. As a further guide to conditions in the heavily obscured HII regions they also observed Br γ emission, requiring an input $> 4 \times 10^{54}$ s^{-1} ionising photons to maintain the HII regions in their ionised state.

2.2 The Galactic Centre

The line of sight from the Earth to the centre of our Galaxy is heavily obscured at visible wavelengths ($A_v \sim 27$mag - Becklin, Gatley and Werner 1978); infrared and radio studies have therefore proved crucial in exploiting the opportunity to obtain a close-up view of events in a galaxy nucleus. Time-variable positron annihilation radiation (511keV) has been detected from the general direction of the Galactic Centre (e.g. Lingenfelter and Ramaty 1982). A non-thermal radio source is found just to the SW of the infrared source IRS 16 (e.g. Lo 1982; Tresch-Fienberg et al. 1984; Marcaide, this conference).

Far-infrared photometry (30-100µm) by Becklin, Gatley and Werner (1982) indicates that dust in the central regions is confined chiefly to the plane of the galaxy, around a central cavity about 4pc diameter; inside the 'bubble', the density of dust is much lower ($A_v \ll 1$). IRS 16 lies at the centre of the cavity, where the distributions of total far-infrared luminosity and colour temperature reach their maxima. Other high-density clumps of plasma are found within the central cavity, and together with IRS 16 make up the region known as Sgr A (West); IRS 16 has infrared colours bluer than these surrounding clumps. IRS 16 shows no evidence of CO bands; all other infrared sources observed in the region are found to show prominent CO bands, indicating the presence of late-type luminous stars in all the infrared sources except IRS 16 (see, e.g., the spectra of IRS 16 and IRS 7 presented by Hall, Kleinmann and Scoville (1978)). Genzel et al. (1984) have used the [OI] fine-structure line at 63µm to measure the rotation of gas around the galactic centre; they find that the rotational velocity at a radius $\sim$1pc corresponds to a mass within the central parsec of $\sim 3 \times 10^6$ M$_\odot$. These observations of a non-thermal source radiating $> 10^7$L$_\odot$ in the centre of a cavity 4pc across, embedded within a region of high dust density led Gatley (1984) to the conclusion that mass loss from the central object creates a bubble in the interstellar medium.

Recent advances in the spectroscopic equipment at the 3.8 metre U.K. Infrared Telescope (UKIRT) have made it possible to carry out a number of crucial spectroscopic tests of this conclusion. Richard Wade's

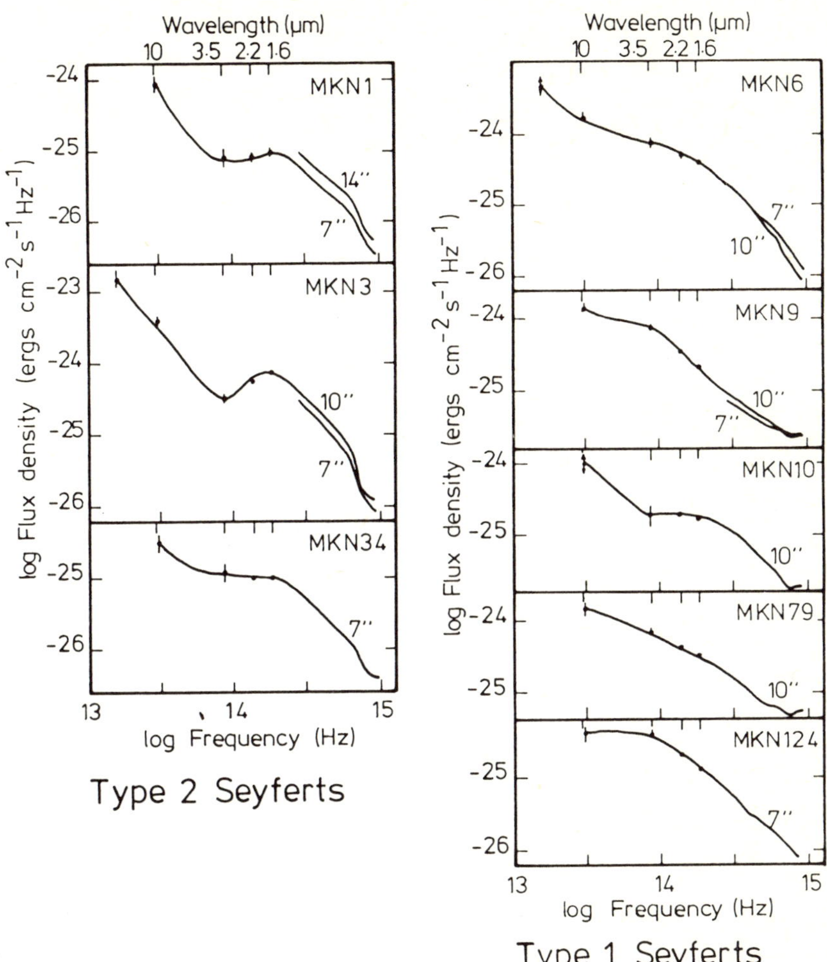

Figure 2:-   The near-infrared spectra of Seyfert galaxies adapted from Neugebauer et al. 1976; (see also the more up-to-date classification plots in the paper by Lawrence at this conference or in Lawrence et al. 1984).

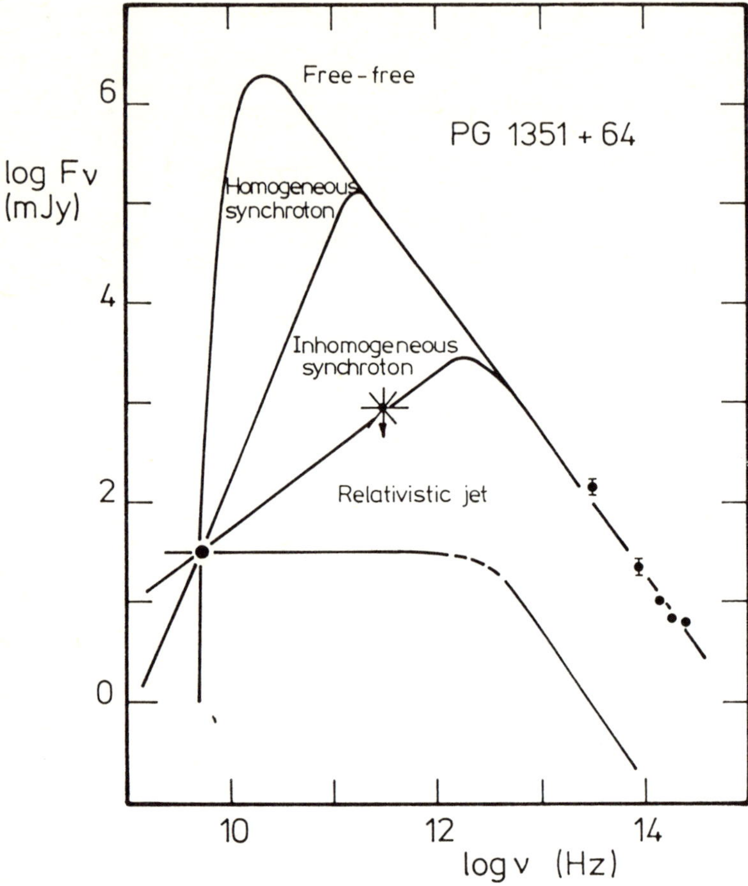

Figure 3:-   The radio-to-optical continua for various single-component emission mechanisms; these show how even an upper limit at millimetre wavelengths (the asterisk) can be of crucial importance in distinguishing between various basic models. Illustration adapted from Ennis, Neugebauer and Werner (1982).

CGS2, a 7-element, InSb, 1-5µm, cooled grating spectrometer was used to discover and study molecular hydrogen in the Galactic centre and to follow up Hall, Kleinmann and Scoville's discovery of broad λ 2.06µm He 2^1P-2^1S emission.

Emission of vibrationally excited molecular hydrogen has been observed in many parts of the Milky Way, and is found to be associated with star-forming regions and with supernova remnants. Excitation of this H_2 is thought to be by shocks (see, e.g., the review by Shull and Beckwith 1982). Given the possibility that mass loss from IRS 16 might stimulate the formation of shock fronts in the dense region surrounding the central bubble, Gatley et al. (1984) observed the 2.12µm v = 1 $\rightarrow$ 0 S(1) line of H_2 in and around the central region. Gatley (1984) concluded that "The strength of the emission from the shocked gas peaks sharply and symmetrically at 40" ($\sim$ 2pc) on either side of IRS 16. This is <u>not</u> an artifact of extinction, for the extinction in the ring itself is only $A_v \sim 1$.... and the foreground extinction is fairly uniform.... Rather, the distribution of shocked gas is symmetric around IRS 16, and forms a thin ring.... a mass-loss rate of > 10^{-4} M_o yr^{-1} is required to account for the strength of the observed molecular hydrogen emission" (c.f. Hutchings 1976; Lacy, Townes and Hollenbach 1982; Genzel et al. 1984).

The next step was to examine the evidence for mass loss in the spectrum of IRS16 itself. As stated above Hall, Kleinmann and Scoville (1978) had already observed very broad HeI λ 2.06µm emission in IRS16. (Some confusion had arisen from their report of <u>narrow</u> Bγ emission, which led to speculation about a vast overabundance of He in the broad-line region; the CGS 2 observations at UKIRT included checking the line profile of hydrogen Bα, which was found to have a FWOI $\sim$ 1,400 km s^{-1}, essentially identical to that of the He line. Furthermore, no He 5$\rightarrow$4 transition was detected in this spectral region). Geballe, Gatley, Wade and their team obtained line profiles of the λ 2.06µm line in the region containing IRS16 (Geballe 1984). The UKIRT group found that the He λ 2.06µm line profile was widest near IRS16; CID observations in the 8-13µm region by Tresch-Fienberg et al. (1984), images at 1.2µm and 2.2µm by Storey and Allen (1983; see also Allen, Hyland and Jones 1983) and CCD observations near 1µm by Becklin and by Storey and Allen (1983) serve to emphasise that the structure of this region is quite complex. The wide line profile could not be fitted with a single gaussian - broad and narrow components were required. The narrow component was found to be consistent with the motions observed by the Berkeley group (Lacy et al 1980) in the [Ne II] 12.8µm line.

The only major part of Gatley's original speculations which some might consider has not been fully confirmed is whether or not there is conclusive evidence for radial outflow (see, e.g. Lo and Claussen 1983). Included in his argument is the assumption that the shocked H_2 <u>is</u> caused by such outflow from the central non-thermal engine. The shocks could instead be formed as part of the general dissipation of non-circular motions in the ring; Genzel et al. (1984) for example conclude that mass loss from a single driving source at the galactic centre is not likely to be sufficient to explain the observed [OI] λ 63µm emisssion.

It has been argued that the He λ2.06µm line profile indicates organised motion (Geballe 1984). The infrared 2.06µm transition and the λ 584Å Lyα resonance transition occur from the same 2 ^1P upper level and the branching ratio is roughly 1000:1 in favour, of course, of the λ 584Å line. Geballe argues that because the λ 2.06µm line is so strong the gas must be very optically thick in the λ 584Å line, so

that many resonance transitions can occur back to the 2 ^{1}P level; the
degree to which two-photon emission can produce λ 2.06μm emission was
not discussed. Now if the rapid gas motions were locally turbulent,
relatively few resonance scatterings would occur as a result of the
relative Doppler shift of the λ 584Å absorption and emission; the gas
would become optically thin to resonance scattering, photons near
λ 584Å would escape and very few λ 2.06μm photons would be emitted. On
the other hand, if the motion is organised (such as in radial motion),
plenty of suitable local He atoms, of velocity similar to that emitting
the resonance line, are available for resonance scattering and
subsequent emission of λ 2.06μm photons. However if small quiescent
clouds are sufficiently optically thick to resonance scattering, they
may be in turbulent motion relative to each other yet still produce
significant λ 2.06μm emission (Carswell personal communication). Nevertheless the two estimates of mass-loss rate from the central object -
based on the observations of H_2 and on the line-profile observations of
He - are in encouragingly good agreement, so the picture of radial
outflow may after all, be correct.

Although we see clear evidence for non-thermal activity and a
central engine in the nucleus of our own galaxy, it should be realised
that (a): this activity is confined to the central parsec of the
galaxy and (b): the total luminosity of ∿ 10^7 L$_\odot$ is about two orders
of magnitude less than the flux from the surrounding central star-
formation regions - 230pc across (Gatley and Becklin 1981; see also
Telesco et al 1984) - that one might study when comparing our Galaxy
with others placed several Mpc or more away from us; see, e.g., the
low-resolution IRAS images presented by Gautier et al. (1984) and the
spectroscopic study of NGC 253 by Beck, Beckwith and Gatley (1984). It
is of interest that M31 has a Low Ionisation Nuclear Emission Region
(LINER)-type nucleus (Keel, as quoted by Pagel at this conference) but
is also a very weak infrared emitter; it also has a central peak of
infrared radiation, seen in the 60μm IRAS image presented by Habing et
al. (1984). They conclude that "the radiation from the central part is
due to hot dust heated by red giants in the bulge.... the co-incidence
with the non-thermal radio radiation from the center is remarkable, but
probably fortuitous". Viewed from another galaxy, the star-formation
activity and associated dust in the central part of the Milky Way would
be noted in a survey like that by Becklin et al. (1984b), but the
non-thermal activity described above would escape detection at 10μm.
The obvious unanswered questions include the following: how common is
the existence of such non-thermal activity in other galaxy nuclei? -
is the nucleus of our galaxy in the quiescent stage of some much more
spectacular history? - is the frequent co-incidence of nuclear non-
thermal activity and central star-formation activity, that is being
highlighted with the examples in this section, (see also Thuan 1984),
really fortuitous?

2.3 <u>M82 and other starburst galaxies</u>

M82 is a powerful infrared source. There is so much dust in the
galaxy that it is able to absorb most of the 2 x 10^{53} Lyman continuum
photons s^{-1} input from young stars (as measured by the Bα and Bγ line
strengths) and re-radiate much of this energy in the far-infrared
region of the spectrum; the bolometric luminosity is ∿ 4 x 10^{10}L$_\odot$.
This scenario is, of course, based on indirect evidence for the young,
hot stars. Until recently, the most direct evidence for their presence
was the existence of emission lines of ionised gas and a corresponding
level of free-free radio (continuum) emission at 3mm (see, e.g., the
typical starburst continuum spectrum shown in Figure 4).

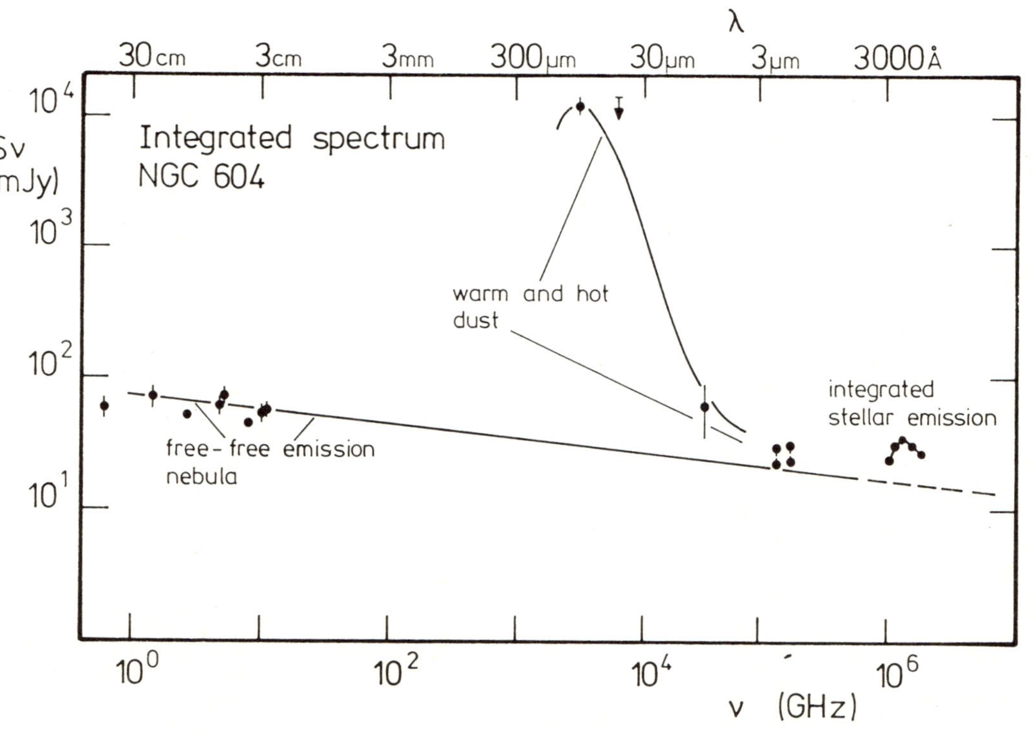

Figure 4:-   The spectrum of starburst activity over 6 decades in frequency - illustration adapted from Israel et al. 1982.

Working through detailed starburst models (e.g. Rieke et al. 1980; Weedman et al. 1981; Gehrz, Sramek and Weedman 1983; Campbell and Terlevich 1984)), one finds it necessary to invoke a high level of supernova activity in order to explain the observed levels of radio and x-ray emission (see also Harwit and Pacini 1975; Biermann 1976; Hummel 1980; Condon 1980; Condon et al 1982). For example Gehrz, Sramek and Weedman expect several supernovae yr^{-1} in NGC 3690; in NGC 7714, Weedman et al predict the existence of $\sim 10^4$ supernova remnants in a volume of $\sim 280pc$ radius with a supernova rate approaching one per year. Similarly, in M82 and many other starburst nuclei, supernova rates $\sim 1 yr^{-1}$ are expected. M. Phillips (personal communication) has started using an optical CCD to make regular checks for supernovae in a number of the 'starburst' nuclei, even though he realises that dust associated with the starburst will hinder his efforts quite severely. Laques et al (1980) have already reported brightening of one of the 'hot spots' in NGC 2903 while van den Bergh (1980) has commented on the incidence of two type 1 supernovae in NGC 5253 within less than a century. Searches for direct evidence of large numbers of supernovae are obviously important if we are to understand galaxies such as M82 in terms of simple starburst models; in view of the extinction problems, it was obviously better to use radio wavelengths for the search. M82 has, once again, revealed the most significant recent results. Kronberg's talk at this conference was illustrated with remarkable data from the VLA showing a large number of apparently transient radio sources lying in strings roughly along the major dimension of the galaxy. The behaviour of these sources is entirely consistent with identification as supernovae and their remnants; the associated supernova rate in M82 is estimated as $\sim 0.2 yr^{-1}$.

These observations appear to be a significant success for the simple models of Rieke et al. (1980). These models were based on a power-law initial mass function (IMF) $dN/dm \propto m^{-\alpha}$ ($\alpha = 2.35$ is the Salpeter solar neighbourhood solution) and an exponentially decaying star formation rate (SFR, the number of stars formed per unit time interval, $\propto e^{-t/to}$). The models were successfully constrained to fit the total bolometric luminosity of M82, the Lyman continuum flux, the absolute magnitude at K (-23.2) and the observed CO index (an indicator of the presence of late-type giant stars - see, e.g. Ridgway 1984; Campbell and Terlevich 1984). However, in order that matter can be successfully recycled through several generations of star formation without becoming locked up in low-mass stars, it is found that the formation of low-mass stars ($<3M_o$) must be strongly suppressed relative to the IMF in the solar neighbourhood.

The resulting starburst continuum spectrum (see, e.g., Figure 4) has the following components (Rieke et al 1980, Rieke 1981):-

i)   Non-thermal radio flux at several centimetres: supernova remnants.

ii) Thermal free-free (continuum) flux at several millimetres: ionised gas.

iii) Excess in mid and far i.r.: dust heated by young, hot stars.

iv) Near infrared flux: direct radiation from luminous red stars formed at an earlier phase of the starburst.

v)   Inferred ultraviolet flux: the young hot stars.

vi) X-radiation: stellar and supernova remnants.

A period of rapid star formation, lasting 10^7 to 10^8 years can plausibly explain most of the observed characteristics of M82. As discussed in section 1.2, Terlevich and Melnick have recently developed

a more sophisticated version of this general type of starburst model to
link all the different classes of narrow-line galaxies - starburst
nuclei, LINERS and Seyfert 2 nuclei.

2.4 NGC1097 and NGC1365

The "hot-spot" nuclei of these barred-spiral galaxies (Figure 5, 6)
were first described in detail by Sersic and Pastoriza (1965) - see
also the excellent photograph of NGC1097 by Rickard (1975). Similar
structure has recently been detected in NGC 1097 at 6 and 20cm
(Wolstencroft, Tully and Perley 1984). Early speculations by the
Soviet Byurakan group (e.g. Tovmassian 1966) suggested that the cluster
of bright spots found in these nuclei represent the active remains of a
'split nucleus'. These speculations appeared to have been laid to rest
when Osmer, Smith and Weedman (1974) reported that the nuclear region
contains short-lived massive HII regions photoionised by OB stars
created in a recent burst of star formation. There was no sign of any
non-thermal activity from their data, but they had used large circular
diaphragms for their observations, up to 40" diameter. The signal
their scanner received was thus the integrated light of all the spots
in a given nucleus. More recently, Telesco and Gatley (1981) have made
scans of the nuclear region at 10µm using a 5" beam diameter. Like
Osmer, Smith and Weedman, they concluded that "the extended clumpy
visible structure is almost certainly a complex of HII regions powered
by early-type stars". They estimated a bolometric luminosity $\sim 10^{11}$ $L_\odot$
for the central 30^n of NGC 1097. The 20cm radio emission was found by
Wolstencroft, Tully and Perley to lie in a ring "co-incident with the
optical hotspot and 10-µm emission"; though non-thermal, they find
that the 20cm radiation can be interpreted in terms of supernova
remnants associated with the star formation activity.

Osmer, Smith and Weedman (1974) included several of these hot-spot
nuclei in their study and found the spectra were all broadly similar to
each other; they went on to make the apparently reasonable assumption
that all the spots in a given nucleus had similar properties. This was
in fact a mistake, for we already knew that the central spot in NGC
1097 had a spectrum quite different from the kiloparsec outer ring of
spots (Smith 1972). When superior spatial and spectral resolution
became available at the Anglo-Australian Telescope (AAT), detailed
studies of the individual spots were undertaken.

(a) NGC 1097. Meaburn et al. (1981) noted that the central spot,
the true nucleus of NGC 1097, emits [NII] emission with FWOI $\sim$ 800
km s^{-1} and suggested a Seyfert 2 classification; Keel (1983) has since
identified the emission as being that of a "Liner" - see Heckman (1980)
and the paper by Diaz and Pagel at this conference. Radio maps made
with the VLA reveal a weak, compact, flat-spectrum source at the
position of the central spot. Talent (1982) has observed a strong
optical continuum from this position, dominated by late-type stars.
Phillips et al. (1984) have made a detailed study of the optical
spectrum and, by appeal to the [NII]λ 6584/Hα vs [OIII]λ 5007/Hβ
diagram of Baldwin, Phillips and Terlevich (1981), confirm Keel's
conclusion that the true nucleus is a classic Liner. They note that
while the ring spectrum shows strong emission lines of uniformly low
ionisation, the nuclear spectrum has much weaker emission lines but
with a wide range of ionisation; they do however confirm the observation of Meaburn et al. that the nuclear emission lines are broad. The
Liner type of spectrum was originally thought to arise from $\sim$ 100
km s^{-1} shocks (Heckman 1980; Draine 1980 has discussed the nature of
infrared emission to be expected from dust in shocked gas). However,

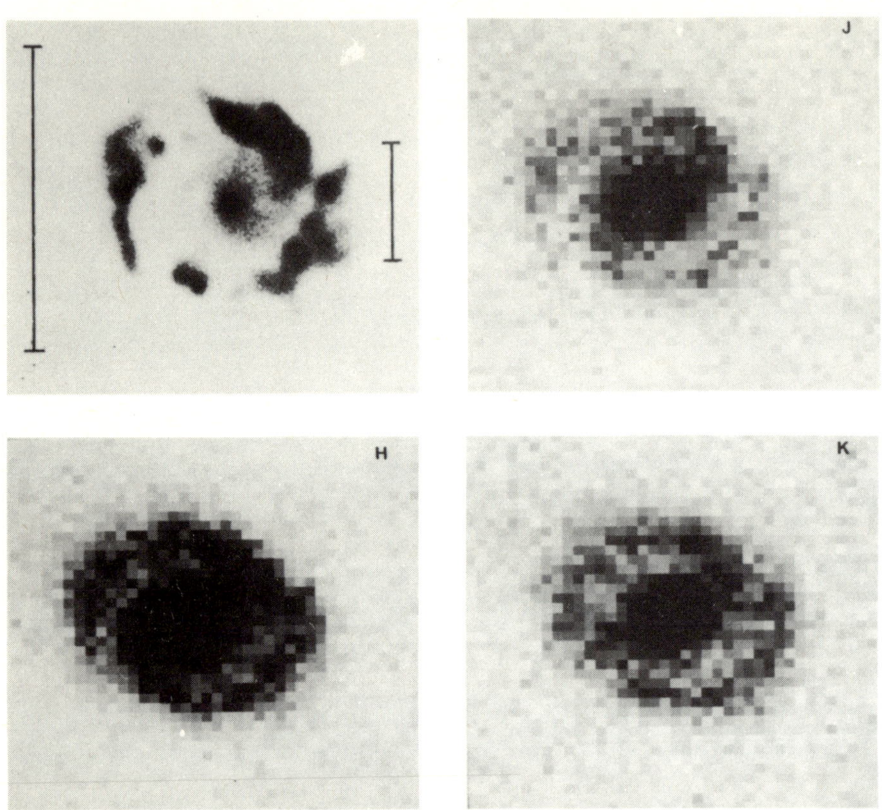

Figure 5:-   The nucleus of NGC 1097, as viewed in the visible and near-infrared regions. The central spot, the true LINER nucleus of NGC 1097 (Pagel and Diaz, this conference), becomes gradually more dominant as the wavelength increases. Illustration provided by Dr. Martin Ward, based on work by him in collaboration with Drs. Axon, Bailey and Hough. The near-infrared images were taken at the AAT using J. Bailey's mapping software and D. Allen's infrared photometer-spectrometer. Optical photo taken by the author, using the Tololo 1.5 metre telescope at f13.5. The long bar corresponds to 30", the short bar to 1kpc at the assumed distance of 17 Mpc.

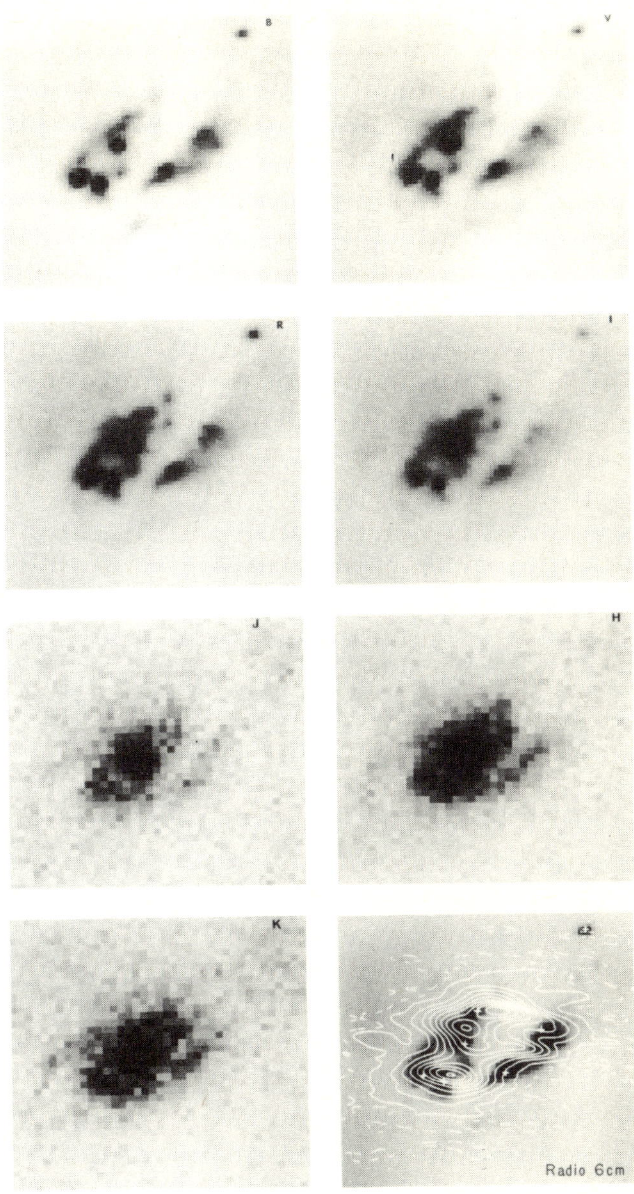

Figure 6:-   AAT images of the nucleus in NGC 1365. The true nucleus, of Seyfert 1 type (Edmunds and Pagel 1982), is seen to dominate at longer wavelengths. Illustration kindly provided by Dr. Martin Ward. Details of infrared images as for Figure 5. Optical images made using a direct CCD camera. The 6cm contour overlay is taken from Figure 4b of Sandqvist, Jörsäter and Lindblad 1982; the value of the outermost solid contour is 1mJy per beam and the contour interval is 1mJy.

recent work (Ferland and Netzer 1983; Keel 1983; Halpern and Steiner 1983; Pequignot 1984) has shown that the Liner spectrum can arise from photoionised gas. A discussion of the source of ionising photons - which tackles the problem of the different relative intensities of the optical λ 4686Å line from one object to another - has been given at this conference by Diaz and Pagel; this discussion has been general- ised further by Terlevich and Melnick (1984), as discussed in Section 1.2.

In summary we have, in the central regions of NGC 1097, a huge region of star-formation activity accompanied, apparently, by a marginal level of non-thermal activity, as evinced by the very weak, flat-spectrum, compact radio source.

(b) NGC 1365. In Figure 6 we see that one of the hot spots in the central region of NGC 1365 is also clearly distinguishable from its neighbours on infrared images. Edmunds and Pagel (1982) found the spectrum of this particular hot spot - the true nucleus of NGC 1365 - shows a very broad asymmetric component on both the $H\alpha$ and $H\beta$ emission lines. The FWOI is at least 4,000 km s^{-1}, and may be as high as 16,000 km s^{-1}. The infrared observations of Glass (1973) and Phillips and Frogel (1980) imply the presence of extended emission from hot dust. All the observations are consistent with a picture involving a non-thermal nucleus surrounded by a region of prolific star-formation activity. The non-thermal nucleus is somewhat more obvious in NGC 1365 than in NGC 1097, but had been missed in the early optical surveys. It was the use of X-ray techniques (e.g. Ward et al. 1978) which first drew attention to the possibility of significant non-thermal activity within this highly reddened nuclear region. This fact, and the images in Figures 5 and 6 combine to emphasise the point made earlier that existing surveys for non-thermal Seyfert-like activity are heavily biassed against the discovery of reddened nuclei. Searches for Seyfert nuclei are biassed against finding them when associated with intense star-formation activity and its associated dust. Such an association may be the rule, rather than the exception, at least at low non-thermal luminosities; thus Markarian-type Seyferts could be a tiny minority. (Curiously, the Seyfert 1 optical nucleus appears to have no signifi- cant corresponding radio source - Sandqvist, Jörsäter and Lindblad 1982. The properties of galaxies with more powerful compact nuclear radio sources have been described recently by Heckman et al. 1983 and by Hummel, van der Hulst and Dickey 1984).

2.5 The "mini-Seyferts" NGC 1667, NGC 5135 and IC 5135
Thuan (1984) has recently reported that the nuclei of these three galaxies show "Seyfert 2-like nuclear activity, one to two orders of magnitude smaller than that of the 'typical' Seyfert 2 galaxy.... Besides showing Seyfert-like characteristics such as broad, low contrast wings in the $Ly\alpha$ profiles and the unusual strengths of the emission lines of very high ionisation species such as HeII, NGC 5135 and IC 5135 also show starburst characteristics in their ultraviolet spectra.... They may be galaxies undergoing the transition from the starburst stage to the mini-Seyfert stage". He suggests that sub- stantial amounts of dust may exist close to the nucleus. Infrared observations of these objects are clearly needed.

2.6 NGC 1068
This Seyfert galaxy is especially suitable for detailed infrared studies as it emits at least 3×10^{11} L$_\odot$ in the wavelength range from 1 to 300μm (Telesco, Harper and Loewenstein 1976; Telesco and

Harper 1980), yet is only ∿18 Mpc distant from us. Telesco et al. (1984) have recently completed a series of infrared photometric observations of the central regions of NGC 1068. They mapped the distribution of 10μm emission from the central 40″, made scans of the galaxy at 60μm and 158μm and performed multi-aperture photometry at these same far-infrared wavelengths. They find that NGC 1068 emits extended 10μm and 60μm radiation, corresponding to a "disc" ∿ 3kpc across. At far infrared wavelengths, the extended disc dominates. Even in the visible, the ratio of integrated fluxes from the nucleus and disc is only ∿ ½. At 10μm however, the nucleus dominates, its integrated flux being ∿ 10 times that of the surrounding disc (Telesco et al. 1984). One is led to imagine a longer-wavelength version of the sequence of images in Figures 5 and 6; 10μm and 20μm images would show the nucleus of NGC 1068 clearly, but by 100μm the nucleus would have become lost in the extended radiation from the starburst disc. It is this combination of extended emission which peaks at ∿ 100μm and nuclear emission peaking near 20μm which leads to the complex shape of the continuum. Telesco et al. note that their data "permit for the first time a comparison of the large-scale spatial distributions of infrared, visual, molecular and radio emission in NGC 1068 and provide a firm basis for the separation of the energy distributions of the nuclear and extended components". They found that roughly half of the <u>total</u> infrared luminosity originates in the Seyfert (i.e. non-thermal) nucleus and half in a 3kpc (35″) diameter disc surrounding it. They conclude that the disc component corresponds to a heavily obscured, extended starburst. The disc is so luminous because it is big, not because it has particularly high surface brightness. Its surface brightness is comparable to that of the 230pc star-formation region of the Milky Way. The corresponding region of NGC 1068 is 3kpc across, which results in a total power level two orders of magnitude greater.

It is still proving quite difficult to sort out the various contributions to the polarisation in NGC 1068, which has already been detected from the ultraviolet to 13μm. At short wavelengths, the polarisation has been attributed to dust scattering and absorption (Angel et al. 1976) along with a possible non-thermal component (McLean et al. 1983; Miller and Antonucci 1983; Antonucci 1983; Schmidt and Miller 1984; see also the paper by Martin at this conference). Here, the major problem is the need to correct properly for the effects of dilution by starlight. Once this is done, the remaining continuum polarisation is essentially independent of wavelength. Even then, three possible mechanisms could be invoked to explain this: (a) non-thermal radiation, direct from the source - the least plausible option because the observed position angle of the short-wavelength polarisation (102°) is observed to be perpendicular to the direction of the radio jet. (b) electron scattering of the non-thermal radiation in an accretion disc, (c) an unusual distribution of grain sizes which results in polarisation which happens to be independent of wavelength. In the near-infrared, Lebofsky, Rieke and Kemp (1978) have attributed the polarisation to a combination of dust scattering and non-thermal radiation (see also Rudy et al. 1982b). At wavelengths λ > 5μm, thermal dust emission masks the non-thermal radiation. Knacke et al (1974) deduced that in the 10μm region, the polarisation arises either from differential dust absorption in the two planes of polarisation or from non-thermal emission. Aitken et al (1984) used a multi-channel 8-13μm spectropolarimeter to show that the polarisation in this wavelength region is at most only weakly dependent on wavelength - unlike the polarisation seen in the Becklin-Neugebauer (BN) object and

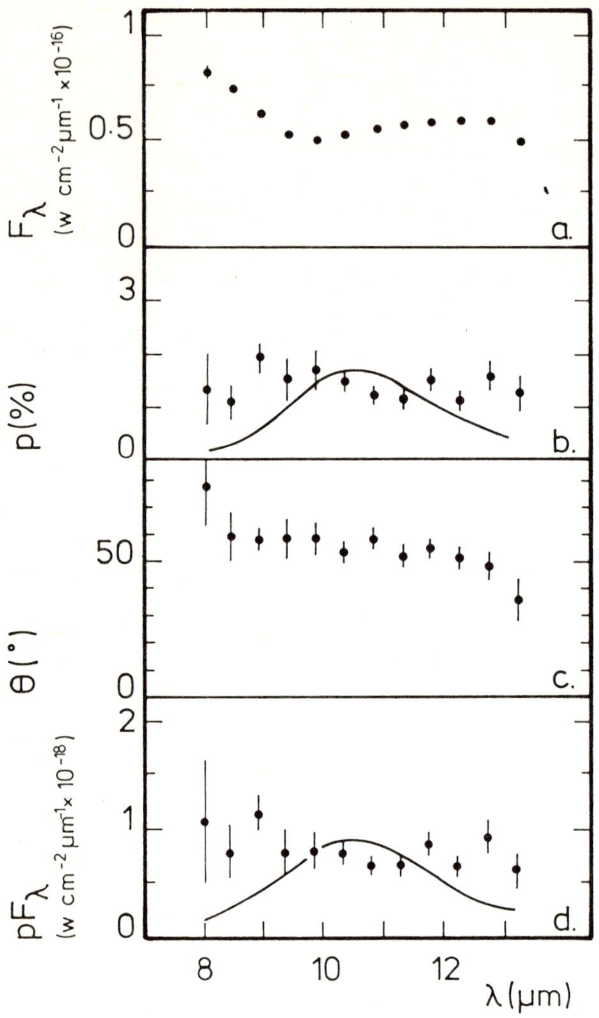

Figure 7:-   The 8-13µm polarisation spectrum of NGC 1068 shows little dependence of polarisation on wavelength, unlike the polarisation from well-studied observed sources in the Milky Way (examples given by the solid curves). Illustration adapted from Aitken et al. (1984); quantities plotted are (a) Flux, F_λ; (b) Polarisation, p; (c) Position angle, θ; (d) polarised intensity, $p\,F_\lambda$.

other well-studied obscured sources where the polarisation appears to
be peaked at the 10μm silicate feature. Fitting the BN polarisation
curve to the data from NGC 1068 (Figure 7), Aitken et al. were able to
show that less than 18% of the polarisation can be due to aligned
grains like those found in BN. They also found the position angle of
polarisation (55°) is essentially independent of wavelength, which is
why they believe that only a single polarisation mechanism operates in
the 8-13μm region. The polarisation angle is almost exactly perpendicular to the position angle of the rotation axis (145°). This,
coupled with weak dependence of polarisation on wavelength, could
indicate scattering of unpolarised thermally-emitted flux by other
grains which are themselves aligned parallel to the galactic plane
(McLean, personal communication; see discussion by Knacke and Capps
1974 of relative position angles for different polarisation
mechanisms). On the other hand, there is some suggestion of possible
variability in the position-angle data (Aitken et al 1984), which would
indicate the presence of a variable, non-thermal infrared component.

2.7 3C Galaxies

Rudy et al. (1984) have recently completed optical polarimetry and
infrared photometry of the broad-line radio galaxy (BLRG) 3C109. They
found that the Hα emission line as well as the adjacent continuum was
7% polarised. No intrinsically polarised mechanism for producing
emission lines is known, so this result is interpreted as evidence for
extinction by aligned dust grains. This interpretion in terms of dust
is supported by the steep rise in the continuum from 1.25 to 20μm,
similar to (radioquiet) Seyfert galaxies with strong thermal emission
(see also Elvis et al. 1984; Lawrence et al. 1984). In the BLRG, the
ratio Hα/Hβ $\sim$ 10 is too large to be explained purely in terms of
radiation transfer effects (Kwan and Krolik 1981); the 1.25-10μm
slopes are similar to those in the optical. Rudy et al. find that,
as a class, the broad-line regions of BLRGs are more heavily reddened
than those of either the Seyfert 1 galaxies or quasars. They also
suggest that there is a gradual decrease in the dust content of these
three classes with increasing source luminosity.

Lilly and Longair (1984) have recently completed JHK photometry of
81 3CR galaxies with $0 < z < 1.6$ selected from a statistically complete
sample of 90 radio galaxies. Though they find bursts of star formation
which they consider may be associated with the radio activity, they
believe that this early star formation activity is not concentrated in
the nuclear region. Lilly, Longair and McLean (1983) have obtained
deep BVRI images of 3C352 (z = 0.81) and find that the blue and ultraviolet excess is distributed throughout that galaxy (although the
[OIII] λ 5007 emission may indeed be associated with the nucleus).
Their principal reason for rejecting the notion of a major non-stellar
component to the i.r. flux is the observation that the images of many
of the galaxies on the CCD exposures have resolved non-stellar
profiles. In particular 3C267, 280 and 324 - which are amongst the
bluest objects on the (r-K) diagram - are as extended as would be
expected for giant elliptical galaxies at $z \sim 1$. They find no galaxy
significantly redder than a passively evolving galaxy, and there is a
significant scatter of colours blueward from this model. They find the
ultraviolet excess is correlated with the line strength of the narrow
emission lines. In their view, this could be a result of the star
formation in the galaxy producing gas which can flow into the nucleus,
producing non-thermal radio activity and a correlation between W_λ [OII]
and Δ (r-K). In view of the fact that dust, as well as blue stars, is

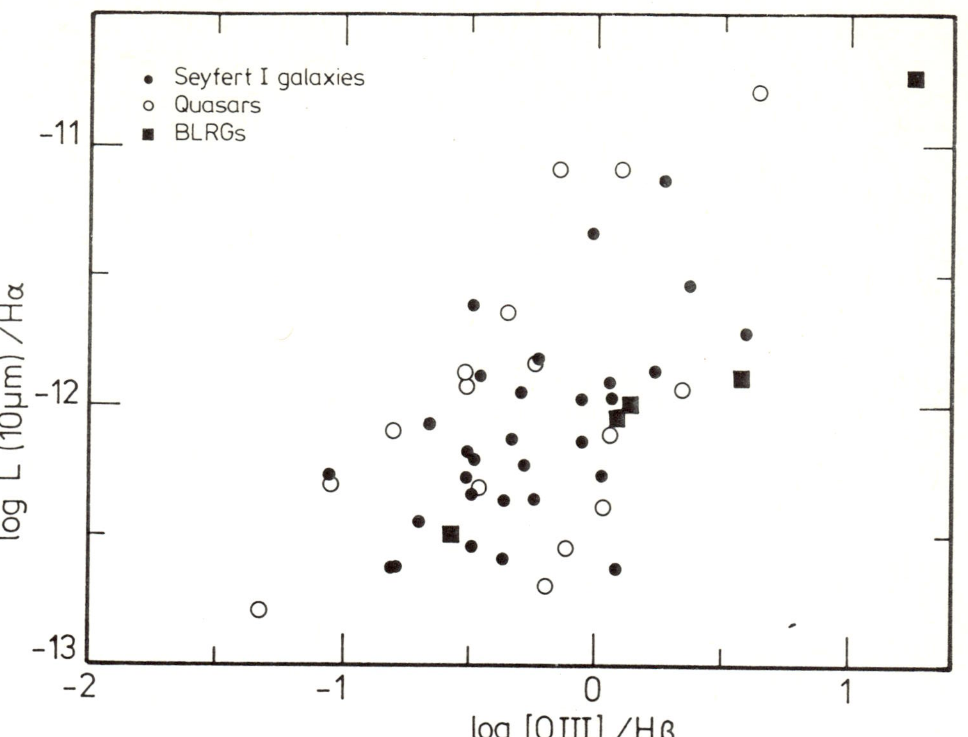

Figure 8:-   The correlation between 10µm emission and the strength of [OIII] found by Rudy (1984). Dust emission is strong in objects with a strong narrow-line region. Such a region becomes weaker in more luminous objects; dust is more difficult to detect in quasars than in Seyfert galaxies.

associated with star formation episodes, it would be wise to check these galaxies at longer wavelengths. This process has been started by Lilly and Longair (personal communication) who find, in 3CR galaxies with emission lines, that 3.5μm excess over starlight from a giant elliptical is common - contary to the initial indications of work by Elvis et al. (1984). Both groups agree that the 1-2.2μm radiation from these galaxies is indistinguishable from starlight; as these are the most distant stellar systems which have been systematically catalogued, the 3CR galaxies without broad emission lines are ideal for studies of the early evolution of stars in giant elliptical galaxies (Lilly and Longair 1982, 1984).

2.8 NGC7469 and the overall occurrence of dust in AGN.
Star formation activity is still detectable in type 1 Seyferts, where the non-thermal component is often thought to dominate (see, e.g. Smith et al 1983). In the type 1 Seyfert galaxy NGC 7469, a source less than $2''$ (480pc) in diameter does indeed dominate the near-infrared continuum. Cutri et al. (1984) have however found that nearly 80% of the $5 \times 10^7 L_\odot$ emitted in the 3.3μm dust feature (Rudy et al 1982a) arises in an extended region, between 240pc and 730pc from the central source. They use the high observed temperature (> 300°K) of the dust to conclude that it has most likely been heated in situ by young hot stars in circumnuclear HII regions. Cutri et al. suggest that "the regions around the nucleus of NGC7469 may support star formation complexes as large as those in M82", and propose several observational tests of their hypothesis for NGC7469 and other active galaxies in order "to reveal whether the phenomenon of large-scale star formation in active galactic nuclei is a rare or common occurence".
Rudy (1984) has recently completed a major study of the effects of dust on the properties of active galaxies with broad emission lines. The bias inherent in selecting Seyferts by means of their ultraviolet continuum has already been emphasised in section 1. Rudy points out that this bias can also affect our perception of the most common polarisation mechanism in Seyferts. Selection by UV excess will ensure that among dust - induced polarisations, the one involving least extinction will be selected - namely scattering.
Rudy's principal result is illustrated in Figure 8. Dust, responsible for much of the 10μm emission, is likely to be present in substantial amounts in AGN if the object supports a strong narrow-line region (NLR). He finds this result holds for quasars as well as for the lower-luminosity objects, and that there is no evidence that the dust-to-gas ratio in the NLR differs significantly between the Seyfert 1 galaxies and the quasars. In an earlier summary (Smith 1984), I presented a fairly detailed review of the situation concerning dust in Seyferts and quasars. Dust is very common in Seyfert 2 galaxies, is present in most Seyfert 1 galaxies, but is hard to detect in quasars - see also Soifer and Neugebauer (1981); Puetter et al (1981); Rudy and Puetter (1982); Smith et al. (1983); LeVan et al (1984); Rudy et al. (1984). Rudy's explanation is that the NLR weakens with increasing luminosity (but see also Gaskell 1983, 1984). He intends to try to confirm this result for quasars using objects with strong [OIII]. Because the optical spectral index and optical polarisation of radio-loud and radio-quiet quasars are similar (Richstone and Schmidt 1980; Stockman, Moore and Angel 1984), Rudy also concludes that dust is unlikely to play much of a role in luminous quasars (see also Neugebauer et al 1984). Aitken (personal communication) and his collaborators now have 8-13μm spectra of 32 galaxy nuclei, and their results

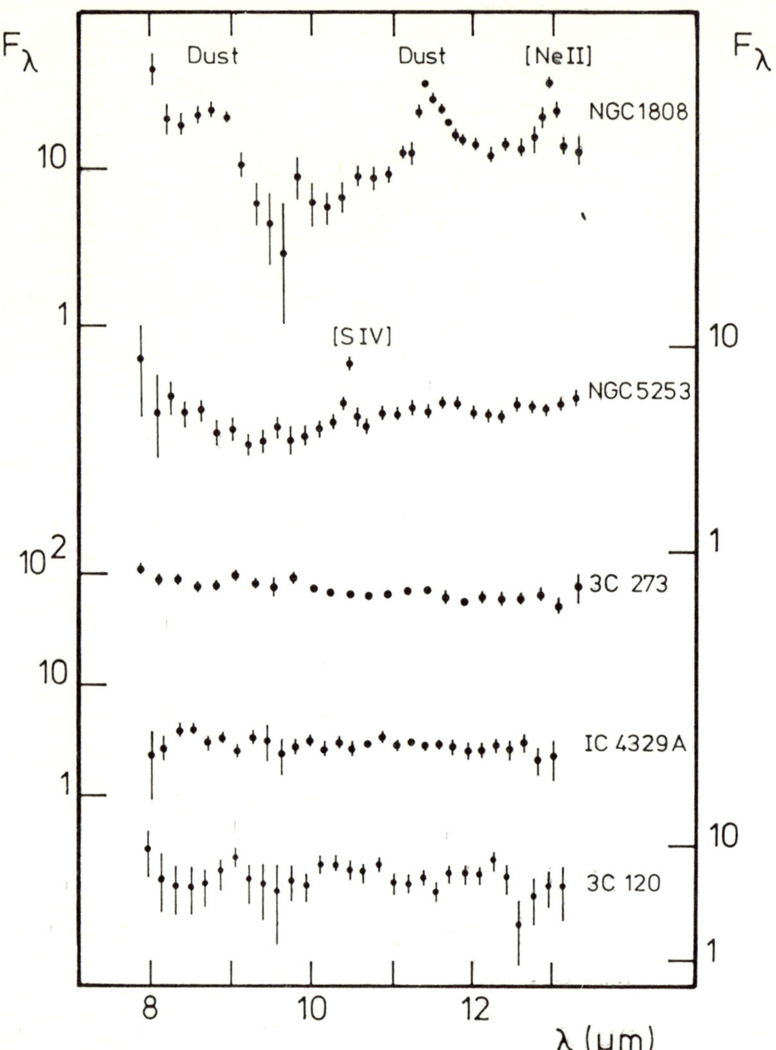

Figure 9:-   Representative 8-13μm spectra of narrow-lined galaxies
(NGC 1808, NGC 5253), a Seyfert galaxy (NGC 4329A), a
broad-lined radio galaxy (3C120) and a quasar (3C273) as
obtained by Aitken, Roche, Phillips and collaborators.
Their latest statistics are summarised in Table 1.

are summarised in Table 1.   8-13µm dust emission features are not usually seen in spectra of QSOs, BLRGs and Seyferts - whereas they are seen easily in most starburst nuclei;   some typical spectra are presented in Figure 9.

Table 1:   The 8-13µm spectra of galaxy nuclei*

Object type	Number Observed	Spectral class	Number in spectral class
Quasars	2	Power-Law.	2
Seyfert nuclei	18	Pure power law.	13
		Power law + silicate absorption.	3
		Unidentified dust emission features (NGC 7469; NGC 7582).	2
Starburst nuclei	12	Prominent dust emission features, usually accompanied by [NeII] λ 12.8µm emission.	10
		Smooth, possible silicate absorption. [SIV] emission (IIZw40; NGC5253).	2

*See also Figure 11 and section 2.7.   Data kindly supplied by Dr. David Aitken.   See also Phillips, Aitken and Roche (1984);   Roche et al. (1984).

Although dust is normally hard to detect in very luminous objects, counter examples can be found - particularly if one is prepared to go to objects as bizarre as Mrk 231.   Cutri, Rieke and Lebofsky (1984) have concluded that "with the exception of Mrk 231, strong silicate absorption seems to be typical of starburst, not Seyfert galaxies [See Figure 9].   It is therefore attractive to model the spectrum of Mrk 231 as a combination of the two types of activity....   The luminosity of $1-2 \times 10^{12}L_\odot$ .... is then divided roughly equally between starburst and Seyfert activity.... the starburst has a luminosity 15 to 20 times as great as that of M82 and the Seyfert activity is about 100 times more energetic than in NGC 4151".   Note, however, that Aitken et al. (personal communication) have now observed silicate features in 3 Seyfert galaxies (see Table 1 and also Frogel, Elias and Phillips 1982).   IRAS observers (e.g. Joseph 1984; Rowan-Robinson 1984) are finding other galaxies with infrared luminosities $\sim 10^{12}L_\odot$; NGC 6240 and Arp 220 appear to be interacting or merging galaxies having far-infrared luminosities within a factor 2 of Mkn 231.   Rowan-Robinson (1984) suggests that Arp 220 may be a quasar which is just switching on inside its cocoon of dust; its IR spectral ratio F100µm/F12µm is much greater than that for either Mkn 231 or for M82.   Hyland and Allen

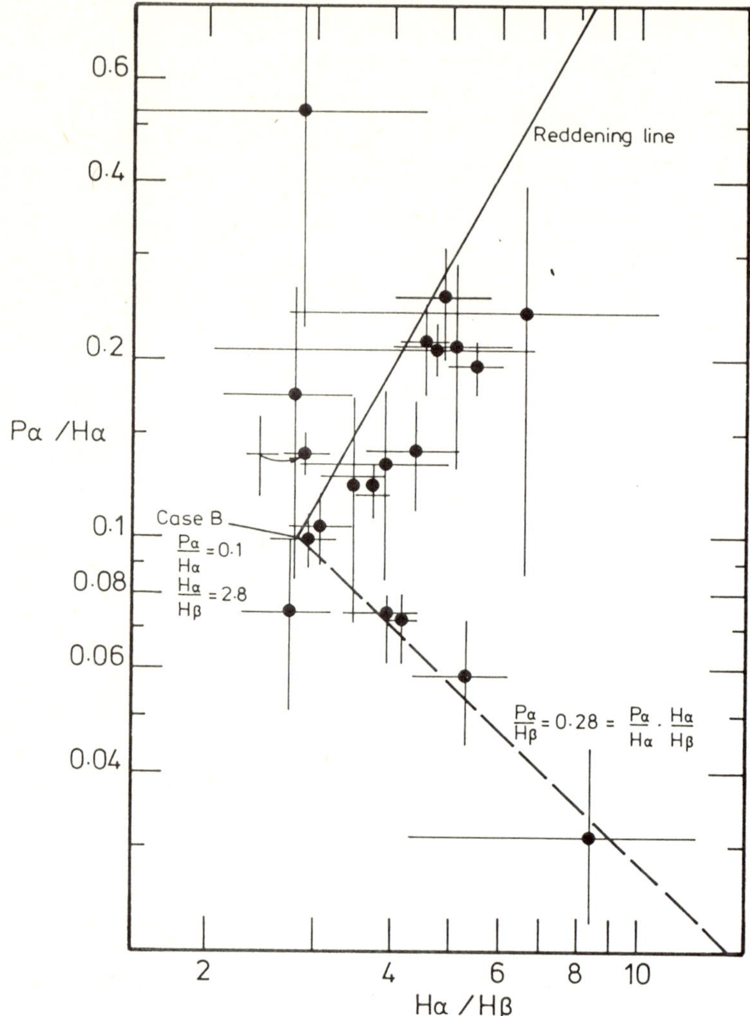

Figure 10:-   Seyfert galaxies are distributed either along a reddening line or along a line of constant Pα/Hβ, as shown in this illustration adapted from Lacy et al. (1982). Quasars (not plotted here) are found spread out along the Pα/Hβ = 0.28 line only, and no evidence is found in these diagrams for significant amounts of dust reddening in quasars. The bottom right of the diagram corresponds to enhanced Hα (large Hα/Hβ and small Pα/Hβ).

(1982) have concluded from high-quality near-infrared broad-band photometry that thermal emission from hot dust (T $\sim$ 1200°K) is common in
quasars, and exists mainly in the broad-emission-line region. Similar
conclusions have also been reached for mildly variable, low polarisation QSOs by Cutri and Wisniewski (1983). Finally, Neugebauer et al.
(1984) report an indication of excess 100μm emission over the presumed
power-law continuum component in two radio-quiet QSOs.
   Studies of the atomic hydrogen spectrum in the UV, optical and
infrared (UVOIR) region have yielded convincing evidence that dust is
present in both type 1 and type 2 Seyfert galaxies. Lacy et al (1982)
showed that the Seyfert galaxies were distributed along a reddening
line, or along a line of constant Pα/Hβ (Figure 10; see also Rudy and
Willner 1983; Puetter et al. 1981). On the other hand Soifer et al.
(1981) showed that quasars were spread out along the Pα/Hβ = 0.28 line
only; and no evidence was found by them for significant amounts of
dust reddening in quasars.
   Now, although the Pα/Hβ value for quasars is close to the Menzel and
Baker Case B, radiative-recombination-cascade line-flux ratio (e.g.
Osterbrock 1974), Hα is enhanced relative to Pα and Hβ; Hα/Hβ ratios
are often large while Pα/Hα ratios are correspondingly small. Reddening will clearly not explain such observations; high densities and
large optical depths appear to be required, which selectively populate
hydrogen energy levels from the ground up, increasing the intensity of
some of the redder lines (see discussion by Carswell at this
conference). At high optical depths, Lyα resonance trapping maintains
a significant population of hydrogen atoms at the n=2 level. Balmer
and even higher series transitions can become optically thick; note
that Case B calculations assume that while the Lyman lines are
optically thick, all other higher-series lines are optically thin. An
advantage of the infrared is that one can observe lines with lower
levels n>2 so that one can begin to test these ideas using the ratios
of lines from a common upper level (e.g. Pβ: Hγ:Lyδ).
   LeVan et al (1984) have recently studied the λ1.083μm HeI line in
quasars and Seyfert galaxies. The intensity ratio R (HeI) = λ1.083μm/
λ0.5876μm is a sensitive measure of the physical conditions found in
AGN. High values of R (compared with theoretical reference values Ro
for a hypothetical dustless Galactic gaseous nebula) indicate the
presence of reddening, presumably by dust. Low values of R/Ro indicate
that the λ0.5876μm line has probably been enhanced by radiative
transfer processes in a high-density, optically thick gas (as described
in more detail in the paper by LeVan et al). For QSOs and Seyfert 1
galaxies, they found values R/Ro $\lesssim$ 1/3, equivalent to optical depths
τ $\gtrsim$ 10^3, but little or no dust extinction appears necessary.

2.9 <u>3C273</u>
   Lee et al. (1982) searched carefully for evidence of 3.28μm dust
emission in 3C273, and Roche et al. (1984) made a similar search for
the dust emission features at 8.65μm and 11.25μm; none of these
features was found. Sellgren et al. (1983) have recently published
observations of the line profile of Paschen alpha in 3C273; they find
"the observed similarity of the line profiles Pα, at 1.875μm, and CIV,
at 1550Å, leads us to conclude that there is no differential reddening
due to dust, between clouds moving at different velocities within the
broad line region of 3C273". Finally, Clegg et al. (1983) presented
the continuum spectrum of 3C273 from 1.2μm to 20cm, which showed no
evidence for dust emission - an essentially perfect straight-line fit
$F_\nu \sim \nu^{-0.7}$ was obtained from 3μm to 5mm, i.e. over more than 3 decades

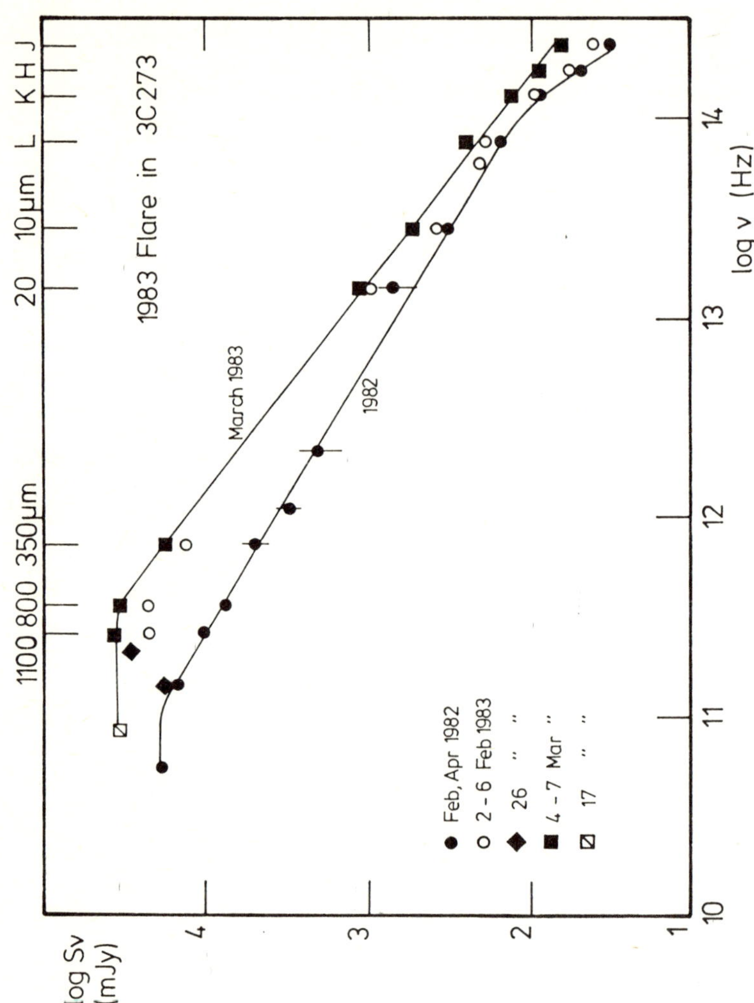

Figure 11:- Flaring activity in 3C273, adapted from Robson et al. 1983 (see also the paper by Robson at this conference).

in frequency; during a recent flare, this entire section of spectrum increased in flux - as shown in Figure 11 and described later in the conference by Robson (see also Robson et al. 1983). Because the flare was concurrent in the IR-to-near-millimetre spectrum of 3C273 it is likely that the emission over this range of frequency originates in the same very compact region of the source. The time scale of the flare ($\sim$ 2-4 months) is suggestive of an origin within the central 0.1pc of the source; the flare propagated to longer wavelengths, while decaying at shorter wavelengths, consistent with the canonical relativistic expanding cloud models of, for example, Pauliny-Toth and Kellerman (1966) - see also the paper presented by O'Dea at this conference. If 3C273 remains quiescent for a long period, it may be possible to trace the effect of this continuum flare on the emission lines in a few years' time (see, e.g., the reviews by Penston and by Carswell at this conference).

At these luminosities, the non-thermal component seems at first sight to have become totally dominant over any star-formation activity. Objects such as Arp 220, the "cocoon quasar", may however provide a significant set of exceptions (Rowan-Robinson 1984; see also section 2.8). Broad emission lines testify to the presence of a few solar masses of dense photoionised gas around the non-thermal nucleus; the physics of the emission-line region have been outlined for us here already by Dr. Carswell.

2.10 <u>OJ287 (0851+202) and other BL Lac objects</u>

In the BL Lac objects, such as OJ287, even the evidence for gas essentially disappears; all signs of nuclear starburst activity, including dust, are overwhelmed by the scale of non-thermal power-law activity. In the absence of infrared emission lines, multifrequency broad-band photometry and polarimetry dominate the study of infrared and millimetre-wave emission from these objects (see, e.g., Bregman et al. 1984) who conclude that in 0735 + 178 "rapid and dramatic variations evident at infrared and optical wavelengths are absent at radio and X-ray frequencies. These observations support a picture where the IR-UV flux emanates from a small region, while the X-rays are produced by the inverse-Compton process in the radio emitting region (partly opaque synchrotron emission gives rise to the radio flux)". BL Lac objects vary rapidly, so the multifrequency work has to be organised so that the various spectral points are all obtained within a short time interval. Holmes et al. (1984<u>b</u>), for example, report a 7% change in flux in 1 hour from OJ287, though most of the recent work has been aimed at time scales of days and months. Note for example that half a light day corresponds to the Schwarzschild radius, for a $4 \times 10^9 M_\odot$ black hole (Lynden-Bell 1978).

In their study of OJ287, Holmes et al. (1984<u>b</u>) report broad-band photometric observations made at UBVRIJHK on 7, 8, 9 and 10th January 1983. Similar observations of OJ287, also made in January 1983 have been reported by Balonek et al. (1983). As explained in more detail at this conference by Brand, the wavelength dependence of the polarisation could not be reproduced by either a single synchrotron component model, or a two-component model with spectral cut offs. They found that a superposition of two, optically thin polarised components (p, θ independent of frequency) was sufficient to explain the gross behaviour of the data. They also noted that the polarisations of each component are very similar, and tend to vary in unison from night to night - suggesting a physical connection between the two components. The very rapid variability noted above indicates that "the infrared luminosity is near the Eddington limit for accretion onto a massive black hole....

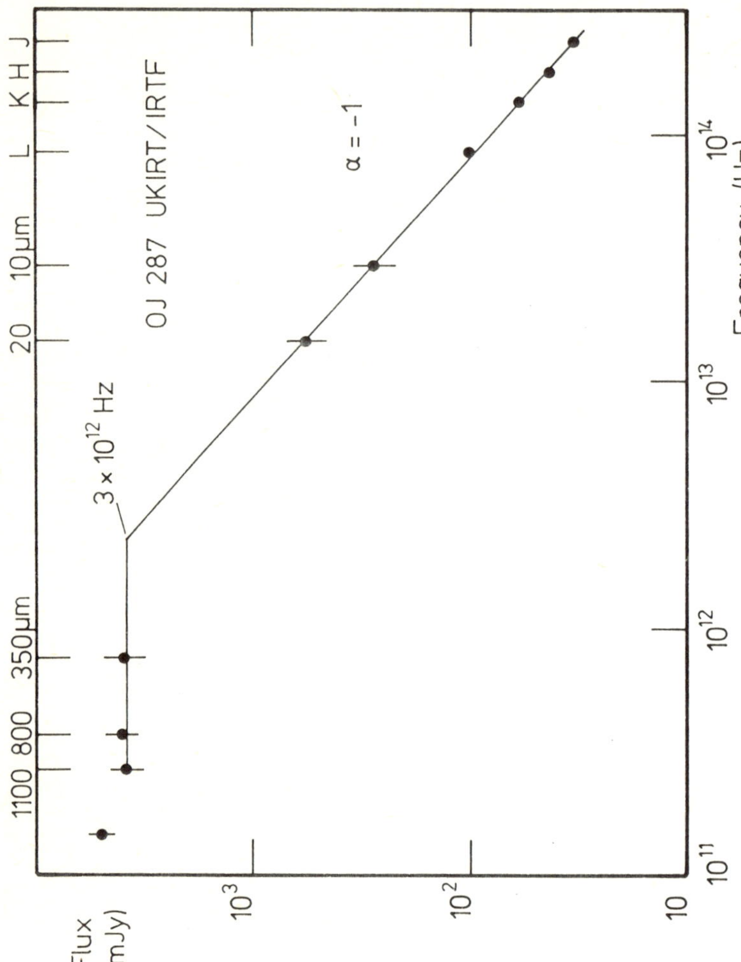

Figure 12:-   The infrared and high-frequency radio spectrum of OJ287 (see also the paper by Gear at this conference).

which strongly suggests the presence of a jet (Rees 1978a,b). The observed radiation can then be understood in terms of a cloud of emitting electrons being accelerated in the jet (e.g., Blandford and Konigl 1979, Marscher 1980)".

These results for OJ287 have been extended to a total of 30 BL Lacs by Holmes et al (1984a), who conclude that the two-component model with relativistic beaming yields a good description of BL Lacs as a class. Relativistic beaming was invoked to explain the observed violation of the Elliot-Shapiro (1974) variability-luminosity criterion

$$\log t_{min} \text{ (secs)} > \log L \text{ (erg s}^{-1}) - 43.1$$

where the minimum timescale of variability is assumed to be the light travel time across the Schwarzschild radius and the main interaction between radiation leaving and matter infalling is assumed to be electron scattering.

Some of these BL Lacs have also been studied at infrared, sub -millimetre and millimetre wavelengths, as explained in more detail later in this conference by Gear (see also Gear et al. 1984a,b). These broad-band continuum data are also broadly consistent with synchrotron emission from relativistic jets aligned close to the line of sight. The spectrum of OJ287 shown in Figure 12 was taken during a four day interval 3 months later than the observations by Holmes et al. The slope of the spectrum at frequencies $\nu < 3 \times 10^{12}$ Hz is close to zero; Gear et al. have found this to be a very common property of BL Lacs and OVV quasars (in agreement with the earlier conclusions of Jones et al. 1981 and of Ennis, Neugebauer and Werner 1982) and is likely to indicate the radial dependence of magnetic field in these objects.

The section of spectrum at $\nu > 3 \times 10^{12}$ Hz is straight, with slope $\alpha = 1.0$ ($F_\nu \alpha \nu^\alpha$). There is no sign of any dust emission. Indeed, there is little sign of any gas either in these objects. It is of interest to note that the QSO 1156 + 295 (Wills et al. 1983; Glassgold et al. 1983) has recently exhibited a 5-magnitude optical outburst. In the 'low' state, it shows an emission spectrum similar to that of a normal QSO. When bright, the optical spectrum of 1156 + 295 was essen-tially featureless, like that of a BL Lac object.

We turn finally to the continuum shape for the very red radio sources - the empty-field objects first discussed by Rieke, Lebofsky and Kinman (1979). In their original survey, they found six objects (at 2.2µm) at the positions of flat-spectrum radio sources, that had either no optical identification at all or had been identified with very faint red sources. The infrared-to-optical spectra of these sources have power-law spectra of index -3 ($S \alpha \upsilon^\alpha$), and this extreme redness suggested that a distinct class of active galaxy had been discovered. Though this slope is much steeper than the spectra of previously known QSOs, a follow-up study by Impey and Brand (1981) showed that such sources are in fact just the tail of the normal distribution of quasar colours, that they are relatively common, and that a considerable fraction of these sources will have properties similar to the BL Lac objects (see also Beichman et al. 1981a,b).

As mentioned earlier, Rieke and Lebofsky (1980) show that optical identification programmes for radio-source samples produce samples biased against very red sources, even if optical colour was not used as a criterion in making the identifications. Furthermore Lebofsky, Rieke and Walsh (1983) have shown that compact steep-spectrum sources are as detectable at 2µm as are compact flat-spectrum sources, although such a sample of steep-spectrum sources does contain a large fraction of high-redshift galaxies. Axon, Bailey and Hough (1982) have reported an extremely red nucleus in the radio elliptical IC 5063 which could be either an $\alpha = -4.5$ non-thermal source, or the Wien tail of a 650°K blackbody.   They point out that steep-spectrum, low-luminosity sources like this would have escaped detection in existing surveys.

As pointed out explicitly by Beichman et al. (1981b), "the importance of these sources is that their extreme nature forces a close examination of the emission mechanism thought to be operating in the whole class".   This fact, and the very red colour of these sources naturally led to searches for these objects at longer wavelengths. Beichman and collaborators (1981a,b) reported detections of 1413 + 135 first at 10μm and then at 1.0 millimetre.   Summaries of the multi-frequency spectrum of 1413 + 135 were given by Beichman et al. 1981a and by Bregman et al. 1981.   Both papers agree that the cause of the steep infrared continuum is an extremely sharp high-energy cutoff in the distribution of synchrotron-emitting electrons (see also Rieke, Lebofsky and Wisniewski 1982, who point out that the cutoff is nearly as sharp as is theoretically possible for synchrotron emission). Bregman et al. also showed that this cutoff is unlikely to be caused by internal reddening.   The other properties of 1413 + 135 are generally similar to the BL Lac objects.   If the emission at 1mm is due to incoherent synchrotron radiation, then the source is very compact and a large magnetic field ($\sim$ 10G) is required.   The X-radiation cannot then be primary radiation, but rather inverse Compton upscattering with Lorentz factor $\gamma \sim 300$.   The shape of the spectrum is attributed to synchrotron radiation from a relativistic jet, whose axis makes a small angle to the line of sight (see, e.g. Ennis, Neugebauer and Werner 1982b).   Rieke, Lebofsky and Wisniewski stress that the abruptness of the   inflection   in   the   infrared   spectrum   means   that   if   normal synchrotron   emission   is   responsible,   then   the   sources   must   have remarkably   uniform   magnetic   fields,   and   the   electron   acceleration process   in   some   of   these   sources   may   produce   a   nearly monoenergetic electron spectrum.

3. CONCLUSIONS AND FUTURE WORK

Examples   of   mixed   non-thermal   and   starburst   activity   include galaxies at all luminosities - the non-thermal component can appear at levels   differing   by   more   than   5   orders   of   magnitude   in   starburst galaxies,   ranging   from   the   activity   in   the   centre   of   our   own   Galaxy ($\sim 10^7$ L$_\odot$) to that observed in Mrk 231 ($\sim 10^{12}$L$_\odot$).   At the low luminosity end of the nuclear non-thermal activity (e.g. L $\sim 10^7$L$_\odot$ in our galaxy - see section 2.2), that activity is dwarfed by the burst of star formation activity in the central region.   At the high end, in objects like the flaring QSO 1156 + 295, the star-formation activity, if any, has not been detected.

In order to establish a firm observational link between non-thermal and star formation activity a more systematic approach is now possible in observations of the nuclei of galaxies.   It is important to study well defined samples of objects initially selected from surveys in a number of different wavelength regions.   Suitable samples include for example, the brighter Virgo galaxies, the 3CR, the Piccinotti - Elvis X-ray sample (Piccinotti et al. 1982;   Elvis et al 1984) and the sample of bright Sbc galaxies described at this conference by Davies.   A detailed theoretical model for a link between the starburst and non-thermal nuclei - which includes testable predictions - will, I suspect, also prove to be a fascinating challenge in the next few years.

4. ACKNOWLEDGEMENTS

I wish to thank Ian Gatley, Tom Geballe, Richard Wade, Ian McLean, Walter Gear and Ian Robson for many useful discussions. I am particularly grateful to Martin Ward and Rick Rudy for supplying material which stimulated the approach adopted in this review. Mark Sykes went to extraordinary lengths to obtain preprints for me. Tim Hawarden and Ann Savage kindly provided comments on an initial draft of the manuscript. Linda Taylor handled the typing with her customary efficiency. Finally, I would like to acknowledge my wife Anamaria and daughters Paulina and Carolita - nearly all this review was written at home and at night, which would have been impossible without their support.

REFERENCES

Aitken, D.K., Bailey, J.A., Briggs, G., Hough, J.H. and Roche, P.F. 1984 Nature, in press. "Infrared Spectropolarimetry of NGC 1068".

Allen, D.A., Hyland, A.R. and Jones, T.J. 1983 M.N.R.A.S. 204:1145.

Angel, J.R.P., Stockman, H.S., Woolf, N.J., Beaver, A.E. and Martin, P.G. 1976 Ap.J. (Letters) 206:L5.

Antonucci, R.R.J. 1983 Bull. Am. Astr. Soc. 15:976.

Arens, J.F., Lamb, G.M., Peck, M.C., Moseley, H., Hoffmann, W.F., Tresch-Fienberg, R. and Fazio, G.G. 1984 Ap.J, in press (April 15). "High Spatial Resolution Observations of NGC 7027 with a 10 micron array camera".

Arp, H. 1966 Ap.J. Suppl. 14:1.

Axon, D.J., Bailey, J. and Hough, J.H. 1982 Nature 299:234.

Bailey, M.E. 1982 M.N.R.A.S. 200:247.

Baldwin, J.A., Phillips, M.M. and Terlevich, R. 1981 P.A.S.P. 93:5.

Balonek, T.J., Smith, P.S., Zeilik, M., Burns, J.O., Puschell, J.J., Schmidt, G., Elston, R., Heckert, P., Impey, C., Barvainis, R. and Kenney, J. 1983 Bull. Am. Astr. Soc. 15:977.

Balzano, V.A. 1983 Ap.J. 268:602.

Balzano, V.A. and Weedman, D.W. 1981 Ap.J.243:756.

Beck, S.C., Beckwith S. and Gatley I 1984 Ap.J. in press (April 15). "Observations of Infrared Hydrogen Recombination Line Emission from External Galaxies".

Becklin, E.E., Gatley, I. and Werner, M.W. 1982 Ap.J. 258:135.

Becklin, E.E., Depoy, D.L., Heasley, J.N., Lester, D.F. and Wynn-Williams, C.G. 1984a Bull. Am. Astr. Soc. 15:914.

Becklin, E.E., Devereux, N.A., Capps, R.W., Scoville, N.Z. and Young, J. 1984b, Bull. Am. Astr. Soc. in press (proceedings of 162nd AAS meeting, Minneapolis). "A 10μm Survey of Activity in Galactic Nuclei: The 100 Brightest Virgo Galaxies".

Beichman, C.A., Neugebauer, G., Soifer, B.T., Wooten, H.A., Roellig, T. and Harvey, P.M. 1981a Nature 293:711.

Beichman, C.A., Pravdo, S.H., Neugebauer, G., Soifer, B.T., Matthews, K., and Wootten, H.A. 1981b Ap.J. 247:780.

Bierman, P. 1976 Astron. and Ap. 53:295.

Blandford, R.D. and Könnarion, A. 1979 Ap.J. 232:34.

Bothun, G.D. and Schommer, E.R. 1982 Astron.J. 87:1368.

Bregman, J.N., Lebofsky, M.J., Aller, M.F., Rieke, G.H., Aller, H.D., Hodge, P.E., Glassgold, A.E. and Huggins, P.J. 1981 Nature 293:714.

Bregman, J.N. and 20 co-authors 1984 Ap.J. 276:454.
Campbell, A.W. and Terlevich R. 1984 M.N.R.A.S., in press. "Studies of
    Violet Star Formation - III. Origin of the Infrared Luminosity".
Carroll, T.J. and Kwan, J. 1983 Ap.J. 274:113.
Clegg, P.E., Gear, W.K., Ade, P.A.R., Robson, E.I., Smith, M.G.,
    Nolt, I.G., Radostitz, J.V., Glaccum, W., Harper, D.A. and Low, F.J.
    1983 Ap.J. 273:58.
Condon, J.J. 1980 Ap.J. 242:894.
Condon, J.J. and Dressel, L.L. 1978 Ap.J. 221:456.
Condon, J.J., Condon, M.A., Gisler, G. and Puschell, J.J. 1982 Ap.J.
    252:102.
Cunningham, C.T., Robson, E.I., Ade, P.A.R. and Radostitz, J.V. 1984
    M.N.R.A.S., submitted. "The Submillimetre and Millimetre Spectrum
    of NGC 5128".
Cutri, R.M., Rudy, R.J., Rieke, G.H., Tokunaga, A.T. and Willner, S.P.
    1984 Ap.J., in press (May 15). "The Spatial Extent of the 3.3μm
    Emission Feature in the Seyfert Galaxy NGC 7469".
Cutri, R.M. and Wisniewski, W.Z. 1983 Bull. Am. Astr. Soc. 15:958.
Cutri, R.M., Rieke, G.H. and Lebofsky, M.J. 1984 Ap.J. (submitted).
    "Markarian 231:A Seyfert Nucleus in a Giant Elliptical Galaxy".
Dahari, O. 1983 Bull. Am. Astr. Soc. 15:987.
Draine, B.T. 1981 Ap.J.245:880.
Edmunds, M.G. and Pagel, B.E.J. 1982 M.N.R.A.S. 198:1089.
Elliot, J.L. and Shapiro, S.L. 1974 Ap.J. (Letters) 192:L3.
Elvis, M., Willner, S.P., Fabbiano, G., Carleton, N.P., Lawrence, A.
    and Ward, M. 1984 Ap.J., in press. "1-20μm Infrared Photometry of
    3CR Radio Galaxies".
Ennis, D.J., Neugebauer, G. and Werner, M. 1982 Ap.J. 262:460.
Feldman, F.R., Weedman, D.W., Balzano, V.A. and Ramsey, L.W. 1982 Ap.J.
    256:427.
Ferland, G.J. and Netzer, H. 1983 Ap.J. 264:105.
Fischer, J., Simon, M., Benson, J. and Solomon, P.M. 1983 Ap.J.
    (Letters) 273:L27.
Frogel, J.F., Elias, J.H. and Phillips, M.M. 1982 Ap.J. 260:70.
Gaskell, C.M. 1983 Bull Am. Astr. Soc. 15:988.
Gaskell, C.M. 1984 Ap.Letters. 24:43.
Gatley, I. 1984 In:Kessler, M.F. and Phillips, J.P. (eds) Galactic and
    Extragalactic Infrared Spectroscopy, p351. D. Reidel Publishing
    Company, Dordrecht, Holland.
Gatley, I. and Becklin, E.E. 1981 In:Wynn-Williams, C.G., and
    Cruikshank, D.P. (eds) Infrared Astronomy p281. D. Reidel
    Publishing Company, Dordrecht, Holland.
Gatley, I., Jones. T.J., Hyland, A.R., Beattie, D.H. and Lee, T.J.
    1984, preprint. "Shocked Molecular Hydrogen Emission from the
    Center of the Galaxy".
Gautier, T.N., Hauser, M.G., Beichman, C.A., Low, F., Neugebauer, G.,
    Rowan-Robinson, M., Aumann, H.H., Boggess, N., Emerson, J.P.,
    Harris, S., Houck, J.R., Jennings, R.E. and Marsden, P.L. 1984,
    Ap.J. (Letters) 278:L57.
Gear, W.K., Robson, E.I., Ade, P.A.R., Smith, M.G., Clegg, P.E.,
    Cunningham, C.T., Griffin, M.J., Nolt, I.G. and Radostitz, J.V.
    1984a Ap.J in press (May 1). "Millimetre-wave Observations of
    Flat-Spectrum Radio Sources".
Gear, W.K., Ade, P.A.R., Griffin, M.J., Robson, E.I., Brown, L.,
    Smith, M.G., Nolt, I.G. and Radostitz, J.V. 1984b Preprint.
    "Multifrequency Observations of Blazars: The 1μm to 2mm Continuum".

Geballe, T.R. 1984. Talk given at the Royal Observatory Edinburgh.
Gehrz, R.D., Sramek, R.A. and Weedman, D.W. 1983 Ap.J. 267:551.
Genzel, R., Watson, D.M., Townes, C.H., Dinerstein, H.L.,
    Hollenbach, D., Lester, D.F., Werner, M. and Storey, J.W.V. 1984
    Ap.J. 276:551.
Glass, I.S. 1973 M.N.R.A.S. 164:155.
Glass, I.S. 1984 Preprint "JHK colours of 'Ordinary' galaxies".
Glassgold, A.E. and 21 co-authors 1983 Ap.J.274:101.
Goebel, J.H., Witteborn, F.C., McCreight, C.R., Stafford, P., Jared, D.
    and Wisniewski, W. 1984 Bull. Am. Astr. Soc. 15:999.
Gunn, J.E. 1979 In: Hazard, C. and Mitton, S. (eds) Active Galactic
    Nuclei, p213. Cambridge University Press, Cambridge, England.
Habing, H.J., Miley, G.K., Young, E., Baud, B., Boggess, N., Clegg, P.,
    de Jong, T., Harris, S., Raimond, E., Rowan-Robinson, M. and
    Soifer, B.T. 1984, Ap.J. (Letters) 278:L59.
Hall, D.N.B., Kleinmann, S.G. and Scoville, N.Z. 1982 Ap.J. (Letters)
    260, L53.
Halpern, J.P. and Steiner, J.E. 1983 Ap.J. 269:L37.
Harwit, M. and Pacini, F. 1975 Ap.J. (Letters) 200:L127.
Heckman, T.M. 1980 Astron. and Ap. 87:152.
Heckman, T.M., Lebofsky, M.J., Rieke, G.H. and van Breugel, W. 1983
    Ap.J. 272:400.
Hildebrand, R.H., Whitcomb, S.E., Winston, R., Stiening, R.F.,
    Harper, D.A. and Moseley, S.H. 1977 Ap.J. 216:698.
Holmes, P.A., Brand, P.W.J.L., Impey, C.D. and Williams, P.M. 1984a
    M.N.R.A.S., in press. "Infrared Polarimetry and Photometry of BL
    Lac Objects. Data paper III".
Holmes, P.A., Brand, P.W.J.L., Impey, C.D., Williams, P.M., Smith, P.,
    Elston, R., Balonek, T., Zeilik, M., Burns J., Heckert, P.,
    Barvainis, R., Kenney, J., Schmidt, G. and Puschell, J. 1984b
    M.N.R.A.S., in press. "A Polarisation Flare in OJ 287".
Hummel, E. 1980 Astron. and Ap. 89:L1.
Hummel, E., van der Hulst, J.M. and Dickey, J.M. 1984 preprint.
    "Central Radio Sources in Spiral Galaxies".
Hutchings, J.B. 1976 Ap.J. 203:438.
Hutchings, J.B. and Campbell, B. 1983 Nature 303:584.
Hyland, A.R. and Allen, D.A. 1982 M.N.R.A.S. 199:943.
Impey, C.D. and Brand, P.W.J.L. 1981 Nature 292:814.
Israel, F.P., Gatley, I., Matthews, K. and Neugebauer, G. 1982 Astron.
    and Ap. 105:229.
Jones, T.W., Rudnick, L., Owen, F.N., Puschell, J.J., Ennis, D.J., and
    Werner, M.W. 1981 Ap.J. 243:97.
de Jong, T. 1984 Observatory, in press. "IR Excess in Galaxies as
    derived from IRAS data" - talk presented at Specialist Discussion
    Meeting of the RAS "The First Results from IRAS" - May 1984.
de Jong, T., Clegg, P.E., Soifer, B.T., Rowan-Robinson, M., Habing, H.,
    Houck, J.R., Aumann, H.H. and Raimond, E. 1984 Ap.J. (Letters)
    278:L67.
Joseph, R.D. 1984 Observatory, in press. "IRAS Observations of
    Interacting Galaxies" - talk presented at Specialist Discussion
    Meeting of the RAS. "The First Results from IRAS" - May 1984.
Joseph, R.D., Meikle, W.P.S., Robertson, N.A. and Wright G.S. 1984
    M.N.R.A.S., in press. "Recent Star Formation in Interacting
    Galaxies".
Keel, W.C. 1980 Astron. J. 85:198.
Keel, W.C. 1983 Ap.J. 269:466.

Klein, U. 1984 Observatory 104:58.
Knacke, R.F. and Capps, R.W. 1974 Ap.J. 192:L19.
Kwan, J. and Krolik, J.H. 1981 Ap.J. 250:478.
Lacy, J.H., Townes, C.H., Geballe, T.R. and Hollenbach, D.J. 1980 Ap.J
    241:132.
Lacy, J.H., Townes, C.H. and Hollenbach, D.J. 1982 Ap.J. 262:120.
Lacy, J.H., Soifer, B.T., Neugebauer, G., Matthews, K., Malkan, M.,
    Becklin, E.E., Wu, C-C., Boggess, A. and Gull, T.R. 1982 Ap.J.
    256:75.
Laques, P., Nieto, J.-L., Vidal, J.-L., Augé, A. and Despiau, R. 1980
    Nature 288:145.
Larson, R.B. and Tinsley, B.M. 1978 Ap.J. 219:46.
Lawrence, A., Ward, M., Elvis, M., Fabbiano, G., Willner, S.,
    Carleton, N. and Longmore, A. 1984, M.N.R.A.S. in press. "1-20μm
    Observations of Low Luminosity Active Galaxies".
Lebofsky, M.J., Rieke, G.H. and Kemp, J.C. 1978 Ap.J. 222:95.
Lebofsky, M.J., Rieke, G.H. and Walsh, D. 1983 M.N.R.A.S. 203:727.
Lee, T.J., Beattie, D.H., Gatley, I., Brand, P.W.J.L., Jones, T. and
    Hyland, A.R. 1982 Nature 295:214.
LeVan, P.D., Puetter, R.C., Smith, H.E. and Rudy, R.J. 1984 Preprint
    "HeI λ10830 Emission in Seyfert Galaxies and QSOs".
Lilly, S.J. and Longair, M.S. 1982 M.N.R.A.S. 199:1053.
Lilly, S.J., Longair, M.S. and McLean, I.S. 1983 Nature 301:488.
Lilly, S.J. and Longair, M.S. 1984, preprint. "Infrared Photometry of
    Distant Radio Galaxies".
Lingenfelter, R.E. and Ramaty, R. 1982 In:Riegler, G.R. and
    Blandford, R.D. (eds) The Galactic Center, p83. AIP Conference
    Proceedings #83.
Lo, K.Y. 1982 In:Riegler, G.R. and Blandford, R.D. (eds) The Galactic
    Center, p1. AIP Conference Proceedings.
Lo, K.Y. and Claussen, M.J. 1983 Nature 306:647.
Longmore, A.J, Sharples, R.M., Tokunaga, A.T., Rudy, R.J.,
    Robson, E.I., Ade, P.A.R. and Radostitz, J. 1984 M.N.R.A.S., in
    press. "Continuum Emission from the Nucleus of NGC 1275".
Lynden-Bell, D. 1978 Physica Scripta 17:185.
McAlary, C.W., McLaren, R. A., McGonegal, R.J. and Maza, J. 1983 Ap.J.
    Suppl. 52:341.
McLean, I.S., Aspin, C., Heathcote, S.R. and McCaughrean, M.J. 1983
    Nature 304:609.
Marscher, A.P. 1980 Ap.J. 235:386.
Meaburn, J., Terrett, D.L., Theokas, A and Walsh, J.R. 1981 M.N.R.A.S.
    195:39.
Miller, J.S. and Antonucci, R. 1983 Ap.J. (Letters). 271:L7.
Moneti, A., Forrest, W.J., Pipher, J.L. and Woodward, C.E. 1983 Bull.
    Am. Astr. Soc. 15:914.
Moorwood, A.F.M. and Glass, I.S. 1982 Astron. and Ap. 115:84.
Morris, M. and Rickard, L.J. 1982 Ann. Rev. Astron. and Ap. 20:517.
Neugebauer, G., Becklin, E.E., Oke, J.B. and Searle, L. 1976 Ap.J.
    205:29.
Neugebauer, G., Soifer, B.T.., Miley, G., Young, E., Beichman, C.A.,
    Clegg, P.E., Habing, H., Harris, S., Low, F. and Rowan-Robinson, M.
    1984, Ap.J. (Letters) 278:L83.
Norman, C. and Silk, J. 1983 Ap.J. 266:502.
Osmer, P.S., Smith, M.G. and Weedman, D.W. 1974 Ap.J. 192:279.
Osterbrock, D.E. 1974 Astrophysics of Gaseous Nebulae p64. Freeman.
Osterbrock, D.E. 1984 Q.J.R.A.S. 25:1.

Pauliny-Toth, I.K.K. and Kellerman, K.I. 1966 Ap.J. 146:634.
Péquignot, D. 1984, Astron. and Ap. 131:159.
Phillips, M.M. and Frogel, J.A. 1980 Ap.J. 235:761.
Phillips, M.M., Charles, P.A. and Baldwin, J.A. 1983 Ap.J. 266:485.
Phillips, M.M., Aitken, D.K. and Roche, P.F. 1984 M.N.R.A.S. 207:25.
Phillips, M.M., Pagel, B.E.J., Edmunds, M.G. and Diaz A. 1984
    M.N.R.A.S., in press. "Nuclear Activity in Two Spiral Galaxies with
    Jets:NGC 1097 and NGC 1598".
Piccinotti, G., Mushotsky, R.F., Boldt, E.A., Holt, S.S., Marshall,
    F.E., Serlemitsos, P.J. and Shafer, R.A. 1982 Ap.J.253:485.
Puetter, R.C., Smith, H.E., Willner, S.P. and Pipher, J.L. 1981 Ap.J.
    243:345.
Rees, M.J. 1978a Nature 275:516.
Rees, M.J. 1978b Observatory 98:210.
Richstone, D.O. and Schmidt, M. 1980 Ap.J. 235:361.
Rickard, J.J. 1975 Astron. and Ap. 40:339.
Rickard, L.J. and Harvey, P.M. 1984, Ap.J., in press. "Far Infrared
    Observations of Galactic Nuclei".
Ridgway, S.T. 1984 In:Kessler, M.F. and Phillips, J.P. (eds) Galactic
    and Extragalactic Infrared Spectroscopy, P 309. D. Reidel
    Publishing Company, Dordrecht, Holland.
Rieke, G.H. 1978 Ap.J. 226:550.
Rieke, G.H. 1981 In:Wynn-Williams, C.G. and Cruikshank, D.P. (eds)
    Infrared Astronomy, p 317. D. Reidel Publishing Company, Dordrecht,
    Holland.
Rieke, G.H. and Lebofsky, M.J. 1979 Ann. Rev. Astr. and Ap. 17:477.
Rieke, G.H., Lebofsky, M.J. and Kinman, T.D. 1979 Ap.J. (Letters)
    232:L151.
Rieke, G.H. and Lebofsky, M.J. 1980. In: Abell, G.O. and Peebles,
    P.J.E. (eds) Objects of High Redshift, Proc, IAU Symp No. 92, p263.
    D. Reidel Publishing Company, Dordrecht, Holland.
Rieke, G.H., Lebofsky, M.J., Thompson, R.I., Low, F.J. and
    Tokunaga, A.T. 1980 Ap.J. 238:24.
Robson, E.I., Gear, W.K., Clegg, P.E., Ade, P.A.R., Smith, M.G.,
    Griffin, M.J., Nolt, I.G., Radostitz, J.V. and Howard, R.J. 1983
    Nature 305:194.
Roche, P.F., Aitken, D.K., Phillips, M.M. and Whitmore, B. 1984,
    M.N.R.A.S. 207:35.
Roos, N. 1981 Astron. and Ap. 95:349.
Rowan-Robinson, M. 1984 Observatory, in press. "Arp 220" - talk
    presented at Specialist Discussion Meeting of the R.A.S. "The First
    Results from IRAS" - May 1984.
Rudy, R.J. 1984. Ap.J. in press. "Effects of Dust on the Infrared
    Emission, Selected Line Ratios, and Polarisation of Seyfert 1
    Galaxies, Broad-Line Radio Galaxies, and Quasars".
Rudy, R.J. and Puetter, R.C. 1982 Ap.J. 263:43.
Rudy, R.J., Jones, B., LeVan, P.D., Puetter, R.C., Smith, H.E.,
    Willner, S.P. and Tokunaga, A.T. 1982a Ap.J. 257:570.
Rudy, R.J., LeVan, P.D., Puetter, R.C., Smith, H.E. and Willner, S.P.
    1982b Ap.J. 253:53.
Rudy, R.J. and Willner, S.P. 1983 Ap.J. (Letters) 267:L69.
Rudy, R.J., Schmidt, G.D., Stockman, H.S. and Tokunaga, A.T. 1984 Ap.J.
    in press (March 15). "The Dusty, Luminous Broad-Line Radio Galaxy
    3C109".
Sandqvist, A., Jörsäter, S. and Lindblad, P.O. 1982 Astron. and Ap.
    110:336.

Schmidt, G.D. and Miller, J.S. 1984 Ap.J. in press. "Spectropolari-
    metry of Seyfert Nuclei".
Schweizer, F. 1980 Ap.J. 237:303.
Scoville, N.Z., Becklin, E.E., Young, J.S. and Capps, R.W. 1983 Ap.J
    271:512.
Sellgren, K., Soifer, B.T., Neugebauer, G. and Matthews, K. 1983
    P.A.S.P. 95:289.
Sérsic, J.L. and Pastoriza, M. 1965 P.A.S.P. 77:287.
Shull, J.M. and Beckwith, S. 1982 Ann. Rev. Astr. Ap. 20:163.
Smith, H.A., Lada, C.J. Thronson, H.A., Jr., Glaccum, W., Harper, D.A.,
    Loewenstein, R.F. and Smith, J. 1983 Ap.J. 274:571.
Smith, M.G. 1972 Bull. Am. Astr. Soc. 4:237.
Smith, M.G. 1984 In: Kessler, M.F. and Phillips, J.P. (eds) Galactic
    and Extragalactic Infrared Spectroscopy, p369. D. Reidel Publishing
    Company, Dordrecht, Holland.
Soifer, B.T. and Neugebauer, G., 1981 In:Wynn-Williams, C.G. and
    Cruikshank, D.P. (eds) Infrared Astronomy, IAU Symposium 96, p329.
    D.Reidel Publishing Company, Dordrecht, Holland.
Soifer, B.T., Neugebauer, G., Oke, J.B. and Matthews, K. 1981
    Ap.J.243:369.
Soifer, B.T., Rowan-Robinson, M., Houck, J.R., de Jong, T.,
    Neugebauer, G., Aumann, H.H., Beichman, C.A., Boggess, N., Clegg,
    P.E., Emerson, J.P., Gillett, F.C., Habing, H., Hauser, M.G., Low,
    F., Miley G. and Young, E. 1984 Ap.J. (Letters) 278:L71.
Stein, W.A. and Soifer, B.T. 1983 Ann. Rev. Astr. and Ap. 21:177.
Stocke, J.T., Leibert, J., Gioia, I.M., Griffiths, R.E., Maccacaro, T.,
    Danziger, I.J., Kunth, D. and Lub, J. 1983 Ap.J. 273:458.
Stockman, H.S., Moore, R.L. and Angel, J.R.P. 1984 Ap.J., in press
    (April 15). "The Optical Polarisation Properties of 'Normal'
    Quasars".
Stockton, A. 1982 Ap.J. 257:33.
Stockton, A. and MacKenty, J.W. 1983 Nature, 305:678.
Storey, J.W.V. and Allen, D.A. 1983 M.N.R.A.S. 204:1153.
Talent, D.L. 1982 P.A.S.P. 94:36
Telesco, C.M. and Gatley, I. 1981 Ap.J. (Letters) 247:L11.
Telesco, C.M. and Harper, D.A. 1980 Ap.J. 235:392.
Telesco, C.M., Harper, D.A. and Loewenstein, R.F. 1976 Ap.J. (Letters)
    203:L53.
Telesco, C.M., Becklin, E.E., Wynn-Williams, C.G. and Harper, D.A. 1984
    Ap.J., in press (July 15). "A Luminous 3 Kiloparsec Infrared Disk
    in NGC 1068".
Terlevich, R. and Melnick, J. 1984 in preparation. "Warmers:The
    Missing Link Between HII Regions and Seyfert Galaxies".
Thuan, T.X. 1984 preprint "Ultraviolet Observations of Starburst and
    Mini-Seyfert Galactic Nuclei".
Toomre A. and Toomre J. 1972 Ap.J. 178:623.
Tovmassian, G.M. 1966 Astrofizika 2:419.
Tresch-Feinberg, R., Fazio, G.G., Gezari, D.Y., Lamb, G.M., Shu, P.,
    Hoffman, W.F., McCreight, C.R. and Gatley, I. 1984, Bull. Am. Astr.
    Soc. 15:992.
Van den Bergh, S. 1980 P.A.S.P. 92:122.
Van der Hulst, J.M., Crane, P.C. and Keel, W.C. 1981 Astron. J.
    86:1175.
de Vaucouleurs, G., de Vaucouleurs, A. and Corwin, H.G. 1976 "Second
    Reference Catalogue of Bright Galaxies". University of Texas Press,
    Austin.

Ward, M.J., Wilson, A.S., Penston, M.V., Elvis, M., Maccacaro, T. and
    Tritton, K.P. 1978 Ap.J. 223:788.
Ward, M., Allen, D.A., Wilson, A.S., Smith, M.G. and Wright, A.E. 1982
    M.N.R.A.S. 199:953.
Weedman, D.W. 1983 Ap.J. 266:479.
Weedman, D.W., Feldman, F.R., Balzano, V.A., Ramsey, L.W., Sramek, R.A.
    and Wu, C.-C. 1981 Ap.J. 248:105.
Willner, S.P., Ward, M., Longmore, A., Lawrence, A., Fabbiano, G., and
    Elvis, M. 1984 P.A.S.P. 96:143.
Wills, B.J. and 33 co-authors 1983 Ap.J. 274:62.
Wolstencroft, R.D., Tully, R.B. and Perley, R.A. 1984 M.N.R.A.S.
    207:889.
Young, J.S., Gallagher, J.S. and Hunter, D.A. 1984 Ap.J. 276:476.

1-20μm observations of active galaxies

A. Lawrence

Royal Greenwich Observatory, Herstmonceux Castle
Hailsham, East Sussex, UK

SUMMARY

I will discuss some highlights of a continuing IR study of ~60 galactic nuclei of many different kinds. Normal nuclei, starburst nuclei and Type 1 Seyferts have well defined IR spectral shapes, whereas Type 2 Seyferts seem to be a mixture of things. Three broad line radio galaxies were observed to have very steep power law continua throughout 0.3-20μm, raising problems with photo-ionisation models.

Normal galactic nuclei

Substantial JHK large aperture studies of elliptical and spiral galaxies have been carried out by Aaronson (1977), Persson et al (1979) and Griersmith et al (1982). We have made a small study (5 galaxies) extending this work in several ways –
(a) small apertures, 5-8", (b) galaxies selected for having no, or very weak, emission lines, (c) measurements in L and M bands.
An example is shown in Fig. 2. All the galaxies have the same colours within ± .05 mag. Agreement with large aperture studies at JHK is good. The colours are close to those of an M giant. Thus we seem to have an old stellar population of high and uniform metal abundance, and little radial effect within ~ 1 Kpc. Results are discussed in more detail in Willner et al (1984).

Type 1 Seyferts

Rieke (1978) studied Type 1 Seyferts from 1 to 10 μm. We have concentrated on pushing out to 20 μm, and collecting data for a complete X-ray selected sample. Analysis of these data is still in progress (Ward et al 1984 in preparation). Here I simply wish to point out the typical form for the more luminous Seyferts – a power law of slope 1-1.5. An example is shown in Fig. 2.

Radio galaxies

We looked at four Narrow Line Radio Galaxies (NLRG) and three Broad Line Radio Galaxies (BLRG). The results on the NLRG were unexciting – upper limits only at 10 and 20μm, and JHKL colours consistent with old stars, after K correction. The lack of IR excess is marginally consistent with the hypothesis that NLRG correspond to (radio-quiet) Type 2 Seyferts, i.e. in having similar IR/optical/X-ray properties. Observations a few times more sensitive are needed.

The BLRG we studied were all unusual in having very steep optical spectra (α ~2.5). The main result is that this slope continues out to 20μm, suggesting that it is a real non-thermal power law (Fig. 1). However, there are problems. Extrapolating into the UV, we can count

BROAD LINE RADIO GALAXIES

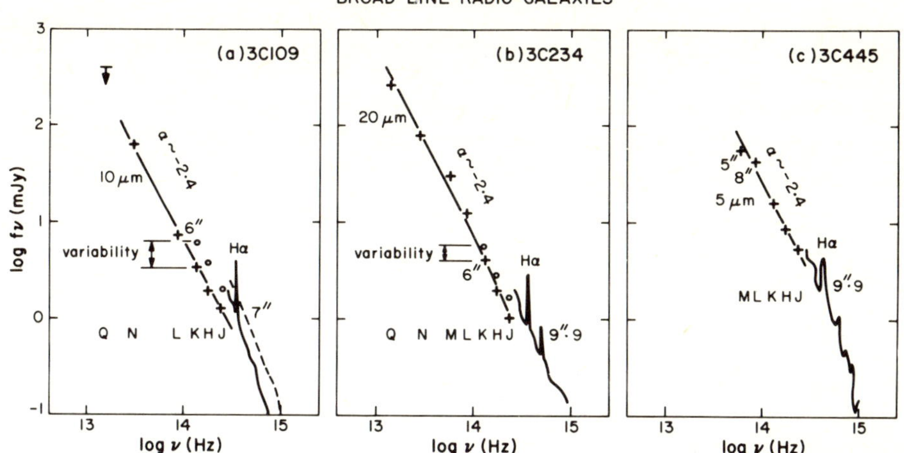

FIG. 1. Flux density distributions of radio galaxies. Optical data
and extra JHK data (circles) from Yee and Oke (1978) and Lilly and
Longair (1982)

the number of Lyman continuum photons, and ask if it is enough to
produce the observed Hα flux by photo-ionisation, assuming a covering
factor of unity ? The answer is no in all three cases, by at least an
order of magnitude. Furthermore, in the case of 3C 234, the He II
λ4686/Hβ ratio implies a slope of ~1.5 for the ionising continuum.
Simple extrapolation also fails to explain the observed X-ray flux.

All of these objects have large Balmer decrements. An alternative
explanation for the steep continua is large reddening plus dust re-
radiation – then the true (pre-extinguished) continuum may be consistent
with the Hα and He II/Hβ data. However, a conspiracy between reddening
and re-radiation is required to simulate a power law over 2 decades of
frequency. Furthermore, comparing our data with earlier optical and JHK
data indicates variability without a change of slope, strongly
suggesting a non-thermal source.

One thing is clear. In all these galaxies, <u>most</u> of their emitted
energy must emerge (originally or eventually) in the far-infrared.
Further discussion can be found in Elvis <u>et al</u> (1984).

<u>Low luminosity active nuclei</u>
Optical spectroscopy reveals three kinds of low-level "activity" in
spiral galaxy nuclei – roughly, Type 2 Seyferts, LINERS, and starburst
nuclei. Is this diversity reflected in the IR? We should be aware, as
Malcolm Smith has stressed in his review, that non-thermal and star-
formation activity may often co-exist (there may even be a causal
connection). Fig. 2 displays three "templates" for the components
likely to be present in a galactic nucleus – non-thermal power law, old
stars, and the averaged spectrum of a star-forming complex. (The only
appropriate comparison available for the star-forming complex is the
giant HII region NGC 5461, in a spiral arm of M101. The flat JHK is
presumably recombination radiation, and the steep NQ due to dust heated
to ~180 K).

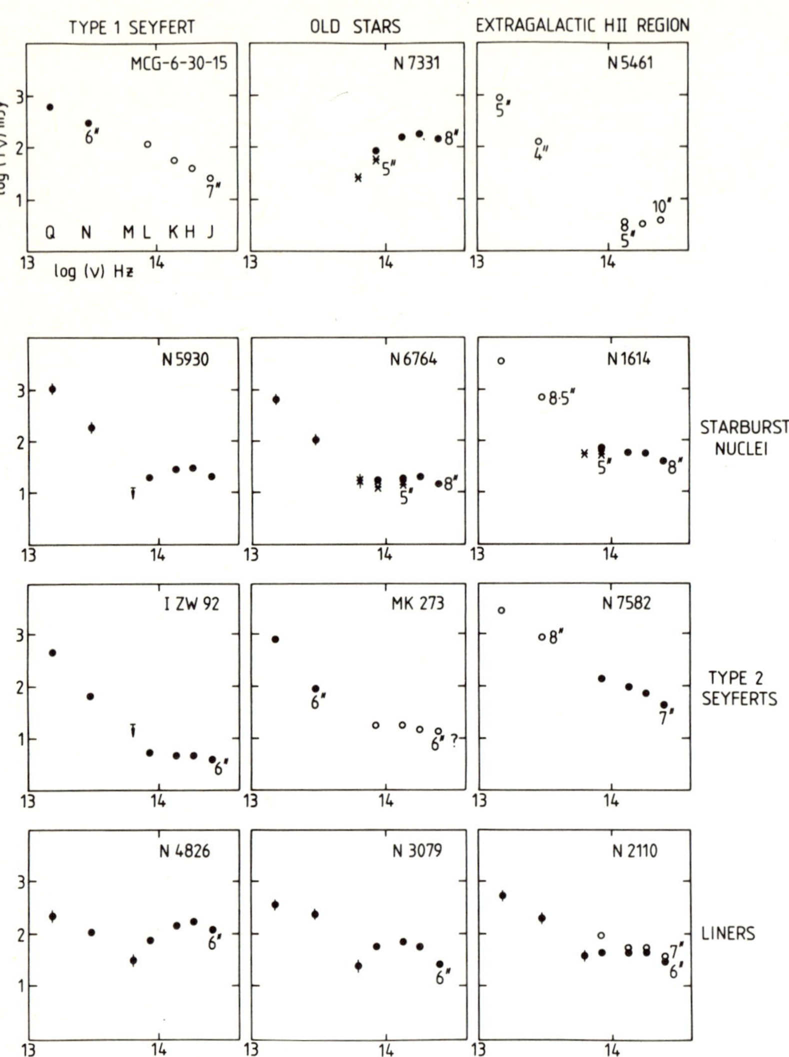

FIG. 2.   Top: examples of the three components likely to be present in low luminosity active nuclei.   Below: examples of the three main groups (as defined spectroscopically) of low luminosity active galactic nuclei.   Extra data (open circles) from: Blitz et al (1981);   Ward et al (1982);   Lebofsky and Rieke (1979);   Rieke (1978);   Frogel et al (1982)

Now let us look at some galactic nuclei.   The starburst nuclei indeed look roughly similar to N5461, except that a large old stellar component is apparent, especially in N5930.   However, after subtraction of the largest possible stellar component, the relative fluxes in the L band and in the Balmer lines are not consistent with recombination − another component is necessary, perhaps very hot dust.

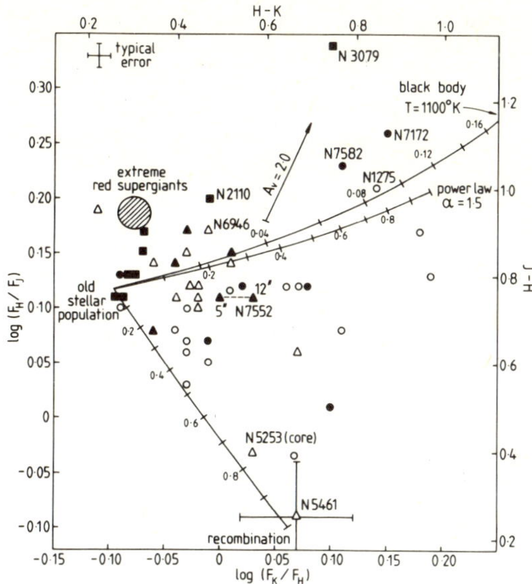

FIG. 3. Flux ratio diagram, showing mixing curves between various expected components. Triangles: starburst nuclei. Squares: LINERS. Circles: Type 2 Seyferts. Open symbols are extra data from: Balzano and Weedman (1981); Rieke (1978); Moorwood & Glass (1982); Blitz et al (1981); Rudy et al (1983).

The LINERS all have relatively flat NQ excesses, similar to Type 1 Seyferts. However, curvature is required to meet the M band measurements, implying a thermal origin for the emission, with temperature ~250 K. The JHKL data for LINERS are generally pure stellar (after a reddening correction in the case of N3079 - it is edge on). The exception is N2110, which spectroscopically is a link case between LINER and Type 1 Seyferts. Again, the NQ component must be thermal, but at JHKL there is variability on a 2 year timescale. Perhaps the best explanation for the IR emission from LINERS is dust heated by a central source, rather than by star formation.

Now I put on my revolutionary hat. I don't believe in Type 2 Seyferts. What I really mean is that they're not a well defined class, but often a dumping ground for several quite different kinds of object. A variety of shapes is seen in the IR. N7582 shows a power law slope, and is probably an obscured Type 1 Seyfert. Many others look similar to starburst nuclei.

Fig. 3 shows the three groups on a JHK diagram, with mixing curves between the important components. The starburst nuclei lie between the recombination point and red supergiants (rather than red giants), consistent with recent (10^7 years) massive star formation. The Type 2 Seyferts are scattered over a large area. While some are probably non-thermal sources of some kind, many lie along the line between red supergiants and recombination radiation, but further along than the starburst nuclei. Perhaps these are very strong, high excitation, starbursts of some kind?

Further discussion of low luminosity active galaxies can be found in Lawrence et al (1984).

Acknowledgements

This work is a collaboration with Martin Ward, Martin Elvis, Andy Longmore, Steve Willner, Pepi Fabbiano, and Nat Carleton. The revolutionary hat was borrowed from Roberto Terlevich, who doesn't believe in LINERS either.

References

Aaronson M, PhD thesis, Harvard University

Balzano VA, Weedman DW, 1981 Astrophys J, 243:756

Blitz L, Israel FP, Neugebauer G, Gatley I, Lee TJ, Beattie DH, 1981 Astrophys J, 249:76

Elvis MS, Carleton N, Fabbiano G, Lawrence A, Ward MJ, Willner SP, 1984 Astrophys J, in press

Frogel JA, Elias JH, Phillips MM, 1982 Astrophys J, 260:70

Griersmith D, Hyland AR, Jones TJ, 1982 Astron J, 87:1106

Lawrence A, Ward M, Elvis M, Fabbiano G, Willner S, Carleton N, Longmore A, 1984 Astrophys J, sumbitted

Lebofsky MJ, Rieke GH, 1979 Astrophys J, 229:111

Lilly SJ, Longair MS, 1982 Mon. Not. R. astr. Soc. 199:1053

Moorwood AFM, Glass IS, 1982 Astron Astrophys 115:84

Persson SE, Frogel JA, Aaronson M, 1979 Astrophys J supp, 39:61

Rieke GH, 1978 Astrophys J, 226:550

Rudy RJ, LeVan PD, Rodriguez-Espinosa JM, 1982 Astron J, 87:598

Ward M, Allen DA, Wilson AS, Smith MG, Wright AE, 1982 Mon. Not. R. astr. Soc, 199:953

Willner SP, Ward M, Longmore A, Lawrence A, Fabbiano G, Elvis M, 1984 Publ.Astron.Soc.Pacif., 96:143

Yee HKC, Oke JB, 1978 Astrophys J 226:753

A Flare in the Infrared to Radio Continuum Spectrum of 3C273.

E.I.Robson[*], W.K.Gear[*], P.A.R.Ade[+], & I.G.Nolt[#].

[*] Division of Physics & Astronomy, Preston Polytechnic, Preston PR1 2TQ

[+] Dept. of Physics, Queen Mary College, London E1 4NS

[#] Dept. of Physics, University of Oregon, Eugene, OR 97403, U.S.A.

SUMMARY

We have observed a well defined outburst in the continuum spectrum
of the quasar 3C273. The flare propagated from the submillimetre region
to the radio while at the same time decaying in the near and
far-infrared. Subsequent monitoring has shown that within the space of
a year the source has attained its quiescent continuum level at all
wavelengths shorter than 20 μm. The near and far infrared spectral
shape of the flare was similar to the quiescent spectrum but a
pronounced peak was observed in the short millimetre region. This peak,
compatible with synchrotron self-absorption, was seen to propagate to
longer wavelengths over a time of two months. The timescale and
energetics of the flare suggest a major outburst of energy from a
region within a parsec of the nucleus. Precise details   depend
critically on whether isotropic or relativisitic beaming are assumed.
The data are consistent with a relativistic beaming model.

INTRODUCTION

3C273 was the first quasar for which spectroscopic data revealed the
extragalactic nature of this class of object (see introductory review
by Hazard, this volume). Being close and bright it has been the object
of continuous study over many wavelengths. However until recently, the
continuum spectrum of 3C273 in the region 40   μm to 700   μm was
unknown. This far-infrared/submillimetre region is extremely difficult
to investigate because between 40 μ and 300 μm the earth's atmosphere
is totally opaque from all ground-based observatories. With new
facilities such as UKIRT, the world's largest infrared telescope,
projecting 4.2 km above sea-level, and the Kuiper Airborne Observatory
this unexplored region of the spectrum can be investigated for the
first time. The region is crucial for studying the continuum processes
pertaining to the source (see eg. review by Robson (1982)). Some
galaxies such as M82 and NGC1068, have submillimetre/far-infrared
emission from thermal re-radiation of dust which dominate their
bolometric luminosity (see review by Smith, this volume). In a
programme of observations of active galaxies (Gear et al 1984) using
UKIRT and the QMC/Oregon helium-three-cooled photometer (Ade et al
1984) we obtained the first ground-based submillimetre detection of a
quasar, 3C273. Combined with data from the Kuiper Airborne Observatory
and NRAO the complete infrared to radio continuum spectrum of 3C273 was
presented by Clegg et al (1983) who also discussed models decribing
possible emission mechanisms. Relativistic beaming was taken as the

underlying physical process because of the observed superluminal motion in this quasar (Pearson et al 1981).

We subsequently began a programme of single-epoch (snapshot) multifrequency observations of a sample of sources which were believed to be powered by beaming from relativistic jets (see Gear et al, this volume). Naturally 3C273 was included in this sample. Our main observational tool was the UKIRT facility: this is ideal because as well as a large collecting area on the highest site, it has a suite of common-user photometers covering the 1.25 μm to 35 μm region. With our photometer all ground-based accessible atmospheric windows from 1.25 μm to 1200 μm are available. Furthermore, the telescope is provided with a rotating dichroic which can feed the infrared radiation from the source into the photometers positioned at the Cassegrain foci. The optical radiation from the source is fed through the dichroic to a T.V. camera for precise acquisition and guiding. Multiband photometry over a wavelength range of 1000 is therefore possible from a single telescope in a single observing period which can be as short as three hours. As an example we recently made 1.25 μm to 1200 μm observations simultaneous with the EXOSAT X-ray satellite.

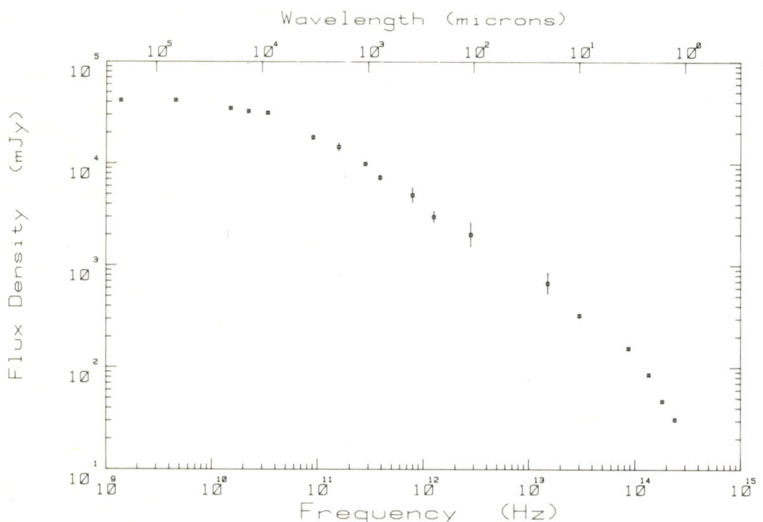

Fig. 1.   The 'quiescent' continuum spectrum of 3C273. The source of the data points are given in Clegg et al 1983 and Robson et al 1983.

OBSERVATIONS

During the programme of UKIRT multifrequency monitoring we were fortuitous in that 3C273 underwent a dramatic variation. Details of photometers and calibration procedure for the data presented below are found in Robson et al (1983). Our observations during the first week of February 1983 showed that 3C273 had increased in flux from its quiescent level at all our observing wavelengths. An observing session using the NRAO 12-m telescope at Kitt Peak on February 26th indicated a rise in flux density at 1.3 and 2.0 mm consistent with the UKIRT 1.1 mm data of three weeks previously. These data when combined with the UKIRT fluxes suggested a broad peak in the spectrum between 0.8 and 1.3 mm. This was confirmed by data from the 22-m radio telescope of the Crimean Astrophysical Observatory (Zinchenko, private communication). This showed that the flux at 3.39 mm remained steady throughout February and below the 1.3 mm value determined on February 26th. Within the first week in March we again observed 3C273 at wavelengths from 1.2 µm to 1.1 mm and discovered that the flux had increased further at all wavelengths. The 0.8 and 1.1 mm fluxes had increased by about 60% in 30 days. The broad peak appeared to have moved to longer wavelengths and this was confirmed within two weeks when 3.35 mm data were obtained from NRAO. The flux at this wavelength had increased from its February value by about 35%. On March 25th UKIRT 1.25 µm to 20 µm observations showed that the IR flux had decreased to a level approximately the same as in the first week of February. The data give some suggestion of a spectral flattening between 10 µm and 20 µm. Between April 9th to 11th 1.1 mm, 1.3 mm and 2.0 mm data from NRAO revealed that the peak in the spectrum had propagated to longer wavelengths, the 2.0 mm datum having the highest flux. A 3.35 mm measurement from the same telescope within two weeks strengthened the propagation hypothesis, the peak now being between 2.0 and 3.35 mm. In the last two days of April and first two days of May, UKIRT data revealed that the near infrared fluxes had faded below the February values, the 800 µm flux was marginally higher but the 1.1 mm point was still 50% higher. On May 20th and 22nd 1.1 mm, 1.3 mm and 2.0 mm observations showed that by now these fluxes were also in decline. This steady decrease was monitored during June and into August before 3C273 became unobservable. In February 1984 the infrared fluxes had attained their quiescent values within the errors of observations but the millimetre and submillimetre fluxes were still high. UKIRT infrared service observations on March 8th showed that there had been no change in the 1.25 µm to 3.5 µm data; the flare had certainly ended in the infrared. If the maximum development of the flare in the infrared is postulated to be that of the early March data then the infrared fluxes had risen by just over 100%. The millimetre flux however had risen by at least 230% above its quiescent value.

DISCUSSION

Figure 1 presents the quiescent spectrum and Figure 2 shows the observations within the period of February 26th to March 7th. The inset in the latter shows the NRAO data between April 9th to 11th. Our data alone are sufficient to give a crude description of the spectral development of the flare. The spectral index, α, $(S_\nu \propto \nu^\alpha)$ of the source in the quiescent state is: -1.1 between 0.5 µm and 3.5 µm, -0.7 between 3.5 µm and 5 mm and -0.1 for longer wavelengths until it again reverts to -0.7. During the outburst the infrared fluxes between

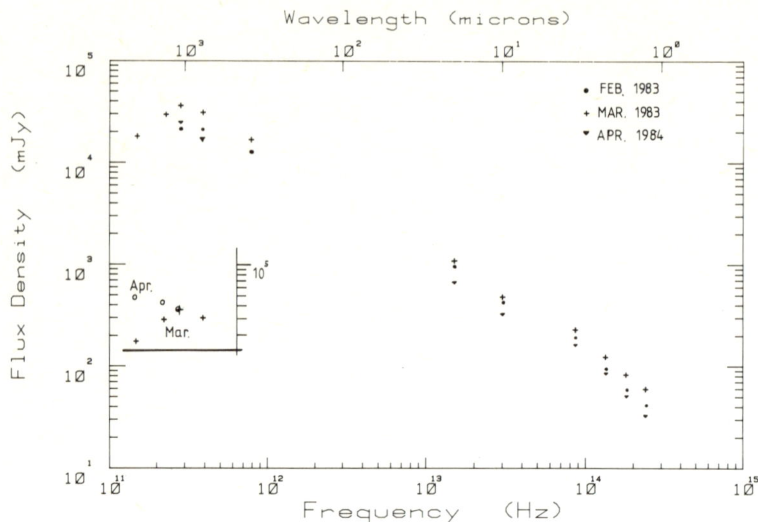

Fig. 2.   Spectral development of the flare for three epochs. The
inset shows the millimetre progression during a time interval of one month. The crossed points clearly show the
millimetre turnover indicative of synchrotron self-absorption. Flux values can be found from Robson et al 1983.

1.25 μm and 20 μm increased at all wavelengths, however the 1.25 μm
flux increased most and the result was a single power law from 1.25 μm
to 3.8 μm with a flatter slope. Between 10 and 20 μm the fluxes merely
increased   preserving   the   initial   slope   within   the   errors   of
observation. At millimetre wavelengths a pronounced peak, suggestive of
a turnover from synchrotron self-absorption was observed: this turnover
was seen to propagate to longer wavelengths. Our observations only
allow a rough guide to the progression which we observed to move from
around 800 μm to 1100 μm in about 30 days. We are currently searching
for 3.35 mm and longer millimetre data from April 1983 onwards and near
infrared data during the latter months of 1982 and January 1983 to
define more accurately the flare development.

   After subtracting the quiescent continuum spectrum, the spectrum of
the flare can be described by $S_\nu(Jy) \sim 6000/\nu(GHz)$. The total flux in the
observers frame between 1 μm and 1 mm is of order $4 \times 10^{-13} Wm^{-2}$. In
the same frame, the total luminosity of the flare is of order 3×10^{12} $L_\odot$
using a value for the Hubble constant of 100 km s^{-1} Mpc^{-1}. However this
assumes isotropic emission and as mentioned previously 3C273 is
observed to have superluminal motion, thereby indicating relativistic
motion and beaming a most likely source parameter. When this is taken
into account using a value for the bulk relativistric motion of 5.3
(Pearson et al. 1981), the luminosity in the rest frame of the
relativistic jet is reduced by nearly a factor of 1000. Consideration
of the timescale of the flare suggests that for isotropic emission the
event occured over a period of 2 to three months, limiting the size of
the emission region to about a 0.1 pc in extent. If relativistic
beaming is the cause, the timescale in the frame of the beam is
lengthened by the bulk factor of 5.3. The turnover frequency can also
be used to give some idea of the scale of the emitting region.

Following the discussion given in Burbidge et al (1974), the size of a self-absorbed synchrotron emitting region is given by:

$$R = 2 \times 10^{14}\, S_m^{\frac{1}{2}}\, \nu_m^{-1}\, L_D\, (1+z)^{-2} \quad m$$

where the flux S_m is in Janskys, the turnover frequency ν_m is in units of 10^{12} Hz , the luminosity distance L_D is in Gpc and z is the redshift. This gives values of R around 4×10^{16} m (1 pc).

CONCLUSIONS

The ability to obtain 'snapshot' multifrequency observations over a wavelength range of 1000 has been shown to produce valuable data for investigating the spectral and temporal behaviour of particular extragalactic objects. In the case of many such objects the submillimetre/far infrared dominates the bolometric luminosity (in the absence of a strong X or γ-ray fluxes). We have shown the overall behaviour of a single well-defined flare for the quasar 3C273. The luminosity of the flare is of order $3 \times 10^{12} L_\odot$ assuming isotropic emmission but is greatly reduced if relativistic beaming is in operation. It is clear that from the timescale of variability and the very high turnover frequency the emission most probably comes from a region extremely close to the central power source of the quasar. The availability of data from other observers would give a more detailed picture of the precise temporal progression of the flare. Using the data base above we are currently exploring models of relativisitic jets to explain the flare observed in 3C273; the formation of knots within the beam seeming to be a promising line of attack (Marscher et al in preparation).

REFERENCES

Ade PAR, Griffin MJ, Cunningham CT, Predko S, Nolt IG 1984 Infrared Phys
   (in press)
Burbidge GR, Jones TW, O'Dell SL 1974 Ap.J. 193:43
Clegg PE, Gear WK, Ade PAR, Robson EI, Smith MG, Nolt IG, Radostitz JV,
   Glaccum W, Harper DA, Low FJ 1983 Ap.J. 273:58-63
Gear WK, Robson EI, Ade PAR, Smith MG, Clegg PE, Cunningham CT, Griffin
   MJ, Nolt IG, Radostitz JV 1984 Ap.J. 280:102-106
Pearson TJ, Unwin S, Cohen MH, Linfield R, Readhead ACS, Seilstad GA,
   Simon RS, Walker RC 1981 Nature 290:365-368
Robson EI 1982 in Proc ESA Workshop Noordwijkerhout, ESA SP-189, 47-52
Robson EI, Gear WK, Clegg PE, Ade PAR, Smith MG, Griffin MJ, Nolt IG,
   Radostitz JV, Howard RJ 1983 Nature 305 194-196

The Infrared to Millimetre Continuum Emission of Blazars

W.K.Gear[#], E.I.Robson[#], P.A.R.Ade[*], I.G.Nolt[+]

[#] Division of Physics & Astronomy, Preston Polytechnic, Preston PR1 2TQ

[*] Dept of Physics, Queen Mary College, London E1 4NS

[+] Dept of Physics, University of Oregon, Eugene OR 97403, U.S.A.

SUMMARY

Observations at wavelengths between 1 micron and 2 millimetres of the continuum emission from a sample of Blazars (BLLacs and OVV quasars) show that most of these objects have very flat milllimetre/ submillimetre spectra, even up to 800 GHz. Two sources however, do exhibit spectral breaks close to 300 GHz. These sources must all therefore be very compact. All the objects have near-infrared (1-4 micron) spectra steeper than -1.0 and several show evidence of breaks around 10 microns, suggestive of energy losses. The overall continuum shape may be explained in terms of synchrotron radiation from an isothermal relativistic jet aligned close to our line of sight.

INTRODUCTION

The wealth of results and interpretations presented at this conference reflect the problems associated with   attempting to understand any of the complicted processes occuring in Active Galactic Nuclei (AGN).

We have attempted a slightly simpler problem by considering only continuum emission, the obvious objects to study then being those which only have continuum, namely the BLLacs (plus the OVV quasars which, although they do show strong line emission, have continuum properties identical with the BLLacs ; Angel and Stockman 1980) We also take an unusual viewpoint by measuring the previously undetermined submillimetre and far-infrared properties of these objects.

The far-infrared/submillimetre region of the spectrum has only recently been opened up for astronomical observations (see e.g. Robson 1982), the major reason being the opacity of the atmosphere. The development of new detector technology and the use of high, dry sites such   as Mauna Kea have enabled the atmospheric windows at 10, 20, 30, 350, 450, 800 and 1100 microns to be utilized. For measurements between   30 amd 300 microns astronomers have to resort to airborne or satellite observatories. This paper discusses the results of a   ground based   programme of measurements of Blazars in the atmospheric windows 1-20 microns and 300-2000 microns.

MOTIVATION

There are several good reasons why we wished to observe Blazars in this very difficult region of the spectrum, apart from the obvious one of "you never know what you might find until you look !".

1) Firstly, a look at the known overall continuum shape of a typical Blazar (see Fig 1) shows that there must be a transition from the flat radio spectrum to steep optical/ infrared power law which must occur in the large gap in the spectrum between $\sim 10^{11}$ Hz and 10^{13} Hz.

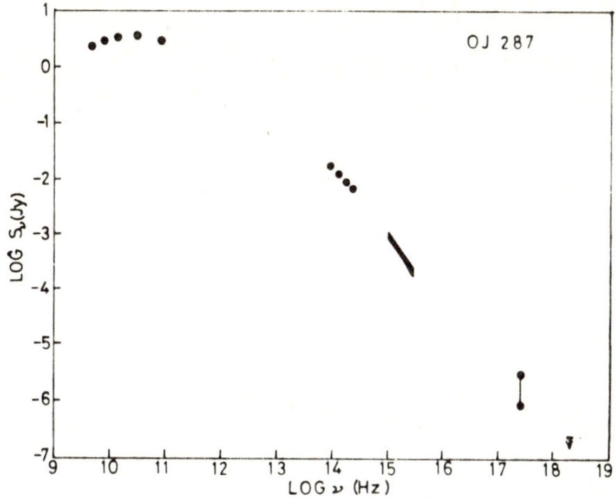

Figure 1: The radio through X-ray spectrum of 0851+202 (OJ287) from Worral et al (1982), showing the unexplored submillimetre and far-infrared region.

2) If the radio through infrared emission is incoherent synchrotron radiation, then the spectral break in the submillimetre is almost certainly due to self-absorption. Determining the position of this break gives us information on the compactness and magnetic field strength in the source.
3) It is crucial to know the emission properties in the region of the spectrum where the bulk of the energy output occurs. Figure 2 shows that for the Blazars this occurs in the far-infrared.

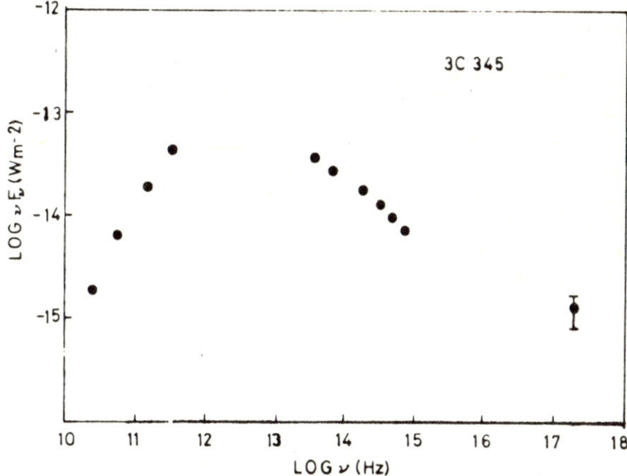

Figure 2: A plot of Log ν versus log $\nu\,F_\nu$ (energy per decade of frequency) for the source 1641+399 (3C345).

4) Radio outbursts in compact sources seem to propagate from high to low frequencies (see paper by O'Dea, this volume). In order to fully determine the evolution of such events it is therefore necessary to observe them at the highest radio frequencies.
5)Determining the spectral index of the optically thin emission immediately above the self absorption turnover tells us the electron energy distribution index in the source, unmodified by radiation or expansion losses.
6) If the overall radio through X-ray continuum of Blazars is to be understood in terms of a synchrotron self-Compton model, then the X-ray properties are very sensitive to the properties at the peak of the synchrotron spectrum.

RESULTS

        Figure 3 shows the shape of the 1 micron to 2 millimetre continuum of a sample of 12 Blazars. The 400, 800 and 1100 micron measurements were obtained from UKIRT using the QMC/Oregon He-3 photometer (Ade   et al 1984) and the 1300 and 2000 micron measurements were obtained from the NRAO 12m dish using the NRAO continuum photometer (Radostitz et al 1984). The 1-20 micron observations were made using the UKIRT and IRTF common user instrumentation.

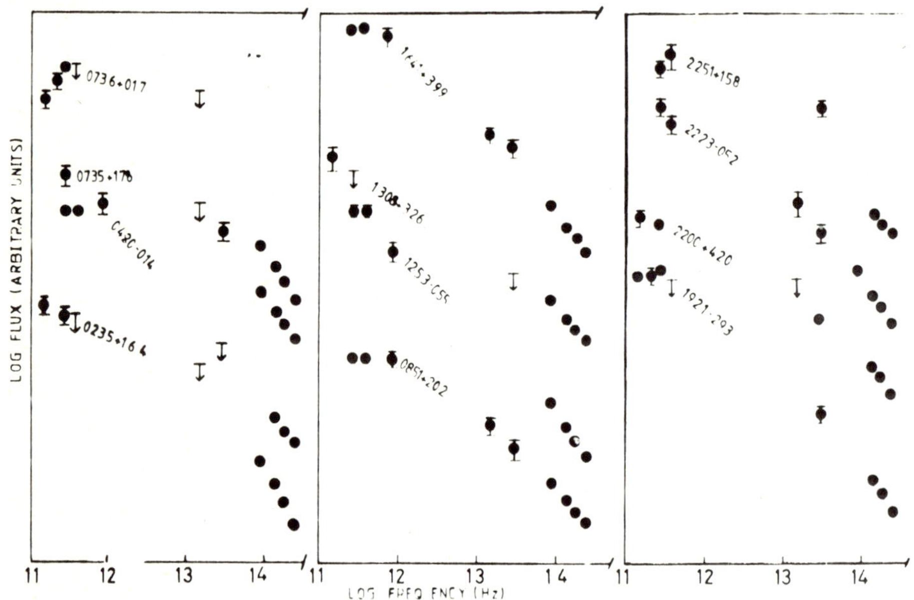

Figure 3: The millimetre to infrared spectra of a sample of 12 blazars.

        The observational results may be summarized as follows:
1) The millimetre wave spectra of all the sources are flat up to at least 300 GHz (1mm). Four sources; 0420-014, OJ287, 1413+135 and 3C345 are all flat up to 800 GHz (400 um), while two; 3C279 and 3C446 have turnovers close to 1mm.
2) The 1-4 µm spectra are all steeper than -1.0, the steepest being -1.6. Several sources show spectral breaks around 10 µm.

3) The 1 μm to 2mm luminosities, (estimated by integrating the observed spectra assuming isotropic emission, and drawing a straight line connecting the highest observed submillimetre frequency and the lowest observed infrared frequency) vary between 2×10^{11} (BLLac) and 4×10^{14} (3C446) $h^{-2} L_\odot$.

DISCUSSION

The fact that these objects have such high turnover frequencies means that they must necessarily be very compact. A simple calculation using the formalism of Burbidge, Jones and O'Dell (1974) gives sizes ~ 10 milliarcseconds and magnetic fields ~ 10^{-2} Gauss.

This calculation treats the submillimete and infrared emission as arising in a single homogeneous region, however such a region should have an optically thick slope of 2.5 whereas the observed slope is zero. In the case of one source (0735+178) it has been shown that the flatness of the radio spectrum may be due to the superposition of a number of such sources (see also review by Porcas, this volume). However, a flat spectrum arises more naturally from an inhomogeneous model (e.g. Marscher 1977) especially if one includes relativistic beaming in order to explain the observed apparent superluminal motion in some of these sources. In the relativistic jet model developed by Königl (1981), a flat optically thick spectrum arises if the jet is isothermal, i.e. the radiative and expansion losses are replaced by some reacceleration mechanism, such as a shock wave. It is then tempting to regard the millimetre and submillimetre emission as partially opaque emission from a smoothly inhomogenous jet, with the VLBI structure arising further out where the jet has become "clumpy" but still with the overall continuum adding to give a flat spectrum.

This model is also consistent with the fact that on the higher frequency VLBI maps a greater proportion of the total flux arises in an unresolved core, since at higher frequencies one should see further into the jet. Emission at the turnover frequency then arises at the most compact region of the source, the nozzle of the postulated jet.

We consider a slightly modified version of the model presented in Königl (1981). In this model synchrotron emission arises from an unresolved, conical, inhomogeneous relativistic jet of opening angle ($\phi \sim 0.1$) inclined at an angle θ ($> \phi$) to the line of sight. The jet has a constant velocity $\beta_j = v_j/c$ at all radial distances R from the apex and thus a constant Doppler factor $D_j = \Gamma_j^{-1}(1-\beta_j \cos\theta)^{-1}$ where $\Gamma_j = (1-\beta_j^2)^{-1/2}$ is the bulk Lorentz factor. The relativistic electrons are assumed to have a power law distribution of energies $n(\gamma_e) = K\gamma_e^{-5}$. The electrons are assumed to be continually reaccelerated so that the flow is isothermal, and $K = K_1 R^{-2}$; $B = B_1 R^{-1}$ where R is in parsecs and $K_1 = K(1pc)$, $B_1 = B(1pc)$.

The optically thin emission will be dominated by the radius R_m at which the self absorption turnover frequency becomes comparable to the synchrotron loss break frequency with slope $\alpha = -s/2$ if s>2 and $\alpha = 1-s$ if s<2. Since all the sources observed have a 1-4 μm slope steeper than -1 we take former case.

From the equations in Königl (1981) we obtain the following expressions

$$B_1 = a(s)[z^2 D_j \operatorname{cosec}\theta \, S_m \Gamma_j^5 \beta_j^5 (1+z)^{-6} \phi^{-1} \nu^{-5}]^{1/7} \quad \text{Gauss} \qquad (1)$$

$$R_m = b(s)[\Gamma_j \beta_j z^6 \operatorname{cosec}^3\theta \, S_m^3 (1+z)^{-11} \phi^{-3} \nu_m^{-8} D_j^{-1}]^{1/7} \quad \text{pc} \qquad (2)$$

$$K_1 = c(s)[\nu_m^{3+2s}(1+z)^s(\mathrm{cosec}\,\theta)^s\,z^{2(5+s)}\,S_m^{5+s}\,D_j^{-11-5s}(\Gamma_j\beta_j)^{-3-2s}\phi^{-12-s}]^{1/7}$$

$$\mathrm{cm}^{-3} \quad (3)$$

where a,b and c are functions of s and S_m is the flux in janskys at ν_m
Thus with knowledge of the observables ν_m, S_m, s ,z and Γ_j we can derive values for R_m , B_1 and K_1 and hence $B(R_m)$ and $K(R_m)$.

Applying this model to our observations, taking a typical value of $\Gamma_j \sim$ 5 we obtain R $\sim 10^{-4}$ to 10^{-3} parsecs (corresponding to the expected radii of the accretion disk around a 10^8 to 10^9 $M_\odot$ Black Hole, see review by Blandford, this volume). We also derive magnetic fields at this radius typically $\sim$ 10 Gauss and $K(R_m)$ $\sim 10^8$ cm.$^{-3}$(Note that the actual electron number density depends on the lower cutoff γ_c to the electron energy distribution, for $\gamma_c \sim 100$ we obtain n $\sim 10^5$ cm^{-3}.)

In order to test this model we need to monitor the sample for variabiity timescales and also to measure the X-ray emission. If the X-rays have a self-Compton origin then they will be scattered far-infrared photons (although see paper by Fabian, this volume). We are currently involved in both projects, preliminary results on 1921-293 (OV236) and the monitoring of a flare in 3C273 have been reported by Gear et al (1983) and Robson et al (1983). When millimetre VLBI becomes rgularly available it will be a valuable tool in resolving the regions discussed here.

ACKNOWLEDGEMENTS

We would like to thank all those who made these observations possible, especially J.V.Radostitz, M.J.Griffin, M.G.Smith, G.Veeder and L.Lebofsky. We also thank D.G.Vickers, C.Bissick and S. Predko for technical support and Dr. P.E.Clegg for useful discussions.

REFERENCES

Ade,P.A.R et al 1984, Infr. Phys., in press.
Angel,J.R.P. and Stockman,H.S. 1980, Ann.Rev.Astr.Ap. 18,321.
Clegg,P.E. et al 1983, Ap.J. 273,58.
Gear,W.K. et al 1983, Nature, 303,406.
Gear,W.K. et al 1984a, Ap.J. in press.
Gear,W.K. et al 1984b, Ap.J. submitted.
Konigl,A. 1981, Ap.J. 243,700.
Marscher,A.P. 1977, Ap.J. 216,244.
Radostitz,J.V. et al 1984, in preparation.
Robson,E.I. 1982, in "Scientific Importance of Submillimeter Observations" Proc. ESA conference Noorwjikerhout (ESA sp-189) p.147.
Robson,E.I et al 1983, Nature, 305,194.
Worrall,D.M. et al 1982, Ap.J. 261,403.

EMISSION LINE REGIONS OF ACTIVE GALACTIC NUCLEI AND QUASARS

R.F. Carswell

Institute of Astronomy, Madingley Road, Cambridge CB3 OHA, England.

SUMMARY

The observational constraints on the conditions in the gas which give rise to the emission lines seen in the spectra of quasars and active galactic nuclei are summarized, with particular attention being paid to the region responsible for the broad lines. Some general requirements on the physical conditions and geometry are described, and the emission line region of the quasar 3C273 is used to illustrate the successes, and shortcomings, of photoionization models. Comparisons of the emission line profiles of different ions show that there must be some radial flow, and obscuration, in the region of the emission line material.

INTRODUCTION

The defining feature of an active galactic nucleus is the presence of a powerful compact source of non-thermal continuum radiation. In many cases the conditions in the gas in the neighbourhood of this source are such that the object spectrum shows relatively broad emission lines covering a wide range of ionization. The emission line material is sufficiently far from the continuum source that it provides few direct clues to the nature of the power source, but it is important to understand the behaviour of the outlying material if we are to obtain a complete picture of quasars and Seyfert galaxies. There is perhaps only one important (and obvious) quantity determined from the emission lines which is of crucial importance in considering the continuum source, and that is the redshift. This is the main (and often only) indicator of the distance to the object, knowledge of which is necessary for consideration of the energy requirements.

Quasars and Seyfert 1 galaxies show very broad permitted lines, with velocity widths typically about 5000 km s^{-1} (FWHM), and narrower forbidden lines with widths of up to about 500 km s^{-1} are usually also present. Seyfert 2 galaxies are characterized by showing no distinguishable broad line component, and all lines have widths typical of the forbidden lines in Seyfert 1 galaxies and quasars. The most reasonable interpretation of the widths of the lines is that they are Doppler broadened by mass motions, and the clear separation in line widths suggests that the broad permitted lines and narrow lines arise in physically distinct regions. These are referred to as the Broad-line region (BLR) and narrow-line region (NLR). Here we shall deal in a little detail with the constraints that the observations place on the conditions in the BLR, and merely sketch some of the NLR results.

While the picture of the emission line region is based on the common features of observations of many objects, it can be helpful to have one specific example to illustrate some of the more important points. The object chosen for this purpose here is the redshift $z = 2.81$ quasar Q0207-398 (Fig. 1). The measured line widths in this object are about 2500 km s^{-1} (Wilkes, 1984), which makes them rather narrower than usual, but we are able to see quite easily lines from a wide range of ionizations, OIλ1305 to NVλ1240. Other features in this spectrum are the obviously asymmetric CIVλ1549 line profile, and the presence of large numbers of absorption lines. The absorption features almost invariably arise outside the quasar emission regions, and so are not relevant here.

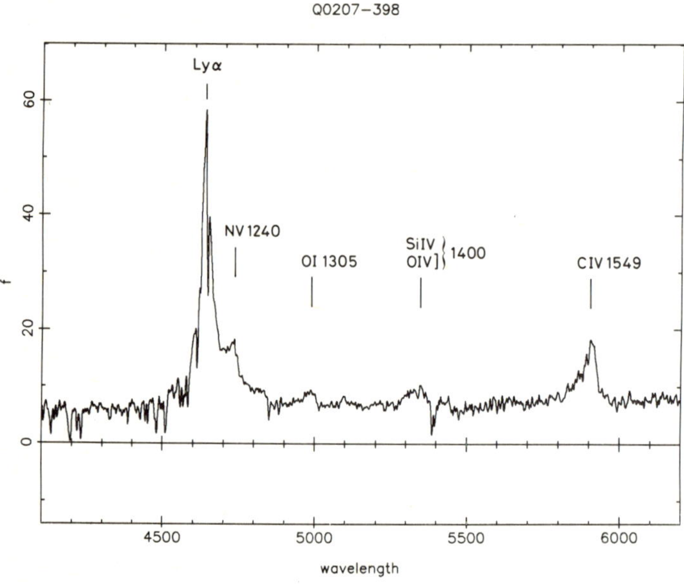

FIG. 1. A spectrum of the quasar Q0207-398, with the prominent emission lines marked. The flux scale is proportional to photons/sec/A.

In the following sections we shall first consider the implications of the observed emission line spectra on our interpretation of the ionization in the BLR material, and then turn to constraints on the geometry and kinematics of the BLR. For a much more comprehensive review of the properties of the emission line region than is given here, see Davidson and Netzer (1979), and for good general reviews of the topic, see Smith (1981) and Osterbrock (1984).

IONIZATION CONDITIONS IN THE BROAD-LINE REGION.

Since we know that the gas is in the neighbourhood of a powerful source of ionizing radiation, it is natural to suppose that the ionization levels in the cloud are dominated by this continuum source. Other mechanisms have, of course, been considered, but these are generally found to be inappropriate for quasars and AGN. For a detailed discussion of the possibilities, and details of the derivation of many of the results presented here, see Davidson and Netzer (1979).

In most AGN the broad emission lines are either permitted or semi-forbidden. There are no signs of a broad component to the forbidden lines. If the absence of these lines is due to collisional de-excitation, then the electron density n_e must be greater than 10^7 cm^{-3}. The strength of CIII]λ1909 relative to the other lines suggests that collisional de-excitation of the upper level is not important for this line, and so the electron density n_e can not be much greater than 10^{10} cm^{-3}. More refined estimates come from photoionization model considerations using the relative line strengths, and these suggest that n_e is nearer the upper end of the range, and is of order $10^{9.5}$ cm^{-3} (Davidson and Netzer, 1979).

There are few observable lines from which to estimate a typical temperature for the line-emitting gas. One pair which is usable in principle consists of the CIIIλ977 and CIII]λ1909 lines, but the CIIIλ977 line is difficult to measure in the confusion of the Lyman absorption lines in quasar spectra and at too low a wavelength to be measurable in Seyfert galaxies using IUE. There are also some complications because the CIII]λ1909 line depends on the density, but generally limits on the λ977 strength suggest temperatures $< 25000K$ are typical. Models, where the energy gain is balanced against the energy loss rates, give temperatures of order 2.10^4K.

While it is not appropriate to detail the model calculations, it is useful to have an idea of some of the considerations involved. Since the gas is photoionized, the relative numbers of ions of a given element can be described in terms of an ionization parameter, U, defined as the ratio of the ionizing photon number density (which determines the photoionization rate) to the electron density (which determines the recombination rate). Thus, for a single central continuum source, the ionization parameter for clouds at distance R and with electron density n_e is $U = Q(H)/(4\pi R^2 cn_e)$. Here Q(H) is the number of photons per second emitted by the continuum source which are capable of ionizing hydrogen from the ground state, i.e. with energies above 13.6eV. Strictly speaking only the relative number of HII to HI is proportional to U (along with some dependence on the temperature). For other ions the shape of the continuum could be important, though for smooth continua and ionization potentials near that of hydrogen any corrections will be small.

However, it is not the relative ionizations which we observe, but relative fluxes from emission lines from those ions. In principle, and after some computation, though, we may use the relative strengths of lines from different ions such as CIII]λ1909/CIVλ1549 to estimate an effective value for the ionization parameter, U, in any object where we can measure the relative fluxes. The practicalities are not quite as straightforward, since strong lines from neighbouring ions such as CIV and CIII do not often occur, and interpretation of the the CIII/CIV pair itself is complicated by the fact that the CIII]λ1909 line flux is dependent on the density. One approach is to use, for example, the Lyα/CIV and CIII/CIV line ratios together to estimate an effective U and n , the hydrogen number density, as has been done by Mushotzky and Ferland (1984). Fig 2, which is an adaptation of Mushotzky and Ferland's figure 2, shows curves of constant density and constant ionization parameter in a Lyα/CIV and CIII/CIV flux ratio plane for dust-free models with solar abundances. Following Mushotzky and Ferland a two component continuum was used, with energy fluxes $F(\nu) = F_1(\nu) + F_2(\nu)$. $F_1(\nu) \propto \nu^{-1.2}\exp(-\nu/\nu_c)$, where $\nu_c = 200eV$, and $F_2(\nu) \propto \nu^{-0.7}$. The ratios of the two components were set by requiring $Q_1(H)/Q_2(H) = 6$.

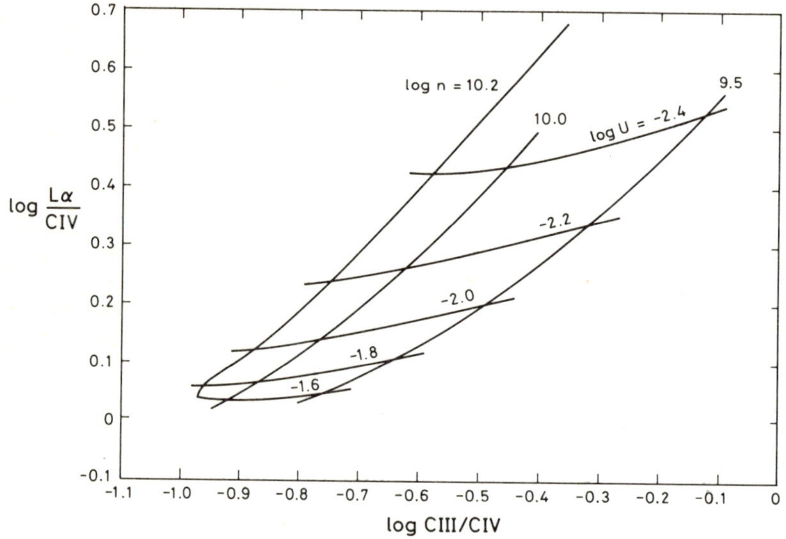

FIG. 2. The dependence of ionization parameter, U, and hydrogen density, n, on the Lyα/CIVλ1549 and CIII]λ1909/CIVλ1549 line flux ratios. For details see text.

Using these two line ratios, Mushotzky and Ferland (1984) have determined effective ionization parameters for a number of quasars and AGN, with luminosities over the whole range differing by a factor of about 10^5. The range of inferred ionization parameters is much smaller, with $-1.2 > \log U > -2.7$. As had been known from individual studies, the (lower luminosity) Seyfert galaxies have higher ionization parameters than the quasars. Mushotzky and Ferland were able to establish that, for the available data, $U \propto L^{-0.25}$, where L is the luminosity of the object. They show that, as a consequence, for high luminosity objects the CIVλ1549 equivalent width is roughly inversely proportional to the continuum luminosity. This is known to hold for flat spectrum radio sources (Baldwin, 1977a; Baldwin et al., 1978; Wampler et al.,1984).

So far we have considered only the high ionization lines which must arise in, or near, that part of the emission line region where most of the hydrogen is ionized. The presence of lines from species with low ionization potentials, such as MgIIλ2798 and OIλ1305, shows that there must be significant HI regions as well. Thus there is a third parameter which affects the fluxes of the low ionization lines, the Lyman limit optical depth, τ_L, of the emission line region. An upper limit for τ_L is about 3.10^5 (depending on the model), since otherwise the CII]λ2326 line should be stronger than the observations allow (e.g. Kwan and Krolik, 1981). This corresponds to a maximum thickness of order 5.10^{12} cm. For the models described here, a value $\tau_L=10^5$ was chosen arbitrarily.

A less direct indication of high Lyman optical depths comes from the Lyman to Balmer line ratios. Baldwin (1977b), from a composite quasar spectrum, showed that the Lyα/Hβ intensity ratio in quasars is not about 40 as predicted from simple recombination theory, but is somewhat closer to 3. Subsequent observations, using IUE at low redshifts (e.g. Wu et al., 1983), and infrared CVF spectroscopy of Hα at redshifts around 2 (Hyland, Becklin and Neugebauer, 1978; Puetter et al., 1981; Soifer et al., 1981; Allen et al., 1982), have confirmed this result. There is some scatter, but generally Lyα/Hβ is about 7, and Lyα/Hα is about 2 - 3. The most likely explanation for these line ratios (Weisheit, Shields and Tarter, 1981) is that most of the Balmer line radiation arises in an optically thick region where the X-rays from the continuum source give rise to suprathermal electrons. These electrons thermalize to give an extended warm zone in the quasar cloud, and, since the Lyα is effectively trapped, ionizations from the n=2 level in hydrogen are quite common. Every time this occurs, a potential Lyα photon is lost, and there is the possibility again for Balmer, and higher order, lines to be emitted. In a crude sense then, a fraction of the X-ray energy is going in to e.g. Hα and Hβ photons; many of the Lyman continuum photons are converted to Lyα photons. If this is correct, then (ignoring questions of the variation of cloud geometry and optical depth with luminosity) the Lyα flux should correlate with the ultraviolet luminosity, and the Balmer line fluxes with the X-ray luminosity. Presumably the luminosities in these ranges correlate with the optical luminosity, and so we might expect the line fluxes to correlate with the continuum. Such a correlation has been found by Yee (1980) for the Balmer lines.

With this general background, we now turn to a particular case to see how well the models compare with the observed line flux ratios. The best case for making a detailed comparison is for 3C273, where the line strengths are known over a wide wavelength range (Ulrich et al., 1980), and the ionizing continuum is best known of all quasars (see Bezler et al., 1984), so uncertainties in its shape should affect our results less than for other objects. There is an important gap in the observations of the 3C273 continuum flux, however, spanning the region from the Lyman limit to the soft X-rays (at a few kilovolts), and it is difficult to guess how the spectrum behaves there. For the illustrative models presented here we have used log linear interpolation across this frequency region.

The ionization parameter and (hydrogen) density were set by comparing model predictions for the Lyα/CIV and CIII/CIV line ratios with those observed for the object (Table 1). These yield, for a region with constant density throughout, $\log U = -2.2$ and $\log n_H = 10$. If we then construct a model with $\tau_L = 10^5$, we find that CII]λ2326 and MgIIλ2798 are too strong, so instead we have truncated the model emission zone at $\log \tau_L = 3.5$ to compute Model 1 in Table 1. Thus the Lyα, CIV, CIII and MgII lines have been used to define the parameters, so comparison with reality must be based on the other lines. Generally the (few) low ionization lines are within an acceptable factor of two or so of the predictions, but the Balmer lines and the high ionization lines agree very poorly.

TABLE 1.   Emission line ratios

Line	3C273	Model 1	Model 2	Model 3	Q0207-398
OVI 1035	1.0:	0.02	0.02	0.07	1.11
Lyα	2.17	2.12	2.26	2.18	4.80
NV 1240	0.46	0.03	0.03	0.05	0.48
OI 1305	0.11				0.22
SIV/OIV]1400	0.30	0.14	0.14	0.09	0.33
CIV 1549	1.00	1.00	1.00	1.00	1.00
HeII 1640	0.15	0.10	0.11	0.10	
OIII]1663	0.15	0.14	0.14	0.10	
CIII]1909	0.28	0.28	0.27	0.26	0.44
CII] 2326	<0.04	0.01	0.05	0.03	
FeII 2600	<0.07	0.01	0.03	0.01	
MgII 2798	0.15	0.16	0.31	0.15	
HeII 4686	<0.04	0.01	0.01	0.01	
Hβ	0.43	0.04	0.07	0.04	
Hα	1.39	0.34	0.56	0.42	
Pα	0.09	0.03	0.05	0.04	

The Balmer line enhancement mechanism outlined earlier is more effective at higher densities, so we may improve the Balmer line agreement between model and observation by somehow increasing the density in the HI region. An obvious way of doing this without being too arbitrary is to assume that, instead of constant density, the region is at constant pressure throughout. The neutral, cooler, zone must then have a higher density to maintain pressure equilibrium with the HII region. Model 2 has been computed assuming pressure equilibrium, $\log U = -2.2$ and $\log n_H = 10$ on the inner side of the cloud, and $\log \tau_L = 5$. The Balmer line agreement is better, but the Balmer decrement is still too high. The OVI and NV line strengths are, of course, barely changed. In principle, it is possible to modify the continuum in the region where there are no observations to constrain us, but even an increase to $F(\nu)$ = constant between the Lyman limit and the beginning of the observed X-ray continuum (with a very steep cutoff at low X-ray energies to force agreement with the observations) leaves both NVλ1240 and OVIλ1035 too weak by a factor of about three.

All the above discussion has been based on models calculated assuming that the heavy element abundances are solar. However the diagnostic for the ionization parameter, the Lyα/CIV ratio, is also sensitive to the abundances. If we attempt to construct a model using the 3C273 continuum with, for example, 1/10 solar abundances, we find that we cannot reproduce the line ratios observed. The Lyα/CIV ratio always exceeds 2.5 for a reasonable range of parameters, and the CIII/CIV ratio is less   than 0.1 usually. Just to illustrate the differences, Table 1 also contains the results of a calculation based on 0.5 times solar abundances. To reproduce the Lyα/CIV and CIII/CIV ratios the ionization parameter and the density had to be changed considerably. Model 3 is for constant pressure, log U = -1.90, log n_H = 9.45 and log $\bar{\tau}_L$= 5.0. It is an important consequence of this sort of calculation that the abundances in quasars are generally quite close to solar, but that the inferred densities and conditions depend quite sensitively on the abundances chosen. For a more general discussion of the possible range, and effects, of the heavy element abundances see Davidson and Netzer (1979).

3C273 is not anomalous in having much stronger high ionization lines than the models allow. The line ratios for the high redshift quasar Q0207-398, which was used to provide a sample spectrum earlier, are also given in Table 1. There are some detailed differences between this object and 3C273, but the high ionization lines are still strong relative to CIVλ1549. Log U = -2.6 and log n = 10.2 adequately fit the Lyα/CIV and CIII/CIV line ratios, given the standard continuum used to obtain Fig. 2. Then, for a one component model, the predicted OVIλ1035/CIVλ1549 flux ratio is 0.008, and that for NVλ1240 is 0.022.

The absence of sufficiently strong high ionization lines is a feature of the models which has led to suggestions that a single cloud picture is too simplified (e.g. Davidson and Netzer, 1979). A second high ionization component, or a range of conditions in a population of emission line clouds, provide the most plausible explanation. There have also been suggestions that there must be a denser, physically distinct region to that given by the conventional models to explain the Hα/Hβ ratio observed (Collin-Souffrin et al., 1982). Many of the low ionization lines, such as MgIIλ2798 and the FeII lines (Joly, 1982) would also arise predominantly in this region. Thus, to explain the emission line strengths that we see, we are being forced to consider a situation where a range of sizes, densities and conditions apply. While this is computationally inconvenient, and difficult to use to predict line strengths since we have little idea of what these ranges should be, it is surely more plausible than a situation in which all the material is in uniformly sized clouds all at the same density and distance from the QSO.

However, before abandoning the single component models we should consider one further complication, if only to sweep it under the carpet - and that is dust. Dust absorption and scattering in our Galaxy, and possibly in the intergalactic medium towards the quasar, will reduce the observed ultraviolet line fluxes more than the optical ones. Absorption within the galaxy containing the active nucleus, and within the emission line region, will have the same effect. Thus the models we have set up to explain the observed line ratios may fail to do so adequately simply because we are not measuring the correct, intrinsic, line fluxes.

It is possible to construct illustrative models which give most line ratios to have values typical of those observed, including the high ionization lines, using a single component cloud population and dust obscuration (see Davidson and Netzer, 1979). The observed Lyα/Hα ratio, in particular, is readily explained through the loss of trapped

Lyα photons on dust in the emission line region clouds. If the densities and optical depths are sufficiently low that there is no significant collisional excitation from the n=2 level in hydrogen, then the observed Balmer decrement need then be only a little steeper than the traditional case B value, consistent with the observations. However, the Pα/Hβ ratio should then be greater than the case B value (about 0.28), while in 3C273 it is found to be somewhat smaller (0.21). Thus the hydrogen line fluxes must be modified by some process other than dust reddening, presumably involving optical depth and collisional effects. Unfortunately, this implies that the Balmer line ratios will also be affected, and so the Balmer decrement will not give a reliable estimate of the amount of dust reddening. We might suppose that, since the Hα/Hβ ratio is increased both by dust and, apparently, n=2 collisional excitation, the dust component is likely to be small, but in the absence of self-consistent models this is no more than a guess. On the other hand, there is the possibility that the continuum polarization in many QSOs is primarily due to dust (Stockman et al., 1984) which is likely to affect the emission lines as well, and small amounts of dust do provide reasonably internally consistent line ratios in some well studied Seyfert galaxies (e.g. Ward and Morris, 1984).

DISTRIBUTION OF THE BLR MATERIAL

The total amount of emitting gas, and its distribution in the region of the continuum source is clearly of great importance, and is one of the things we know very little about. While the details of the conditions in the line emitting regions are subject to some uncertainty, the simple models may still give us reasonable estimates for some of the global parameters, such as overall size, and mass, of the material involved. The mass of gas in the emission line region may be estimated by measuring the emitted flux in a line, and so, through the models, the total numbers of the relevant ion present. The mass estimates that result are a little uncertain, but quantities of order 10 - 100 M$_\odot$ result. A typical distance from the continuum source comes from knowledge of the ionization parameter and the density, and, through direct measurement, the luminosity of the object. Then, from the equation defining the ionization parameter (U = Q(H)/4πR^{2}cn$_e$ for a single small continuum source) given in the previous section, it is a simple matter to estimate a typical distance R. The numbers here range from about 0.1 - 1 pc for the AGN to about 1 - 10 pc for quasars.

From the distance, density, and mass estimates we find that the fraction of the volume filled by the ionized gas is about 10^{-4}. If the size upper limit applies to all dimensions of the clouds (so they are assumed not to be pancake - like), then there must be at least 10^{10} of them, and possibly a considerably larger number. The space between the clouds is unlikely to be empty, and is possibly filled with a diffuse inter-cloud medium at a temperature of perhaps 10^{8}K (Krolik, McKee and Tarter, 1981).

Attempts have been made to count the number of clouds directly in a few bright Seyfert galaxies by cross-correlating the Hα and Hβ profile structure (Atwood et al., 1982). The idea here is that if there are few enough clouds then it should be possible to see velocity structure on scales comparable with the thermal widths of individual clouds. No such structure has been found, and so Atwood et al. inferred that the number of clouds emitting Balmer lines must exceed 3.10^4. The technique they used relies on the effects of a finite number of clouds exceeding those of a finite signal-to-noise ratio in the observations, and so it is unlikely that there will be an improvement of more than another factor of ten, at best. Thus we are a long way from observationally testing the rough estimates of cloud numbers.

A second important quantity is the fraction of the continuum source which is covered by the emission-line clouds. This is determined by comparing the number of photons in a recombination line, such as Lyα, with the number of ionizing photons available in the Lyman continuum. The covering factors obtained in this way for quasars are usually quite small, typically about 10% if all the Lyα is from recombination (Baldwin and Netzer, 1978). Direct observations of the Lyman limit region in high redshift quasars confirm directly that the fraction of the sky as seen from the continuum source that is covered by optically thick clouds is less than about 15% (Smith et al., 1981; MacAlpine and Feldman, 1982; Oke and Korycansky, 1982).

The optical depth considerations described above do not affect the estimate of the covering factor deduced from the Lyα equivalent width very much, since most of the recombination radiation will arise in the HII zone. However, using Hα will give misleadingly high estimates, since the process described earlier enhances the Balmer line flux. For some quasars the covering factors deduced from Hα are greater than unity, and so it is quite likely that the covering factors of order unity derived for Seyfert galaxies are similarly misleading. Wu et al. (1983) give IUE measurements of Lyα equivalent widths for Seyfert galaxies which allow approximate covering factors to be determined. Generally they are a little higher than in quasars, with a mean of 17%. As with the quasars, this quantity should be multiplied by some number of order 0.5 - 0.8 to allow for collisionally excited Lyα. Note though that if there is a significant contribution to the Lyα flux from optically thin material, then the method outlined here would overestimate the fraction of the continuum source covered by optically thick clouds.

The Lyman limit observations of quasars could also give us an indication of the sizes of the clouds compared with the continuum source. If, as we expect, the clouds are much smaller, then there should be a lower apparent continuum, by an amount corresponding to the covering factor (after some allowance for any optically thin component to the cloud population), at wavelengths shortward of the emission Lyman limit. To test this requires good continuum observations of quasars with large Lyα equivalent widths. These are almost invariably quite faint, so this test has not yet been done. If the clouds are larger than the continuum source, then in a fraction of QSOs, corresponding to the mean covering factor, there will be a strong Lyman discontinuity at a redshift near the emission redshift, which should be seen to have an anomalously weak corresponding Lyα absorption line. The last point arises simply because an emission line cloud is unlikely to cover the entire Lyα emission region as seen from the Earth. No such system has yet been found.

BROAD LINE REGION KINEMATICS

Almost by definition, the common feature of the line profiles is their large velocity widths. A typical FWHM is about 5000 km s^{-1}, and the rather difficult to measure full width zero intensity (FWZI) is about 20000 km s^{-1}. The profile shapes range from the smooth (see e.g. the Hβ profile in 3C120, Baldwin et al., 1980) to the asymmetric and lumpy (e.g. Netzer, 1982 in 3C390.3, and others by Gaskell, 1983). There is some tendency for enhanced wings compared with a gaussian, though of course we have no reason to expect a gaussian velocity distribution in any case.

For low redshift objects the broad line profile work has necessarily been restricted to helium and hydrogen lines in the optical region, with a few important results from IUE observations (see Penston, these proceedings). In Seyfert galaxies there is often a small difference in the Hα and Hβ profiles, with a general trend for the Hβ profile to have a slightly larger FWHM than that for Hα (Shuder, 1982). However, it is not hard to find exceptions where the Hα and Hβ profiles are, as far as the measurements allow, identical (Atwood et al., 1982). There is a more pronounced difference between HeIλ5876 and Hα, also in the sense that the HeI line has a larger FWHM. In low redshift quasars the differences between these line profiles are less pronounced than in the (lower luminosity) Seyfert galaxies (Shuder, 1984).

For high redshift quasars there is the possibility of comparing line profiles from a range of ions, though most of the strong lines seen are from relatively high ionization species. Baldwin and Netzer (1978) found that all the strong lines, particularly Lyα, CIVλ1549 and CIII]λ1909 have similar profiles in most cases. There are some detailed differences which occur in some objects (Wilkes, 1984), but the most striking feature is the overall profile similarity in each quasar. There is evidence from a large sample of quasars (33) that the CIVλ1549 profile is generally asymmetric (Young et al., 1982), with an extended tail towards short wavelengths. An example where this asymmetry is quite pronounced is in the redshift z = 2.81 quasar Q0207-398 (Fig. 1). Generally the asymmetry is quite small, as it is for most of the objects chosen by Smith et al. (1981) and Wilkes and Carswell (1982) for comparison of the Lyα and CIVλ1549 profiles, and by Wilkes (1984) for a detailed study of the Lyα, NVλ1240, CIVλ1549 and CIII]λ1909 profiles. In those cases where the CIVλ1549 profile has a pronounced asymmetry the Lyα line usually has a similar asymmetric profile. Unusually, in the example we have used, Q0207-398, the Lyα line probably has an additional relatively narrow component redshifted with respect to the part which matches the CIV profile.

One inference from the profile similarity has been highlighted by Smith et al. (1981). If the Lyα and CIVλ1549 arise predominantly in optically thick clouds, then the absorption and reradiation of Lyα within the cloud will have different characteristics from CIVλ1549 and, particularly, CIII]λ1909. All these lines will be created in the side of each cloud nearest the continuum source and the Lyα must escape almost exclusively towards the continuum source, since the line optical depth outwards must be much greater than the Lyman continuum optical depth, τ_L, which itself is very high. In the cases of CIVλ1549 and

CIII]λ1909 there is no similar barrier to their escape through the HI region, since there most of the carbon will exist as CII, and so there will be no significant absorption (except perhaps on dust) for CIII and CIV line photons travelling in the outward direction. Thus the Lyα photons which we see come predominantly from clouds on the far side of the continuum source, and CIVλ1549 and CIII]λ1909 photons from all clouds which are not occulted by e.g. another cloud or the continuum source.

If we now suppose that there is a net outflow, for example, then the Lyα line will be redshifted with respect to, and narrower than, the other lines from the HII zone. This is contrary to what is seen. Similarly net inflow is ruled out if the lines come predominantly from optically thick clouds, since under these circumstances the Lyα line should be relatively narrower and blueshifted.

Profile comparisons between low ionization (e.g. MgIIλ2798, and the Balmer lines) and high ionization lines are rather rudimentary. This is partly because the strong high ionization lines are in the ultraviolet, and strong low ionization lines are at or near optical wavelengths. It is possible to compare CIII]λ1909 with MgIIλ2798 in the same object, but few spectra with the required S/N and resolution have been obtained at suitable redshifts. Most of the work which has been done hinges on comparing the profiles of weak lines, such as OIλ1305 with Lyα and CIVλ1549 (Wilkes, 1984). The tentative suggestion here is that the low ionization lines are narrower. This is in line with what we might expect from the Seyfert galaxy result that the HeI line is broader than the Balmer lines.

One thing which does appear to be well founded is a differential redshift between low and high ionization lines, amounting to perhaps 750 km s^{-1} (Wilkes, 1984). This was first noticed by Gaskell (1982) in low resolution spectra, and confirmed by Wilkes (1982) from a different low resolution sample. A somewhat more direct demonstration of the velocity difference has been provided by Wilkes (1984), using an adequate number of higher resolution spectra to compare the line profiles and determine relative velocity shifts. Hints of a difference between Lyα and Hα redshifts were noted by Allen et al. (1982).

This leaves us in something of a quandary, having argued against infall or outflow on the basis of the comparison of Lyα profile with CIVλ1549 and the absolute need for considerable systematic gas motions with some form of obscuration if there is a velocity difference of the type described above. How do we reconcile the two requirements?

Since there is a difference, presumably, between the redshifts of the Lyα and Hα lines these are likely to come from different cloud populations. Then the Lyα could well arrive mostly in clouds which are not very optically thick, and so the asymmetry argument against outflow or infall is no longer valid. The Lyα/Hα ratio in the optically thick clouds must then be $\ll 1$, but that should not cause insurmountable problems. Much of the CIVλ1549, NVλ1240 and OVIλ1035 would arise in the optically thin clouds, and the model difficulties in producing sufficiently high fluxes in the last two of these would therefore be overcome.

One of a number of possible pictures which qualitatively satisfy the constraints has been suggested by Heckman (as reported by Weymann et al. 1982).   Outflow material from the quasar is partly occulted by the accretion disk around the black hole (none of these details is essential).   The inner part is presumably more ionized (unless it expands as it flows out) we suppose, and the outer less ionized. The inner region will have much of the radiation occulted by the accretion disk, so all we will see is the material moving towards us, hence blueshifted. The   true redshift would then be given by the low ionization lines, which are in the outer regions. The difference in line widths has important consequences if verified since the flow must decelerate if this picture is correct.

Of course, this is not the only one. It is possible that an internal obscuration, infall model will work in a similar way. All that is required is that the clouds contract so that the density rises more rapidly with decreasing distance than $1/R^2$, and the same model applies. A more extensive discussion of the constraints, and the possibilities, is given by Wilkes (1984). Whatever the reality, it is clear that the kinematic properties and the ionization are linked, and any successful model has to take account of these relationships. Also, differential obscuration will affect the line flux ratios which are used to determine the conditions in the emission line region. This is an area where much work, both observational and theoretical, remains to be done.

CONCLUDING REMARKS

While the emphasis here has been on the broad line region, the conditions in the outlying gas which gives rise to the narrow permitted and forbidden emission lines seen in the spectra of low redshift quasars and Seyfert galaxies is also a fascinating topic in its own right. I do not propose to review this topic here, but refer the reader to reviews by Osterbrock (1978) and Balick and Heckman (1982). As with the broad line clouds we may infer some of the characteristic parameters which apply to the narrow line region (NLR). The presence, and strengths, of the forbidden lines imply that the electron density, n_e, is typically 10^4 cm^{-3}. The Hβ luminosity gives a mass of ionized gas in the region of 10^5 - 10^6 M$_\odot$, in a region of order some hundreds of parsecs across (from ionization balance and, where the object is close enough, direct observation). The line widths are typically less than 500 km s^{-1} (FWHM).

Line profiles have been obtained by Heckman et al. (1981), and subjected to a more comprehensive study by Whittle (1982, 1984). Most of the work has concentrated on the [OIII]λ5007 line, since it is the strongest and so is the easiest to study. These lines are usually found to have stronger wings than a gaussian, and to have a strong asymmetry, almost always towards the blue. As for the broad line material, for which the emission lines have a (weaker) asymmetry in the same sense, this implies that there must be radial gas flows and some source of opacity either within the emitting clouds or in the region containing them. Inferring the direction of the flow is difficult, but a study of the spatially resolved extended emission, and the absorption lines using IUE, strongly suggests that at least for Markarian 509 outflow is occurring (Phillips et al., 1983). However, the relation of the extended emission region to the NLR is not clear, much less the NLR to the broad line region, so it is not necessarily true that outflow dominates everywhere.

The picture which has emerged is one of complexity in both the broad and narrow line regions, and a fuller understanding of the conditions would come most readily from spectroscopic observations with much higher spatial resolution than has been possible so far. We would then be in a position to determine the variations in ionization, obscuration and velocity as a function of the (projected) distance from the continuum source, and relate these to models of the line emitting regions. With the space telescope this should be possible for the narrow line region in nearby objects, but the broad lines will remain beyond its capabilities. For these, however, the distance information may be inferred indirectly by observing the response of the line profiles and fluxes to changes in the continuum flux, as has been done for NGC4151 using IUE (Ulrich et al., 1984; Penston, these proceedings). It is important that such studies be continued and extended.

Acknowledgements:

I am grateful to Gary Ferland for very generously providing the photoionization program which was used to obtain the illustrative models presented here, to Paul Hewett and Malcolm Smith for comments on the manuscript, and to the Radcliffe Trust for support.

REFERENCES

Allen, D.A., Barton, J.R., Gillingham, P.R., and Carswell, R.F., 1982. MNRAS, 200, 271

Atwood, B., Baldwin, J.A., and Carswell, R.F., 1982. Ap.J., 257, 559.

Baldwin, J.A., 1977a. Ap.J., 214, 679.

Baldwin, J.A., 1977b. MNRAS, 178, 67.

Baldwin, J.A., Burke, W.L., Gaskell, C.M., and Wampler, E.J., 1978. Nature, 273, 431.

Baldwin, J.A., Carswell, R.F., Wampler, E.J, Smith, H.E., Burbidge, E.M., and Boksenberg, A., 1980. Ap.J., 236, 388.

Baldwin, J.A., and Netzer, H., 1978. Ap.J., 226, 1.

Balick, B., and Heckman, T.M., 1982. Ann. Rev. Astron. Astrophys., 20, 431.

Bezler, M., Kendziorra, E., Staubert, R., Hasinger, G., Pietsch, W., Reppin, C., Trumper, J., and Voges, W., 1984. Astron. Astrophys. in press.

Collin-Souffrin, S., Dumont, S., and Tully, J., 1982. Astron. Astrophys., 106, 362.

Davidson, K., and Netzer, H., 1979. Rev. Mod. Phys., 51, 715.

Gaskell, C.M., 1982. Ap.J., 263, 79.

Gaskell, C.M., 1983. Proc. 24th Liege Colloq. 'Quasars and Gravitational Lenses', ed. J.P. Swings, Univ. de Liege, Inst. d'Astrophysique, p473.

Heckman, T.M., Miley, G.K., van Breugel, W.J.M., and Butcher, H.R., 1981. Ap.J., 247, 403.

Hyland, A.R., Becklin, E.E., and Neugebauer, G., 1978. Ap.J., 220, L73.

Joly, M., 1982. Astron. Astrophys., 102, 321.

Krolik, J.H., McKee, C.F., and Tarter, C.B., 1981. Ap.J., 249, 422.
Kwan, J., and Krolik, J.H., 1981. Ap.J., 250, 478.
MacAlpine, G.M., and Feldman, F.R., 1982. Ap.J., 258, 11.
Mushotzky, R.F., and Ferland, G.J., 1984. Ap.J., 278, 558.
Netzer, H., 1982. MNRAS, 198, 589.
Oke, J.B., and Korycansky, D.G., 1982. Ap.J., 255, 11.
Osterbrock, D.E., 1978. Phys. Scr., 17, 285.
Osterbrock, D.E., 1984. QJRAS, 25, 1.
Phillips, M.M., Baldwin, J.A., Atwood, B., and Carswell, R.F., 1983.
    Ap.J., 274, 558.
Puetter, R.C., Smith, H.E., Willner, S.P., and Pipher, J.L., 1981.
    Ap.J., 243, 345.
Shuder, J.M., 1982. Ap.J., 259, 48.
Shuder, J.M., 1984. Ap.J., 280, 491.
Smith, M.G., 1981. 'Investigating the Universe', ed. F.D. Kahn
    (Reidel, Dordrecht), p151.
Smith, M.G., Carswell, R.F., Whelan, J.A.J., Wilkes, B.J., Boksenberg,
    A., Clowes, R.G., Savage, A., Cannon, R.D., and Wall, J.V., 1981.
    MNRAS, 195, 437.
Soifer, B.T., Neugebauer, G., Oke, J.B., and Matthews, K., 1981.
    Ap.J., 243, 369.
Stockman, H.S., Moore, R.L., and Angel, J.R.P., 1984. Ap.J., 279, 485.
Ulrich, M.-H., Boksenberg, A., Bromage, G., Carswell, R., Elvius, A.,
    Gabriel, A., Gondhalekar, P.M., Lind, J., Lindegren, L., Longair,
    M.S., Penston, M.V., Perryman, M.A.C., Pettini, M., Perola, G.C.,
    Rees, M., Sciama, D., Snijders, M.A.J., Tanzi, E., Tarenghi, M., and
    Wilson, R., 1980. MNRAS, 192, 561.
Ulrich, M.-H., Boksenberg, A., Bromage, G.E., Clavel, J., Elvius, A.,
    Penston, M.V., Perola, G.C., Pettini, M., Snijders, M.A.J., Tanzi,
    E.G., and Tarenghi, M., 1984. MNRAS, 206, 221.
Wampler, E.J., Gaskell, C.M., Burke, W.L., and Baldwin, J.A., 1984.
    Ap.J., 276, 403.
Ward, M., and Morris, S., 1984. MNRAS, 207, 867.
Weisheit, J.C., Shields, G.A., and Tarter, C.B., 1981. Ap.J., 245,
    406.
Weymann, R.J., Scott, J.S., Schiano, A.V.R., and Christiansen, W.A.,
    1982. Ap.J., 262, 497.
Whittle, D.M., 1982. Ph.D. thesis, Cambridge University.
Whittle, D.M., 1984. MNRAS submitted.
Wilkes, B.J., 1982. Ph.D. thesis, Cambridge University.
Wilkes, B.J., 1984. MNRAS, 207, 73.
Wilkes, B.J., and Carswell, R.F., 1982. MNRAS, 201, 645.
Wu, C.-C., Boggess, A., and Gull, T.R., 1983. Ap.J., 266, 28.
Yee, H.K.C., 1980. Ap.J., 241, 894.
Young, P.J., Sargent, W.L.W., and Boksenberg, A., 1982. Ap.J. Suppl.,
    48, 455.

CHEMICAL ABUNDANCES AND THE IONISATION MECHANISMS IN LINERS

A. I. Díaz
Astronomy Centre, University of Sussex, Falmer, Brighton,
East Sussex, U.K.

B.E.J. Pagel and I.R.G. Wilson
Royal Greenwich Observatory, Herstmonceux Castle, Hailsham,
East Sussex, U.K.

SUMMARY

New near infrared spectrophotometric observations show that the relative strengths of the sulphur emission lines [SIII] $\lambda\lambda$9069, 9532 A with respect to H_α, constitute a powerful diagnostic to distinguish between shock and photoionisation mechanisms in Liners. These observations strongly support the idea, already favored on the basis of abundance arguments, that Liners and Seyferts form a continous sequence, both types of objects being photoionised, but with the ionisation parameter being lower in Liners. This photoionisation could be due either to the presence of a central non-thermal source or to recent star formation activity.

1. Introduction

Nuclear emission in galaxies shows a very wide range of intensities. It is possible, in principle, to distinguish between high excitation and low excitation nuclear emission regions. Seyfert galaxies and many of the emission-line radio galaxies belong to the first category, while the second one includes low-excitation, oxygen rich, nuclear HII regions (non active) and Liners.

The word Liner, first used by Heckman (1980), stands for Low Ionisation Nuclear Emission-line Region and essentially designates an active nucleus whose spectrum is dominated by low-ionisation lines. Liners have moderately strong [OIII], with λ5007/$H_\beta \simeq 3$, compared to a ratio of 10 or more in Seyfert 2's and narrow line regions of Seyfert 1's, and a ratio of less than 1 measured in the high abundance HII regions found in the nuclei of spirals. Liners also have very strong [NII] lines with λ6584/H_α being typically greater than unity. Although the emission-line luminosities of most Liners are much smaller than those of Seyfert galaxies and their featureless continuum much weaker (if at all seen in the blue), Liners share some of the properties of Seyfert galaxies. Often a broad component of H_α is seen and the "narrow" line widths are usually greater than 300 Km/s, they often show X-ray emission and have a tendency to have nuclear compact radio sources. In fact, many Narrow Line Radio Galaxies show optical spectra similar to those of Liners.

Liners are much more common than Seyfert galaxies. Up to one third of all galaxies belong to this group (against the 1% represented by Seyferts, Weedman (1977)) and recent surveys by Stauffer (1982) and Keel (1983) have led to the remarkable conclusion that most, if not all, early-type spirals have emission-line spectra of this type, which provides some suggestion that various forms of nuclear "activity" may be present at a low level in the humblest spirals as well as at a higher level in Seyferts, radio galaxies and quasars.

2. Ionisation mechanisms

The origin of the excitation of the emission lines observed in Liners has been the subject of considerable amount of work over the last few years. The similarities found between Liner optical spectra and those corresponding to supernova remnants led to the initial suggestion that these objects were excited by shock waves (Heckman, 1980; Baldwin, Phillips and Terlevich, 1981) in contrast with the case of Seyferts whose emission-line spectra could be fitted by photoionisation models with a power-law continuum which indeed is often observed at radio, optical and X-ray wavelengths. However, it has recently been pointed out by Ferland and Netzer (1983) and others that the more obvious features of liners can be explained by a power-law continuum similar to that working for Seyferts but with lower ionisation parameter (ratio of ionising photons at the Lyman continuum limit to the electron density), thus suggesting that a continuous sequence of decreasing ionisation parameter can explain the emission-line spectra of broad-line objects, Seyfert 2's and Liners.

Obviously, shock heating in Liners cannot be ruled out on the basis of this argument alone. The unquestionable presence of a non-thermal ionising continuum would be the strongest argument in favor of power-law photoionisation. However, as noted by Heckman (1980), the contribution from a featureless continuum in most Liners does not exceed 10% of the total flux in the blue. In the ultraviolet the situation is also ambiguous, it being difficult to distinguish between the contribution of a weak non-thermal continuum and normal hot stars. Arguments in favor of one mechanism or another have to rely on complementary observations and are quite model dependent (Penston, 1982).

Some weak lines, like [OIII] $\lambda 4363$ and He II $\lambda 4686$ could, in principle, provide crucial tests to distinguish between photoionisation and shock heating mechanisms. Very high temperatures as deduced from the $\lambda 4363/\lambda 5007$ ratio of [OIII] lines would rule out photoionisation models. Unfortunately, two main reasons make the use of this line ambiguous for the discrimination between different kinds of models. First, the difficulty to measure it, due to its intrinsic weakness and to its occurrence adjacent to H_γ and on top of a continuum affected by strong stellar absorption features. Secondly, the fact that in some cases this line may be influenced by high electron densities and hence it cannot be used to determine the electron temperature in the usual way (Carswell et al., 1984).

The strength of the He II line $\lambda 4686$ looks much more promising. Its intensity is essentially dependent on the spectral distribution of the ionising radiation and, in the power-law case, is determined by both the ionisation parameter and the spectral index. Power-law photoionisation models predict a very small range for the He II $\lambda 4686/H_\beta$ ratio. However, the range observed in Liners seems to be much wider (although errors of $\sim 20\%$ are quoted for most of the measurements available). This fact has led Stasińska (1984) to propose a different type of ionising spectrum resembling a black body with an effective temperature higher than 150,000 K.

A link between the presence of this black-body like spectrum and star formation activity in galactic nuclei has recently been suggested by Terlevich and Melnick (1984) based on new studies of the evolution of high mass stars. These stars may suffer very large mass losses at a rate which is strongly dependent on metallicity, ending their evolution as

bare carbon cores and reaching temperatures of about 200,000 K. Models
of photoionisation by star clusters sufficiently old to account for the
presence of an appreciable number of such extremely hot stars, can
adequately reproduce the spectra of Liners and, furthermore, their
predictions for HeII $\lambda 4686/H_\beta$ are compatible with the observed ratios.

Therefore, the definite exclusion of shock heating as the dominant
mechanism acting in Liners would open the possibility that star
formation is responsible for the high excitation spectra observed in
Liners.   This would classify Liners as "non active" objects if the term
activity is used to designate processes which cannot be accounted for in
the context of star formation and evolution.

Another type of plausibility argument has been pointed out by Pagel
(1983) based on abundance considerations.The most striking feature
common to the optical spectra of Liners is the enormous strength of the
[NII] $\lambda 6584$ line.   A plot of [NII] line strength against [OIII] line
strength for Seyferts and Liners (open circles and crosses respectively
in figure 1) shows that power-law photoionisation models place the
various types of objects along a sequence of varying ionisation
parameter with constant and quite plausible abundances (solar or
slightly over solar).   In the case of photoionisation by a star cluster,
this sequence represents an evolution in time with decreasing ionisation
parameters corresponding to increasing cluster ages.   Opposite to the
case of photoionisation, in shock models the nitrogen abundance is
required to increase steadily from high excitation objects to low
excitation objects and an overabundance of nitrogen by a factor of about
three would be indicated for a typical Liner.

In order to investigate if these overabundances would be likely to
exist, some emission-line galactic nuclei were observed with the AAT
using the IPCS detector (by Edmunds and Phillips in 1981 and by Pagel in
1982).   Four of these objects turned out to belong to the Liner family:
NGC 1097, NGC 1433, NGC 1598 and NGC 7590. The observations were made
with a long slit which included both the nucleus and normal galaxy
spectra which in some cases comprised normal HII regions a few hundred
parsecs from the nucleus, thus allowing the determination of the
chemical abundances in these regions.

Figure 1 shows the location in the [NII]/H_α vs [OIII]/H_β diagram of
the points corresponding to our data and a few more objects, M 51, M 81
and NGC 1365, for which similar observations are available.   Several
points for each object are shown, corresponding to the nucleus itself
and to outlying parts of the slit. These data fill the gap between
Seyfert 2's and Liners that appears to exist in the diagram (which, were
it real,would strongly suggest different mechanisms for different
objects) and fit the predicted curve for photoionisation with solar
abundances quite well.   Furthermore, the nuclear abundances are in
agreement with the ones deduced for the nearby HII regions, except
perhaps in the case of M 51 in which the intensity of the [NII] line is
stronger than may be accounted for by models.

All this seems to indicate that the unusual strength of the [NII]
lines in Liners is mainly an excitation effect rather than an abundance
effect.

3.New observations

In recent years the region of the spectrum in the vicinity of 1 μ has
become accessible to spectrophotometric studies through developments in
instrumentation. This allows the use of an extra discriminant that
could be very powerful: the intensities of the far red (near infrared
?) [SIII] lines $\lambda\lambda 9069$, 9532. These lines behave quite distinctly in

shock and photoionisation régimes (figures 2 and 3). Shocks at a reasonable interstellar density of 100 cm^{-3} produce strong [SII] and weak [SIII] compared to typical HII regions, whereas photoionisation predicts rather strong [SIII] in the range of ionisation parameters adequate to fit the [OIII] and [NII] intensities observed in Liners.

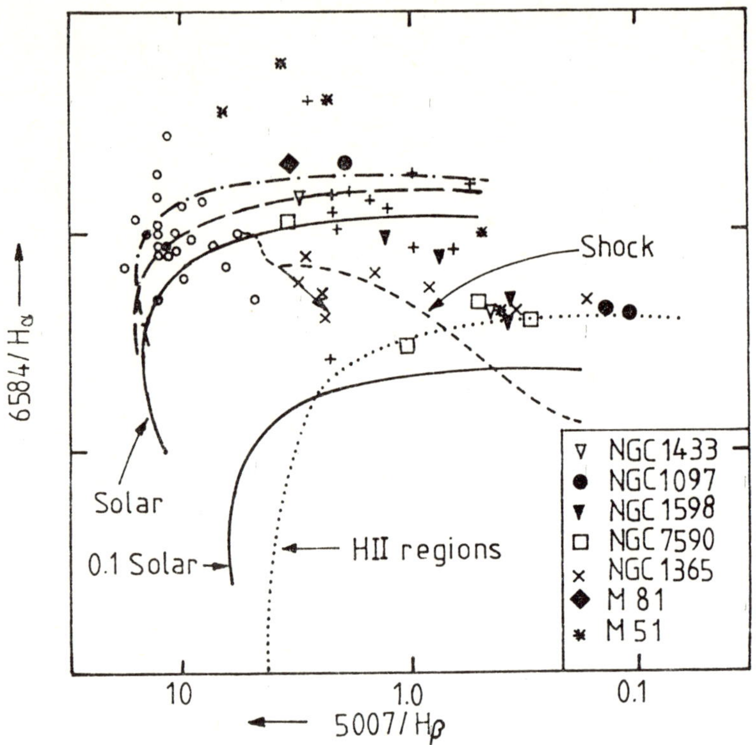

Fig. 1.- [NII]/H$_\alpha$ vs [OIII]/H$_\beta$ diagram. Open circles and crosses represent Seyfert 2's and Liners respectively after Ferland and Netzer (1983). Solid lines represent power-law photoionisation models for the specified chemical abundances. The long dashed line and the dashed-dotted line represent models of photoionisation by a young star cluster in which the contribution of extremely hot stars (see text) is approximated by a black body of Teff= 150,000 K and Teff= 200,000 K respectively. Shock models and the locus for HII regions are indicated.

As part of a wider programme undertaken by Pagel, Díaz and Wilson, some Liners, NGC 1097, NGC 1433 and NGC 1672, and a Seyfert-like galaxy, NGC 1365, have been observed by Wilson in October, 1983 both in the nucleus and in the surrounding HII regions with the Anglo Australian Telescope using the RGO spectrograph and the CCD detector. The data cover the spectral region from 6000 A to 10000 A and have been reduced using the standard STARLINK programs.

The main problem in observing at these wavelengths is the deep and highly structured trough of water vapor absorption which chops out the continuum background to such an extent that the weak lines are barely visible in the raw data. Division by a standard star with a featureless spectrum removes the contribution of this effect to the first order, allowing the [SIII] lines to be detected. Our results are based on both divided and subtracted spectra.

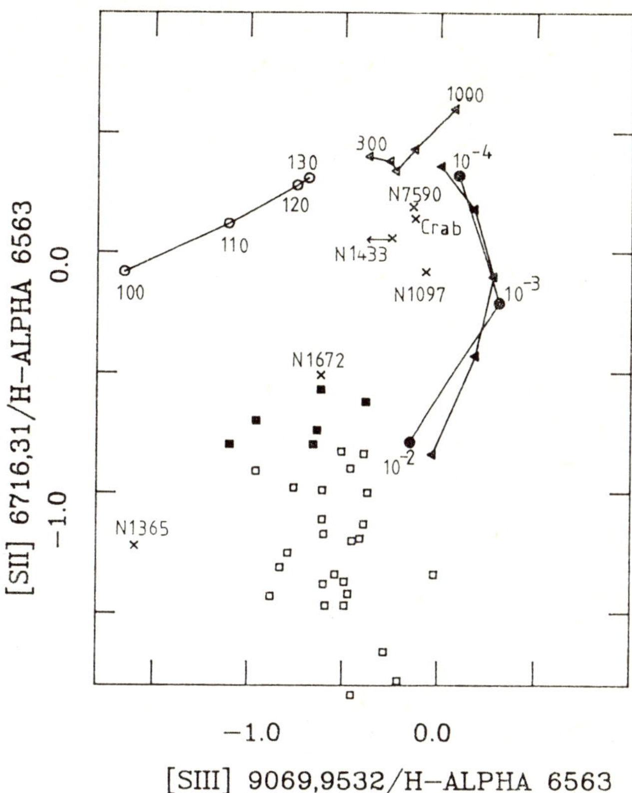

Fig. 2.- Logarithmic line intensity ratios [SII]/H_α vs [SIII]/H_α. Details of the figure are given in the text. The models of photoionisation by a young star cluster correspond to a black body of Teff=200,000 K .

4. Results and conclusions

Figures 2 and 3 show the position of the observed objects on the excitation diagrams [SII]/H_α vs [SIII]/H_α and [SII]/[SIII] vs [OII]/[OIII], together with some theoretical models and the available data on HII regions in the Galaxy and the Magellanic Clouds. The theoretical models are shown as symbols connected by straight lines. Open circles, open triangles, filled circles and filled triangles correspond to "slow shock" models (Dopita et al. 1984), "fast shock" models (Binette et al.,1984), power-law photoionisation models (Stasinska, 1984) and models of photoionisation by a young star cluster (Terlevich, 1984) respectively. Shock models are labeled according to their shock velocities and photoionisation models according to their ionisation parameter. All the theoretical models shown correspond to solar abundances. Our data are shown as filled squares for the nuclear HII regions and crosses for the actual nuclei. Individual objects are labeled. Open squares correspond to normal HII regions. The position of a bright filament in the Crab Nebula studied by Dennefeld and Péquignot (1983) and well fitted by photoionisation models is also shown.

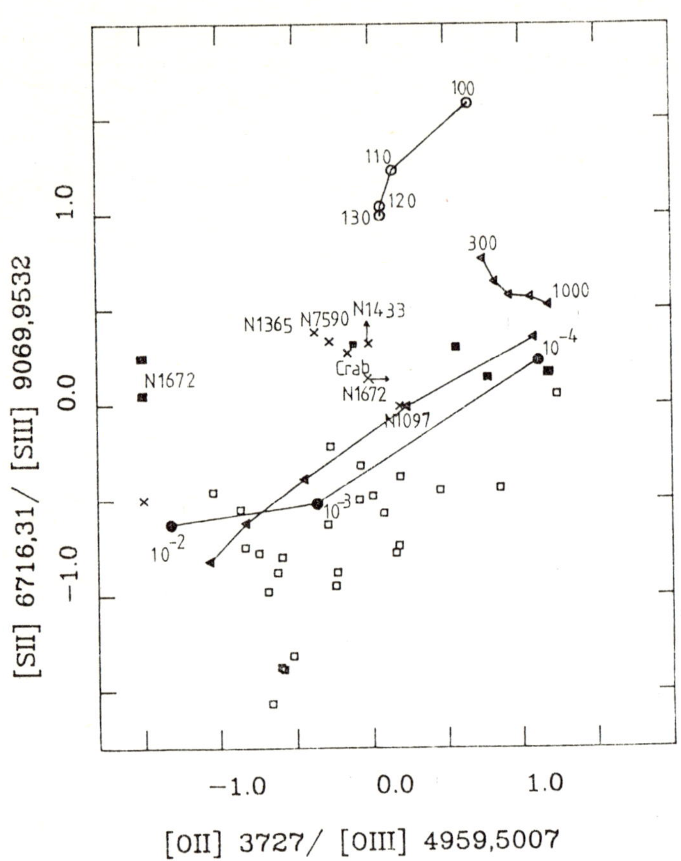

Fig. 3.- Same as Fig. 2 but for [SII]/[SIII] vs [OII]/[OIII].

Perhaps the most remarkable feature of the [SII]/H_α vs [SIII]/H_α diagram (fig. 2) is the clear separation between HII regions and Liner nuclei, except in the case of NGC 1672. This object has a much bluer continuum than the rest of the Liners in our sample and shows strong Balmer absorption features. Bearing in mind the necessary oversimplification of the theoretical models, as evidenced by the position in the diagram of the Crab filament, most of our objects are consistent with photoionisation, although contributions from shocks cannot be ruled out. However, the second diagnostic diagram [SII]/[SIII] vs [OII]/[OIII], (fig. 3), although it does not allow the distinction between different types of photoionisation models (normal hot stars, power-law continuum or young star cluster), seems to be a good discriminant between them and shock models. In fact, all our nuclei lie far from the shock heating zone in this diagram and so does the Crab filament. Thus, we think it is possible to discriminate between shocks and photoionisation if both diagrams are used simultaneously.

Nevertheless, Liners are far from being a homogeneous class of objects. If we assume that several excitation mechanisms are actually at work in Liners, the positions of NGC 1672 and NGC 1433 in the diagnostic diagrams may reflect different relative contributions, a black body contribution being significant for the first and shock heating being important for the second. On the other hand, if the pseudo-activity of Liners is linked just to star formation proccesses, they might as well represent the results of different evolution patterns of the region surrounding the star burst. The evolutionary sequence proposed by Terlevich and Melnick (1984) predicts a large burst of star formation to produce an HII region which would subsequently evolve to a Seyfert 2 type nucleus and later to a blue Liner (as NGC 1672 would be classified), whereas red Liners (e.g. NGC 1433, NGC 1097) would be the result of a smaller star burst.

Acknowledgements

We would like to thank R.J. Terlevich for providing the results of his models in advance of publication.

REFERENCES

Baldwin, J.A., Phillips, M.M. and Terlevich, R.,1981, Pub.
        astr. Soc. Pacific, **93**, 5
Binette, L., Dopita, M.A. and Tuohy, I.R., 1984, Astrophys. J.,
        submitted
Carswell, R.F., Baldwin, J.A., Atwood, B. Phillips, M.M., 1984,
        preprint
Dennefeld, M. and Péquignot, D., 1983, Astron. Astrophys.,
        127, 42
Dopita, M.A., Binette, L., D'Odorico, S. and Benvenuti, P., 1984
        Astrophys. J., **276**, 653
Ferland, G.J. and Netzer, H., 1983, Astrophys. J.,**264**, 105
Heckman, T.M., 1980, Astr. Astrophys., **87**, 152
Keel, W.C., 1983, Astrophys. J. Suppl., **52**, 229
Pagel, B.E.J., 1983, in Formation and Evolution of Galaxies
        and Large Structure in the Universe, Eds. J. Audouze
        and J.T.T. Van, Reidel Pub., p 437
Penston, M.V., 1982, Proceedings of the Third European IUE
        Conference,p. 69
Stasińska, G., 1984, preprint
Stauffer, J.R., 1982, Astrophys. J., **262**, 66
Terlevich, R., 1984, private communication
Terlevich, R. and Melnick, J., 1984, preprint
Weedman, D.W., 1977, Ann. Rev. Astr. Ap., **15**, 69

The Velocity-Field in the Seyfert Galaxy NGC 5643

S. L. Morris and M. J. Ward, Institute of Astronomy, Madingley Rd.,
Cambridge.

A. S. Wilson, Astronomy Program, University of Maryland.

K. Taylor, Royal Greenwich Observatory, Herstmonceux Castle, Hailsham,
Sussex, BN27 1RP.

SUMMARY

We show new TAURUS and VLA data on the Seyfert 2 galaxy NGC 5643,
deriving the detailed velocity-field of the galaxy, and the extended
narrow line emission, also discussing the relation between the optical
and radio emitting plasmas.

Introduction

   The observations described in this paper are part of a programme to
study the environment of active galactic nuclei. The aims of this
collaboration are the following:

   (i)   to study the velocity-field of the parent galaxy of the active
         nucleus so as to investigate the 'feeding' mechanism of the
         central engine driving the activity, and to look for systematic
         differences between 'active' and 'normal' spiral galaxies,

   (ii) to see how the galaxy velocity-field joins onto the narrow line
         region (NLR) velocity-field,

   (iii) to compare the optical morphology and kinematics with the radio
         morphology in order to study the relationship between the
         optical and radio emitting plasmas.

   The galaxy we will be discussing in the paper is NGC 5643, a southern
Seyfert 2 galaxy. The parent galaxy is a nearly face-on (inclination 27°
$\pm 5^{\circ}$, de Vaucouleurs et al. (1976), (RC2)), barred spiral, with a clear
dust lane running along the leading edge of the bar to the east of the
nucleus. It is at a distance of 24 Mpc (H_{o} = 50 km s^{-1} Mpc^{-1}), with no
visible near neighbours on the SERC J survey plate. Its HI 21 cm
line-profile shows the typical double-horned structure expected of a
rotating disk (Reif et al. (1982)). In the nuclear regions, there is
extended high-excitation emission over ~ 1.8 kpc (Whittle (1982)).
Finally, it is also a low luminosity X-ray source (3.10^{40} ergs s^{-1} (0.2
- 4 kev) (Phillips et al. (1983)).

Observations and Reduction

The observations consist of TAURUS (Atherton et al. (1982)) data cubes centred on the Hα and [OIII] λ5007 emission lines. These cubes are seeing-limited Fabry-Perot line-profile maps, covering a 4' x 4' field (28 kpc at the distance of the galaxy) with a velocity resolution of 20 km s^{-1}. The IPCS was used as the detector giving a 50 μm pixel size corresponding to 2" or 240 pc at the distance of the galaxy. The [OIII] data was obtained during poor seeing conditions, but gives results consistent with the Hα, and will not be discussed here.

We also obtained VLA radio maps of the object at both 6 and 20 cm wavelengths with a resolution of 1.3" and 4.2" respectively. A roughly circular beam profile was obtained for this southern object by using the A/B hybrid configuration of the VLA.

The TAURUS data was used to obtain a map of the velocity-field of the gas in the galaxy, fitting gaussian line-profiles to any of the 200 x 200 spectra with sufficient S/N. The resulting maps of gaussian height, position, width and background-level were then used in the following analysis.

In order to study the deviations from simple, circular flow, in the plane of the galaxy, it is necessary to first derive a model for the axisymmetric velocity-field, and then to look at the deviations from this model. A rotation curve was derived from the Hα velocity map, using the method described by Warner et al. (1973). The most uncertain parameter in the fit is the inclination, which could not be determined from the velocity map. The value given in RC2 leads to a maximum rotational velocity rather high for the Tully-Fisher relation given in Bottinelli et al. (1980). Use of this relation, with the observed HI velocity width (208 km s^{-1}, Reif et al. (1982)) and the RC2 absolute blue magnitude, suggests a very large inclination. This problem could be due to contamination of the absolute magnitude by light from the Seyfert nucleus. In order to check the RC2 inclination, a 60 minute IIIaF plate, taken at the AAT by Hanes and Malin was scanned on the APM (Kibblewhite et al. (1984)). The inclination, estimated from faint isophotal contours at major-axis diameters of 31, 38 and 42 kpc, is $23° \pm 3°$, consistent with that in RC2. This value has been used in the following analysis. The derived rotation curve is shown in figure 1. For reasons that will be discussed later, only the eastern half of the bar emission was included in the derived curve. For the region from 3.5 to 10 kpc, in which each radial bin contained 10 or more data points, the error estimate is as shown for a representative point in the centre of the figure. For the other points, the error is uncertain.

A simple model for the observed rotation curve was then fitted, consisting of a solid-body rotation out to the turnover point in the data, and then a power law:

$$V(r) = k \, (r / r_o)^{\alpha}.$$

The parameters found for the fit are: $\alpha = 0.39 \pm 0.1$, $k = 242 \pm 11$ km s^{-1}, $r_o = 6.9$ kpc, and the turnover radius was 1.5 ± 0.3 kpc. These led to the model rotation curve shown in figure 1. This model curve was then used to derive a model velocity map, which was subtracted from the observed map in order to study the deviations.

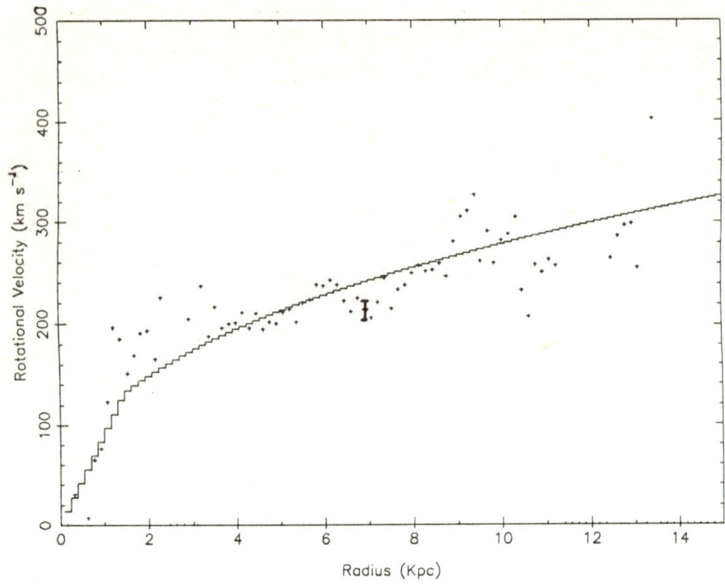

Figure 1 Observed and Model Rotation Curve for NGC 5643.

Discussion

The velocity deviations seen in the galaxy as a whole must be treated
with some caution, due to the uncertainty in the inclination – which
could produce spurious, large-scale red or blue-shifts, but there is a
residual blue-shift of the gas ahead of the end of the bar to the east,
and a similar red-shift ahead of the end of the bar to the west. This
shows up in the observed rotation curve as an excess over the model
curve in the region 2 – 4 kpc.

Independent of the value used for the inclination, there is evidence
for a systematic change from net red-shift to net blue-shift as one
scans from north to south, across the eastern limb of the bar, in the
extended high-excitation gas.

Both of the above phenomena fit in well with current models of gas
flow in barred spiral galaxies, eg Huntley (1980), in which, in the rest
frame of the driving bar potential, gas overtakes the bar from behind at
a fairly obtuse angle, experiences a shock at the leading edge, and then
flows almost parallel to the bar towards the centre. In such a model,
due to the specific orientation of this galaxy, one would also expect
the region just ahead of the end of the bar to show the maximum
projected radial velocity. In broad band pictures of NGC 5643, the
position of the dust lane, which is commonly believed to mark the
position of a shock front, is consistent with this shock lying along the
region of extended line emission, explaining the systematic trend from
north to south.

To move onto the details of the nuclear extended emission, the region
of maximum line-width in the map is also where large deviations from the
predicted radial velocities occur. This region is displaced to the

south-west of the optical nucleus by ~ 3'' (350 pc).
The gas in this region is blue-shifted by ~ 100 km s^{-1}. Figure 2 shows
the morphology of the Hα line and continuum emission (produced from the
TAURUS data cube), and the region of broad, blue-shifted emission is
shown hatched. Also marked are typical line-widths for the various
regions on the map.

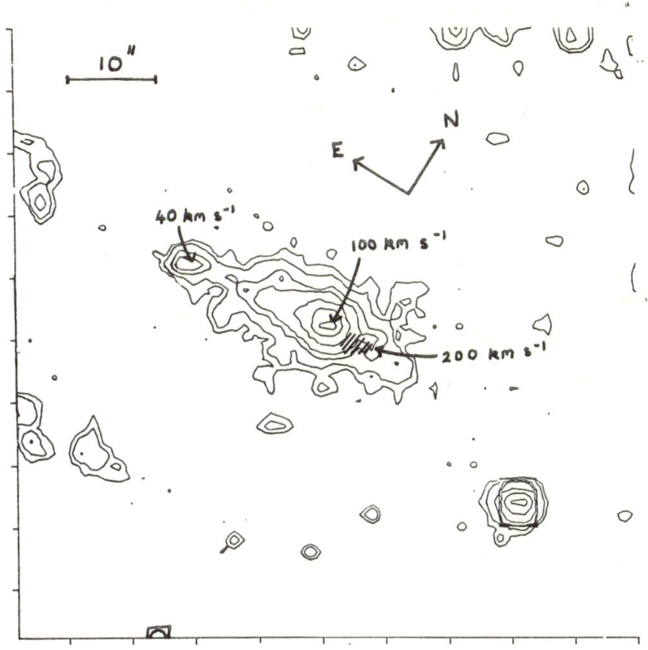

Figure 2 Hα Intensity Map of NGC 5643, result of summing TAURUS cube
in the wavelength direction. Contours 400, 500, 750, 1000, 2000, 5000
10000 counts.

To investigate whether this region of blue-shifted gas is the actual
Seyfert nucleus, obscured by dust, we obtained radio maps of the object.
The 6 cm map obtained (reconstructed with a ~ 4'' beam to show more
extended structure) is shown as figure 3. On it are marked the optical
nucleus, the Hα 'blob' to the east of the extended emission, and the
region of broad, blue-shifted gas. As can be seen, the radio peak
position is consistent with that of the optical. It should also be
mentioned that the galaxy kinematic centre, derived from the rotation
curve fitting, is consistent with the optical peak emission. The radio
structure can be interpreted as a central core, with two lobes extended
at position angle (PA) 87° (± 3°) separated by 17'' (2 kpc). The
spectral indices derived are 0.3 for the core, and 0.6 - 0.7 for the two
lobes (S $\propto \nu^{-\alpha}$). An upper limit to the polarisation of the core
component is ≤ 5 %.

Other position angles to consider are the following:
bar: 90° ± 10°,
projected rotation axis: 46° ± 2°,
optical, core to eastern 'blob': 95° ± 3°.

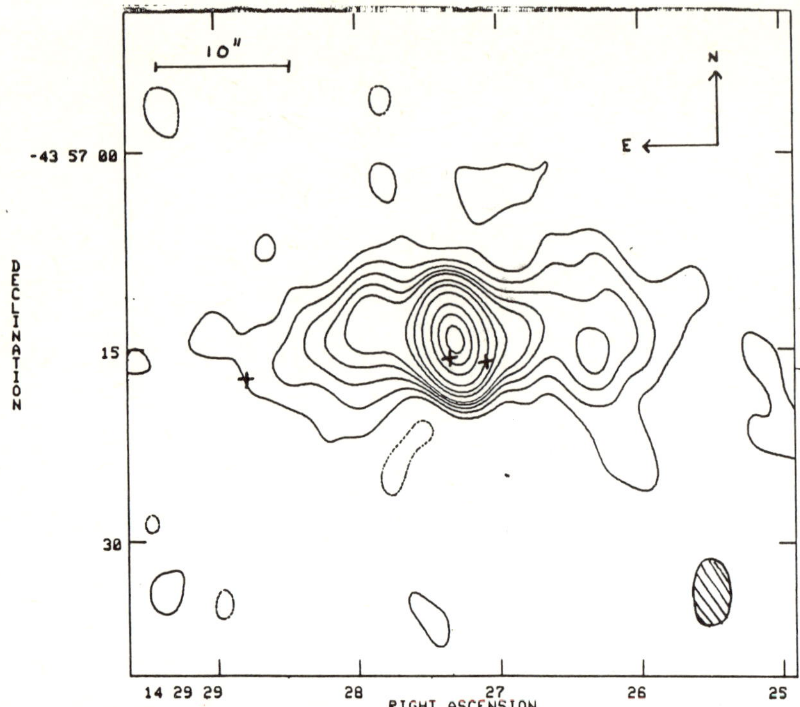

Figure 3 6 cm VLA Map of NGC 5643, beam 5.3 by 3.0 arcseconds. Contours -1, 1, 2, 3, 4, 5, 7.5, 10, 15, 25, 35, 45 times 1.734 10^{-3} Jy/Beam Area.

It now remains to find a consistent scenario to explain the above data. A very simple, and qualitative, attempt is the following. The gas in the galaxy experiences a shock at the leading edge of the bar, at the position shown by the dust lane, and flows down the bar towards the nucleus. This gas then forms a disk, orthogonal to the bar major-axis, as discussed by Durisen et al. (1983). This disk can then both collimate the radio emission, and also obscure the central non-thermal emission to all of the cool gas, apart from that lying along the bar axis. It thus explains both the radio and optical morphology to the east, but not, unfortunately, the blue-shifted gas to the south-west of the nucleus, which seems to require some further ingredient in the recipe. It is of interest that this region also shows an increased $H\alpha$ / $H\beta$ ratio (Whittle (1982)), suggesting increased dust extinction.

The lack of detailed correlation between the radio and optical morphology, except for the common PA of extension, argues against any direct excitation of the optical emission by shocks in the radio 'jets', or the optical velocity-field being driven by these 'jets'.

Acknowledgements

Thanks are due to Mark Whittle, Steve Unger and Alan Pedlar for useful discussion, and also the staff and astronomers at the AAT for their usual high quality assistance with the TAURUS observations. Richard Hook and Joss Bland provided advice and assistance with the

TAURUS data reduction. Dave King provided astrometric positions for stars near NGC 5643, in advance of publication, to allow alignment of the optical and radio data. We are grateful to Dave Hanes and Dave Malin for providing the AAT plate of NGC 5643, and to the APM group, and especially Pete Bunclark, for assistance with scanning this plate on the APM.

References

Atherton, P. D., Taylor, K., Pike, C. D., Harmer, C. F., Parker, N. M. and Hook, R. N., (1982), M. N. R. A. S., 201, 661.

Bottenelli, L., Gougenheim, L., Paturel, G. and de Vaucouleurs, G., (1980), Ap. J., 242, L153.

Durisen, R. H., Tohline, J. E., Burns, J. H. and Dobrovolskis, A. R., (1983), Ap. J., 264, 392.

Huntley, J. M., (1980), Ap. J., 238, 524.

Kibblewhite, E. J., Bridgeland, M. T., Bunclark, P. and Irwin, M. (1984), 'Proceedings of the Astronomical Microdensitometry Conference' (To be Published).

Phillips, M. M., Charles, P. A. and Baldwin, J. A., (1983), Ap. J., 266, 485.

Reif, K., Mebold, U., Goss, W., van Woerden, H. and Seigman, B., (1982), Astron. and Astrop. Supp., 50, 451.

de Vaucouleurs, G., de Vaucouleurs, A. and Corwin, H. G., (1976), "Second Reference Catalogue of Bright Galaxies", Austin, University of Texas, (RC2).

Warner, P. J., Wright, M. C. H. and Baldwin, J. E., (1973), M. N. R. A. S., 163, 163.

Whittle, D. M., (1982), Ph. D. Thesis, Cambridge University.

An Occultation of the Inner Seyfert Nucleus of NGC 4151?

J. Meaburn[1], H. Ohtani[1,2] and C. D. Goudis[3]

1. Department of Astronomy, The University, Manchester M13 9PL, England
2. Kyoto Observatory, University of Kyoto, Kyoto, Japan
3. St. Andrew's University, Patras, Greece

SUMMARY

The brightness of the unresolved nucleus of NGC 4151 was monitored over five months in 1983. Variations of $\cong$ 0.1 mag/day were observed in the U-band and no significant variation was found of the [OIII] 5007 $\overset{\circ}{A}$ emission line. However, an event that was observed on the nights of the 10/11 and 11/12 February 1983 in the continuum around 5672 $\overset{\circ}{A}$ has all the characteristics of an occultation. It is proposed that an inner synchrotron nucleus of $\lesssim$ 3 a.u. diameter was occulted by an opaque cloud $\cong$ 6 a.u. across on those two nights.

INTRODUCTION

NGC 4151 is a galaxy with a type 1 Seyfert nucleus (Khachikian and Weedman 1974) and has unusual importance because of its close proximity for 1 arcsec corresponds to 100 pc over the face of the galaxy for $H = 50$ km s^{-1} Mpc^{-1}.

Its bright nucleus is unresolved at a resolution of 0.08 arcsec (Schwarzchild 1973) and in the U-band varies within a range from 11 – 12.5 magnitudes (Penston et al 1971; Lyuti 1973 and 1977). Indeed flares are observed in this wavelength domain that rise in 10 days and decline over the next 20. These must originate then in regions that are $\lesssim$ 0.01 pc across.

Perola et al (1982) demonstrated that the variations in the ultraviolet continuum are well correlated with those in the optical continuum but that the X-ray emissions vary on a much more rapid time scale. Moreover, ultraviolet flares are followed after 10 – 15 days by enhancements in the brightness of the broad emission lines (Ulrich 1983). The broad emission lines are from ions which can emit where electron densities N_e are $\gtrsim 10^9$ cm^{-3}. These are then from more compact volumes than the regions emitting the narrow [OIII] 5007 $\overset{\circ}{A}$ forbidden line which becomes collisionally de-excited with $N_e \gtrsim 10^7$ cm^{-3}.

The aim of the present work was to observe the brightnesses of the unresolved nucleus of NGC 4151 in the U-band (3900 $\overset{\circ}{A}$ to 3400 $\overset{\circ}{A}$) in the [OIII] 5007 $\overset{\circ}{A}$ line and in a region of continuum, devoid of emission lines, 70 $\overset{\circ}{A}$ wide, at 5672 $\overset{\circ}{A}$. It was hoped that this sequence of measurements would be achieved on as many consecutive nights as possible within the period Feb – June 1983 (inclusive) so that the most rapid variations would be revealed. Incidentally, the continuum at 5672 $\overset{\circ}{A}$ was only monitored to correct the contamination through the [OIII] filter. As it happened, the most interesting phenomenum occurred in this unlikely domain.

OBSERVATIONS AND RESULTS

The observations in February, April and June 1983 were with a single beam photometer (Meaburn 1984) on the 1.9-m Kottamia telescope (Egypt) and in March (1983) on the 1.22-m Kryonerian telescope (Greece) with a similar device (Goudis and Meaburn 1973). Both photometers have EMI 9862 thermocooled photomultipliers as detectors.

The observing apertures were set at 13 arcsec after calibration against double stars to match those used by previous workers. The standard Star 2 of Penston et al (1971), which is only $\cong$ 2 arcmin from the nucleus of NGC 4151 and of similar brightness, was used throughout as a comparison. Star 5 was observed less frequently to check for any variability of Star 2 (none was found and the V, B-V and U-B values for both stars 2 and 5 were found to be within $\pm$ 0.01 magnitudes of those found by Penston et al 1971).

Standard U, B and V band filters were used as well as one of 21 $\overset{o}{A}$ bandwidth centred on [OIII] (Doppler shifted from 5007 $\overset{o}{A}$) and one of 70 $\overset{o}{A}$ bandwidth centred on 5672 $\overset{o}{A}$. The results of these measurements for the unresolved nucleus of NGC 4151 are given in Figures 1 a - e and for reference Star 5 in Figures 1 f - h. Filled circles are for the mean of many measurements (from 2 - 10) throughout one night, whereas open circles are where only one measurement was obtained. Error bars are only shown if the uncertainties, derived from the standard deviation of the means, exceed $\pm$ 0.01 mag.

The relative magnitudes are within the Kottamia system with respect to reference Star 2. The U, B and V magnitudes within the standard system are given on the right hand coordinate.

The open squares in Figure 1b are for the [OIII] measurements after the small contaminations by continuum (estimated from the results in Figure 1c) have been subtracted.

DISCUSSION

In Figure 1a the variations of the U-band within a few days are apparent. Diminutions in brightness of $\lesssim$ 0.1 mag/day were observed in February, April and March while brightenings at a similar rate occurred in March. These are part of the mean flares recognised by Lyuti (1977) which become bright by 0.5 to 1.0 mag in 10 days and then decline in the following 20 - 50 days. The changes in the brightness of the U-band are always greater than those in the B and V bands for these events (Lyuti 1977, and present results).

No significant variations in the U-band brightnesses in many continuous periods of observations throughout 5 hours were detected in the present work.

In Figure 1b it can be seen (open squares) that no significant variations in the brightness of the [OIII] line occurred within this five month period. This constancy, compared with the variability of the broad wings of Hβ, has been pointed out for other Seyfert galaxies by Andrillat and Souffrin (1968), Boksenberg and Netzer (1977) and Tohline and Osterbrock (1976).

However, a most unusual event can be seen in Figure 1c in the continuum from the nucleus of NGC 4151 around 5672 $\overset{o}{A}$. Firstly, the general trend of this light curve follows that shown in Figure 1a for the U-band but, as expected, with an amplitude of the variations being somewhat smaller. However, on the nights of 11/12 and 12/13 February 1983, a diminution of 0.2 magnitudes was observed compared with the brightnesses on both the preceding and following nights. This faintening corresponds to only a minor kink in the U-band curve in Figure 1a.

As a transient event can never be verified by repeated observation (though further similar ones can be searched for) it is most important to appreciate the true significance of the measurements on which it is based. For this purpose all the separate brightness measurements of the nucleus of NGC 4151 that were obtained in the critical nights in February are presented as simple ratios of the corresponding brightnesses of the reference Star 2.

Incidentally, they were all obtained on dark, photometric nights as NGC 4151 crossed the meridian (near the zenith) with the telescope tracking to better than 1 arcsec in 10 minutes. For each value in any one bandwidth the object was measured, followed by the reference star and then the sky background.

It can be seen in Table 1 that the individual U-band and [OIII] + CONT results are very consistent on each night which illustrates the high quality of the sky and technique. Similarly on the nights of the 10/11, 11/12 and 12/13 the 5672 $\overset{\circ}{A}$-band values are within a small range on each night with a very significant jump occurring from the 11/12 to the 12/13. The 5672 $\overset{\circ}{A}$-band was less well observed on the nights of the 8/9 and 9/10 with the very first value of 0.502 being anomalous (cirrus or bad centring?). However, the mean of these three measurements on these two nights is in keeping with the general trend of the 5672 $\overset{\circ}{A}$ brightness over a much larger period (see Figure 1c).

It is concluded that a real event has been observed on the nights of 10/11 and 11/12 February 1983 in the 5672 $\overset{\circ}{A}$-band.

TABLE 1.   The ratios of the brightnesses of the nucleus of NGC 4151 to those of reference star 2 for separate measurements

Date(Feb) 1983	Time (U.T.)	U-band	[OIII]+CONT[*]	5672 $\overset{\circ}{A}$-band
8/9	00:30	2.95	3.69	0.502
	02:09	3.10	3.68	0.600
9/10	22:31	3.00	3.75	0.583
	23:08	2.96	3.82	
	24:00	2.86		
	01:00	2.97		
10/11	22:14	2.74	3.60	0.472
	23:00	2.79	3.49	0.485
	23:50	2.80	3.51	0.480
	01:00	2.78	3.64	0.473
11/12	23:30	2.66		
	23:45	2.71	3.63	0.464
	00:00	2.62	3.62	0.454
12/13	20:56	2.61	3.76	0.540
	21:20	2.59	3.75	0.530
	21:34	2.64	3.75	0.570
	22:17	2.74	3.79	0.553
	22:44	2.61	3.74	0.570

[*] The results are uncorrected for the small contamination by continuum

The shape of this event is most unlike that of the mean flares observed in the U-band and has all the characteristics of an occultation with a decline and rise of $\stackrel{<}{\sim}$ 1 day and duration of $\cong$ 2 days.

In this case the model shown in Figure is proposed as a plausible explanation. Firstly, it is proposed that in the declining phase of a U-band and UV continuum mean flare that the U-band emission is composed largely of atomic recombination continuum emitted by a volume of $\stackrel{<}{\sim}$ 0.01 per diameter with Ne > 10^9. In this case an inner compact continuum (synchrotron?) source could have much more contrast at 5672 $\overset{\circ}{A}$ where the recombination continuum will be very low. The event shown in Figure 1c on the nights of the 10/11 and 11/12 Feb could be the consequence of the occultation of this central source by an orbiting opaque cloud. For a decline in brightness in one day and a typical cloud velocity of 5000 km s^{-1} the inner nucleus would have to be $\stackrel{<}{\sim}$ 3 a.u. across and for a duration of the occultation of 2 days the cloud would be 6 a.u. in extent.

If this interpretation is correct this sequence of observations could have yielded an accurate upper limit to the size of the accretion disk which surrounds the driving force of a Seyfert nucleus.

It would seem most important to continue these observations in an attempt to measure the shape of the declining or increasing light curve of an occultation to permit more accurate estimations for this size.

Acknowledgements

These observations were performed on the Kottamia Telescope under the Anglo-Egyptian agreement and the programme of observations of Seyfert nuclei (started in 1982) on this telescope was generously sponsored by the Science and Engineering Research Council in grants to JM.

JM and HO are grateful to the staff at the Kottamia Observatory for their assistance in 1983 and CG and JM to the staff of the Kryonerian Observatory in 1983.

JM acknowledges receipt of a Senior SERC Fellowship during this work.

REFERENCES

Andrillat Y, Souffrin S 1968 Astrophys Letters <u>1</u> 111
Boksenberg A, Netzer H 1977 Astrophys J <u>212</u> 37
Goudis CD, Meaburn J 1973 Astrophys Space Sci <u>20</u> 149
Kachikian EY, Weedman DM 1974 Astrophys J <u>192</u> 581
Lyuti VM 1973 Soviet Astron <u>16</u> 763
Lyuti VM 1977 Soviet Astron <u>21</u> 655
Meaburn J 1984 in preparation
Penston MJ, Penston MV, Sandage A 1971 Publ Astron Soc Pacific <u>83</u> 783
Perola GC, Boksenberg A, Bromage GE, Clavel J, Elvis M, Elvis A, Gondhalekar PM, Lind J, Lloyd C, Penston MV, Pettini M, Snijders MA, Tanzi EG, Tarenghi M, Ulrich MH, Warwick RS 1982 Mon Not R astr Soc <u>200</u> 293
Schwarzchild M 1973 Astrophys J <u>182</u> 357
Tohline JE, Osterbrock DE 1976 Astrophys J Letters <u>210</u> L117
Ulrich MH 1983 ESO Scientific Preprint No273 (10th Texas Symposium on Relativistic Astrophysics 1982)

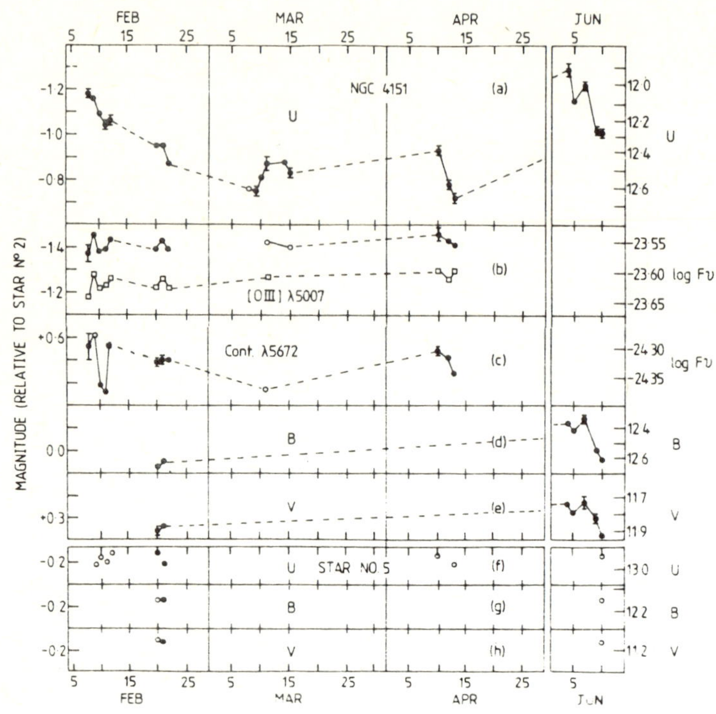

FIG 1.   Light curves for NGC 4151

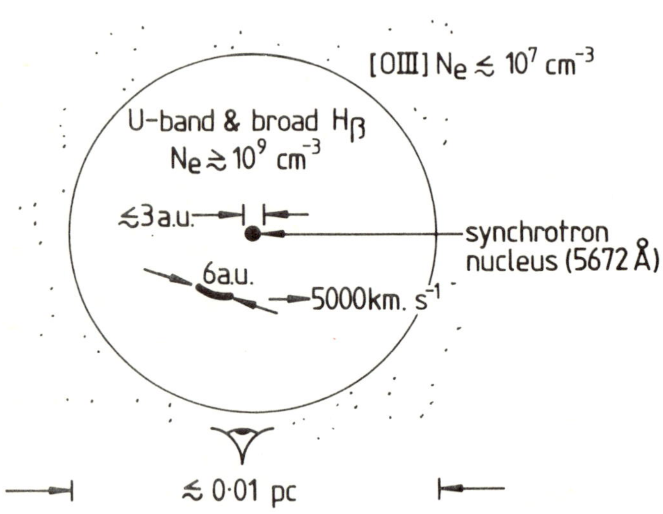

FIG 2.   A model of the nucleus

MR 2251-178: Gravitational Interaction Associated with Quasar Activity

M.A.C. Perryman

Astrophysics Division, Space Science Department of ESA, ESTEC,
Noordwijk, The Netherlands

SUMMARY

The possibility that quasars are activated by interactions between
galaxies has received attention recently following the discovery of
extended emission surrounding many low and moderate redshift quasars,
and the distorted morphological appearances of some of these extensions.
We have recently obtained narrow-band optical images of the field of
the low-redshift quasar MR 2251-178 which reveal, in addition to
nebulosity closely associated with the quasar, diffuse line-emitting
regions separated by up to 100 kpc from the nucleus but kinematically
associated with it. We attribute these regions to density perturbations
of the gaseous envelope or disc now known to rotate about this quasar,
and it appears plausible that these density enhancements have been
caused by a tidal interaction between the quasar and a nearby active
galaxy in the same cluster. Limits of between 2-4 x 10^8 yrs can be
placed on the time elapsed since this interaction. In this paper I
present the arguments leading to these conclusions, and consider some of
the implications that these observations may have on our understanding
of active galaxies.

INTRODUCTION

The idea that quasars are activated by interactions between galaxies,
especially within small clusters where stripping and virialization have
occurred more slowly than in large clusters, has been discussed by
Stockton (1982), Bothun et al. (1982), Hutchings & Campbell (1983),
Stockton & MacKenty (1983), and others.
The nearby quasar MR 2251-178 appears to provide a good example of
such an encounter. The quasar was discovered by Ricker et al. (1978) on
the basis of its X-ray emission, and has been studied since then by
Canizares et al. (1978), Ricker et al. (1979), Phillips (1980), Bergeron
et al. (1983) and Hutchings & Campbell (1983). The quasar (V = 15, z =
0.064, M = -24) is now known to have a significant nebulosity
surrounding the stellar core, weak radio emission and variable optical
emission associated with it, and to lie at the edge of a small irregular
cluster of galaxies.
Detailed long-slit observations of the quasar and its surroundings by
Bergeron et al. (1983) resulted in the discovery of a very large halo of
ionised material surrounding the quasar, traced by the presence of
[O III] 4959/5007 Å emission out to distances of some 200 kpc (H = 50
km/s/Mpc) from the nucleus. This gas was postulated to surround the

quasar in the form of a disc (inferred from the energy balance), and has a rotation curve showing a central velocity gradient of 17 km/s/kpc out to 6 kpc and a small dip near 10 kpc, rising to a broad maximum of about 180 km/s at about 120 kpc to the SE, and reaching a value of about -60 km/s at >25 kpc to the NW. A mass of about 10^{12} $M_\odot$ can then be inferred to lie within a radius of 200 kpc of the quasar.

Observations in [O III] and Hß, made in order to determine the morphology of the line-emitting gas, are described by di Serego, Perryman & Macchetto (1984), and the smoothed [O III]-continuum image is reproduced in Fig. 1. Vertical lines at scan lines 128 and 384, and the series of concentric, roughly parabolic rings visible at the left edge are a result of uncorrected detector imperfections. The Hß-continuum

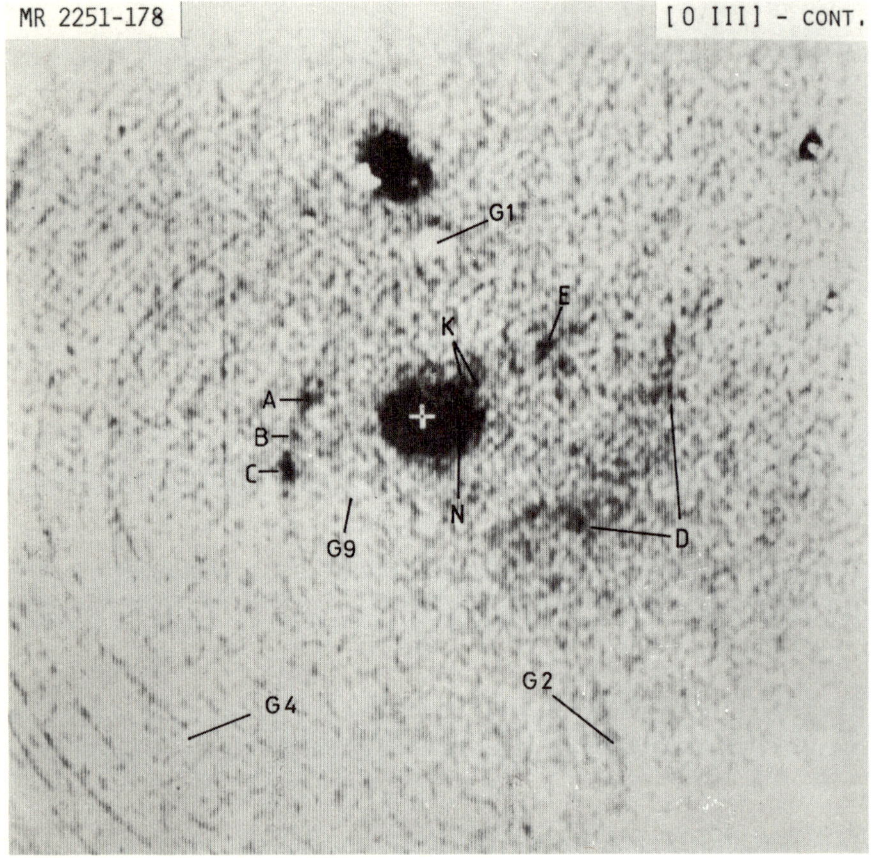

FIG. 1 The field of MR 2251-178 seen as the difference between exposures in redshifted [O III] and scaled line-free continuum. The field is 3.2 x 3.2 arcmin^2, with north to the top and east to the left. The position of galaxies, according to the notation of Phillips (1980) and Bergeron et al. (1983), are marked.

image is quite flat and has not been reproduced here, while the
[O III]-continuum image shows the filamentary nebulosity associated with
the nucleus noted by previous workers (the centre of the nucleus is
marked by a cross), along with regions of extended [O III] emission
further from the nucleus - three prominent regions (A-C) about 25 arcsec
(40 kpc) to the east of the quasar nucleus together forming an 'arm'
some 18 arcsec (30 kpc) in extent, a more diffuse region (D) extending
from 40 arcsec (70 kpc) south-west of the nucleus to some 60 arcsec (100
kpc) to the west of it, and another prominent region (E) some 30 arcsec
(50 kpc) to the north-west of the nucleus.

THE LINE EMITTING FEATURES IN MR 2251-178

In Di Serego, Perryman & Macchetto (1984), we argue that the
inclination angle of the disc structure is likely to lie in the range
$i = 30°-50°$, based upon subjective measurements of the axial ratio of
the extended emitting regions assuming circular outer structure. The
rotation curve derived by Bergeron et al. and corrected for in this way
bears a quantitative similarity to the rotation curves for spiral
galaxies. The maximum diameter of the emitting structure would be some
140 kpc from region C to the westernmost extension of region D - very
large compared with the optical diameters of normal spiral galaxies,
although comparable to the scales of extended H I envelopes known to
exist around some nearby spirals (e.g. Huchtmeier & Richter 1982), and
presumably not inconsistent with the sizes of the giant gaseous halos or
discs inferred to envelop some quasars (e.g. Shaver & Robertson 1983).

Bergeron et al. were able to show that the total [O III] luminosity
and the high-excitation conditions generally prevailing throughout the
extended halo surrounding MR 2251-178 are consistent with
photoionisation by the continuum radiation from the active nucleus. If
we assume a line-of-sight depth of regions A-C equal to their projected
length, the power-law continuum from the quasar nucleus can be shown to
be still capable of supporting the observed [O III] fluxes. On this
assumption, and assuming normal heavy-element abundances, the observed
[O III]/Hß ratios for regions A-C (derived on the basis of rather
marginal detections of Hß) indicate densities in the range
$n = 1-5$ cm^{-3} on the basis of the standard photoionisation models. On
the other hand the linear sizes and separations of regions A-C, their
[O III] luminosities of $5-11 \times 10^{40}$ erg/s, and their [O III]/Hß
ratios are similar to the expected properties of giant H II regions
ionised by hot stars (e.g. Elmegreen & Elmegreen 1983).

Whether or not star formation is occuring within regions A-C is not
central to the question of the origin of the inhomogeneities in the
line-emitting gas, and further observations could certainly clarify this
point. Rather, we wish to stress the evidence in favour of a recent
gravitational interaction between the quasar and galaxy G1, which lies
some 40 arcsec (75 kpc) to the north of it, having occured:
(1) Numerical simulations of galaxy-galaxy interactions by Toomre &
Toomre (1972), Icke (1984) and others, suggest that tidal disturbances
dissipate on rather short timescales. Galaxy G1 has the smallest
projected separation from MR 2251-178, and a radial velocity difference
G1-quasar of +1250 km/s;
(2) Galaxy G1 has both extended radio emission associated with it
(Ricker et al. 1978), and strong low-ionisation optical emission lines
(Bergeron et al. 1983), and therefore is an active galaxy in its own
right;
(3) G1 lies on an extrapolated continuation of the arc formed by

regions A-C;

(4) The models of Icke (1984) have allowed us to estimate velocity
perturbations in the gaseous quasar envelope induced by a prograde,
co-planar interaction between a disc system and a galaxy of comparable
mass, in an attempt to predict the effects of such an encounter. In
these models the perturbations in the gaseous envelope are approximated
by v(pert) = 4π gv $(R/b)^3$ where v is the circular velocity in the
perturbed galaxy, b is the impact parameter, R is the characteristic
radius of the galaxy at which $M(R) = 0.5$ $M(\infty)$, and the dimensionless
constant g = 0.39. We have taken a lower limit of R = 20 kpc,
v = 200 km/s, and arbitrarily taken b = 50 kpc, giving v(pert)=60 km/s,
considerably larger than the expected local sound speed. We have no
satisfactory way of estimating b, but we conclude that shocks may have
occured in the gaseous disc assuming plausible parameters for an
interaction between the quasar and galaxy G1. It should be noted that
the geometry of the system satisfies Icke's model requirements;

(5) Interaction models frequently predict countertidal features on
the far sides of the interacting nuclei, and this could explain the
diffuse emitting regions seen to the SW of the nucleus;

(6) On the basis of the ionisation conditions within galaxy G1 we
have derived a lower limit to the time elapsed since the proposed
interaction of 2×10^8 yr, and on the basis of the assumed cluster size,
an upper limit to this time of 4×10^8 yr. A recent burst of star
formation within galaxy G1 associated with the tidal encounter (e.g.
Larson & Tinsley 1978) may be consistent with this elapsed time.

TIDAL ENCOUNTERS AND NUCLEAR ACTIVITY

Precisely how such an encounter could initiate the quasar activity is
not well understood, it being rather unlikely that an interaction of the
type considered for MR 2251-178 could produce sufficient disturbances in
the inner regions to be responsible for activating the nucleus directly.
It is important to bear in mind that, even if a large fraction of
quasars are eventually found to lie within small clusters in which
gravitational interactions have left prominent evidence, a causal
connection between such encounters and the Seyfert or quasar activity
must still be demonstrated.

A model for the flow of material into the nuclear regions of barred
spiral galaxies has been proposed by Tubbs (1982). In this model gas is
channelled into the galactic centre along the bar, following tidal
interactions and distortions in the outer regions. Further observations
could decide whether such a model could apply in the case of
MR 2251-178. We have evidence for knots and filamentary structure
extending to the NW of the nucleus itself, reminiscent of the
instability condensations seen around dominant galaxies in X-ray
clusters (e.g. Heckman 1981) although condensation of the line-emitting
gas may have been triggered by inflowing material associated with the
underlying galaxy rather than with the surrounding cluster.

Models of Seyfert activity have also considered the possibility of an
interaction between the nucleus and the associated galaxy disc, largely
inspired by the peculiar appearance of certain classical Seyferts,
especially the frequent presence of faint annular structures (Adams
1977, Heckman, Balick & Sullivan 1978, Simkin, Su & Schwartz 1980).
For MR 2251-178 the full width at zero intensity of the Balmer lines
measured by Canizares et al. (1978) of between 18 000 km/s and 23 000
km/s for Hß and Hα respectively, would be consistent with the DR
(inner disc and outer ring) morphological classification of Su & Simkin

(1980) using their reported correlation between the full-width at zero
intensity of the Balmer lines and disc structure found for Seyfert 1
galaxies. This correlation has been interpreted by Su & Simkin as an
indication that the line-broadening is related to the mechanism of
material infalling into the nucleus. The high X-ray luminosity of
MR 2251-178, Lx(2-10 keV) = 10^{45} erg/s (Ricker et al. 1979), is also
consistent with this DR structure on the basis of the correlation found
between morphological type and X-ray intensity by Su & Simkin (1980).

DISCUSSION

Whether MR 2251-178 is classified as a Seyfert galaxy or as a quasar,
further studies of this object should, because of its proximity, tell us
much more about the phenomenon of nuclear activity.

It will be of considerable interest to establish whether the type of
velocity field seen in the gas surrounding the nucleus of MR 2251-178 is
typical of that surrounding higher redshift quasars. The form of the
rotation curve of MR 2251-178, being similar in many respects to that of
nearby spiral galaxies, suggests that there are at least two distinct
components to its mass distribution - a disc (already suspected on the
basis of the inferred covering factor, and to some extent confirmed by
the [O III] morphology) and a halo. One implication of this geometrical
picture is that the spectrum of the quasar nucleus will be sensitive to
the inclinition of such a disc to the line of sight. A second concerns
the possible use of galaxy rotation curves as a standard metric. With
the luminous active nucleus photoionising and illuminating the extended
gas, such a rotation curve may be measurable at considerably larger
redshifts than that of MR 2251-178. Understanding the evolution of the
mass distribution necessary to interpret a high redshift rotation curve
will, of course, be an additional challenge.

REFERENCES

Adams TF 1977 Ap. J. Suppl. 33:19
Bergeron J, Boksenberg A, Dennefeld M, Tarenghi M 1983 Mon Not R Astr
     Soc 202:125
Bothun GD et al. 1982 Astron. J. 87:1621
Canizares CR, McClintock JE, Ricker GR 1978 Ap. J. Letters 226:L1
Elmegreen BG, Elmegreen DM 1983 Mon Not R Astr Soc 203:31
Heckman TM 1981 Ap. J. Letters 250:L59
Heckman TM, Balick B, Sullivan WT 1978 Ap. J. 224:745
Huchtmeier WK, Richter OG 1982 Astron. Astrophys. 109:331
Hutchings JB, Campbell B 1983 Nature 303:584
Icke V 1984 Preprint
Larson RB, Tinsley BM 1978 Ap. J. 219:46
Phillips MM 1980 Ap. J. Letters 236:L45
Ricker GR et al. 1978 Nature 271:35
Ricker GR et al. 1979 X-Ray Astronomy p281. Baity WA, Peterson LE (eds)
     Pergamon, Oxford
Serego S di, Perryman MAC, Macchetto F 1984 Ap. J. Submitted
Shaver PA, Robertson JG 1983 Ap. J. Letters 268:L57
Simkin SM, Su HJ, Schwarz MP 1980 Ap. J. 237:404
Stockton A 1982 Ap. J. 252:33
Stockton A, MacKenty JW 1983 Nature 305:678
Su HJ, Simkin SM 1980 Ap. J. Letters 238:L1
Toomre A, Toomre J 1972 Ap. J. 178:623
Tubbs AD 1982 Ap. J. 255:458

Optical and infrared polarization of active galactic nuclei

P.G. Martin

Canadian Institute for Theoretical Astrophysics, University of Toronto,
Toronto, Ontario, Canada M5S 1A7

SUMMARY

   Since the review by Stockman and Angel (1980) under a similar title
there have been some significant developments. Among the more striking
is extensive evidence for the alignment of the electric vector of
optical polarization with respect to other indicators of source
structure, providing basic information on the geometry deep in the
nucleus. Important to interpretations of such alignments is
identification of the mechanism which produces the polarization. There
is agreement that the high and variable polarization of the so-called
blazars is of synchrotron origin, but there is a divergence of opinion
as to whether polarization that appears to be caused by scattering
should be (or is readily) attributed to dust or electrons. This paper
discusses the basis for such interpretations, stressing both the types
of observation that can be made to distinguish between the alternatives
and the sources of confusion and ambiguity. The potential value of
considering polarized flux as well as degree of polarization is pointed
out. While it is natural to seek unifying (simplifying and satisfying)
concepts, a summary of available observations shows that the
individualities and complexities of different active galactic nuclei
have to be recognized too. It is clear that several polarization
mechanisms are to be found in practice, sometimes within the same
object.

INTRODUCTION

   What is unknown about active galactic nuclei that can be further
elucidated by studies of polarization in the optical and infrared?
Perhaps the most basic information provided is geometrical. The very
existence of polarization indicates a lack of either complete symmetry
or complete disorder, and the orientation of the electric vector, $\underline{E}$,
should reflect any basic geometrical organization, like axisymmetry or
an ordered magnetic field. Indeed, recent work has shown intriguing
alignments of $\underline{E}$ with respect to other geometrical indicators, such as
radio structure. Another area that polarization data impact is
certainly the emission mechanisms responsible for the observed
continuum; polarization can be useful in identifying and even
separating various components that might be present. Also, it has
become clear that the environment and geometry in the nucleus, including
the regions producing broad and narrow emission lines (BLR and NLR
respectively), can be probed. One potential polarizing agent in this
environment is dust, which at the same time will affect the emission
line ratios and shape of the observed continuum through its extinction

and will affect the observed infrared continuum through its thermal
radiation. Polarization is therefore involved in a number of
interconnected ways with an overall understanding of the nature of
activity in galactic nuclei.

Underlying any discussion and interpretation is of necessity an
identification of the mechanisms by which the observed polarization is
produced and modified. Herein lies some of the current difficulty in
making full use of the available observations; different mechanisms
appear to operate in different objects, and perhaps even within a single
object. While there are some controversies with respect to
interpretation, placing the field in what might be dubbed a state of
polarized flux, it is possible to look forward to their resolution by
appropriately refined observations. Therefore, this paper will address
the potential, as well as the actual, results of the techniques.

The next section is devoted to the types of observation that can be,
and often have been, done, in order to emphasize some of the principles
underlying the interpretations, and also some of the complications that
can arise. This overview benefits from hindsight, since many of the
phenomena have been observed. The subdivisions are for pedagogical
purposes, and naturally in interpreting particular active galactic
nuclei all possible data should be considered in concert. Following are
two sections, one on what actually has been accomplished, and one
briefly summarizing by extension some future prospects.

Another aspect of an overview is to provide bibliographic material,
for there is no real substitute for reading the original papers. This
aspect is made easier in this instance by earlier reviews which cover
various facets of this subject: Angel (1974), Visvanathan (1974a),
Angel and Stockman (1980) and Ulrich (1984). The references here to the
literature can be said to be least incomplete for the period since 1980,
and naturally they reflect this author's interests and ignorance.

The notation follows what has become standard (e.g. Clarke 1974).
Polarized light is described fundamentally in terms of the Stokes
parameters: total flux I, linearly polarized flux Q and U (referred to
generically as P), and circularly polarized flux V. In a source with
multiple components the Stokes parameters for the individual components
are additive. The direction of vibration of the electric vector, $\underline{E}$, of
the linearly polarized component is described by a position angle, θ.
The fractional linear polarization (P/I), called the degree of
polarization, is designated p and is often expressed in per cent;
unlike P, p must be corrected for dilution by unpolarized components (a
recurring theme).

TYPES OF OBSERVATION

<u>Surveys</u>
The most basic type of observation is establishing whether the
various classes of active galactic nuclei are characterized by
measurable polarization. A number of surveys, to be summarized below,
have been carried out. Data from these surveys provide the basis for
statistical studies (e.g. typical polarization for each class) and for
correlations (e.g. degree of polarization with respect to physical
properties of the source that might be related to the polarization
mechanism, and orientation of $\underline{E}$ with respect to geometrical features of
the source).

Variability

Polarization observations of active galactic nuclei were first motivated by the possibility of confirming the synchrotron origin of the featureless continuum. One identfying feature of a compact source would be rapid variability of polarization, in degree and/or position angle. This indeed is a characteristic of some active galactic nuclei. Subsequently it has been discovered that (non-variable) polarization produced by scattering (dust or thermal electrons) or on transmission through aligned dust can also be present. Therefore, many of the other types of observation are aimed at distinguishing between the possible origins of the polarization so that the deeper significance of the polarization in interpreting the nature of the source can be realized.

Wavelength dependence of polarization

Ideally, the most fundamental parameter, polarized flux, would be reported. However, many polarimeters yield only the degree of polarization through a differential measurement which does not require absolute measurements of flux. This is still useful if the effects of dilution by unpolarized light can be assessed. The wavelength dependence of θ (rotation of $\underline{E}$ with wavelength) is also of interest.

Some characteristics of the various polarization mechanisms do differ. A simple <u>synchrotron</u> component would have constant p and θ, and so P would have the same power-law dependence (spectral index) as I. <u>Electron scattering</u> would also produce constant p and θ; P would have the same spectrum as the unscattered source, which in some cases might be a power-law. Thus, given this signature, the alternatives of electron scattering and synchrotron polarization would generally need to be distinguished by some further considerations, such as timescale of variability, magnitude of p, and polarization of emission lines. Even these tests might not be definitive. For example, polarization of emission lines like the continuum could be taken to support electron scattering, but lack of polarization in the lines would not rule out electron scattering; it would only place restrictions on the placement of the scatterers relative to the regions emitting the continuum as opposed to the line radiation. This illustrates the obvious point that it is essential to consider as many diagnostics as possible. It must also be realized that the above-mentioned signature might not be immediately evident because of dilution by unpolarized light. If, as is likely, the amount of dilution is dependent on wavelength, p would become wavelength dependent and the net spectral shape, I, would be altered (see the subsection on contamination below). However, θ and P would remain as useful diagnostics. Contamination by light with a different polarization is the worst possibility. One clue to multiple components of polarization is a wavelength dependence of θ, but this clue could be absent if the polarizing mechanisms had the same wavelength dependence and/or shared the same geometrical orientation ($\underline{E}$s orthogonal or parallel).

Scattering by <u>dust</u> introduces further possibilities. What distinguishes dust from electron scattering is wavelength dependence in several key ingredients: the scattering cross-section, σ_s, increases to the blue, except for grains that are large compared to the wavelength for which σ_s is fairly constant; the intrinsic polarization of the scattered light, say p_s, is wavelength independent for small particles, as for electrons, but decreases as the size becomes larger than the wavelength; the phase function for small particles is also like that for electrons, but becomes increasingly forward peaked for larger particles, thus altering the effective scattering geometry. For the most general problem one should consider the matrix elements for producing scattered polarized flux, rather than σ_s', the phase function and p_s separately (Martin 1978). For this exposition call such an element M_s. What are the possible outcomes?

When grains are <u>smaller</u> than the wavelength, M_s increases to the blue, the increase in σ_s' more than compensating for any decrease in p_s. Thus, relative to I, P will be enhanced to the blue (e.g. P would be concave relative to an incident power-law); this is the signature to be sought by measuring P (even this can be confused if there is a complex incident spectrum or several components of polarization). Suppose that only p is considered. There are three possibilities. If direct light from the nucleus is so completely obscured (e.g. nucleus surrounded by a disk seen edge on) that the light is predominantly scattered, then $p \approx p_s$, which in the case of small particles is quite independent of wavelength. This result could therefore be confused with electron scattering or synchrotron polarization, unless use were made of the spectral shape of P. If on the other hand there is substantial direct light, then the scattered light will be relatively enhanced in the blue, and so p will rise to the blue. This is the striking effect usually cited as evidence for dust, but from the constraints just described it is clearly not the only possible signature of dust scattering, only a conspicuous one. The third possibility is an alteration of either of the first two by the addition of diluting unpolarized light. In a common situation in which there is relatively more dilution in the red, p will rise to the blue in both cases. This illustrates that a rise in p to the blue does not immediately signify dust scattering; first the effects of dilution must be removed.

This can be reinforced by working the analysis backwards. Suppose p is observed to rise to the blue. This might indicate dust, or be a spurious result of dilution. If removal of dilution is possible and p still rises to the blue, then dust is more clearly indicated. If p is wavelength independent, either dust scattering with no direct light, or electron scattering (with or without direct light), or synchrotron polarization are possiblities. That is why the use of the dilution-independent quantity P can be so important. Another diagnostic to check the analysis is emission line polarization, since emission lines are less strongly affected by dilution.

This does not exhaust the possible outcomes. When the grains are <u>larger</u> than the wavelength, then M_s is no longer a steeply rising function of frequency; it might even fall. This then has obvious implications for the three possibilities discussed for small grains. It is important to observe that the maximum p possible is less than for small grains or electrons, since p_s is less, especially if a range of scattering angles is considered. This fact might be used to rule out large dust particles in objects with moderately large p.

A final complication for either large or small grains is optical depth. Multiple scattering can alter the wavelength dependence of p and P beyond easy recognition.

In an optically-thin medium, θ would normally be constant with wavelength, but even in this case θ might vary if dust properties (e.g. size) depend systematically on position. If the medium has appreciable optical depth, then the effective scattering geometry might be wavelength dependent, resulting in a wavelength dependence of θ as well as an altered wavelength dependence of both p and P. Such effects are seen in thick circumstellar dust envelopes. Note, however, that an observation of a wavelength dependence of θ does not by itself signify an asymmetric optically-thick dust cloud. As mentioned above, the mixture of polarization sources can produce the same phenomenon.

If light from the source propagates through a medium of
aligned grains, polarization as well as extinction will result. This is
the phenomenon causing interstellar polarization in the Galaxy. In the
Galaxy, p has a maximum near 5500Å, but it does not follow that the
wavelength dependence of p would be the same in other galaxies unless
the aligned particles were the same (depends on size and composition).
For example, particles smaller than the wavelength produce p rising to
the blue, providing yet another alternative explanation of such an
observation. If there is non-uniform alignment coupled with a change in
particle characteristics along the line of sight, then θ might depend on
wavelength (e.g. Martin 1974).

Geometry
    The geometry is relevant to the magnitude of p and the orientation of
E, and can in principle be deduced from observations. The orientation
of E can be compared to other indicators of source structure: on a
large scale (and possibly unrelated to structure in the nucleus),
orientation of the optical image of the host galaxy, of dust lanes, and
of extended radio structure (e.g. as measured at the VLA); on a small
scale, VLBI radio structure (comparable size to the BLR and continuum
emitting regions); E of the integrated radio polarization (which for
useful analysis has to be identified with the large or small scale radio
emission). The alignments that have been discovered underline the
necessity of understanding how the basic symmetry in the sources relates
to the production of polarization.
    In the case of optically-thin synchrotron radiation, E is orthogonal
to the projection of the ordered component of the magnetic field on the
plane of the sky. Note that order is a requirement, and in the most
favorable geometry p can be as high as 70%.
    Polarization by transmission through dust depends on alignment, E
ordinarily being orthogonal to the long axis of the projected grain
profile and, in the case of magnetic alignment like in the Galaxy,
parallel to the projection of the ordered component of the magnetic
field on the plane of the sky. Normally p increases with the degree of
alignment and with optical depth. High polarization is therefore also
accompanied by high extinction.
    Scattering. Polarization by scattering depends on there being an
asymmetry of the scattering medium, the extreme being a single
off-centre cloud. For a single cloud, E is orthogonal to the projected
radius vector from the cloud to the central source. A more realistic
geometry might be an oblate spheroid (a flat disk in the extreme). In
the optically-thin case, E is parallel to the projection of the axis of
symmetry of the oblate spheroid on the plane of the sky (E can be
orthogonal to the above for large dust particles, but then the
polarization would be small.) When the optical depth is significant,
there can still be net polarization of the spheroid emission from
scattering in the "atmosphere"; E is then orthogonal to the spheroid
axis. Another distinct possibility in a flattened system is for the
optical depth to be large in the plane but small towards the poles.
This produces an anisotopic illumination, and so even if there were an
isotropic distribution of scatterers surrounding the object, the
scattered light would be polarized, again with E orthogonal to the axis
of symmetry. A spatially-resolved example of this geometry is provided
by bipolar nebulae (e.g. Schmidt, Angel and Beaver 1978).
    The magnitude of p also provides restrictions on the optical depth
(total mass) of the scatterers and/or geometry (particularly important
for high p). The intrinsic polarization of the scattered light, p_s, can
be 100% for electrons and small particles; however, in the oblate

spheroid geometry all scattering angles are not so favorable and also
some cancellation of polarization would occur, reducing the maximum
integrated p to ~20%. For large particles the intrinsic polarization is
smaller, as noted. The observed p is generally much less because of
dilution by light direct from the source and unfavorable orientation (no
polarization at all when viewed along the axis of symmetry); p is also
smaller when there is multiple scattering. One way in which p could be
enhanced is if direct unpolarized light from the nucleus is obscured,
leaving undiluted polarized scattered light (this might occur in the
anisotropic illumination model if the flattened system were viewed
nearly edge on).

In models in which the scatterers are distributed over a considerable
volume, the <u>mass of scatterers</u> required to produce sufficient scattered
light is proportional to the square of the distance from the source (it
takes more material to produce a given covering factor with a fixed
optical depth) and is inversely proportional to the scattering
cross-section per unit mass. The latter is about four orders of
magnitude larger for submicron-sized dust particles than for electrons
(the effective mass of the electron is the hydrogen mass for this
calculation), which gives an strong advantage to dust even after
allowance for heavy element abundances. Of course, dust must be able to
form and survive in the proposed environment; there is certainly at
least circumstantial evidence for dust in active galactic nuclei (see
e.g. Rudy 1984). Schmidt and Miller (1984) point out that the mass
derived from emission line strengths in either the BLR or the NLR is
inadequate to support significant electron scattering, for typical
parameters. This conclusion can be checked for individual objects for
which a more specific geometry (and scale) is proposed. If the BLR
clouds are confined by a hot diffuse gas, then the contribution of the
associated electrons to the polarization might be significant; however,
the thermal velocity dispersion of the hot electrons ($>10^4$ km s^{-1} for
$T>10^7$K) will smear out any emission lines in scattered light, and so
this model would not be appropriate for those objects in which polarized
line emission appears. (The same phenomenon blurs out the Fraunhofer
spectrum in the solar K corona.) Another location for electrons is a
"corona" surrounding the central source (accretion disk, radiation
torus, etc.); there would be no optical line emission from this dense
hot gas. Thus electron scattering cannot be ruled out a priori, and
each object should be considered on an individual basis.

The polarization of emission lines relative to the continuum is
obviously intimately tied to the relative geometrical configurations,
and is discussed separately.

<u>Circular polarization.</u> Intrinsic circular polarization is unlikely
to be important in optical non-thermal emission. However, circular
polarization can be produced by dust, through linear to circular
conversion. The required linear polarization might arise in an
independent process, or in the case of dust scattering might be produced
in the first step(s) of multiple scattering. Linear to circular
conversion by transmission through aligned grains can occur if the
incident <u>E</u> is neither orthogonal nor parallel to the direction of grain
alignment; this mechanism explains the circular polarization of the
Crab Nebula (Martin and Angel 1974). Circular polarization by
scattering also requires a breaking of axial symmetry; multiple
scattering in an asymmetrical circumstellar envelope is the proposed
explanation for circular polarization of late-type stars (e.g. Angel and
Martin 1973). The wavelength dependence of V depends on both that of P
(incident) and that of the linear-to-circular conversion efficiency.
Because the process producing V can contribute to P, the observable

quantity V/P gives only a first approximation of the wavelength
dependence of the conversion efficiency. The other observable, V/I,
depends as well on dilution.

Polarization of emission lines
   This is very useful in identifying the mechanism of polarization, or
more generally in recognizing the presence of multiple sources of
polarization. Measurements can be made using interference filters, or
preferably with a spectrograph (spectropolarimetry). It is important to
be able to distinguish between the apparent polarization at the
wavelength of the line (p_T) which includes light in the continuum, and
p_L of the line itself (the effect of the underlying continuum, with
polarization p_c, being removed by formal subtraction of Stokes
parameters).
   For a polarized <u>synchrotron</u> continuum source with no scattering, p_L=0
and p_T<p_c. The emission line spectrum would not appear in P and would
be in absorption in p. The same could be true in the case of
scattering, if the geometry were such that the lines, which come from a
more extended region than the continuum, were unpolarized. (The
scatterers are said to be interior to the BLR and NLR; a hybrid
situation, with lines from the BLR polarized similar to the continuum,
and lines from the NLR less polarized, would indicate that the
scatterers were interior to the NLR but exterior to or mixed with the
BLR, etc.)
   Detection of definite polarization in emission lines is important
because it establishes the presence of <u>scattering</u>, or transmission
through <u>aligned grains</u>. If p_L were measured for lines from the same
region, at a number of different wavelengths, then it might be possible
to distinguish between electron scattering and polarization by dust
(with complications as in the continuum). Also, because of the high
velocity dispersion, very hot electrons would smear out the scattered
emission lines.
   Emission line polarization can be presented in a number of ways. An
emission line polarized by scattering in the same amount as the
continuum (p_L=p_c=p_T) would have the same equivalent width in P as in I,
except when there is diluting unpolarized continuum, in which case it
would be larger. With such dilution, p_c<p_T<p_L. The position angles
would be the same, however; this point is generally useful in
establishing whether the continuum polarization arises from the same
mechanism. Cases in which lines are polarized with both p and θ
different than the continuum indicate a different scattering geometry,
or the presence of a polarized synchrotron continuum.
   Polarization through emission lines also provides information on the
velocity field (see below).

Complications
   Often the situation might not be straightforward. Addition of
multiple synchrotron components with different spectra can introduce a
wavelength dependence in p and θ. Complicated non-thermal sources with
curved spectra can have frequency dependent p too. Variabilty in p and
θ can be enhanced if a source with relativistic bulk motion is seen
nearly end on.
   <u>Contamination.</u> A serious contaminant in many Seyfert and radio
galaxies studied (and in some low luminosity BL Lacs) is light from
stars in the host galaxy. This contamination depends on aperture size,
seeing, relative brightness of the nucleus, and redshift. (In
spectropolarimetry, atmospheric refraction can be a further nuisance if
the nucleus is not well centred in the small slit at all

wavelengths.) Therefore, the most useful measurements are those which
yield the dilution-independent parameter P; the most common
measurements are of p. If unpolarized, this contaminating starlight
causes a wavelength dependent dilution of the nuclear source. Because
the starlight is typically redder than the nuclear light (e.g. bulge
population), the dilution is least in the blue, causing a spurious rise
in p (not P). There would be no effect on θ.

Photometric decompositions of the spectrum are not always reliable,
and in any case would usually be based on measurements through a
different aperture in different conditions of seeing. When there is
simultaneous spectrophotometry through the same aperture, techniques of
spectral decomposition can be used. Antonucci (1984) describes in
detail the fitting of a late-type galaxy spectrum plus an assumed
power-law component to the observed spectrum. This can be checked using
high resolution spectrophotometric observations including absorption
lines to establish the amount of starlight in the same aperture
(complications are that the metallicity of the stars must be known, and
that it is difficult to remove an early-type stellar component). Note,
however, that this decomposition cannot be used to correct the
polarization for dilution unless some further assumption about the
polarization of the starlight component is made. Some of the starlight
might be polarized by dust and/or electrons, but even then the
polarization might be different because of the different geometry.
Usually it is assumed to be unpolarized, but in discussing Mrk 3 Schmidt
and Miller (1984) have also entertained the possibility that the
starlight is polarized, thus preserving the evidence for dust.

An assumption about the starlight polarization can be avoided if
spectropolarimetry through stellar absorption lines is obtained. There
will be less dilution in the observed stellar absorption lines;
therefore, if the starlight is unpolarized, p (not P) will be enhanced
in the lines (p shows an inverted stellar spectrum). The amount of
contaminating light can then be assessed and removed (again it is
difficult to remove a hot stellar component). Such complete data are so
far only available for NGC 1068 (see below); good signal-to-noise in
absorption lines is difficult to obtain in the fainter galaxies. A more
useful alternative for the fainter objects (Miller 1984) may be to use
breaks or slope changes in the spectrum, since these are useful
diagnostics of the amount of starlight. These continuum features would
appear in the spectrum and would be inverted in p; however, they would
not be seen in P (unless there were also scattered starlight),
emphasizing the importance of examining both p and P.

There are other sources of contaminating light. At 3000-4000 Å there
is excess emission that has been attributed to the dense gas in the BLR;
polarization of this light would then follow that of the BLR emission
lines (see NGC 4151 below). In QSOs an apparently thermal excess
(30000 K) is redshifted into the optical window; its polarization can
be compared to that of the featureless power-law continuum. In the
infrared, thermal emission by hot dust might be present. This might
become polarized by scattering (the optical depth is much less if the
scattering is by dust). If the dust particles are aligned, the emission
is intrinsically polarized with $\underline{E}$ parallel to the projection of the long
axis of the grain profile on the plane of the sky. Unpolarized thermal
emission acts to dilute the polarized emission from other components, as
described above for starlight.

<u>Aligned dust outside the nucleus.</u> Aligned interstellar dust in the
plane of the host galaxy, well away from the nuclear region, could
polarize all nuclear components equally (at a given wavelength); there
is still a potential problem of contaminating starlight falling in the

aperture. This polarizing mechanism offers little insight into the
nature of the nucleus but has to be recognized so that unwarranted
interpretations are not made; it would also indicate significant
foreground extinction.

Finally, interstellar polarization from the Galaxy could be present.
This component of polarization will have a wavelength dependence given
by the Serkowski formula, peaking in the visual passband. The apparent
dependence of p on wavelength is altered, and a wavelength dependence of
θ can be introduced. For statistical studies, an allowance can be made
using the known average dependence of polarization (extinction) on
Galactic latitude. However, for an individual galaxy the foreground
should be measured directly by observing faint stars within a few arc
minutes of the object. Interstellar polarization is patchy. Even faint
SAO stars are not always reliable, for they might be neither
sufficiently close to the object on the sky nor sufficiently far away
(Schmidt and Miller 1984).

ACTUAL RESULTS

Surveys and variability

BL Lacs. Of the active galactic nuclei, BL Lac objects have the
largest and most variable polarization. A compilation, including other
properties of the sources, and a discussion is given by Angel and
Stockman (1980). Two major reports on monitoring are by Angel et
al. (1978) and Impey et al. (1982). Polarization as high as 43% has
been found. Strong variability on all timescales down to 4 h has been
observed, but not all objects are variable. Optical and infrared
polarization normally vary in step, although there have been a few
reported exceptions. In some objects the orientation of E, while
variable, shows a preferred direction, while in others there is random
orientation.

BL Lac itself was the subject of a multisite campaign to provide 24-h
coverage (Moore et al. 1982), and results from a subsequent campaign
have not yet been analysed (Axon 1984). The photometric and
polarimetric fluctuations have the appearance of noise, and have been
interpreted in terms of a multicomponent structure. For models of BL
Lacs in general a most challenging result is that large changes in flux
can occur with little change in p, and vice versa. This can be
considered profitably in the context of beaming models, in which a jet
with relativistic motion is seen nearly end on (e.g. Blandford and
Königl 1979). Such models are already suggested by the inferred high
brightness temperatures, and by superluminal expansion of radio
components in some objects.

QSOs. Stockman, Moore and Angel (1984) present the final results for
142 objects in the "bright QSO" survey, and mention other surveys in
progress. They find a distinction between a minority displaying high
polarization (>3%) and the rest for which <p>=0.6%. The former are
referred to as HPQs, and are further discussed by Moore and Stockman
(1981, 1984) who extended the QSO survey to include 239 objects. Of 186
radio-selected objects, 20 are HPQs. On the other hand,
optically-selected QSOs are not HPQs. (There are objects with broad
absorption lines (BALQs) of which 6 of 30 show high polarization
(Stockman, Moore and Angel 1984); the polarization of BALQs is not
highly variable and so it is suggested that they are a distinct
class.) Except for their strong emission lines, HPQs are similar to BL
Lacs. HPQs show large amplitude, rapid optical variability (optically
violent variables), steep, smooth infrared-optical continua, and strong,
compact, flat-spectrum radio sources, sometimes with low frequency

variability or superluminal motion. Measurements of a few objects
(Moore and Stockman 1981) indicate that HPQs do not have significant
optical circular polarization (V/I<0.5%).

On the other hand LPQs have low polarization (p=0-2%); they are
discussed by Stockman, Moore and Angel (1984). The polarization is
quite constant in p and θ. Contamination by interstellar polarization
from the Galaxy is present, but on average is not overwhelming; it has
been analysed only statistically. In the probability distribution of
polarization, there is no distinction between radio-selected and
optically-selected LPQs (unlike for HPQs).

Polarization of optical jets. The knots in the jet of M87 are
strongly polarized, with changes from knot to knot in the orientation of
E tracking those of the radio polarization (Schmidt, Peterson and Beaver
1978). One knot in 3C 277.3 shows similarly high polarization, with E
perpendicular to the jet (Miley et al. 1981). Roeser (this volume) has
recently reported detection of polarization in the jet of 3C 273. This
evidence for synchrotron emission away from the nucleus provides strong
evidence for in situ acceleration of the electrons.

Seyferts. Martin et al. (1983) surveyed 99 objects, 67 being of type
1; ten type 1 and four (of 32) type 2 Seyferts have p>2.5%. Thus the
success rate for finding highly polarized objects is similar to that for
QSOs (if Seyfert nuclei did not suffer dilution from the host galaxy,
the rate would be even higher). However, the polarization seems to be
qualitatively different. Among the Seyferts, only NGC 1275 shows rapid
polarization variability like the HPQs (e.g. Angel et al. 1978); this
object is in many other respects not typical of Seyferts. Data on the
(lack of) variability of the polarization of other Seyferts are provided
in Maza (1979) and the survey, and subsequent papers dealing with
spectropolarimetry. NGC 4151 is well known for its variability in p (θ
also varies, but rather than being intrinsic to the nuclear component
this might arise from different relative contamination by the
line-polarizing component as the nuclear flux and p vary). 3C 390.3 and
3C 120, which are also classified as BLRGs (see radio galaxies below),
appear to be variable.

Seyferts with low polarization are typically somewhat less polarized
than the LPQs, and contamination by Galactic interstellar polarization
is also present. However, after correction for dilution by the host
galaxy, typical Seyfert galaxies probably would have higher residual
polarization than LPQs. The measured polarization does not appear to
correlate with a number of other properties of the sources (Maza 1979);
this does not preclude searching for such correlations among the more
highly polarized objects.

Circular polarization has been searched for in a few objects with
strong linear polarization (Maza 1979 and Table 1); only in NGC 1068
was it detected (Angel et al. 1976).

Radio galaxies. Rudy et al. (1983) surveyed 13 broad line radio
galaxies (BLRGs), finding two highly polarized objects (p>7%); of
these, 3C 234 had been reported in a small survey by Antonucci (1982).
Unlike in HPQs, the polarization in these two objects is not strongly
variable. For the rest, the polarization is generally larger than in
LPQs or Seyfert 1 galaxies. Antonucci (1984) reports on a larger sample
of radio galaxies (some overlap with the above and some NLRGs). Only
3C 332 is added as a new highly polarized source (p>2.5%). Of 22
objects with polarization detections (not seriously contaminated with
Galactic interstellar polarization), about half show evidence for
variability.

Wavelength dependence of polarization

Most information concerns the continuum, since emission line measurements are either not sufficiently precise or do not span a wide range in wavelength. Except for spectropolarimetry of the brightest and most highly polarized objects, data are from broad band polarimetry.

<u>BL Lacs.</u> For these variable objects it is important to have simultaneous observations. Bailey, Hough and Axon (1983) summarize new and previous evidence comparing the near-infrared and optical polarization (also further evidence on variability). The basic result is that both p and θ are constant with wavelength, which is taken, along with the variability, as evidence for synchrotron emission. A important refinement is suggested in the data of BL Lac: at times when the polarization is high (p>10%), the optical polarization is larger (factor 1.2) than in the infrared. In other highly polarized objects, the optical polarization is also somewhat larger than the infrared. Another refinement is the presence of occasional wavelength dependence in some objects (e.g. Puschell et al. 1983); understanding this unusual behaviour is important in distinguishing between models for the variability.

The polarization of low luminosity BL Lacs can be diluted significantly by the host galaxy; if the intrinsic polarization is wavelength independent (as above), then p will rise to the blue, as observed. Maza, Martin and Angel (1978) demonstrate how such information can be useful in separating the non-thermal spectrum from the starlight, obtaining spectral indices similar to those found from other independent analyses. Because of uncertainties in the spectral shape of the starlight to be subtracted, observations of polarized flux would be preferable for establishing the spectral index of the non-thermal, polarized component; although the result is not displayed, Antonucci (1984) comments that that the polarized flux (in wavelength units) rises to the blue in Mrk 501.

<u>HPQs.</u> Moore and Stockman (1981) show that there is little wavelength dependence across the optical spectrum (9 objects), although in three objects, the polarization in the red is somewhat larger than in the blue. In these cases p<10%, and the trend is consistent with that described above for BL Lacs.

For one significantly polarized member of the distinct class of BALQs, PHL 5200, there is spectropolarimetry by Stockman, Angel and Hier (1981), who find that the polarization arises in the continuum and is not variable. Their data indicate a slight increase in polarization to the blue, while spectropolarimetry by Miller (1984) shows no wavelength dependence. Data on the emission lines is mentioned below.

<u>LPQs.</u> Two-colour observations of a subset (15 brighter and more highly polarized objects) suggest a mild wavelength dependence, with the polarization somewhat higher in the blue (Stockman, Moore and Angel 1984). The result should be considered tentative because the observational uncertainty is significant, because the effects of contamination by Galactic interstellar polarization (which might bias the data in this direction) have to be considered, and because, in constrast to this, unpublished Lick spectropolarimetry of 6 high-latitude aligned LPQs (Miller 1984) shows no wavelength dependence in p. Because of the wide range in redshift, significantly different portions of the intrinsic spectrum are sampled with such optical measurements. Thus, in carrying out statistical analyses one should not lose sight of the possibility of different components in the spectrum with potentially different polarization.

Two objects which do stand out in the Stockman, Moore and Angel list are 1004+130 and the BALQ 1246-057; they are not among those measured by Miller and are obviously worthy of further investigation.
  <u>Seyferts.</u> The coverage here is the most extensive because of the relative brightness of these objects. The types of measurement that have been made, including emission line polarization, are summarized in Table 1. The papers cited contain extensive discussions of individual objects which are not easily summarized here. Much of the interpretation hinges on the relationship of the line polarization to that of the continuum, which is the topic of the next subsection. A common, though not universal, property found in the optical is a rise in p to the blue. This has often been cited as evidence for scattering by small particles of dust (e.g. Angel and Stockman 1980). However, as mentioned above, the effects of contamination (dilution) by the host galaxy can be substantial for this class of active nucleus, producing a spurious result, and so each galaxy has to be examined on an individual basis. If there is evidence for low dilution, the issue of modification of the wavelength dependence of p is unimportant; Mrk 486 is one such case which provides good evidence for dust. The most complete data available are for <u>NGC 1068</u> (Miller and Antonucci 1983), and so that application is of fundamental interest. Dilution is very important and the starlight appears to be unpolarized (stellar absorption features appear inverted in p), accounting for much of the apparent rise in p to the blue. The deduced "intrinsic" polarization of the power-law component is high (16%) and quite independent of wavelength. The basic conclusion about the importance of dilution is supported by an independent analysis by McLean et al. (1983) which did not use absorption line information and assumed the intrinsic polarization was independent of wavelength. The latter results, however, included the possibility of reddening and low polarization of the stellar component, and thus arrived at somewhat different properties of the components. This emphasizes that the actual decompositions based on a limited portion of the optical spectrum are uncertain in detail.
  It is noted in the discussions that a wavelength-independent p is immediately suggestive of either electron scattering or a synchrotron origin, but as emphasized above this might also be interpreted as small dust particles if the observed light is largely scattered (even in an electron scattering model, absence of direct light also helps explain the high "intrinsic" polarization and the fact that extension of the observed power-law predicts too few photons to ionize the observed line-emitting gas; Miller and Antonucci 1983). To distinguish between these alternatives, observations of P are clearly important.
  The polarized flux, which is less dependent on the decompostion of the spectrum, is not shown by Miller and Antonucci (1983), but from the stated flat intrinsic polarization and the quoted power-law spectrum (spectral index -2 for frequency units), P is expected to drop to the blue (also Miller 1984). On the other hand, Ward et al. (1984) find that P is rather flat through the optical spectrum (U through I). Thus there appears to be some unresolved discordance in the basic data to be addressed in the forthcoming papers. Extending the wavelength coverage, the latter authors find a rise in P in the near-infrared from J through 10 μm (p shows a peak at K as well as in the blue). This spectral shape of P is thus concave relative to a falling power law and is suggestive of dust. The position angle changes somewhat from J to L, and then is markedly different at 10 μm. If there are several components contributing to the polarization, then interpretations based on optical data alone are obviously less certain. And if the diluting starlight is somewhat polarized, then aperture effects can influence P too.

TABLE 1. <u>Optical polarization in Seyfert and radio galaxies</u>

Object	Type	p %	σ_p %	Continuum BB	S	Lines F	S	V/P	$\sigma_{V/P}$
All Seyfert galaxies known to have p≥1.5%									
Mrk231	1	2.87 ± 0.08		TL,TM	TL,SM	TL	TL,SM	-0.010 ± 0.011	
Mrk376	1	4.38	0.14	MS	MS		MS	0.015	0.023
Mrk486	1	3.40	0.14	TM	St,SM		St,SM		
Mrk704	1	3.58	0.17	TM		TM			
Mrk1239	1	4.09	0.14						
NGC931	1	3.	0.3	SM					
NGC3227	1	1.77	0.09	TL	TL,SM	TL	TL,SM		
NGC6814	1	1.77	0.16		SM				
IC4329A	1	5.74	0.23	MS,W	MS		MS	-0.009	0.013
3C390.3	1	1.8(typical)		R		R			
ESO140G43	1	5.47	0.19	TM,W		TM			
ESO141G55	1	1.50	0.08	TM		TM			
MCG-6-30-15	1	4.65	0.13	TM,W		TM			
TOL1351-375	1	4.11	0.18	TM					
3A0557-383	1	4.63	0.17						
Mrk3	2	1.79	0.07	TL,TM	TL,SM	TL	TL,SM	-0.028	0.016
Mrk348	2	1.79	0.22	TM					
Mrk463E	2	4.18	0.16						
NGC1068	2	2.11	0.02	A,W	A,MA,MAH	A	A,MA,MAH	-0.027	0.003
NGC1275	2	3.49	0.10	M				0.003 ± 0.005	
NGC2992	2	3.32	0.18	TM			TM		
NGC5506	2	3.10	0.16	TM			TM		
NGC7674	2	1.88	0.14						
ESO185IG13	2	1.50 ± 0.30							
Low polarization Seyferts for which line polarization is measured									
Mrk509	1	1.03 ± 0.05		TM			TM		
NGC3516	1	0.79	0.06	TL	TL		TL		
NGC4151	1	1.(typical)		T	T,S	T	T,S		
NGC5548	1	0.72	0.10	TM					
NGC7582	2	1.03 ± 0.12		TM			TM		
Radio galaxies for which line polarization is measured									
3C109		7.1 ± 0.7		RS			RS		
3C120		0.9	0.15		An		An		
3C227		1.6	0.44		An		An		
3C234		10.2	0.6		An		An		
3C382		1.4	0.15		An		An		
3C445		1.4	0.2		An		An		
Mrk501		3.(typical)		MM	An	(no lines)			
1831+731		1.0 ± 0.1			An		An		

Linear polarization. BB: Broadband, F: Filter, S: Spectropolarimetry.
Circular polarization: from M, except NGC 1068 (A); there are
additional observations of Mrk 231, NGC 1068 and NGC 1275.
Bibliographic key. A: Angel et al. (1976), An: Antonucci (1984),
M: Maza (1979), MA: Miller and Antonucci (1983), MAH: McLean et
al. (1983), MM: Maza, Martin and Angel (1978), MS: Martin et
al. (1982), R: Rudy et al. (1983), RS: Rudy et al. (1984),
S: Schmidt and Miller (1980), SM: Schmidt and Miller (1984),
St: Stockman (1976), T: Thompson et al. (1979), TL: Thompson et
al. (1980), TM: Thompson and Martin (1984), W: Ward et al. (1984).

The evidence for dilution by unpolarized starlight weakens the conclusion by Angel et al. (1976) that the continuum polarization in NGC 1068 is due to scattering by dust. However, even with dilution dust is probably not ruled out. Evidence for at least some scattering comes from emission line polarization (see below), but this of course does not mean that all of the polarization arises from this component. It has to be admitted that the structure of these sources is complex.

The presence of <u>circular polarization</u> shows that some of the scattering is by dust. Angel et al. (1976) attribute the circular polarization to multiple scattering in an asymmetric dust cloud. This is difficult to model quantitatively, but it is noted that the same type of geometry might exist in circumstellar envelopes of late-type stars where polarization of the same magnitude is observed. In the interpretations in which there is high continuum linear polarization, linear to circular conversion would be associated with dust producing the relatively low ($p \approx 1\%$) linear polarization of the emission lines. The observed ellipticity, V/P, a quantity relatively independent of the details of dilution, is a few percent, falling to the blue. If the low linear polarization is produced by aligned grains, the circular polarization might be attributed to the accompanying birefringence, as advocated by McLean et al. (1983). However, the actual birefringence is probably too weak to explain the observed ellipticity. Scattering of the continuum linear polarization is the other possibility. This can be evaluated by examination of the relative sizes of two scattering matrix elements, that for production of linear polarization (the low line polarization) and that for linear to circular conversion (Martin 1978). Such a mechanism could produce the observed ellipticity; however, there are some restrictions on the properties of the dust. If the size is small, then the conversion efficiency is low, especially for non-absorbing materials; if the size is large, polarization can suffer from cancellation if there is a range of size and scattering angle; the optimum size is comparable to the wavelength. In these models the intrinsic circular polarization of the emission lines would be zero. Only $H\alpha$ has been observed and the signal-to-noise is not sufficient to be definitive.

In any case, the production of circular polarization is not trivial and the question arises whether the special geometry is unique to NGC 1068. No other Seyfert galaxy shows circular polarization (Table 1), but none has been measured to the same low level of ellipticity (except NGC 1275 which is probably irrelevant to this discussion).

Another well studied object, <u>NGC 4151</u>, illustrates some other interesting points. It is the only Seyfert for which plots of polarized flux have been published; P is consistent with a power-law in frequency with spectral index -0.35. If the starlight that is also present is unpolarized, then this can be taken as the spectral index of a synchrotron component (recall that the polarization is also variable). This flat power-law combined with late-type stellar light can explain the visible spectrum (Schmidt and Miller 1980), but the relationship of the power-law to the infrared and ultraviolet spectrum (seen with IUE) is not clear. The overall spectral index is steeper than inferred from this limited portion of the visible spectrum.

Another interesting feature of the spectrum is a broad emission hump to the blue of 4000Å. A reduction in the apparent polarization at the hump indicates that the excess emission has low polarization. If the emission is attributed to blended Balmer lines and Balmer continuum emission, then the BLR emission lines must also have low polarization; this has been measured to be the case, and so the interpretation is consistent.

In only one galaxy with appreciable polarization, NGC 6814, was foreground polarization in the Galaxy suspected (Martin et al. 1983). This has been confirmed by extensive observations by Schmidt and Miller (1984).

Radio galaxies. Data available for several objects have been interpreted in the papers cited in Table 1. Contamination by the host galaxy can again be significant, removing most evidence for polarization rising to the blue. The most highly polarized object, 3C 234, has p and θ fairly independent of wavelength; after correction for starlight, p might even rise somewhat in the red. The high polarization of 3C 109 is similarly constant. Further discussion is deferred until the end of the next section on emission line polarization.

Polarization of emission lines

No data are available for BL Lacs, which have weak lines if any. Published observations deal with the brighter Seyferts and some BLRGs (Table 1). This work offers great potential, but the richness of polarization behaviour has sometimes only added to the complexity of the interpretation by revealing the presence of multiple components causing polarization. The following subsections deal with some Seyferts; BLRGs and QSOs are mentioned last.

Permitted lines. In NGC 1068, for example, both p and θ of the permitted lines are close to the continuum values, suggesting a common origin. However, if dilution of the continuum is important, as described, then the agreement of p is just a coincidence. The agreement of θ is still interesting, and would indicate that the different polarization mechanisms reflect the same basic symmetry in the nucleus. Mrk 3 shows the same phenomenon, but there the data are not as precise and it is not known definitely that the starlight has low polarization. If, as it seems, these are coincidences, one still can ask whether there might be objects in which p is really the same for the permitted lines and the nuclear light, but where p in the continuum appears lower because of stellar dilution. ESO 140 G43 and MCG-6-30-15 show just this behaviour at Hα; so does the BLRG 3C 227.

Notable cases in which lines and continuum have the same high polarization are Mrk 231, Mrk 376, and IC 4329A. In these objects p rises steeply to the blue, which can be taken to indicate scattering by dust. Another possibility, at least for IC 4329A, is suggested by the alignment of E with the dust lane of this edge-on galaxy; transmission through aligned, small dust particles is therefore an alternative explanation.

There are also objects in which the permitted-line polarization is zero, or at least much less than the continuum (e.g. NGC 4151, NGC 5548, and Mrk 509). This does not necessarily indicate synchrotron polarization, but if a scattering interpretation is advanced the scatterers must be appropriately placed, probably well inside the BLR.

Forbidden lines. Usually just [O III] 5007Å has been measured adequately. In NGC 4151, the forbidden lines and the narrow components of the permitted lines have a small polarization distinct from the continuum. The broad component of the permitted lines appears to be unpolarized. Thus the NLR and BLR are separate regions, as in most theories.

In Mrk 3 and NGC 1068 the polarization is less than in the continuum and at a different position angle; in Mrk 3 the residual might be purely from foreground Galactic polarization. Again, in scattering models for the continuum polarization, this places the scatterers inside the NLR. The forbidden-line polarization is also different than the polarization of the permitted-line emission, which in these type 2

objects is a narrow-line component. This shows directly that there are
distinct zones having different densities even within what might be
called the NLR (Schmidt and Miller 1984). This does not contradict the
general model by Lawrence and Elvis (1982) that type 2 Seyferts are just
type 1 Seyferts in which the BLR is obscured (the distinction being
velocity dispersion, rather than density).

In IC 4329A the polarization is high and equal to the permitted-line
and continuum polarization, consistent with a model in which
polarization is produced by aligned grains well outside (and thus
probably unrelated to) the nucleus. In NGC 2992, both the permitted and
forbidden line polarizations are higher than the continuum and at a
somewhat different position angle. As in IC 4329A, $\underline{E}$ is aligned roughly
with a dust lane along the major axis of the galaxy. In fact, alignment
with large-scale optical structure is a common property (see below).

Polarized satellites. This is a curious phenomenon first seen in
NGC 1068. These appear as enhancements in p about one linewidth redward
of the Balmer lines. According to Miller and Antonucci (1983), they are
produced by low flux, high polarization satellites; Antonucci (1984)
comments that these satellites have the same high polarization as the
continuum corrected for dilution, which would argue for a scattering
origin for both. The redshift of the satellite can arise in two
components: motion of the scatterers and/or emitting gas away from the
observer, which breaks the symmetry in the nucleus, or relative motion
of the scatterer away from the gas producing this Balmer emission. If
this component of the gas has on average the motion of the continuum
source, then the scatterers would form an expanding system (not
necessarily spherically symmetric). Mrk 231 shows asymmetry in p
through H$\propto$ which also suggests expansion (Schmidt and Miller 1984). On
the other hand, the BLRG 3C 382 shows a blue "phantom" in the broad
component of H$\propto$, suggesting systematic infall, while in 3C 234 the red
and blue halves of H$\propto$ have the same polarization (Antonucci 1984).

The ingredients underlying this phenomenon are not without parallel
in a more familiar (and spatially resolved) setting. Diffuse and
filamentary light in the halo of M82 is polarized, and probably arises
from scattering of disk and nuclear light by dust (e.g. Schmidt, Angel
and Cromwell 1976). Both H$\propto$ and [N II] are polarized like the continuum
(Visvanathan 1974b), and are therefore also scattered light.
Consequently, the complex velocity field shown by the emission lines is
probably strongly affected by the motion of the scatterers (Sanders and
Balamore 1971).

Radio galaxies. Only a few of the many interesting objects will be
mentioned here; the original papers contain extensive discussion. H$\propto$
in 3C 234 shows the remarkably high polarization of 13%, close to the
continuum value in p and θ. Antonucci (1984) favours scattering by
electrons to produce this high polarization in the red. To explain the
high p here and in NGC 1068, and the perpedicular alignment discussed
below (these are the prototypes), it may be advantageous to appeal to an
edge-on geometry in which the direct nuclear light is highly obscured
and there is effectively anisotropic illumination of the scattering
medium. In that case the possibility of dust scattering can be
reintroduced and so the polarized flux should be examined carefully.
The [O III] line is unpolarized, showing the separation of the
line-emitting regions.

In the other high polarization object, 3C 109, H$\propto$ is also polarized
like the continuum; Rudy et al. (1984) favour transmission by dust,
illustrating the different interpretations that are made of similar
data.

In 3C 382, the Hα line is polarized, with p comparable to the continuum, but $\underline{E}$ orthogonal (unlike the blue phantom just mentioned); the line polarization appears to be variable.

The polarization in 1831+731 rises to the blue, providing the best case for dust scattering. An unusual feature to be confirmed is that the [O III] line is more highly polarized than the continuum at the same θ; Hβ, while probably polarized, does not show such an excess.

QSOs. Data for these fainter objects are scarce, but are becoming available (Miller 1984). In the HPQ 3C 446 the lines are unpolarized, as might be expected. Hβ polarization has been detected in a few normal QSOs. There is evidence also that the 3000Å bump is unpolarized, as in NGC 4151.

Polarization of the lines and absorption troughs is important to understanding the geometry of the unusual BALQ systems. In PHL 5200, Stockman, Angel and Hier (1981) find the S IV 1397Å emission to be significantly less polarized than the continuum (essentially unpolarized), and suggest that the other lines could be unpolarized too. Miller (1984) finds the C IV 1549Å emission to be significantly polarized. Because of ionization stratification, polarization differences from line to line are conceiveable, and would clearly be interesting to confirm.

Alignments

Optical structure. Such an analysis is available only for Seyfert galaxies. In separate discussions, several objects have been noted in which $\underline{E}$ is closely parallel to a dust lane and/or the major axis: IC 4329A, NGC 931, NGC 2992, MCG-6-30-15, Mrk 376, and Mrk 486. Ward et al. (1984) find similar preferential alignment among additional objects with p>3% (Mrk 463, Mrk 704, 3A0557-38, NGC 5506, and ESO 140 G43), but random alignment among 8 objects with lower, but definite, polarization. The high polarization group has systematically higher Balmer decrements too. This is not without complication though. For example, $\underline{E}$ swings systematically through 40° from B to H in IC 4329A (Ward et al. 1984); there is a definite swing in the optical in NGC 5506 too. Where line polarization is available, p and θ are often consistent with the continuum values, but sometimes are not (see examples above in the subsections on emission lines). All of this is suggestive of either grains aligned in the galactic disk or organization of the scatterers in the nuclear region in a manner related to the large-scale structure of the galaxy.

In NGC 1068 and NGC 4151, $\underline{E}$ of the low [O III] polarization is along the minor axis.

The strong polarization of Mrk 231 shows no preferential orientation with respect to the optical structure (Mrk 231 has a "difficult" radio map too). An unrelated, but interesting, feature of Mrk 231 is the strong Na D absorption in the spectrum. Polarization in the line provides a probe of the geometrical placement of the interstellar absorbing cloud in the nucleus.

Radio structure. This is of interest because it is believed that the structure of radio sources (double lobes and jets) is rooted in a basic axial symmetry in the nucleus, the axis being defined by the angular momentum of the central object. Optical polarization, which probes closer to the nucleus, might reveal this same symmetry.

The first striking evidence was that $\underline{E}$ in LPQs is systematically aligned parallel to the direction of extended radio structure (Stockman, Angel and Miley 1979). An updated histogram of $\Delta\theta$ (Moore and Stockman 1984) hints at the possibility of a bimodal distribution, some objects having perpendicular orientation.

HPQs and BL Lacs have variable polarization, but Wardle, Moore and
Angel (1984) have found that those with a relatively weak radio core
tend to show a preferred direction for $\underline{E}$, and that this tends to be
aligned with the axis of radio structure (5 objects).

Ulrich (1984) has considered whether there is alignment of $\underline{E}$ with
respect to VLBI structure. Here the number of objects in the test is
limited because of the small intersection of the two sets, those with
measured polarization and those with measured VLBI structure. The
objects tend to be the highly polarized BL Lacs and HPQs. For the few
(7) with a preferred orientation of $\underline{E}$, there is a tendency for $\underline{E}$ to be
parallel to the radio axis. One of the best alignments is 3C 273, which
has only low polarization. On the other hand, Mrk 348, a Seyfert 2
galaxy not in Ulrich's list, has $\underline{E}$ orthogonal to the VLBI structure.
Rusk (1984) has enlarged this sample to 28, and finds that there is a
strong preference for alignment too. This alignment with VLBI structure
is perhaps not surprising, given the alignment of $\underline{E}$ with large scale
structure in LPQs and the fact that large scale structure tends to be
aligned with VLBI structure; however, the tests are independent because
the two samples of objects are not the same. The alignment with large
scale structure indicates that the source has a persistent geometry
(axisymmetry) for long periods of time; on the other hand, the VLBI
structure is more closely coeval with the optical emission being
observed, and is therefore most valuable in sources with bent or more
complex structure.

If the ordered component of the magnetic field runs along the radio
axis, as is often assumed (see also radio polarization), and if the
optical synchrotron radiation is optically thin, then $\underline{E}$ would be
orthogonal to the radio axis, contrary to what is seen. One of the
stated assumptions must be incorrect. In the case of the low constant
polarization in LPQs there is the additional alternative that the
polarization arises from scattering; an optically-thin scattering disk
in a plane perpendicular to the radio axis (a reasonable geometry) would
give polarization of the right sense. Whether the variations that are
seen in the polarization of some of the VLBI objects could be explained
in a scattering model would have to be considered.

For Seyferts, the distribution in $\Delta\theta$ appears to be bimodal. One of
the most interesting results is a separation of the
spectroscopically-defined type 1 (parallel orientation) and type 2
(perpendicular orientation) galaxies (Antonucci 1983 and updated in
1984). Antonucci (1982, 1984) suggests a bimodal distribution for radio
galaxies too. In finding these alignment effects, Antonucci emphasizes
that sources affected by variability or contamination by Galactic
interstellar polarization must be avoided. It can also be noted that
alignment might not be perfect if more than one contribution to
polarization is present; in the prototype for perpendicular
orientation, NGC 1068, θ (as well as p) is a complicated function of
wavelength in the optical and infrared.

Radio polarization. Usually radio polarization orientation is
compared to radio source structure, and so the orientation relative to $\underline{E}$
in the optical is only by association (it could obviously be compared
directly). Again, the overlap between lists of sources having suitable
data is small. There is evidence that $\underline{E}$ of the integrated intrinsic
radio polarization (i.e. corrected for Faraday rotation) is orthogonal
to the VLBI structure (Rusk 1984), indicating an alignment of the
magnetic field in the radio-emitting region along the structure if the
synchrotron emission is optically thin. This is in agreement with the
study of 10 sources by Browne et al. (1982) in which it was found that
VLBI and MERLIN scale structures were aligned, and both orthogonal to $\underline{E}$

of the integrated intrinsic radio polarization.

It is not generally known whether the polarized radio flux is
associated with the VLBI components or the structure on some larger
scale; Browne et al. (1982) favour polarization from jet structure
rather than the core. A sample can be constructed of those sources
unresolved by the VLA, to rule out the largest scales, and the same
preferred alignment occurs (Rusk 1984). Polarization has actually been
mapped at VLBI resolution in one source, 3C 454.3 (Cotton et al. 1983);
only a small fraction (0.2) of the polarized flux was resolved away.
Assuming the ordered field to lie along the jet, it appears that near
the core the radio emission is optically thick, making a transition to
optically thin along the jet; the orientation of $\underline{E}$ of the integrated
polarization has the sense of the presumed optically-thin component,
i.e. orthogonal to the source axis. Because the scales are still
different, the magnetic field inferred from VLBI measurements is not
necessarily the same as in the optically-emitting region. In 3C 354.3
the optical polarization is variable, showing no preferred θ.

FUTURE PROSPECTS

The detail in which the polarization has become known in many objects
is such that a single unifying theory or mechanism seems unlikely; the
objects appear to have their own individualities which defy simple
classification. It should go without saying that interpretations of
individual objects draw simultaneously on the many types of observation
that have been parcelled separately here for the purposes of
presentation. There is not space to describe further the varied
analyses that have often been made accompanying the presentation of the
observations, but attention can again be drawn to a few recent papers as
examples of the directions of ongoing research.

For the highly polarized variable objects, synchrotron polarization
seems to be accepted, but the detailed mechanism by which the
polarization changes needs to be found. Puschell el al. (1983) have
emphasized how the interpretation of the occasional occurences of
wavelength dependence might be the most rewarding.

How the high polarization and variability relate to relativistic
beaming and radio properties is the subject of continuing investigation
(e.g. Wardle, Moore and Angel 1984). It is interesting to ask if there
are radio-quiet BL Lacs or blazars. Beaming models predict none, and
this is supported by the QSO surveys in which the HPQs were identified;
the models do predict some optically quiet ones (3C 120?). Craine,
Duerr and Tapia (1978) have shown how the characteristic high optical
polarization can be used to find BL Lacs in fields identified in the
radio, but do not recommend the technique as efficient in searching
blank fields. Candidates identified using Schmidt plates might provide
better opportunities.

In comparing the properties of HPQs and LPQs, Moore and Stockman
(1984) conclude that anisotropic emission seen from different aspects is
not sufficient to explain the differences, making it likely that there
is a fundamental difference in their physical structure. Stockman,
Moore, and Angel (1984) consider the constraints on mechanisms to
explain the polarization (or lack thereof) of "normal" quasars.

Polarization in many of the other non-variable objects appears to be
due to scattering, by electrons or dust. Rudy (1984) has written on the
different phenomena related to dust in active galactic nuclei.
Antonucci (1984) presents arguments for electron scattering. As well as
identifying the scattering agent, high resolution spectropolarimetry has
the potential of sorting out the relative locations of scattering,

emission (lines, featureless continuum, and starlight) and absorption.

Blandford (this volume) has suggested how different rates of accretion, relative to the Eddington limit, combined with different masses of the central object, and different viewing angles might account for the different classes of active galactic nucleus. The central region is quite different in the various accretion regimes (ion torus, thin disk, radiation torus and cauldron), and the implications for polarization should be investigated. Symmetry in the nucleus, suggested by the polarization together with the evidence for alignment, should be related to such organization of material and magnetic fields in the nucleus (and to other properties), as begun by Antonucci (1984).

Often a paper like this would end with a recital of observations that would resolve the outstanding issues. In this case, however, it might be better just to call a moratorium on further observations while stock is taken of what the available observations mean! Usually it is hoped that by broadening the wavelength range over which a phenomenon is studied, a clearer understanding will be reached. This hope may yet be realized here, but it must be appreciated that a clear understanding can also be complex! New observations in the infrared and ultraviolet can be awaited with anticipation, since new spectral components can be examined: in the infrared, thermal emission from dust and non-thermal continuum, and in the ultraviolet, emission humps and the 30000 K radiation. Spacecraft observations using small apertures could minimize some of the uncertainties associated with dilution by starlight in the host galaxy. The limited number of infrared observations available so far do appear to add to the complexity, and perhaps the ultraviolet data expected from WUPPE and ST will do so too. It might be some time before theory fully catches up.

ACKNOWLEDGEMENTS

Preparation of this paper was facilitated by the kind cooperation of several individuals. Preprints and reprints of relevant work were made available by R.R.J. Antonucci, R.J. Rudy, G.D. Schmidt and H.S. Stockman. R.E. Rusk supplied results of his investigations of VLBI-related matters in advance of publication. Stimulating discussions of work in progess were held during this meeting with R.R.J. Antonucci, D.J. Axon and M.J. Ward, and later with J.S. Miller. This work is supported by NSERC of Canada.

REFERENCES

Angel JRP 1974 In: Gehrels T (ed) Planets, Stars and Nebulae Studied with Photopolarimetry 54-63. University of Arizona Press, Tucson
Angel JRP, Boroson TA, Adams MT, Duerr RE, Giampapa MS, Gresham MS, Gural PS, Hubbard EN, Kopriva DA, Moore RL, Peterson BM, Schmidt GD, Turnshek DA, Wilkerson MS, Zotov NV, Maza J, Kinman TD 1978 In: Wolfe AM (ed) Pittsburgh Conference on BL Lac Objects 117-148. Pittsburgh University Press, Pittsburgh
Angel JRP, Martin PG 1973 Ap J Lett 180:L39-L41
Angel JRP, Stockman HS 1980 Ann Rev Astr Ap 8:321-361
Angel JRP, Stockman HS, Woolf NJ, Beaver EA, Martin PG 1976 Ap J Lett 206:L5-L9
Antonucci RRJ 1982 Nature 299:605-606
Antonucci RRJ 1983 Nature 303:158-159
Antonucci RRJ 1984 Ap J 278:499-520
Axon DJ 1984 private communication
Bailey J, Hough JH, Axon DJ 1983 MNRAS 203:339-344

Blandford RD, Königl A 1979 Ap J 232:34-48
Browne IWA, Clark RR, Moore PK, Muxlow TWB, Wilkinson PN, Cohen MH,
    Porcas RW 1982 Nature 299:788-793
Clarke D 1974 In: Gehrels T (ed) Planets, Stars and Nebulae Studied with
    Photopolarimetry 45-53. University of Arizona Press, Tucson
Cotton WD, Geldzahler BJ, Marcaide JM, Shapiro II, Sanroma M, Ruis A
    1983 preprint
Craine ER, Duerr R, Tapia S 1978 In: Wolfe AM (ed) Pittsburgh Conference
    on BL Lac Objects 99-116. Pittsburgh University Press, Pittsburgh
Impey CD, Brand PWJL, Wolstencroft RD, Williams PM 1982 MNRAS 200:19-40
Lawrence A, Elvis M 1982 Ap J 256:410-426
Martin PG 1974 Ap J 187:461-472
Martin PG 1978 Cosmic Dust. Oxford University Press, Oxford
Martin PG, Angel JRP 1974 Ap J 193:343-351
Martin PG, Stockman HS, Angel JRP, Maza J, Beaver EA 1982 Ap J 255:65-69
Martin PG, Thompson IB, Maza J, Angel JRP 1983 Ap J 266:470-478
Maza J 1979 Polarization of Seyfert Galaxies and Related Objects.
    University of Toronto, Ph.D. thesis
Maza J, Martin PG, Angel JRP 1978 Ap J 224:368-374
McLean IS, Aspin C, Heathcote SR, McCaughrean MJ 1983 Nature 304:609-611
Miley GK, Heckman TM, Butcher HR, van Breugel WJM 1981 Ap J Lett
    247:L5-L9
Miller JS 1984 private communication
Miller JS, Antonucci RRJ 1983 Ap J Lett 271:L7-L11
Moore RL, McGraw JT, Angel JRP, Duerr R, Lebofsky MJ, Rieke GH,
    Wisniewski WZ, Axon DJ, Bailey J, Hough JM, Thompson I, Breger M,
    Schulz H, Clayton GC, Martin PG, Miller JS, Schmidt GD, Africano J
    Miller HR 1982 Ap J 260:415-436
Moore RL, Stockman HS 1981 Ap J 243:60-75
Moore RL, Stockman HS 1984 Ap J 279:465-498
Puschell JJ, Jones TW, Phillips AC, Rudnick L, Simpson E, Sitko M, Stein
    WA, Moneti A 1983 Ap J 265:625-631
Rudy RRJ 1984 Ap J in press
Rudy RJ, Schmidt GD, Stockman HS, Moore RL 1983 Ap J 271:59-64
Rudy RJ, Schmidt GD, Stockman HS, Tokunaga AT 1984 Ap J 278:530-535
Rusk RE 1984 in preparation
Sanders RH, Balamore DS 1971 Ap J 166:7-12
Schmidt GD, Angel JRP, Beaver EA 1978 Ap J 219:477-486
Schmidt GD, Angel JRP, Cromwell RH 1976 Ap J 206:888-897
Schmidt GD, Miller JS 1980 Ap J 240:759-767
Schmidt GD, Miller JS 1984 preprint
Schmidt GD, Peterson BM, Beaver EA 1978 Ap J Lett 220:L31-L36
Stockman HS 1976 private communication
Stockman HS, Angel JRP, Hier RG 1981 Ap J 243:404-410
Stockman HS, Angel JRP, Miley GK 1979 Ap J Lett 227:L55-L58
Stockman HS, Moore RL, Angel JRP 1984 Ap J 279:485-498
Thompson IB, Landstreet JD, Angel JRP, Stockman HS, Woolf NJ, Martin PG,
    Maza J, Beaver EA 1979 Ap J 229:909-916
Thompson IB, Landstreet JD, Stockman HS, Angel JRP, Beaver EA 1980 MNRAS
    192:53-74
Thompson IB, Martin PG 1984, in preparation
Ulrich M-H 1984 In: Fanti R, Kellermann KI, Setti G (eds) VLBI and
    Compact Radio Sources (IAU Symp. 110) in press. Reidel, Dordrecht
Visvanathan N 1974a In: Gehrels T (ed) Planets, Stars and Nebulae
    Studied with Photopolarimetry 1059-1083. Univ Arizona Press, Tucson
Visvanathan N 1974b Ap J 192:319-324
Ward MJ, Axon DJ, Hough JH, Bailey J 1984 in preparation
Wardle JFC, Moore RL, Angel JRP 1984 Ap J 279:93-111

Infrared and Optical Photopolarimetry of Blazars

P.W.J.L. Brand

Department of Astronomy, University of Edinburgh, Royal Observatory,
Edinburgh EH9 3HJ

SUMMARY

An extended programme of photopolarimetry of blazars in the near
infrared is revealing significant correlations between intrinsic
properties: (1) significant fraction of all objects observed have
provided evidence of beamed emission; (2) the most luminous sources do
tend to have the most variable polarization position angles; (3) the
sources most luminous at outburst tend to be those which show the
highest polarization; (4) we confirm the correlation, found by Bailey,
Hough and Axon, between degree of polarization and the strength of
wavelength dependence of polarization. Detailed observations of the
blazar OJ287 are modelled by a simple two-component model, which gives
some clues to the nature of the blazar emitting region.

INTRODUCTION

Of all active galactic nuclei, blazars (Wolfe 1978) are the objects
most likely to reveal the workings of the ultimate power source. The
high observed polarization (>40% in some cases, eg Impey, Brand and
Tapia 1982) strongly implicates synchrotron radiation as the emission
mechanism. This is corroborated by the power-law-like spectrum which
is observed — in those cases where truly simultaneous measurements
have been made — to be continuous from ultraviolet through radio
frequencies (Bregman et al 1982, contributions by Robson, Gear, Sch-
lickheiser, this volume). Flux changes of 100% per day are not unusual
in blazars (Impey et al 1984), which imply emission regions no bigger
than a light day; and taken with the isotropic luminosities derived
from redshifts (where known), this suggests that only black holes and
their attendant accretion discs can be the source of power.
These severe constraints, of size and luminosity, can be relaxed if
relativistic beaming of the emission is involved (Begelman, Blandford
and Rees 1984). However, most such schemes also require the presence
of a black hole which is therefore the weapon of choice in the attack
on the problem of blazar energetics.
To some extent separable from that issue is the question of how the
emission of radiation takes place. It is to solve this problem that
our blazar-monitoring programme was devised. Since the end of 1979 we
have monitored the photometric and polarimetric properties of 35
blazars in the near infrared with the UK Infrared Telescope and, in
collaboration with optical polarimetrists, simultaneously in the
near-UV through visible wavebands (Impey, Brand and Tapia 1982, Impey
et al 1982, Impey et al 1984, Holmes et al 1984abc). We use, on UKIRT,
a rotating HR polaroid followed by a Lyot depolariser (kindly

loaned by H M Dyck) in the beam of the near IR InSb common user photometer. Residual instrumental polarization in JHK wavebands is less
than a few tenths of a percent, so we are photon noise limited.

STATISTICAL PROPERTIES

The largest steady isotropic luminosity that a body can produce is
the Eddington luminosity, proportional to the mass of the emitter. A
limit can be placed on the mass of the body using the light travel time
and a super-Eddington luminosity implies a violation of steadiness or
isotropy (Elliot and Shapiro 1974). In the non-steady case, the
relevant constraint is that derived by eg Guilbert, Fabian and Rees
(1983). However, it can be shown in this case that even if the luminosity change is small, but the polarization properties change significantly, then the Eddington limit violation implies non-isotropic
emission.

We have observed 5 instances where the limit is violated, and
another 10 where the limit is approached. We believe that this is
prima facie evidence for beaming — relativistic or otherwise — of the
energy transport in these sources.

Secondly, we have investigated correlations between luminosity and
the polarization properties of our sample (Holmes et al 1984b).

There exists an intrinsic correlation, at the 92% confidence level,
between maximum observed source luminosity, L_{max}, and the observed
range of source polarization position angle, $\Delta\theta$. Of course there is
risk in correlating ranges of properties, since observed ranges of
stochastically varying properties will increase with the number of
observations. This effect has now been carefully calculated, and
properly discounted in these correlations, using Spearman partial rank
correlations between the various parameters including the number of
observations made of a given object.

The correlation between L_{max} and $\Delta\theta$ has an interesting explanation
in the context of a beam model, since an end on view will result in
perspective amplifications of small changes in beam direction while the
end on aspect causes a relativistic enhancement of an otherwise lower
apparent luminosity.

Thus this correlation can be regarded as secondary confirmation of
the importance of relativistic beaming in blazars (Blandford and Rees
1978).

Between luminosity and degree of polarization the following correlations have been found. The maximum degree of polarization P_{max}
correlates with L_{max} at the 1 percent significance level. However,
L_{mean} does not correlate with P_{mean}. An implication is that the most
intrinsically luminous sources, when in outburst, are also the most
highly polarized. It may be that the process of the outburst 'irons
out' the magnetic field structure, producing a higher observed polarization.

Wavelength dependence of polarization

The degree of polarization of the radiation from a power law distribution of relativistic electrons isotropically moving in a magnetic
field, is independent of frequency (a classical synchrotron radiator).
This behaviour is observed in the majority of blazars whose non-thermal
radiation is not diluted by galactic starlight. However, in a small
number of cases, this is not so (eg Impey, Brand and Tapia 1982), and

wavelength dependent polarization is observed. Usually, although not always, members of this subset are more highly polarized at higher frequencies. The shape of the p(λ) curve rules out any simple Faraday depolarization model.

Bailey, Hough and Axon (1983) discovered an intriguing correlation between the strength of wavelength-dependence of polarization and the degree of polarization in a sample of blazars. This correlation shows strongly in our data also (Fig 1), and seems to offer a clue to the emission process (see below).

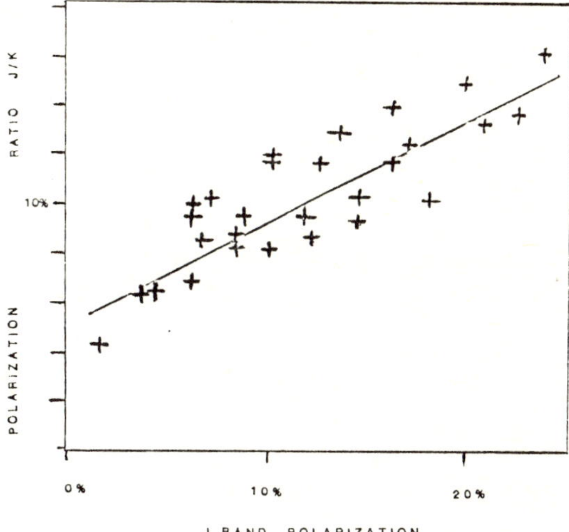

FIG. 1. The ratio of degree of polarization plotted against the polarization at 1.25µm, showing a strong correlation

OJ287

We have studied wavelength-dependent polarization behaviour in an individual object, OJ287, during an observing run in January 1983 (Holmes et al 1984c). The observations were over four nights, and spanned near-UV to 2.2µm wavelengths. The data are summarised in figures 2(a), (b), (c). It can be seen that there is strong wavelength dependence of both position angle and degree of polarization, and that these dependences change dramatically with time. Simultaneously the flux density changes. Indeed, on the night of 1983 Jan 8 we observed a change of 7% in one hour at the K waveband. The shape of the spectrum did not change markedly, however, and is not consistent with either the propagation of a region of unit optical depth in frequency space, or with rapid energy losses in a high-energy tail.

The totality of this behaviour can be modelled summarily in the following way. There are two separate classical synchrotron radiators being observed simultaneously. Each has a convex spectrum, generally similar in shape to the combined spectrum we observe but one is steeper everywhere by a spectral index difference of a half. Each radiator has a degree of polarization of approximately 15%. One of the radiators rotates physically with respect to the other. Then the changes we observe are due to the superposition of the two sets of Stokes parameters. Various parts of this model are critical. In order to produce the dramatic wavelength dependence, the two sources must have closely equal degrees of polarization at one wavelength, and have different

spectral slopes, so that each source can dominate at one wavelength
regime. There has to be a physical rotation of the magnetic field
geometry in at least one source, and each source must have separately
and similarly curved spectra in order to achieve overall spectral
curvature.

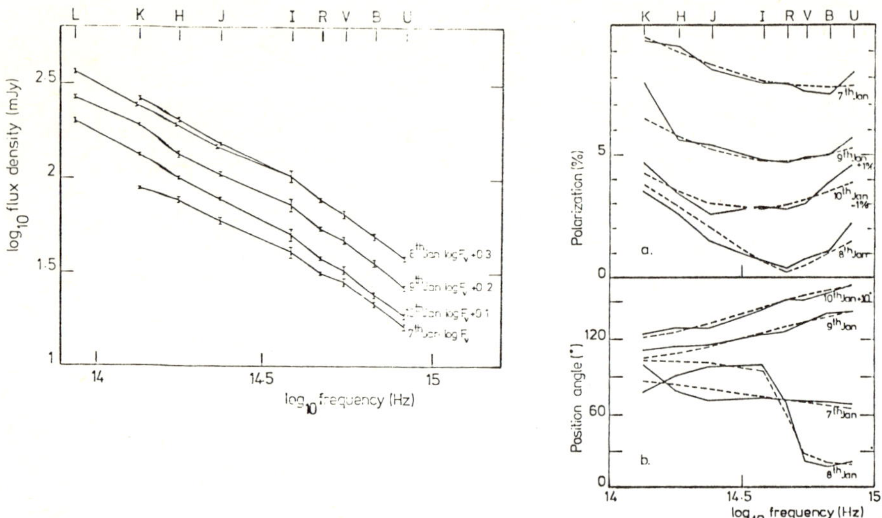

FIG. 2. OJ287 (a) the flux density as a function of frequency on
each of four nights, with variability observed on one night
(b and c) the degree and position angle of polarization at the same
times.
Model fits are shown as dashed lines.

   The model fits based on these precepts are shown in figures 2(abc),
and are excellent.
   The suggestions we make following this modelling are:
   (1) since the intensity variations in each source are correlated,
although not due to rapid energy loss or optical depth effects, the
relativistic electrons orginate from outside of the magnetic field
involved in the emission process.
   (2) by a similar token, the spectral curvature is imprinted on the
electron energy distribution before the electrons arrive at the radiating region.
   (3) the variations in intensity, both increase and decrease, are
most readily explained by variations in the number of electrons arri-
ving at the emitting region.

(4) the difference in spectral index between the two components is due to synchrotron losses nearly instantaneously modifying the electron energy distribution (steepening it by unity in the energy distribution power law in one component, which implies a stronger magnetic field in that component.

(5) Since it is the low magnetic field component that rotates, while the other remains relatively static, we speculate that the lower magnetic pressure permits the ram pressure associated with the arrival of the electron stream more readily to move that component, and to rotate it with respect to the higher magnetic field region.

(6) This series of conclusions fits together well in a beam or jet model of energy generation. A stream of relativistic electrons, possibly generated at an accretion disk, impinges on a density enhancement (a 'bump' on the jet wall or a self-generated density enhancement Marschner 1980)), and the lower density, lower magnetic field regions are pulled off, naturally rotating as they do so.

CONCLUSION

Returning to the phenomenology, we believe that this dramatic behaviour in OJ287 has enabled us to produce an explanation for wavelength-dependence in blazars, coupled with the explanation for the mysterious polarization versus wavelength dependence of polarization correlation. Only a two-component model will produce the variability we observe, including wavelength dependence. Only a model with a few (two?) components contributing can, on a statistical basis, have a high degree of polarization. We obviously and urgently must bear down on these insights with new observations.

REFERENCES

Bailey J, Hough J, Axon D 1983 Mon Not R ast Soc 203, 339

Begelman MC, Blandford RD, Rees MJ 1984 Revs Mod Phys 56, 255

Blandford RD, Rees MJ In: Pittsburgh Conference on BL Lac objects (ed) AM Wolfe

Bregman JN, Glassgold AE, Huggins PJ, Pollock JT, Pica AJ, Smith AG, Webb JR, Ku WHM, Rudy RJ, LeVan PD, Williams PM, Brand PWJL, Neugebauer G, Balonek TJ, Dent WA, Aller HD, Aller MF, Hodge PE, 1982 Astrophys J 253, 19

Elliot JL, Shapiro SL 1974 Astrophys J 192, L3

Guilbert PW, Fabian AC, Rees MJ 1983 Mon Not R astr Soc 205, 593

Holmes PA, Brand PWJL, Impey CD, Williams PM 1984a Mon Not R ast Soc, in press

Holmes PA, Brand PWJL, Impey CD, Williams PM 1984b Mon Not R ast Soc, submitted

Holmes PA, Brand PWJL, Impey CD, Williams PM, Smith P, Elston R, Balonek T, Zeilik M, Burns J, Heckert P, Barvainis R, Kenney J, Schmidt G, Puschell J 1984c Mon Not R ast Soc, submitted

Impey CD, Brand PWJL, Tapia S 1982 Mon Not R ast Soc 198, 1

Impey CD, Brand PWJL, Wolstencroft RD, Williams PM 1982 Mon Not R ast Soc 200, 19

Impey CD, Brand PWJL, Wolstencroft RD, Williams PM 1984 Mon Not R ast Soc 208, ___

Marschner AP 1980 Astrophys J 239, 296

Wolfe AM 1978 Pittsburgh Conference on BL Lac Objects, University of Pittsburgh

UV OBSERVATIONS OF ACTIVE GALACTIC NUCLEI

M. V. Penston

Royal Greenwich Observatory, Herstmonceux Castle,
Hailsham, East Sussex, U.K.

The full text of this lecture is as given in the Proceedings of the
3rd IUE Conference, Madrid, 1982 (ESA SP-176, p.69).

X-RAYS FROM ACTIVE GALACTIC NUCLEI

A.C. Fabian

Institute of Astronomy, Madingley Road, Cambridge, CB3 OHA, England.

INTRODUCTION

Observations made with satellites and balloons have shown that a large fraction of the bolometric luminosity of many Active Galactic Nuclei emerges as X- or soft gamma-rays. The spectral luminosity appears to peak around 1 MeV. X-ray variability has been reported on timescales ranging from minutes to days and years. The emission region is therefore compact in the sense that the dimensionless ratio

$$\ell = \frac{L}{R} \frac{\sigma_T}{m_e c^3}$$

may exceed unity. (L and R are the luminosity and radius of the source and σ_T and m_e are the Thomson cross-section and mass of the electron, respectively). Understanding the source of the X-radiation should be a significant step toward discovering the mechanisms by which the 'central engine' works.

Active Galactic Nuclei are the most efficient 'steady' sources of luminosity known. It is usually considered therefore that the 'central engine' is an accreting black hole, but it is worth recalling that <u>direct</u> observational evidence for such a hypothesis is slight. There is, however, a strong spectral similarity between the X-ray properties of many Active Galactic Nuclei and the Galactic black-hole candidates such as Cygnus X-1. The behaviour of these sources may therefore help our understanding of Active Galactic Nuclei.

It has recently become apparent that mildly relativistic electron-positron pairs dominate the production and transfer of the observed X- and gamma radiation in some, if not all, Active Galactic Nuclei. These pairs are mostly created by photon-photon interactions and may produce a scattering atmosphere throughout the heart of the source and influence the behaviour of radiation of all wavelengths. Radiation pressure may blow the outer edge of this atmosphere away and generate an electron-positron wind.

The picture that is emerging for many Active Galactic Nuclei is one in which the 'central engine' is embedded in an electron-positron gas which is then surrounded by a two-phase medium. The existence and behaviour of the environment is controlled by the X- and gamma-ray emission.

Seyfert Galaxies as X-Ray Sources

Most Seyfert galaxies are found to have comparatively hard X-ray spectra, well fit by a power-law with an energy index of about 0.62 over the range of 0.7 keV to more than 120 kev (Mushotzky 1982, Rothschild et al. 1983, Petre et al. 1984). There are a few notable exceptions, known as the 'Soft Seyferts' (e.g. Pravdo et al. 1981; Pravdo & Marshall 1984), which have an energy index >1 below about 2 keV. As the spectral slope in the optical and ultraviolet wavebands is similar, but that joining them to the X-ray band is at least 1.3 (α_{ox}; Kriss & Canizares 1982), there must be a steep drop in the spectrum of all Seyfert galaxies in the extreme ultraviolet-soft X-ray bands. Perhaps this part of the spectrum is shifted to higher energies in the 'Soft Seyferts' and they are really little different from the rest.

Intrinsic photoelectric absorption is found to be relatively common in those Seyfert galaxies which have 2-10 keV luminosities less than about 5×10^{43} erg s^{-1} (Mushotzky 1982, Culhane 1984, Petre et al. 1984). The absorbing column in the more luminous galaxies is typically less than 2×10^{21} cm^{-2}. The absorption has been well studied in NGC4151 (Barr et al. 1977; Holt et al. 1980) and it has been suggested that clouds in the Broad-Line Region are responsible. The absorbing column in this case is about 10^{23} cm^{-2} and corresponds to a thickness of 10^{13} cm for a cloud density of 10^{10} cm^{-3}. The absorption in NGC 4151 is not totally 'black'; some residual flux is seen below 2 keV. This could be due to the clouds only partially covering the X-ray source (Holt et al. 1980) or to bremsstrahlung from the intercloud medium (Guilbert, Fabian & McCray 1983).

A careful study of the HEAO-1 data for variability by Tennant (1983), (see also Tennant & Mushotzky 1983) has shown that variations on time-scales of less than a few hours are uncommon. The most remarkable exceptions are NGC 6814, which was found to vary in less than 100 seconds (Tennant et al. 1981), and NGC 4051 which varies in less than 1000 seconds (Marshall et al. 1983). Many reports in the literature suggest that variability is common on timescales longer than about a day (e.g. Marshall, Warwick & Pounds 1980), but it is not yet clear how important 'confusion noise' (i.e. that due to unresolved sources) might be in mimicking variability from scans in different directions. Nevertheless, there are cases where we can be fairly certain that the variations were in the source. NGC 4151 is variable on timescales of days (Lawrence 1980). Its spectrum has sometimes been detected out to an MeV (Perotti et al. 1981). Hard X-ray variability has been reported from HEAO-1 data (Baity et al. 1984) which shows no change in the spectral shape below about 100 keV, but a marked steepening occurred above that energy as the source luminosity increased.

X-ray luminosity functions have been constructed from the hundred or so Seyfert and Seyfert-like galaxies found to be X-ray sources from HEAO-1, the Einstein Observatory and earlier satellite surveys (Piccinotti et al. 1982, Maccacaro et al. 1983). The contribution of Seyfert galaxies to the X-ray background can then be determined. Assuming no source evolution, this turns out to be about 10 percent and extrapolates to explain the 'MeV bump' in the soft gamma-ray background, if the lowest luminosity considered is 10^{43} erg s^{-1} (Rothschild et al. 1983). Whether this is a coincidence or not, the steepness of the gamma-ray background testifies that the spectra of all Active Galaxies must turnover above about an MeV.

X-Rays from QSOs

   Most QSOs are too faint at X-ray wavelengths to have yet been detected above a few keV. Below that energy the X-ray telescopes on the
Einstein Observatory are a hundred to one thousand times more sensitive
than ordinary collimated proportional counters. Many QSOs have been
detected by this Observatory and have shown that the ratio of X-ray to
optical luminosity of QSOs is similar to that of Seyfert galaxies (Kriss
& Canizares 1982; see also Cheng et al. 1984). Broadband X-ray spectra
of QSOs are very few. 3C273 has an energy index that varies between 0.2
and 0.7 in the 1-200 keV range (Primini et al. 1979; Worrall et al. 1979;
Bezler et al.1984) and the radio-quiet QSO 0241+622 has an index of 0.9
(Worrall et al. 1980). Similar values have also been obtained from
Einstein Observatory Solid-State Detector data (Petre et al. 1984).
Halpern & Grindlay (1982) and Worrall & Marshall (1984) have compared
0.5-3 keV Einstein fluxes with those in the 2-10 keV band to obtain a
mean index of about 0.8, with a relatively large error. Imaging
instruments with a wide energy range are needed to study the X-ray
spectra of QSOs further.
   If they do constitute the X-ray background (see e.g. Giacconi et al.
1979; Setti & Woltjer 1982) then their mean spectrum must mimic that of
40 keV thermal bremsstrahlung (Marshall et al. 1980). This cannot occur
without some complicated spectral evolution (De Zotti et al. 1982). The
contribution of QSOs to the X-ray background has been extensively
discussed (Tananbaum et al. 1979; Zamorani et al. 1981; Ku et al. 1980;
Cheney & Rowan-Robinson 1981; Avni & Tananbaum 1982; Kembhavi & Fabian
1982; Reichert et al. 1982; Chanan 1983; Anderson & Margon 1983; Stocke
et al. 1983; Gioia et al. 1983; Marshall et al. 1983. 1984; Kriss &
Canizares 1982; Cheng et al. 1984). It is usual to refer the Einstein
flux to a spectral luminosity at 2 keV in the rest-frame of the source.
Deep surveys (Giacconi et al. 1979; Griffiths et al. 1983) suggest that
about 20 (±10) percent of the X-ray background at that energy has been
resolved, or to be more correct, that percentage of the background
observed at higher energies extrapolated down to 2 keV, for the spectrum
is uncertain below 3 keV.
   Radio-loud QSOs appear to be more X-ray luminous than radio-quiet
ones. Since they are the minority population of all QSOs, the contribution of radio-loud QSOs to the (extrapolated) X-ray background is only
about 15 percent. The contribution of radio-quiet QSOs is very uncertain
as only about 15 percent of those radio-quiet QSOs deliberately observed
were actually detected. There could, for example, be a population of
X-ray-quiet QSOs similar to the situation at radio wavelengths. Much
further work is needed to sort out this problem and determine whether it
is a selection effect due to short observing times.
   Variability has been observed in some QSOs (Zamorani 1984; Zamorani
et al. 1984).

X-Ray Emission from BL Lac Objects

   The X-ray spectrum from BL Lac objects is steep below photon energies
of a few keV, making them prominent soft X-ray sources. PKS 2155-304,
for example, is fit by an energy index of about 1.8 over the range of
0.1-10 keV. Above this last energy the spectrum appears to flatten and
may have a hard tail (Urry & Mushotzky 1982; but Rothschild (private
communication) reports that the index is 0.7). A particularly interesting spectral feature has been reported from this BL Lac object by
Canizares & Kruper (1984) using the Objective Grating Spectrometer on

the Einstein Observatory. They find an absorption feature around 600 eV
which is probably due to OVIII, either very close to the source or in
the intergalactic medium. The redshift of PKS 2155-304 is 0.117
(Bowyer et al. 1984).
    Schwartz et al. (1983) report that BL Lac objects are more X-ray
variable than QSOs. This could in part be due to small changes in the
very steep spectral index and does not necessarily mean large changes in
total luminosity.

Low Luminosity AGN and Radio Galaxies

    X-rays are observed from a variety of active galaxies apart from
those mentioned above. The nearest radio galaxy, Cen-A, has long been
recognised as a variable X- and gamma-ray source (Terrell 1984). Many
other radio galaxies have been detected (see e.g. Miller et al. 1984).
    The nucleus of our Galaxy (Watson et al. 1981) and that of several
other nearby galaxies are weak X-ray sources (Long & Van Speybroek 1984).
Starburst galaxies and many other sites of activity are strong sources
of X-radiation with luminosities ranging up to about 10^{41} erg s^{-1}
(Fabbiano, Feigelson & Zamorani 1982). It is possible that we are
observing the cumulative emission from many X-ray binaries in some cases
rather than any coherent object.
    Elvis, Soltan & Keel (1984) have estimated the luminosity function of
low luminosity active nuclei. It is flatter than that for Seyfert
galaxies and, again if there is no evolution, they contribute about $\sim$20
percent of the X-ray background.

The Significance of $m_e c^2$

    The X-ray spectra of individual Active Galactic Nuclei and the
constraints from the X-ray background strongly suggest that a typical
spectrum is hard and extends to about 1 MeV. An energy index of 0.6
means that most of the luminosity is emitted at the highest energies.
Whatever mechanism controls the MeV turnover must therefore be an
important one.
    X-radiation can generally be considered as produced by two separate
pathways (Fig 1). One is from a thermal plasma (in which the electrons
and ions have an approximately Maxwellian distribution) and the radiation
is generated by bremsstrahlung or is scattered up to X-ray energies by
Comptonization. The other major pathway is from relativistic electrons
(which are usually assumed to have a power law distribution) and the
processes are synchrotron or inverse Compton scattering. Both pathways
can give a natural upper energy limit around an MeV due to the production
of electron-positron pairs.
    Pairs act as a thermostat in a thermal plasma heated above 10^9 K
(Bisnovatyii-Kogan, Zeldovich & Sunyaev 1971; Gould 1982; Lightman 1982;
Svensson 1982). Once pairs begin to be generated by particle and photon
collisions, they generate more photons and the pair production rate
increases. Eventually, all the energy put into the gas goes into pairs
and not into increasing the temperature. The maximum temperature
obtainable is dependent on the optical depth of the source but is always
less than 10^{11} K. This may seem to provide a good explanation for a
spectral cutoff around an MeV in active galaxies, but a problem remains
in the uniformity of the observed spectral shapes. Comptonization
dominates over bremsstrahlung in compact sources (Lightman, Giacconi &
Tananbaum 1978) and the resultant spectrum is very dependent on geometry

and optical depth.

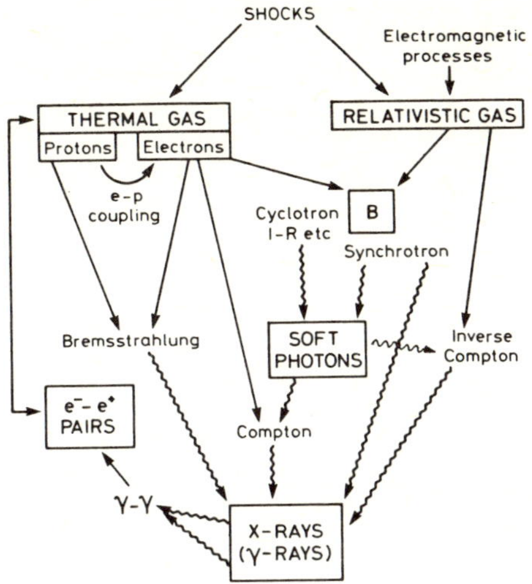

FIG. 1. Radiation pathways for the production of X- and gamma-rays
   (Fabian 1984).

A relativistic plasma provides a better solution if the source is
compact (i.e. $\ell > 1$). The primary X- and gamma-ray production mechanism
will not initially be affected at photon energies above an MeV, but the
photons themselves will be affected if the optical depth through the
source to photon-photon collisions is greater than unity. The cross-
section for this process is approximately similar to that of Thomson
scattering and it comes into operation when the product of the energies
of the colliding gamma-rays and X-ray photons exceeds 2 $m_e c^2$. The cross-
section soon drops off above that threshold. The possible occurrence of
photon-photon collisions in Active Galaxies has been discussed by
(Jelley 1966; Bonometto & Rees 1971; Herterich 1974; Fabian et al. 1976;
Rees 1978; Lightman, Giacconi & Tananbaum 1978; Fabian 1979; Bassani &
Dean 1981; Matteson 1983; Fabian 1983; Kazanas & Protheroe 1983; Salvati
et al. 1983). The important point now apparent is that the flat spectra
imply that most of the luminosity is converted into, or transferred
through, pairs (Guilbert, Fabian & Rees 1983).

The Influence of Electron-Positron Pairs

Photon-photon collisions have a very marked effect upon a source with
a hard primary spectrum when the optical depth in photons of energy
around $m_e c^2$ approaches one. Each primary gamma-ray is then likely to
collide with an X-ray and create a pair. The Thomson optical depth of
such pairs is then also about one, since the cross-section is similar.
The pairs trap the photons and thereby enhance the probability of photons
colliding. More gamma-rays are also radiated by the pairs themselves
since electron-electron and electron-positron Bremsstrahlung become
important. The whole situation is highly non-linear and affects all of

the photons produced in this region. A qualitative description of a
possible outcome is given in Guilbert, Fabian & Rees (1983).

The luminosity at which these effects begin to dominate depends upon
the size of the source and is reached when $\ell > 1$. This can be related to
the Eddington limit, L_{Edd}, in the case of an accreting black-hole since
it involves a similar set of parameters. Generally pair production
starts to become important when the hard luminosity exceeds about one-
thousandth of the Eddington limit and dominates the behaviour when it
exceeds about one percent (Lightman, Giacconi & Tananbaum 1978; Guilbert
Fabian & Rees 1983). This can be seen simply by first recalling that
the Thomson optical depth across a source emitting at L_{Edd} is ~ 1. Each
proton falling into a black hole can lose an energy of up to 30 percent
of its rest-mass which corresponds to (m_p/m_e) gamma-rays of energy $m_e c^2$.
The optical depth in gamma-rays is thus $\sim (m_p/m_e)$, which reduces to unity
if $L=(m_e/m_p)L_{Edd}$.

A possible mechanism for generating the common spectral index in the
X-ray spectra from Active Nuclei derives from the suggestion of Bonometto
& Rees (1971). Consider a continuous source of mono-energetic relativistic electrons distributed through a small region of space. These
electrons lose energy by synchrotron emission or by inverse Compton
scattering lower energy photons (perhaps even those due to synchrotron).
The resultant electron spectrum is a power-law such that the photon
spectrum has an energy index of 0.5, which is fairly close to that
observed although a little too hard. Now suppose that the conditions
are such that the photon spectrum extends beyond $m_e c^2$ and that photon-
photon collisions take place. The pairs also radiate along with the
primary electron spectrum and, as there are more of them at lower
energies, the photon spectrum becomes steeper. Bonometto & Rees (1971)
show that the final output spectrum should have an energy index
between 0.5 and 1.

I have attempted to compute this (non-linear) situation by a more-or-
less brute force method. Electrons and photons are represented by
arrays which are logarithmic in energy. Compton scattering then follows
by a simple linear shift along the photon array. The computation assumes
that mono-energetic, relativistic electrons are injected into a small
region. The photon spectrum is then allowed to evolve with the following
processes being included with varying degrees of sophistication and
repeated at each timestep (which is much less than the light crossing
time of the region): inverse Compton scattering, photon-photon collisions
pair annihilation, radiation transfer and non-relativistic Comptonization. The radiation transfer is very simple and merely account for the
average trapping and leakage of photons. Inverse Compton scattering is
dealt with by assuming that the relativistic electrons introduce a
relative frequency shift to the photon spectrum of $4/3\gamma^2$. Photon-photon
collisions are approximated by adopting the Thomson cross-section and
considering only those X-rays near threshold (cf. Herterich 1974). The
resultant pairs are assumed to have an energy of half that of the gamma-
ray involved (cf. Bonometto & Rees 1971) and are then added into the
electron spectrum for the next timestep. All electrons (and positrons)
with non-relativistic energies are assumed to belong to a Maxwellian
distribution at the Compton temperature, T_c. In order that the acceleration process for the electrons could be ignored, the computation
includes an equal number of positrons (the source would otherwise fill
up with 'dead' electrons). Annihilation then takes place for only the
non-relativistic population, with the Thomson cross-section and within
that time-step. A broadened and shifted annihilation line is added to
the photon spectrum.

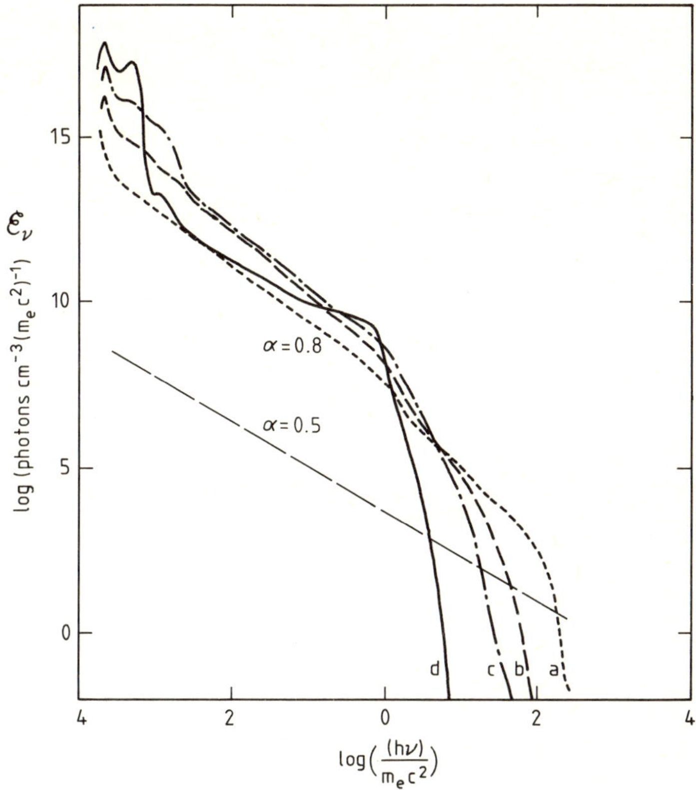

FIG. 2.   Photon density spectra ($cm^{-3}(m_ec^2)^{-1}$) within a region of size 10^{15} cm in which relativistic pairs are continuously injected with a Lorentz – factor $\gamma = 10^3$ at a rate of a) 10^6, b) 10^7, c) 10^8 and d) 10^9 particles per light crossing time. Photons are injected at an energy of 10^{-4} m_ec^2 (i.e. 51 eV). The plots show the spectra after 2 crossing times and subsequently show little change. See text for details of processes considered.

Some results are shown in Figure 2. The Comptonization is difficult to handle when the electrons are non-relativistic and the source becomes optically thick with them. The computations approximate Comptonization by giving photons below the electron temperature a relative upward shift of $(4kT_c/m_ec^2)\tau^2$ and those below that energy a relative downward shift of $(\varepsilon/m_ec^2)\tau^2$, where τ is the Thomson depth through the source. The results shown may only be a rough guide, but the major qualitative properties are understandable (Guilbert, Fabian & Rees 1983). A mean energy index of about 0.8 fits the spectra between about 3 and 300 keV, as expected from the work of Bonometto & Rees (1971). As ℓ exceeds unity the source becomes optically thick with pairs and the non-relativistic Comptonization creates a soft X-ray peak. The radiation temperature, and thus the assumed electron temperature, is then below 1 keV. This is because there is such an abundance of low energy photons from the injection spectrum which are an effective Compton coolant. The shoulder at m_ec^2 is due to the balance between production and trapping

of the annihilation line. Any realistic emergent spectrum would depend
upon boundary effects and would probably include some fraction of the
primary spectrum if injection extends that far.

Nevertheless it does appear that a spectrum somewhat similar to that
of a Seyfert galaxy can be generated by a very simple set of input
parameters. Note that there is a dramatic change at low energies when
the source is optically thick and much of the energy is lost by Comptonization below 1 keV. Pair production has thus had dramatic effect on
the whole spectrum of the source, and would change the optical and
lower frequencies if they had been included. There may be some similarities here to the two-state behaviour of Cygnus X-1. I find the
spectral similarities between that source and other black-hole candidates such as GX339-4 and those of Active Galaxies intriguing and have
speculated that they are not due to coincidence (White, Fabian &
Mushotzky 1984; Figure 3) but that pair production dominates. The soft
Seyferts and/or the BL Lac objects may correspond to the equivalent
'high-states' observed in Cygnus X-1 (Guilbert, Fabian & McCray 1983).

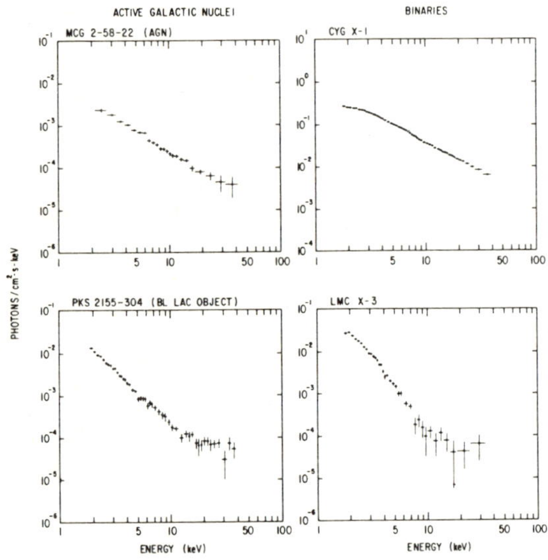

FIG. 3. X-ray spectra of Active Galactic Nuclei compared with those of
   Black-Hole X-ray Binaries. Cygnus X-1 is shown in its 'low-state';
   LMC X-3 represents a 'high-state' spectrum (White, Fabian & Mushotzky
   1984).

A pair atmosphere forms below the Eddington limit for an electron-
proton gas but above that for electrons and positrons alone, which is
about one-thousandth lower. It is therefore possible that pairs can be
blown out of the source, perhaps carrying some more normal matter. There
is, of course, much evidence for outflows in Active Galaxies, both in
the motions within the Broad-Line Region and in the form of jets.

A popular explanation for the power-law spectra in Active Galactic
Nuclei is the Synchro-Self-Compton (SSC) mechanism. Relativistic
electrons with a spectrum compatible with the observations radiate radio,
infrared and optical photons by the synchrotron process and the Compton
scatter them to X- and gamma-ray energies. It should be remembered,
however, that in a compact source the gamma-rays may undergo photon-
photon collisions and produce a mess of pairs, such as described earlier.
Any simple application of the SSC mechanism that predicts an X-ray flux
above that observed (see e.g. Urry & Mushotzky 1982) may also predict
that the source is optically thick to pairs. This is particularly
relevant to those cases, such as the BL Lac objects, where relativistic
beaming is invoked to reduce the predicted X-ray flux to observed
levels. The cooled pairs Compton down-scatter the X-ray photons to
produce a lower X-ray flux without necessarily require relativistic
motion. The 'Compton Catastrophe', in which all of the energy in the
relativistic electrons goes into gamma-rays, may herald the arrival of
a pair plasma.

The Rate of Change of Luminosity

Spectral variations may be an important tool for understanding the
mechanisms underlying Active Galactic Nuclei. Until wide-band X- and
gamma-ray spectroscopy becomes commonplace, we must make do with
luminosity changes. The compactness, L/R and ℓ, or their measured
counterpart, the rate of change of luminosity $\dot{L}$, can be related to
several possible properties (Fabian 1983; see Figure 4).

Bremsstrahlung can still be a dominant process (see e.g. Meszaros
1983) below about 10^{38} erg s^{-2}. Above about 10^{40} erg s^{-2} a hard source
will become optically thick in pairs and above 2×10^{41} erg s^{-2} the source
is radiating too efficiently for simple (e.g. accretion) processes to
apply. A source which is optically thick in pairs will transform much
of its luminosity into lower energies (as in Figure 3) and it need not
be obvious from the observed spectrum that the source is either hard or
compact. Scattering in the pair atmosphere may reduce intrinsic
variations in the primary radiation. The efficiency limit represents
the effects of photon-trapping in the matter assumed to be generating
the luminosity by releasing some fraction (typically less than about
10 percent) of its rest-mass (Fabian 1979).

I thank Roger Blandford, Paul Guilbert, Sterl Phinney and Martin Rees
for discussions.

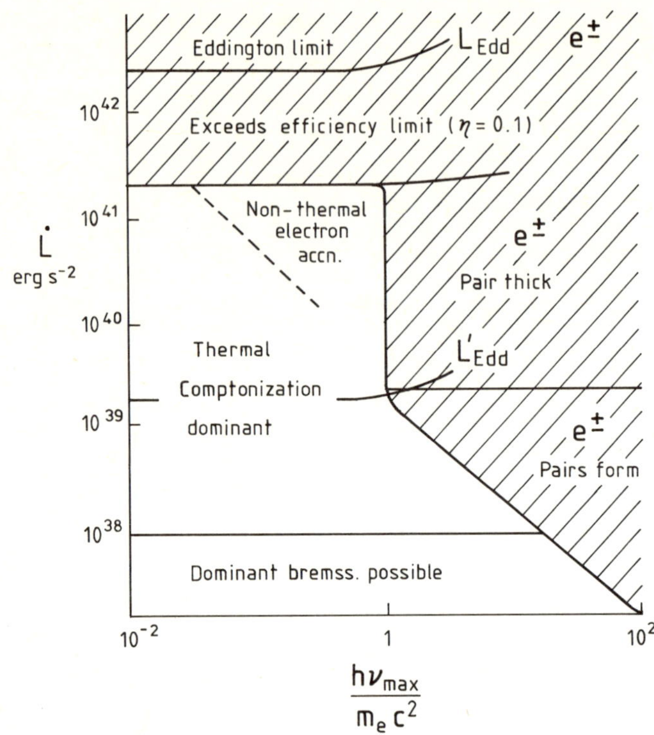

FIG. 4. Schematic diagram showing various regimes for a source with a hard spectrum cutting off above a frequency ν_{max}. Where the source is optically thick in pairs, the entire spectrum is affected and the observed values of $\dot{L}$ (assumed to represent L/R) and ν_{max} will be much reduced.

REFERENCES

Anderson, S.F. & Margon, B., 1983, 24th Liege Intl. Astrophys. Symp., 68.
Avni, Y. & Tananbaum, H., 1982, Astrophys. J., 262, L17.
Baity, W.A., Mushotzky, R.F., Rothschild, R.E., Worrall, D.M., Tennant, A.F. & Primini, F.A., 1984, Astrophys. J., 279, 555.
Barr, P., White, N.E., Sanford, P.W. & Ives, J.C., 1977, Mon. Not. R. astr. Soc., 181, 43P.
Bassani, L. & Dean, A.J., 1981, Nature, 294, 332.
Bezler, M., Kendziorra, E., Staubert, R., Hasinger, G., Pietsch, W., Reppin, C., Trumper, J. & Voges, W., 1984, Astr. Astrophys, in press.
Bisnovatyii-Kogan, G.S., Zeldovich, Y.B. & Sunyaev, R.A., 1971, Sov. Astr., 15, 17.
Bonometto, S. & Rees, M.J., 1971, Mon. Not. R. astr. Soc., 152, 21.
Bowyer, S., Brodie, J., Clarke, J. & Henry, P., 1984, Astrophys. J., 278, L103.
Canizares, C.R. & Kruper, J., 1984, Astrophys. J., 278, L99.

Chanan, G.A., 1983, Astrophys. J., 275, 45.
Cheney, J. & Rowan-Robinson, M., 1981, Mon. Not. R. astr. Soc., 197, 313.
Cheng, F.Z., Danese, L., De Zotti, G. & Lucchini, F., 1983, Mon. Not. R. astr. Soc., 208, 799.
Culhane, J.L., 1984, Phys. Scripta, T7, 134.
De Zotti, G., Boldt, E.A., Cavaliere, A., Danese, L., Marshall, F.E., Swank, J.H. & Symkowiak, A.E., 1982, Astrophys. J., 253, 47.
Elvis, M., Soltan, A. & Keel, W.C., 1984, Astrophys. J., in press.
Fabbiano, G., Feigelson, E. & Zamorani, G., 1982, Astrophys. J., 256, 397.
Fabian, A.C., 1979, Proc. Roy. Soc., 366, 449.
Fabian, A.C., 1983, Proc. Intl. Cosmic Ray Conf., Bangalore.
Fabian, A.C., 1984, Phys. Scripta, T7, 129.
Fabian, A.C., Maccagni, D., Rees, M.J. & Stoeger, W., 1976, Nature, 260, 683.
Giacconi, R., Bechtold, J., Branduardi, G., Forman, J., Henry, J.P., Jones, C., Kellogg, E., Van der Laan, H., Liller, W., Marshall, H., Murray, S.S., Pye, J., Schreier, E., Sargent, W.L.W., Seward, F. & Tananbaum, H., 1979, Astrophys. J., 234, L1.
Gioia, I.M., Feigelson, E.D., Maccacaro, T., Schild, R. & Zamorani, G., 1983, Astrophys. J., 271, 524.
Gould, R.J., 1982, Astrophys. J., 254, 755.
Griffiths, R.E., Murray, S.S., Giacconi, R., Bechtold, J., Murdin, P., Smith, M., MacGillivray, H., Ward, M., Danziger, J., Lub,J., Peterson B.A., Wright, A.E., Batty, M.J., Jauncey, D.L. & Malin, D.F., 1983, Astrophys. J., 269, 375.
Guilbert, P.W., Fabian, A.C. & McCray, R., 1983, Astrophys. J., 266, 466.
Guilbert, P.W., Fabian, A.C. & Rees, M.J., 1983, Mon. Not. R. astr. Soc. 205, 593.
Halpern, J.P. & Grindlay, J.E., 1982, Bull. Am. Astr. Soc., 13, 849.
Herterich, K., 1974, Nature, 250, 311.
Holt, S.S., Mushotzky, R.F., Becker, R.H., Boldt, E.A., Serlemitsos, P.J., Szymkowiak, A.E. & White, N.E., 1980, Astrophys. J., 241, L13.
Jelley, J., 1966, Nature, 211, 472.
Kazanas, D. & Protheroe, R.J., 1983, Nature, 302, 228.
Kembhavi, A.K. & Fabian, A.C., 1982, Mon. Not. R. astr. Soc., 198, 921.
Kriss, G.A. & Canizares, C.R., 1982, Astrophys. J., 261, 51.
Ku, W.H-M., Helfand, D.J. & Lucy, L.B., 1980, Nature, 288, 323.
Lawrence, A., 1980, Mon. Not. R. astr. Soc., 192, 83.
Lightman, A.P., 1982, Astrophys. J., 253, 842.
Lightman, A.P., Giacconi, R. & Tananbaum, H., 1978, Astrophys. J., 224, 375.
Long, K. & Van Speybroeck, L., 1984, in Stellar Mass Accreting X-ray Sources, ed. Lewin, W.H.G. & van den Heuvel, E.A., C.U.P.
Maccacaro, T., Avni, Y., Gioia, I.M., Giannini, P., Griffiths, R.E., Liebert, J., Stocke, J. & Danziger, J., 1983, Astrophys. J., 266,L73.
Marshall, F.E., Boldt, E.A., Holt, S.S., Miller, R., Mushotzky, R.F., Rose, L.A., Rothschild, R. & Serlemitsos, P.J., 1980, Astrophys. J., 235, 4.
Marshall, F.E., Holt, S.S., Mushotzky, R.F. & Becker, R.H., 1983, Astrophys. J., 269, L31.
Marshall, H.L., Tananbaum, H., Zamorani, G., Huchra, J.P., Braccesi, A. & Zitelli, V., 1983, Astrophys. J., 269, 42.
Marshall, H.L., Avni, Y., Braccesi, A., Huchra, J.P., Tananbaum, H., Zamorani, G. & Zitelli, V., 1984, preprint.
Marshall, N., Warwick, R.S. & Pounds, K.A., 1981, Mon. Not. R. astr. Soc., 194, 987.
Matteson, J.L., 1983, Am. Inst. Phys. Conf. Proc., 101, 292.
Meszaros, P., 1983, Astrophys. J., 274, L13.

Miller, L., Fabbiano, G., Longair, M.S., Trinchieri, G. & Elvis, 1984,
    in preparation.
Mushotzky, R.F., 1982, Astrophys. J., 256, 92.
Perotti, F., Della Ventura, A., Villa, G., DiCocco, G., Bassani, L.,
    Butler, R.C., Carter, J.N. & Dean, A.J., 1981, Astrophys. J., 247,
    L63.
Petre, R., Mushotzky, R.F., Krolik, J.H. & Holt, S.S., 1984, Astrophys.
    J., 280, 499.
Piccinotti, G., Mushotzky, R.F., Boldt, E.A., Holt, S.S., Marshall, F.E.,
    Serlemitsos, P.J. & Shafer, R.A., 1982, Astrophys. J., 263, 485.
Pravdo, S.H., Nugent, J.J., Nousek, J.A., Jensen, K., Wilson, A.S. &
    Becker, R.H., 1981, Astrophys. J., 251, 501.
Pravdo, S.H. & Marshall, F.E., 1984, Astrophys. J.
Primini, F.A., Cooke, B.A., Dobson, C.A., Howe, S.K., Scheepmaker, A.,
    Wheaton, W.A., Lewin, W.H.G., Baity, W.A., Gruber, D.E., Matteson,
    J.L. & Peterson, L.A., Nature, 278, 234.
Rees, M.J., 1978, Ann. N.Y. Acad. Sci., 320, 613.
Reichert, G.A., Mason, K.O., Thorstenson, J.R. & Bowyer, S., 1982,
    Astrophys. J., 260, 437.
Rothschild, R.E., Mushotzky, R.F., Baity, W.A., Gruber, D.E., Matteson,
    J.L. & Peterson, L.E., 1983, Astrophys. J., 269, 423.
Salvati, M., Cavaliere, A., Costa, E. & Massaro, E., 1983, Am. Inst.
    Phys. Conf. Proc., 101, 332.
Schwartz, D.A., Madejski, G. & Ku, W.H-M., 1983, in I.A.U. Highlights
    of Astronomy, 6.
Setti, G. & Woltjer, L., 1982, in Proc. Vatican Study Week on 'Cosmology
    and Fundamental Particles'.
Stocke, J.T., Liebert, J., Gioia, I.M., Griffiths, R.E., Maccacaro, T.,
    Danziger, I.J., Kunth, D. & Lub, J., 1983, Astrophys. J., 273, 458.
Svensson, R., 1982, Astrophys. J., 258, 335.
Tananbaum, H., Avni, Y., Branduardi, G., Elvis, M., Fabbiano, G.,
    Feigelson, E., Giacconi, R., Henry, J.P., Pye, J.P., Soltan, A. &
    Zamorani, G., 1979, Astrophys. J., 234, L9.
Tennant, A.F., 1983, Ph.D. Dissertation, Univ. of Maryland.
Tennant, A.F. & Mushotzky, R.F., 1983, Astrophys. J., 264, 92.
Tennant, A.F., Mushotzky, R.F., Boldt, E.A. & Swank, J.H., 1981,
    Astrophys. J., 251. 15.
Terrell, J., 1984, Proc. Munich Conference on X-ray and UV emission from
    AGN, ed. W. Brinkmann & J. Trumper.
Urry, C.M. & Mushotzky, R.F., 1982, Astrophys. J., 253, 38.
Watson, M.G., Willingale, R., Grindlay, J.E. & Hertz, P., 1981,
    Astrophys. J., 250, 142.
White, N.E., Fabian, A.C. & Mushotzky, R.F., 1984, Astr. Astrophys.,
    133, L9.
Worrall, D.M., Mushotzky, R.F., Boldt, E.A., Holt, S.S. & Serlemitsos,
    P.J., 1979, Astrophys. J., 232, 683.
Worrall, D.M., Boldt, E.A., Holt, S.S. & Serlemitsos, P.J., 1980,
    Astrophys. J., 240, 421.
Worrall, D.M. & Marshall, F.E., 1984, Astrophys. J., 276, 434.
Zamorani, G., Henry, J.P., Maccacaro, T., Tananbaum, H., Soltan, A.,
    Liebert, J., Stocke, J., Strittmatter, P.A., Weymann, R.J., Smith,
    M.G. & Condon, J.J., 1981, Astrophys. J., 245, 357.
Zamorani, G., 1984, these Proceedings.
Zamorani, G., Gianini, P., Maccacaro, T. & Tananbaum, H., 1984,
    Astrophys. J., 278, 28.

X-RAY ABSORPTION AND VARIABILITY IN QUASARS

G. Zamorani

Istituto di Radioastronomia, C.N.R.,
Bologna, Italy

SUMMARY

The spectral shape of the X-ray continuum, the presence or absence of intrinsic low energy absorption, the frequency and timescales of observed X-ray variability are different tools at our disposal in the study of the physics of X-ray emission in quasars. In this paper we briefly summarize the present knowledge, from an observational point of view, of some of these aspects. An extremely simplified, but self-consistent picture, in agreement with these data, allows us to derive some order of magnitude estimates of the size of the X-ray emitting region as a function of the X-ray luminosity. Within this picture, it is shown that X-ray variations on timescales of the order of 10^5 s are expected in low luminosity quasars, as a consequence of the passage of one or a few absorbing clouds in front of the emission region.

X-RAY ABSORPTION IN ACTIVE NUCLEI

With respect to the problem of low energy X-ray absorption in active nuclei, there are by now a few interesting results which deserve some attention.

1. It is well known that many low luminosity active galaxies have large measured X-ray absorbing columns, while many high luminosity objects are known not to have such large columns (Mushotzky et al. 1980; Mushotzky 1982; Maccacaro, Perola and Elvis 1982; Petre et al. 1984).

2. In addition to these direct measurements, there is a more indirect, but apparently convincing, evidence in favor of a different importance of the X-ray absorption as a function of the intrinsic luminosity. By collecting the available data on all the Active Galactic Nuclei observed at both "hard" (around 6 KeV) and "soft" (around 1 KeV) X-rays, Lawrence and Elvis (1982) showed that the ratio

of hard to soft X-ray luminosities (HX/SX) appears to be a function of HX. While at high HX (HX > 10^{44} erg s^{-1}) this ratio is approximately constant, near the value expected for a pure power law with an energy index 0.5-0.7, for HX < 10^{44} erg s^{-1} this ratio appears to increase smoothly going to lower intrinsic luminosities. If, as they do, we interpret this effect as due to a deficit in SX and assume that this observed deficit is due to absorption by optically thick clouds (N_H > 10^{22} cm^{-2}) in the broad line region, this result can be used to derive some order of magnitude estimates of the relative sizes of the emitting and absorbing regions. Before doing this, it is interesting to see if the existing EINSTEIN data are consistent with the Lawrence-Elvis effect. Figure 2 in Zamorani (1984) shows the correlation between X-ray and optical luminosities for a large sample of optically selected quasars. From this figure it is clearly seen that some of the low luminosity quasars are indeed consistent with the idea of having been significantly absorbed, i.e. their observed X-ray luminosity is much smaller than what would be expected on the basis of the optical luminosity. On the other hand, there is a good fraction of these low luminosity objects which perfectly follow the correlation defined by high luminosity quasars. From this, we are led to suspect that, even for low luminosity objects, a significant low energy X-ray absorption is present in no more than about one third of the cases.

3. The amount of hydrogen along the line of sight derived from observed absorption in X-ray data is often much larger than that predicted from the reddening of the optical continuum (see Maccacaro, Perola and Elvis 1982).

Many different scenarios can be proposed to explain these observed facts, but perhaps the simplest and still consistent one is the following (see also Reichert et al. 1983 for a similar discussion):

A) "The size of the X-ray emitting region is smaller than the size of the optical emitting region". This is consistent with what is predicted by most accretion models (see, for example, Tucker 1983);

B1) "The size of the X-ray emitting region increases with the mass of the (assumed) central Black Hole". This again is what is expected on a theoretical basis; this size can be reasonably constrained to be in the range 10-50 Schwarzschild radii, 10 R_s being approximately the last stable orbit around a black hole, 50 R_s being the radius where a standing shock can be produced in spherically symmetric accretion (Meszaros and Ostriker 1983);

B2) "The size of the X-ray emitting region increases with the X-ray luminosity". This is a consequence of B1) if the "efficiency" of energy production (that is, the ratio of X-ray luminosity to the Eddington luminosity) is approximately constant for all the objects.

C) "From studies of the emission line spectra of quasars the derived properties of the clouds in the broad line region are: $N_H \gtrsim 10^{22}$ cm^{-2}, $ne = 10^{9-10}$ cm^{-3}, implying sizes for the typical cloud of the order of 10^{13-14} cm (see, for example, Kwan and Krolik 1981 and Krolik, McKee and Tarter 1981)". Note that a single or a few clouds with such a column density ($N_H \gtrsim 10^{22}$ cm^{-2}) can be responsible of the observed low energy X-ray absorption.

D) "The size of the typical absorbing cloud is of the same order as the size of the X-ray emitting region for objects with $L_x = 10^{43}$-10^{44} erg s^{-1}". This derives from the requirement that one or a few clouds can cover significantly the source of the X-ray continuum, thus producing the observed X-ray absorption in low luminosity active nuclei (see points 1 and 2 , above). At the same time, because of A) a much less significant effect would be given by these clouds on the optical continuum (consistently with point 3).

By combining D) and B1) we obtain:

$$10^{13-14} \text{ cm} \simeq (10\text{-}50) \, R_s ,$$

from which we derive that the Schwarzschild radius for these low luminosity active nuclei is of the order of 10^{12} cm, corresponding to a black hole mass of about 3×10^6 M⊙ and to an Eddington luminosity of the order of 3×10^{44} erg s^{-1} . Since the X-ray luminosity of these objects is only a fraction of the total luminosity and L_x/L_{Edd} is already of the order of 0.1, we conclude that, in this scenario, these objects would not be far from being Eddington limited.

E) "In high luminosity quasars ($L_x > 10^{44}$ erg s^{-1}) the size of the X-ray emitting region becomes (much) larger than the typical size of the absorbing clouds". As a consequence, significant absorption, due to the chance passage of one (or a few) clouds along the line of sight, becomes implausible.

X-RAY VARIABILITY

An immediate consequence of the "model" briefly sketched in the previous section is that, in principle, there is the possibility of observing soft X-ray variations (in low luminosity quasars) due to the passage of a thick cloud in front of the emitting region. The expected timescale of variation is given by the ratio between the size (of both the clouds and the emitting region, $10^{13.5}$ cm) and the typical velocity of the clouds. Assuming a velocity of about 3×10^3 km s^{-1} , we obtain an expected timescale of the order of 10^5 s.

A systematic study of the properties of X-ray variability of quasars has been recently performed on a sample of about fifty quasars observed with the EINSTEIN Observatory (see Zamorani et al. 1984). These objects have been analyzed for intensity variations on timescales ranging from a few hundred seconds to 18 months. The main results of this study can be summarized as follows:

a) No object showed significant variability on short timescales, showing that X-ray variability of the order of thirty percent or more is a relatively rare phenomenon on timescales up to a few thousand seconds;

b) On timescales of the order of 10^5 s (about one day) we detected variability in four quasars. Three of these four objects are at the faint end of the X-ray luminosity distribution of the quasars in our sample. Hence, both the timescale and the luminosity of these objects are consistent with what expected on the basis of the model discussed in the previous section;

c) X-ray variability on long timescales (6 - 18 months) has been detected in nine out twelve objects in our sample, showing that it is a very commmon phenomenon on these timescales. However, in contrast with results previously reported for BL Lac objects (Maccagni and Tarenghi 1981; Schwartz, Madejski and Ku 1982), no variation in quasar luminosity larger than a factor of two has been observed.

CONCLUSION

Starting from the existing observational data on the low energy X-ray absorption in active nuclei, and assuming that this absorption is due to optically thick clouds in the broad line region, we have derived order of magnitude estimates of the size of the X-ray emitting region as a function of the X-ray luminosity. The sizes that are derived in this way imply that low luminosity active nuclei (L_x of the order of 10^{43}-10^{44} erg s^{-1}) are not far from being Eddington limited. In these objects there is, in principle, the possibility of observing X-ray variations on timescale of about 10^5 s due to the passage of a cloud in front of the emitting region. Such variations have been actually detected.

REFERENCES

Krolik,J.H., McKee,C.F., and Tarter,C.B.: 1981, Ap.J., 249, 422.
Kwan,J.Y., and Krolik, J.H.: 1981, Ap.J., 250, 478.
Lawrence,A., and Elvis,M,: 1982, Ap.J., 256, 410.
Maccacaro,T., Perola,G.C., and Elvis,M.: 1982, Ap.J., 257, 47.
Maccagni,D., and Tarenghi,M.: 1981, Space Sci. Rev., 30, 55.
Meszaros,P., and Ostriker,J.P.: 1983, Ap.J.(Letters), 273, L59.
Mushotzky,R.F., Marshall,F.E., Boldt,E., Holt,S., and Serlemitsos,P.:
     1980, Ap.J., 235, 377.
Mushotzky,R.F.: 1982, Ap.J., 256, 92.
Petre,R., Mushotzky,R.F., Krolik,J.H., and Holt,S.S.: 1984, Ap.J., in
     press.
Reichert,G., Petre,R., Mushotzky,R.F., and Holt,S.S.: 1983, B.A.A.S.,
     15, 675.
Schwartz,D.A., Madejski,G., and Ku,W.H.M.: 1983, Highlights of
     Astronomy, Vol. 6, Pag. 449.

Tucker,W.: 1983, Ap. J., 271, 531.

Zamorani,G., Giommi,P., Maccacaro,T., and Tananbaum,H.: 1984, Ap.J., 278, 28.

Zamorani,G.: 1984, Proceedings of the IAU Symposium N. 110 on "VLBI and Compact Radio sources", R. Fanti, K. Kellermann and G. Setti (eds), p. 85-94.

EXOSAT Observations of Active Galactic Nuclei

G.Branduardi-Raymont

Mullard Space Science Lab., Holmbury St.Mary, Dorking, Surrey, UK

S.J.Bell Burnell

Royal Observatory, Blackford Hill, Edinburgh, UK

D.Molteni

Istituto di Fisica dell'Universita, Via Archirafi 36, Palermo, Italy

SUMMARY

EXOSAT observations of the Type I Seyfert galaxies ESO141-G55 and NGC5548 are reported. The use of the Low Energy (LE) telescopes in conjunction with the CMA detector and a number of filters allows broad band spectroscopy of these objects in the range .02 to 2.5 keV. While it is difficult to derive information on the shape of the spectrum at the high energy end of the LE experiment (> 500 eV), the soft X-ray response of the CMA permits accurate measurements of the equivalent Hydrogen column density on the line of sight. For NGC5548 we find evidence for a break in the X-ray spectrum at energies below ~500 eV.

INTRODUCTION

Several EXOSAT observations of Active Galactic Nuclei have been carried out to date. The ultimate aim of this programme is the study of their X-ray spectrum simultaneously over the large energy range (.02 to 50 keV) covered by the EXOSAT instrumentation. In this paper we report preliminary results obtained on two objects (ESO141-G55 and NGC5548) using the Low Energy (.02 to 2.5 keV) telescopes (for a description of the instrumentation see de Korte et al. 1981). Broad band spectroscopy was performed with the CMA (Channel Multiplier Array, an imaging detector without intrinsic energy resolution) and a selection of filters which were consecutively positioned in front of it. The filters available for the observations were the 3000 Å thick Lexan, the Aluminium-Parylene and the Boron.

INSTRUMENTAL CHARACTERISTICS AND ANALYSIS PROCEDURE

Fig.1 shows the curves of effective area for the integral instrumentation used, that is the LE telescope, CMA detector and each of three filters (note that the two telescopes onboard EXOSAT have essentially the same effective area, so Fig.1 is representative of both). It is clear that all the curves are very similar at high energies, while they differ substantially below ~.5 keV, in particular due to the edges in the transmission of the different filters. As a consequence, comparing countrates in different filters we will be able to gain information on the spectrum at the low energy end of the LE operational range ($\lesssim$ 500 eV) but it will be more difficult to determine precisely the spectral slope at high energies ($\gtrsim$ 500 eV).

The procedure we adopted in the analysis was as follows: we assumed a power-law spectrum of photon index α = 1.75 for the Seyferts under study, an average value considered 'typical' for AGN (see for instance

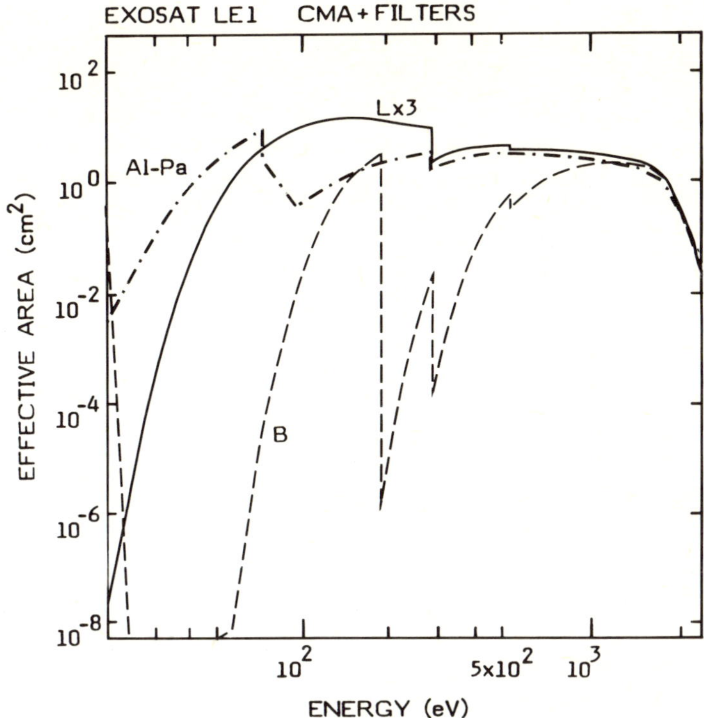

FIG. 1. The effective area of the EXOSAT LE1 telescope in conjunction
with the CMA and various filters.

Mushotzky et al. 1980). For several values of the equivalent Hydrogen
column density N_H we computed the ratio of the counts expected in each
set of two filters which had been used in the observations.   By
comparing the predicted and observed ratios we could then deduce
information about the low energy absorption in the spectrum.

OBSERVATIONS

Table 1 gives a log of the observations and presents the countrates
observed for the two galaxies.   Two observations were made of each
object, one and two months apart for NGC5548 and ESO141–G55
respectively. All data were obtained with the same telescope (LE1) and
filters were alternated in front of the CMA, with the exception of the
first observation of ESO141–G55, which was performed with both LE
instruments and different filters <u>simultaneously</u>.
Fig.2 shows the expected and observed countrate ratio for the first
observation of ESO141–G55. The range of Hydrogen column density
allowed at 1σ is $N_H \geqslant 4.6 \times 10^{20}$ cm^{-2} with a most likely value of 6.7×10^{20}
cm^{-2}. This is marginally higher than the Galactic contribution
$(2–4 \times 10^{20}$ cm^{-2}; Burstein and Heiles, 1982; Ryter et al., 1975, for
$E(B–V)$ to N_H conversion), although it is consistent with it within the
2σ error of the observed ratio. It is not possible to constrain the
Hydrogen column more precisely than this because we have only 2 filter
measurements.   Nevertheless, we can conclude that there is   no
evidence for a low energy excess in the spectrum of ESO141–G55,

TABLE 1.   Details of the observations of ESO141–G55 and NGC5548

ESO141–G55: First and second observation

Date:	4 Sept. 1983		2 Nov. 1983
Telescope:	LE1	LE2	LE1
Filter:	Al–Pa	Lx 3	Lx 3
Exposure:	194 min	243 min	124 min
Counts/s:	.045 + .002	.072 + .003	.052 + .003

NGC5548:

Telescope:	LE1	LE1	LE1
Filter:	Lx 3	B	Al–Pa

First observation (1st February 1984)

Exposure:	47 min	119 min	33 min
Counts/s:	.145 + .008	.010 + .002	.052 + .006

Second observation (3rd March 1984)

Exposure:	238 min	57 min	50 min
Counts/s:	.280 + .005	.023 + .004	.115 + .008

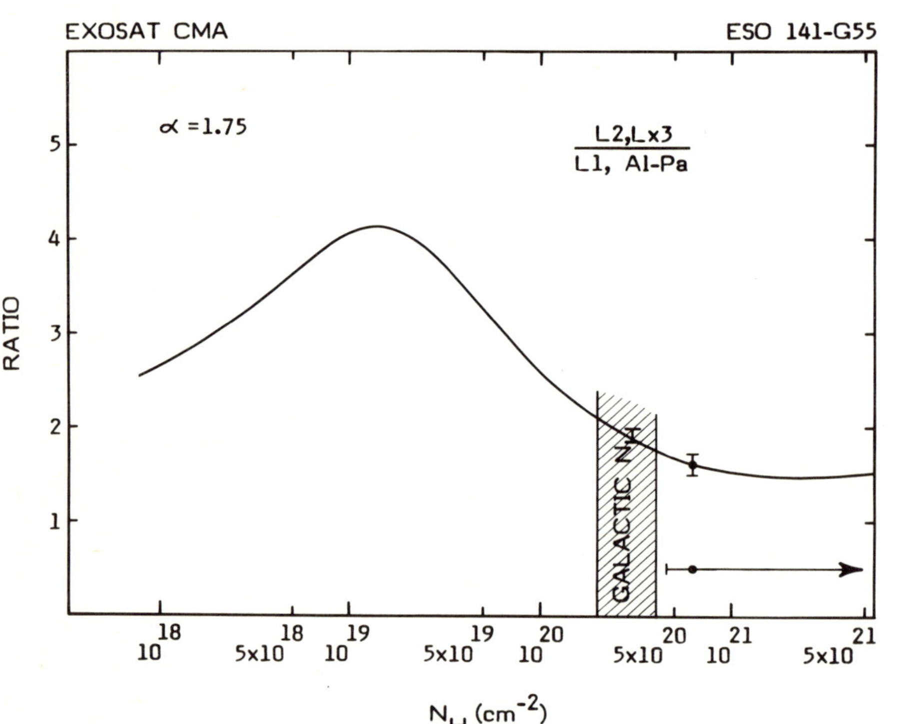

FIG. 2. Expected (curve) and observed (dot and 1σ vertical error bar) countrate ratio for ESO141–G55, first observation. The dot and horizontal error bar indicate the most likely value and error range of the equivalent Hydrogen column density on the line of sight.

in line with the recent findings of Tennant (1983), who has shown the
early report of a steepening of the spectrum at low X-ray energies to
be a spurious result. Moreover, the amount of absorption intrinsic to
this Seyfert must be small, as it is the case for most X-ray AGN.
Comparison of the Lexan 3000 Å countrates during the two observations
(after taking into account the slight difference in effective area of
the LE1 and LE2 telescopes) reveals that a ~20% decrease in the
brightness of ESO141–G55 took place between September and November
1983. For N_H = 6.7x10^{20} cm^{-2} and α = 1.75 the observed countrates
correspond to 4.1 and 3.3x10^{-11} erg cm^{-2} s^{-1} in the .02 to 2.5 keV
range for the first and second observation respectively. These values
convert to a luminosity of 2.5 and 2.x10^{44} erg cm^{-2} s^{-1} for the source.

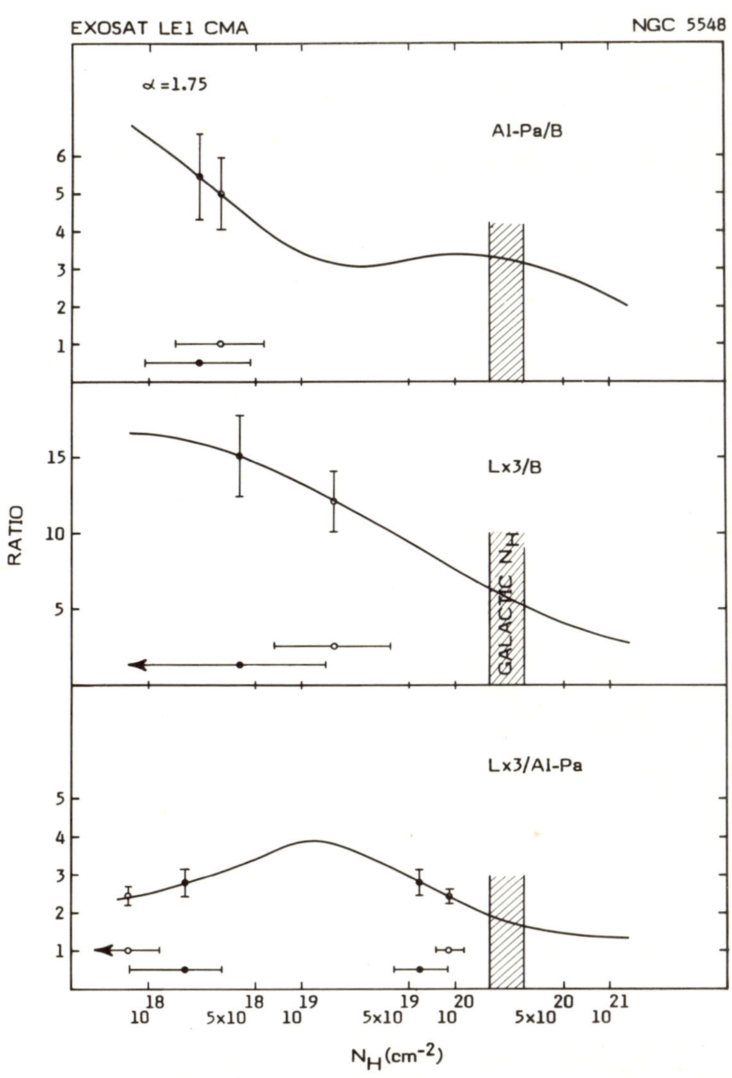

FIG. 3. Expected and observed countrate ratios for NGC5548 (symbols
   as in Fig.2; filled/open dots for first/second observation).

The more interesting case of NGC5548 is illustrated in Fig.3.   For both observations of this Seyfert we were able to use 3 filters in succession, so that we can expect more stringent results in determining its spectrum.   The CMA flux of the source almost exactly doubled between the two observations one month apart, and this happened with all filters, which suggests that there was not significant spectral change despite the large flux variation. Infact Fig.3 shows that the filter ratios for the two observations are rather similar although the implied column density values present large uncertainties, which are due to the poor resolving power of the 'three colour' technique. In particular, the Lexan to Aluminium—Parylene ratio offers two solutions, only one of which is consistent with the Lexan to Boron ratio and only for the first observation. For the second the two filter ratios imply ranges of column density mutually exclusive.   The ratio Aluminium —Parylene to Boron is also shown in Fig.3:   although it cannot add information to that provided by the other ratios, it is useful to prove that the disagreement is not due to some peculiarity of one of the filters.   The remarkable implication of these observations is that the amount of low energy absorption required by the CMA data is about two orders of magnitude lower than that expected from the Galactic obscuration in the direction of NGC5548 (Burstein and Heiles, 1982). For $N_H = 10^{18}$ cm^{-2} and $\alpha = 1.75$ the .02 to 2.5 keV luminosity of the source is 1.8 (first) and 3.5×10^{43} erg s^{-1} (second observation).

DISCUSSION AND CONCLUSION

There are at least two possible explanations of the EXOSAT low energy observations of NGC5548. The first, and most unlikely, is that the flux of the source varied during the course of the observations (our analysis of course assumes that the source is constant while swithching from one filter to another). We can easily exclude this, since the source would have to have varied in almost exactly the same fashion on both occasions. On the other hand, it is much more plausible that the presence of a soft excess in the X—ray spectrum of NGC5548 is responsible for the observations. In turn, this confirms the existence, so far only speculative, of a 'break' in the spectrum, which must occur at or below ~500 eV, where the EXOSAT filters have distinctively different responses. Infact a break is required between the soft X—ray and UV wavelengths to reconcile the flux levels and spectral slopes in the two bands (see for instance Malkan and Sargent, 1982, and Barr et al., 1983). Further observations and detailed fits of the simultaneous ME and CMA data are required before the presence of an excess at low energies in the spectrum of NGC5548 can be properly quantified.   Observations of this source with EXOSAT are continuing. In summary, the results achieved so far show that broad band spectroscopy with the CMA and filters is feasible and promises interesting scientific return.

REFERENCES

Barr P, Willis AJ, Wilson R 1983 MNRAS 203:201-214
Burstein D, Heiles C 1982 A.J. 87:1165-1189
de Korte PAJ et al. Space Science Reviews 30:495-511, 1981.
Malkan MA, Sargent WLW 1982 Ap.J. 254:22-37
Mushotzky RF, Marshall FE, Boldt EA, Holt SS, Serlemitsos PJ 1980 Ap J 235:377-385
Ryter C, Cesarsky CJ, Audouze J 1975 Ap J 198:103-109
Tennant AF 1983 Ph.D.Thesis NASA TM 85101

EXOSAT Observations of Flux Variation from NGC4151

D.R.Whitehouse

Department of Physics and Astronomy
University College London
Mullard Space Science Laboratory
Holmbury St.Mary
Dorking
Surrey, RH5 6NT

SUMMARY

We report the results of a 5-hour observation of the Seyfert galaxy
NGC4151 made with the low energy detectors on EXOSAT. A decrease in
the flux from NGC4151 was detected that was probably due to the
passage of an absorbing cloud of dimensions ~0.1AU along the line of
sight.

Introduction

NGC4151, the X-ray brightest and closest Seyfert, has been
intensively studied at X-ray wavelengths. Ives et al (1976) and Barr
et al (1977) report data taken several months apart with experiment C
on Ariel V. Their observations show no variation in spectral index or
intensity, but do show a variable low energy cut-off due to
photoelectric absorption. This behaviour is in contrast to that of
Cen A where the X-ray slope and absorbing column remain constant and
only the normalisation changes.

The soft X-ray observations require absorption from a large amount
of material (10^{22}–10^{23} atoms cm^{-2}) along the line of sight. This
absorption probably arises from the cold condensed matter that is also
responsible for the broadened optical emission lines. The variation
in the absorbing column would then be due to either the creation and
destruction of clouds along the line of sight or due to orbital
motion.

Observations made with the A-2 experiment on HEAO-1 and the SSS on
HEAO-B confirm this picture with two qualifications. Firstly, the
clouds cannot completely cover the X-ray source; and secondly, that
atoms with $\geq$ have twice the solar abundance. Mushotsky et al (1978)
point out that the limit of the ratio of obscuring filament size, r,
to transverse velocity, V_t, is set at ~10^7 sec by Ariel V.

The Observations

We report the results of a 5-hour observation of NGC4151 made with
the low energy imaging telescopes on board EXOSAT (Taylor et al 1981).
The channel multiplier array detector was used, which is sensitive to
X-rays of energy 0.04–2 keV, but with no intrinsic spectral
resolution. The image is shown in Figure 1 which also shows the

distortion introduced by the partial deployment of the instrument cover.

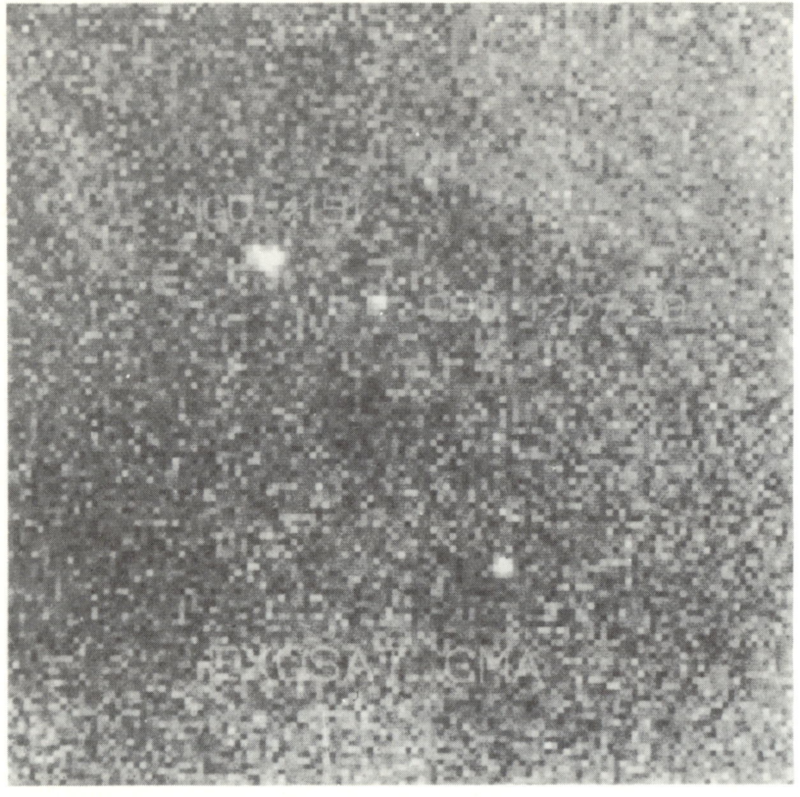

FIG. 1. EXOSAT CMA image of NGC4151 (top left) and the QSO 1207+39
   to its right. The bright spot at the lower right of the image
   is an instrument bright point.

We searched for variability in the flux from NGC4151 by comparing the count rate from the galaxy with that obtained from various regions of the background. Several background regions were selected according to two criteria. Firstly, that they contain no sources above the 3σ level; and secondly, that they contain the same total number of photons as that received from the galaxy.

The timelines from each region were corrected for 'dropouts' and the counts binned on timescales between 1000 and 9500 seconds. The lower timescale is determined by the requirement for a normal distribution of counts in each bin, the upper timescale is merely half the duration of the observation. At intervals of 100 sec a χ^2 comparison was made between the timelines and any values of χ^2 above that expected by the

0.01% level of significance were noted.

Variability was detected from the galaxy on timescales of 1300, 2600 and 3900 seconds. The quasar was analysed in an identical manner and found to be constant. Further investigation revealed that the high values of X^2 were caused by a decrease in the flux from NGC4151 lasting ~2000 seconds. Figure 2 shows this dip which must come from NGC4151 itself as no variation seen in any of the backgrounds could account for it.

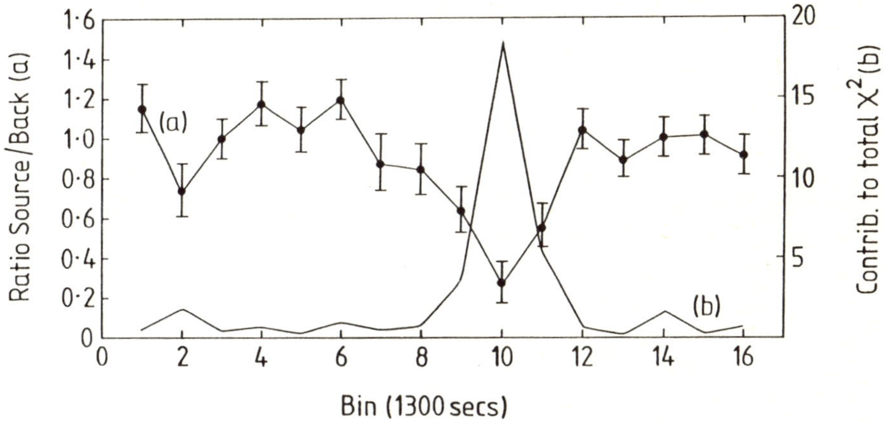

FIG.2. The ratio of counts for NGC4151 to the background and the contribution of each point to the total X^2.

Discussion

There are two possible explanations for this decrease in flux. It could be variation in the continium, or obscuring matter moving across the line of sight. If due to continium variation, a similar event should be visible in data obtained simultaneously by the medium energy (ME) and gas scintillation (GS) detectors which operate at higher energies. Unfortunately, due to instrumental problems, no co-incident ME data is available. Although the GS data shows contamination from solar radiation early in the observation, it is relatively clean at the time of the dip seen in the LE data. It therefore seems likely that this event is due to a to a variation in the absorbing column along the line of sight to NGC4151.

This places a much more stringent limit on r/V_t of ~2×10^3, i.e. 5000 times less than that set by Ariel V. Since we identify the absorbing material with BLR clouds (Sargent 1973) for which we have an indication of the velocity, 10^4 km sec^{-1}, we can estimate the size of the obscuring cloud to be ~2×10^{10} m or 0.1A.U.

References

Barr P. White NE, Sanford PW and Ives JC 1977 Mon. Not. R.A.S. 1981 43p
Ives JC, Sanford PW and Penston MV 1976 Astrophys J. 207, L159
Mushotsky RF, Holt SS and Serlemitsos PJ 1978. Astrophys J. 225, L115
Sargent WL, 1973 in IAU Symposium 55, X-ray and Gamma Ray Astronomy, ed
    RH Bradt and R Giacconi (Dordrecht: Reidel) p 184
Taylor BG, Andresen RD, Peacock A and Zobl R Sp.Sci.Rev. 1981 30,479.

BROAD-BAND CONTEMPORANEOUS STUDIES OF X-RAY BRIGHT AGN

M.J.Coe[1], L.Bassani[2], R.Clement[1,6], A.J.Dean[1], G.Di Cocco[2], B.Elsmore[3],
M. Ferrari-Toniolo[4], F. Giovannelli[4], J.R.Macdougall[5], P.Persi[4] and
L.Spinoglio[1,4].

1. Physics Dept., Southampton Univ., U.K.   2 TESRE-CNR, Bologna,Italy
3. MRAO, Cambridge University, U.K.   4 IAS-CNR, Frascati, Italy
5. Rutherford Laboratory, U.K.   6 Valencia University,Spain

SUMMARY

Multiwaveband contemporaneous observations have been collected for
the three X-ray bright active galactic nuclei: MCG 8-11-11, Mkn 501 and
NGC 1275.
The Seyfert galaxy MCG 8-11-11 and the BL Lac object Mkn 501 were
observed contemporaneously in radio (MRAO, Cambridge,U.K), far-infrared
(IRAS), near-infrared (TIRGO, Gornergrat, Switzerland). and ultraviolet
(IUE); while the peculiar galaxy NGC 1275 was observed at near-infrared
and ultraviolet wavelengths.
The preliminary results of this study are presented.

INTRODUCTION

Contemporaneous observations with a wide frequency coverage are essential for extragalactic objects such as active galactic nuclei. Such
measurements are an important first step to the understanding of the physical mechanism by which the central energy source produces the emitted
radiation. In particular the ultraviolet continuum in the range 1,200-
1,900 A defines the slope of the nuclear power law, while the near infrared (1-5μm) and far-infrared (30-100μm) photometries result in the determination of the stellar and, eventually, dust contributions respectively, coming from the surrounding galaxy.
Due to the sometimes rapid variability in the emission of active galactic nuclei (BL Lac objects have variability timescales as short as one
day), simultaneous, or at least contemporaneous, observations are required.

OBSERVATIONS

Table 1 summarizes the observations of the three galactic nuclei collected during the period September-October 1983: for each observation are
reported the date, the observatory involved, the observing band(s), the
logarithm of the frequency and the correspondent logarithm of the flux
density in mJy ($1mJy=10**(-26)erg\ s^{-1}cm^{-2}Hz^{-1}$).
The far-infrared data collected by the Infrared Astronomical Satellite
of two or the three sources of this study are not yet available at the
time of writing.
The radio observations of MCG 8-11-11 and Mkn 501 were obtained at
the 5km radio telescope of the Mullard Radio Astronomy Observatory
(Cambridge, U.K.) at the frequency of 2.7 GHz. The flux density scale
is based on calibrations made using 3C 48 and 3C 286 for which the adopted values are 9.32 and 9.98 Jy respectively.

TABLE 1. Multifrequency observations of three AGN

Source	Region	Date (U.T)	Observatory	Observing band	Log ν (Hz)	Log S_ν (mJy)
MCG 8-11-11	radio	Sept. 28/Oct 1-3-4	MRAO	2.7 GHz	9.43	2.110 ± 0.010
	far-IR	Sept. 24-25	IRAS	12-25-60-100μm	*	*
	near-IR	Sept. 19	TIRGO	J	14.38	1.430 ± 0.030
				H	14.26	1.680 ± 0.016
				K	14.13	1.815 ± 0.012
				L	13.92	2.030 ± 0.080
	UV	Oct. 6	IUE	1200-1900A	15.29	0.621
NGC 1275	near-IR	Sept. 21	TIRGO	J	14.38	1.633 ± 0.016
				H	14.26	1.521 ± 0.024
				K	14.13	1.530 ± 0.030
				L	13.92	1.658 ± 0.090
	UV	Oct. 6	IUE	1200-1900A	15.20	0.205
Mkn 501	radio	Sept.26-28/Oct	MRAO	2.7 GHz	9.43	3.125 ± 0.012
	far-IR	Oct. 6-7	IRAS	12-25-60-100μm	*	*
	near-IR	Sept. 24	TIRGO	J	14.38	1.425 ± 0.025
				H	14.26	1.609 ± 0.016
				K	14.13	1.522 ± 0.020
				L	13.92	1.64
	UV	Oct. 1	IUE	1200-1900A	15.29	0.333

* data not yet available

The near-infrared photometry of all the three objects has been obtained at the F/20, 1.5m Italian Infrared Telescope on the Gornergrat,
Switzerland (TIRGO) (Citterio et al 1981), using an InSb photometer with
standard J.H.K and L filters. The monochromatic flux for a 0 mag star
is taken to be 1,635 Jy at J (1.25μm), 1,090 Jy at H (1.65μm), 665 Jy at
K(2.2μm) and 277 Jy at L(3.6μm). The beam size used for all the observations was 13 arc sec. The magnitudes of the B.S. standard stars
used during the observations are given by Koorneef (1983). The corrections for reddening have been carried out assuming the values of E(B-V)=
0.173 and E(B-V)=0.385 for NGC 1275 and MCG 8-11-11 respectively (Mc-
Alary et al 1983) and the ratios between the extinction in the different near infrared bands given by Glass et al (1982). No reddening correction is required for Mkn 501, as the adopted value of E(B-V)=0.083
makes them negligible.

The ultraviolet observations have been performed using the short wavelength camera (1150-1950 Å) of the International Ultraviolet Explorer
(IUE), in the large aperture (10x20 arc sec), low resolution ($\Delta\lambda \sim 6$Å)
mode at the ESA Villafranca satellite tracking station (VILSPA). The
three images obtained for Mkn 501, MCG 8-11-11 and NGC 1275 were SWP
21211, SWP 21247 and SWP 21248 with exposure times of 202, 200 and 134
minutes respectively. The ultraviolet data have been de-reddened for
the extinction in the Galaxy, including Mkn 501 in this case, taking the
previous mentioned values of E(B-V). The logarithm of the average flux
densities at 1,550 Å are reported in Table 1.

RESULTS

The spectral energy distributions of MCG 8-11-11, NGC 1275 and Mkn501
are reported in Fig.1. In the following sections we describe the results on each individual source. A more detailed description of the
ultraviolet results is given by Bassani et al (1984).

MCG 8-11-11
During the period of our contemporaneous observations, no rapid variability was detected in the radio emission of MCG 8-11-11 in a one week
time interval at the frequency of 2.7 GHz.
The near-infrared spectrum is well fitted by a power law with spec-
tral index α_{IR} = -1.29 ±0.11, where $S \propto \nu^{\alpha}$. This infrared spectral index is steeper than the value of α_{IR} = -0.86± 0.05 computed from the observations of McAlary etg al (1983), collected with a slightly larger
aperture size (15.8 arc secs).
The slope of the spectrum becomes flatter towards the ultraviolet,
giving a spectral index $\alpha_{UV} \sim -1.10$.
For this Seyfert galaxy the IRAS data will be very important, because
they will cover a region over which any thermal component due to cold
dust surrounding the nucleus will be easily seen.

NGC 1275
The near infrared spectrum observed in NGC 1275 shows two peculiar-
ities:(1) the level of the emission was much lower than previously observed, and (2) the measured infrared colours (J-H)=0.15 and (H-K)=0.55
were unusual not only for this galaxy, but also for active galaxies in
general (see,for example, Ward et al 1982). A possible explanation
of this spectral distribution could be the following: the non-thermal
emission due to synchrotron radiation was so faint, i.e. the nucleus
was "switched off", that it was possible to see in the spectrum a contribution from stars younger and hotter than expected in normal galaxies.
In fact, the black-body radiation arising from these stars would be
peaking at shorter wavelengths ($\lambda < 1.25$μm = J-band), rather than around
the H-band (1.65μm) as expected from a normal galactic population domin-

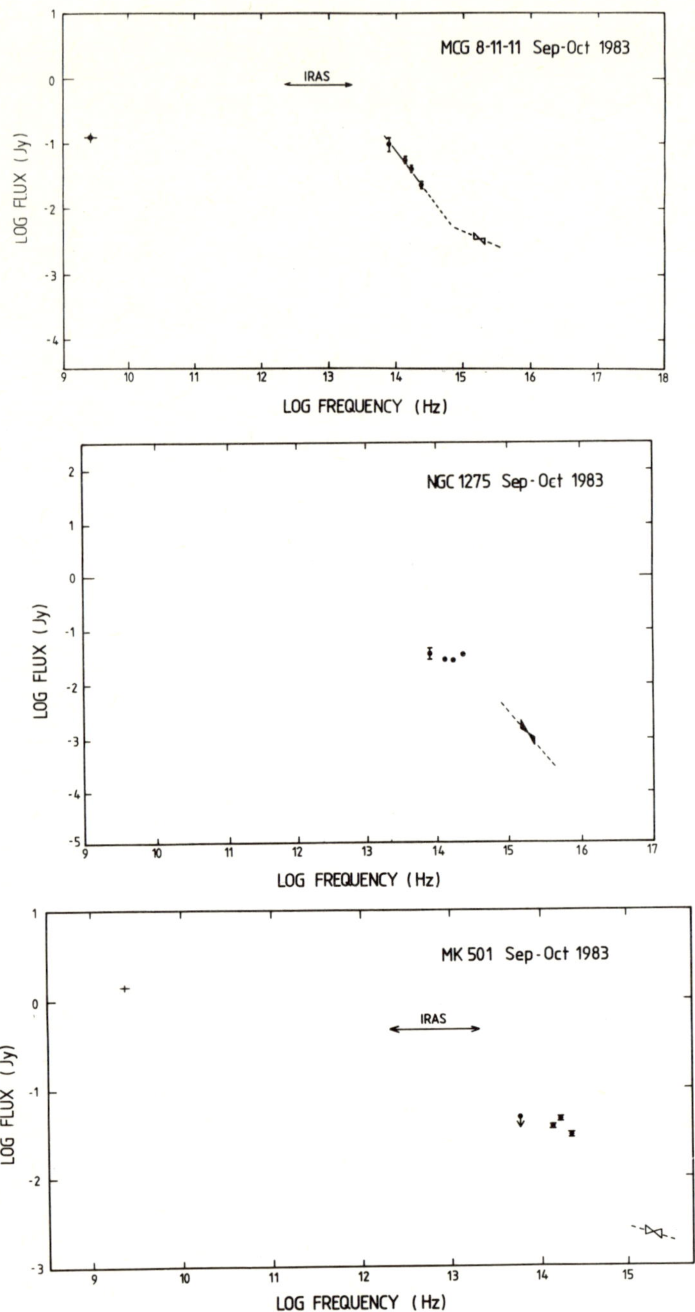

FIG.1. The composite energy distribution of MCG 8-11-11, NGC 1275 and Mkn 501 during the period September-October 1983.

ated by giant late-type stars. This hypothesis is supported by the
optical observations of Malkan and Filippenko (1983), who have measured
roughly the same blue colour either from the nucleus and the nearby star
light of NGC 1275, a colour which is several tenths of a magnitude
bluer than the one of a standard galaxy. They conclude that the additional blue light around the nucleus of NGC 1275 probably comes from a
population of young stars.

However, this particular source needs further detailed studies in the
near-infrared (e.g. monitoring and multiaperture photometry) and will be
one of the targets of our future observations.

The ultraviolet spectrum of NGC 1275 shows a spectral index of the
continuum of $\alpha_{UV} \sim -1.8$, similar to the value of $\alpha_{UV} = -1.6$ given by
Maraschi et al (1983).

<u>Mkn 501</u>

The BL Lac object Mkn 501 does not show rapid radio variability at
the frequency of 2.7 GHz in a one week time interval.

The near-infrared spectrum is characterized by a bump in the H-band,
due to the stellar contribution of the galaxy surrounding the BL Lac object.

The ultraviolet spectrum is fitted by a spectral index of $\alpha_{UV} \sim -0.25$
and is featureless, as expected for BL Lac objects.

For this object the IRAS data, which should be available in May 1984,
will be of great interest, because they will confirm, or otherwise,that
the synchrotron emission is the dominant radiation mechanisms in these
objects.

<u>Acknowledgements</u>

The authors acknowledge the technical staff at the TIRGO Telescope
(CAISMI-CNR), Osservatorio Astrofisico di Arcetri, Florence, Italy) and
the VILSPA staff (ESA, Villafranca, Spain) for their very useful support.

L.Spinoglio would like to acknolwedge the receipt of a SERC research
fellowship at Southampton University, and R. Clement acknowledges the
support by the British Council.

REFERENCES

Bassani L., Coe, M.J., Clement, R., Dean A.J., Di Cocco G., Elsmore B.,
    Ferrari-Toniolo M., Giovannelli F., Macdougall J., Persi P., Perotti
    F., Sembay S., Spinoglio L., Ubertini P.: 1984, Proceedings of the
    Fourth European IUE Conference, 15-18 May, Rome, Italy (In press)
Citterio O., Dilworth C., Iucci N.: 1981, Sky & Telescope, July ,17.
Glass I.S., Moorwood A.F.M., and Eichendorf W.: 1982, Astron.Ap.,<u>107</u>,276
Koorneef J.: 1983, Astron.Ap.Suppl.Series, <u>51</u>, 489.
Malkan M.A., and Filippenko A.V.: 1983, Ap.J.,<u>275</u>, 477
Maraschi L., Tanzi E.G., Treves A.: 1983, Mem.S.A. It., <u>54</u>, 399
McAlary C.W., McLaren R.A., McGonegal R.J., and Maza J: 1983, Ap.J.
    Suppl.Series, <u>52</u>, 341
Ward M., Allen D.A., Wilson A.S., Smith M.G., Wright A.E: 1982, Monthly
    Not.R.astr.Soc., <u>199</u>, 953.

Gamma-rays From Active Galactic Nuclei

L. Bassani[1], A.J. Dean[2], G. Di Cocco[1] and F. Perotti[3]

[1] Istituto TESRE, CNR, Via de' Castagnoli, 1, 40126 Bologna, Italy

[2] Physics Department, Southampton University Highfield, Southampton SO9-5NH, England

[3] Istituto di Fisica Cosmica, CNR, Via Bassini, 15, 20132 Milano, Italy

SUMMARY

In this paper, we review the gamma-ray data on the active galactic nuclei observed to date and we draw some conclusions as to their general characteristics. A number of possible gamma-ray production mechanisms are discussed in light of these measurements. In particular, the relevance to gamma-ray emission of jet structures is presented. The active galaxies contribution to the cosmic diffuse background at gamma-ray energies is estimated. Future prospects for extragalactic gamma-ray astronomy are outlined in view of the coming generation of gamma-ray telescopes.

INTRODUCTION

Extragalactic gamma-ray astronomy is in the early exploration phase. The achievements to date have been limited by the marginal sensitivity of the first wave of effective telescopes and have been restricted to the study of a few objects. However, some extremely significant facts have emerged from this data. Of immediate interest is the discovery that a range of different active galactic types emit gamma-rays. The classes detected at gamma-ray wavelengths include Seyferts, QSO and radio galaxies and are generally represented by the nearest, brightest candidate within the appropriate sub group. This observational situation augers well for the future of extragalactic gamma-ray astronomy since it implies that many more examples of each type emit gamma-rays and, as the sensitivity of gamma-ray telescopes improves, an increasing number of objects will become detectable. Of even greater interest, however, is the clear evidence that Active Galactic Nuclei (AGN) emit the major fraction of their power in the gamma-ray region of the electromagnetic spectrum. From the limited information available it appears that many active galaxies are primarily gamma-ray galaxies. Consequently gamma-ray observations have to be fundamental to the understanding of the nature of these distant objects, a conclusion which is further reinforced by the extreme penetrating power of gamma-ray photons, a factor which enhances their ability to probe deep into the nuclear regions of active galaxies. It is perhaps no accident that the maximum penetrating power of photons through matter is close to 1 MeV and that at this energy active galaxies show a maximum luminosity. The variability timescale of the X/gamma-radiation from AGN ($\sim$ months) may be taken as evidence that the emission is intimately related to the region containing the central power house. If this, as generally accepted, is a massive (M $\sim 10^6$ M$_\odot$) black hole system, then the gamma-emission should permit the study of mildly relativistic plasmas (T $\sim 10^9$ °K) which are thought to form in the inner regions of the associated accretion disks. Since gamma-rays from AGN are generated in a very compact and extremely luminous region, their major cause of opacity (for energies greater than E = $2m_e c^2 \simeq$ 1MeV) is related to collisions with local X-ray photons which lead to the creation of electron positron pairs. In fact, the absorption can be so severe that in order to observe gamma-rays from AGN their emission must be highly collimated into gamma-ray beams which are oriented closely to the direction of the earth. The physical processes associated with the emission and absorption of gamma-rays are very different from those associated with the emission of radio photons from jet structures and thus provide an independent method of determining a unique set of parameters which describes the geometry and kinematics of relativistic jet motion in AGN.

The high gamma-ray luminosity of AGN has also implications with respect to the observed "gamma-ray cosmic diffuse background". On the basis of existing gamma-ray measurements of different types of AGN, it is not only possible to account for the entire cosmic flux but over-subscribe the observed intensity. Time variability and/or restricted angular emission of the gamma-rays may well be needed to reduce the AGN contribution. Whatever the final outcome, it is readily apparent that emission from AGN must provide an important contribution to this cosmic flux. Furthermore, the similarity of the 'bump' in the spectral emission of the cosmic diffuse background close to a few MeV and the peak in the luminosity of AGN at these energies provides another clue to this origin. Consequently, a further insight into the general characteristics of AGN and, possibly into the conditions in the early Universe, may be obtained if one is able to unambiguously associate this diffuse gamma-ray emission with the contribution from a large number of active galaxies.

GAMMA-RAY MEASUREMENTS OF ACTIVE GALACTIC NUCLEI

At X-ray wavelengths a great deal of observational information now exists on the emission from a large number of each of the major classification of active galaxies. On the other hand only few sources have been studied in the gamma-ray region of the spectrum, revealing however, that a good deal of physics can only be learned from observation at E > 0.5 MeV. This discovery phase of extragalactic gamma-ray astronomy is very similar to the early days of X-ray astronomy in the pre-UHURU epoch, before the advent of progressively more sophisticated X-ray satellites, which have established this field as a fully accepted branch of astronomy.

There was a time, however, in which X-ray observations were characterised by a few detections as today happens in gamma-ray astronomy. Table I lists a number of observations of a few extragalactic objects performed by rocket flights in late 1960, and by the UHURU satellite in the early 1970s. As is evident from the Table, the statistical significance of these results were similar to that of present day gamma-ray measurements of the same objects. Then, it is reasonable to expect that improvements in the sensitivity of future gamma-ray detectors if used in conjunction with a generation of gamma-ray satellites, designed to cover all bands from low (few MeV) to high energies (few GeV), will open new opportunities and reveal exciting new discoveries.

TABLE 1. <u>First observations of AGN in X and gamma rays</u>

	X-RAY ASTRONOMY				GAMMA-RAY ASTRONOMY			
Source	Rockets		UHURU		Balloon		COS-B	
	E(KeV)	n σ	E(KeV)	nσ	E(MeV)	nσ	E(MeV)	nσ
3C 273	1–4.5	3 [1]						
	0.25–12	3.1–3.9 [2]	2–6	6 [5]	–	–	50–800	5–7 [6]
CEN A	0.25–12	1–3 [2]	2–6	11 [5]	0.01–2.2	6 [6]	–	–
NGC 1275	0.5–10	2.9–3.8 [3]	2–6	29 [4]	0.02–0.1	5 [6]	–	–
NGC 4151	–	–	2.6	3.8 [4]	0.04–20	6.6 [6]	–	–
					0.02–19	5.2 [6]		

1. Byram et al. (1971)
2. Bower et al. (1970)
3. Fritz et al. (1971)
4. Gursky et al. (1971)
5. Kellogg et al. (1971)
6. References in this work

Now, however, we deal with a limited sample of results and can only draw tentative conclusions as to their possible general characteristics by examining their gamma-ray emission.

The Seyfert Galaxy NGC 4151

At the present time NGC 4151 is not only the best studied extragalactic object at gamma-ray frequencies, but has also been thoroughly investigated at all wavelengths available to the astronomer over a period of years (for a review see Bassani 1981). Its optical spectrum is of basic Seyfert Type I character but with some type II characteristics. Furthermore NGC 4151 is one of the nearest (20 Mpc for $H_o = 50$ km/sec Mpc) and intrinsically weakest (L_x(2–10 KeV) = 7 10^{42} erg/sec) galaxies of the Seyfert class. The overall electromagnetic spectrum appears to contain multiple components (galaxy, nucleus, dust, etc.) having different weights in the different energy bands. However, the high degree of variability observed in the infrared, optical, ultraviolet and X-ray regions of the spectrum (Rieke & Lebofsky 1981, Lyuty 1978, Perola et al. 1982, Lawrence 1980) suggests that the nuclear non-thermal power law component becomes progressively more important going to higher energies until it dominates totally the gamma-ray energy output. NGC 4151 has been observed several times in the low gamma-ray energy range. Table 2 lists these observations.

Originally, the low energy gamma-ray measurements were highly contradictory with detections reported by the MISO team up to 20 MeV (Perotti et al. 1979, 1981a) and by the HEAO 1/A4 group up to few MeV (Baity et al. 1983) and with bracketing upper limits set at much lower flux values by other experimental groups (Meegan & Haymes 1979, Schönfelder 1980, White et al. 1980).

However, it is now evident that all these observations can be explained in terms of a highly variable gamma-ray emission output

TABLE 2. Observation of NGC 4151 in the low energy gamma-ray range

Energy range (MeV)	Experiment	Date	Reference
0.03- 1.2	Ariel V	Dec. 1976	Coe et al. (1981)
0.1 -20	MISO	22 May 1977	Perotti et al. (1979)
1.0 -20	MPI	1 Oct. 1977	Schonfelder (1980)
0.1 -10	Rice U	4 Oct. 1977	Meegan & Haymes (1979)
0.05- 2.0	HEAO A4	Dec. 1977	Baity et al. (1983)
0.05- 2.0	HEAO A4	June 1978	Baity et al. (1983)
1.0 -20	UCR	1 Sept. 1978	White et al. (1980)
0.05- 2.0	HEAO A4	Dec. 1978	Baity et al. (1983)
0.1 -20	MISO	30 Sept. 1979	Perotti et al. (1981a)
0.1 -20	MISO	17 May 1980	Perotti et al. (1983)

over a long timescale (see for example figure 3a, showing the time
history of the NGC4151 flux at 100 keV taken from Baity et al. 1983).
Although the poor statistical significance achieved by the present
generation of gamma-ray telescopes do not allow an accurate determination of the spectral shape above few hundred keV, it is generally
accepted that there must be a spectral change from the lower energy
power law in order to be consistent with the SAS-2 & COS-B 35-200 MeV
upper limits (Bignami et al. 1979, Pollock et al. 1981). There is no
general agreement, however, on the actual location of this spectral
break and/or the spectral shape before the steepening. The early MISO
observations (1977, 1979) favour a power law spectrum having a photon
index α close to 1 which steepens to $\alpha \sim 3$ at few MeV photon energies
(Perotti et al. 1979, 1981a, figure 1). The HEAO 1/A4 measurements
(Baity et al. 1983; figure 2) are more compatible with a single power
law of photon index $\alpha = 1.6$ in the range 2 KeV - 2 MeV for two out of
three observations. The third A4 observation in 1978, when the source
intensity was a factor of two higher, requires a change below 50 keV
from a similar power law to a steeper power law or exponential. As for
the 1980 observations of NGC 4151, the combination of the 2-10 keV
Einstein MPC flux measurement and the hard X-ray MISO data points seem
to indicate a spectrum harder than $\alpha = 1.5$, with a possible break and
steepening at around 50 KeV (Perotti et al. 1983; figure 2). Figure 3
taken from Baity et al. (1983), shows the time history of measured
spectral indices together with the corresponding 100 keV flux values.
When plotted one against the other for the most significant high
energy spectra (figure 4) a weak correlation (3.5 σ) seems to emerge
from the observations, indicating that the spectrum hardens during
flux increases, in a fashion similar to that reported at UV frequencies (Perola et al. 1982). Apart from being variable on a long timescale, the hard X-ray flux shows fluctuations also on a short timescale, typically of the order of days, in phase with the 2-10 keV flux
(figure 5, Baity et al. 1983). At higher energies (> 1 MeV) the
variability implied by the measurements is typically a factor 3-10
over a long timescale (years).

   Furthermore, these variations may occur in step with the 100 KeV
variability. There are positive detections in mid 1977 and late 1979
(MISO data) when the 100 KeV flux was relatively high and there are
upper limits at low flux level in late 1977 and 1978 when the 100 KeV
flux was lower in intensity. In particular the 0.5-5 MeV luminosity
measured by the MISO telescope changed by a factor 4 $\pm$ 2 between May
1977 and September 1979 from $2 \, 10^{45}$ to $5 \, 10^{44}$ erg/sec.

   The shortest timescale of variability in the few MeV range can be

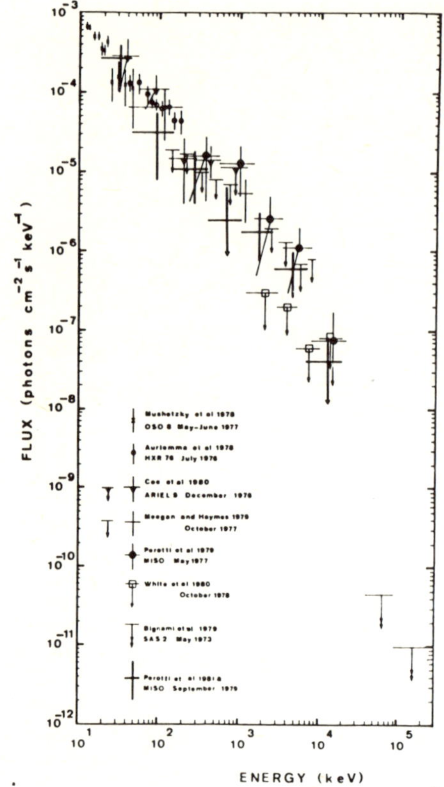

FIG. 1. MISO observations (1977-79) of NGC4151 (Perotti et al. 1981a)

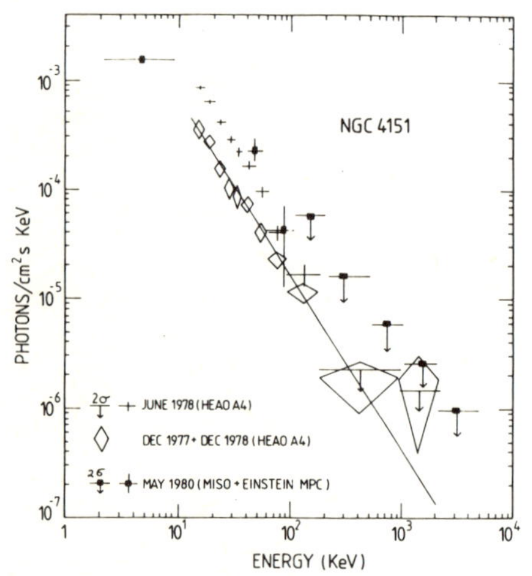

FIG. 2. MISO (1979) and HEAO-A4 observations of NGC4151 (adapted from Baity et al. 1983)

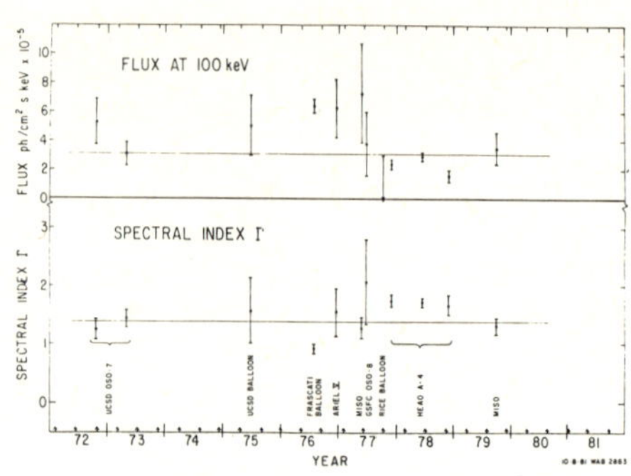

FIG. 3. NGC4151: Time history of 100 keV flux and power law spectral index (Baity et al. 1983)

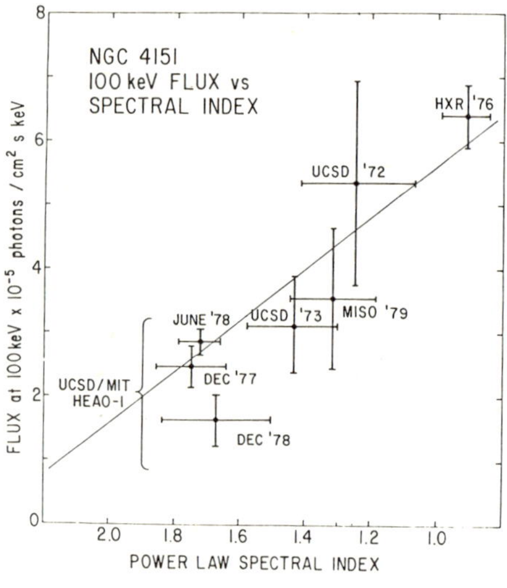

FIG. 4. Correlation between spectral index and 100 keV flux (Baity et al. 1983)

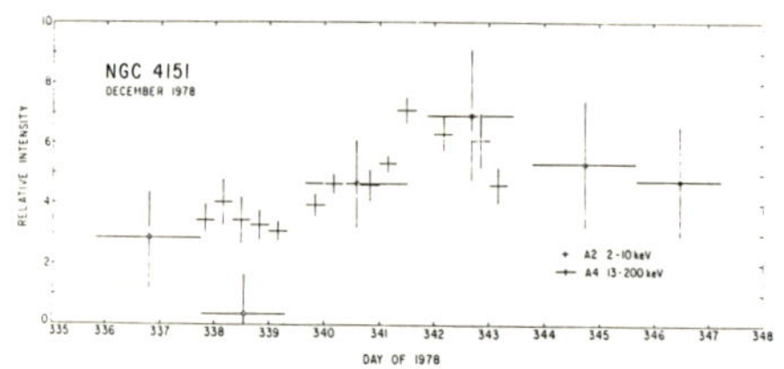

FIG. 5. Short timescale fluctuations of NGC4151 (Baity et al. 1983)

set at $\Delta t \sim 6$ months by comparison with the 1977 observations of the MISO and MPI/Rice University groups.

The Seyfert Galaxy MCG 8-11-11

MCG 8-11-11 is a face on spiral galaxy with a bright Seyfert nucleus. The redshift of z = 0.0205 locates this object at a distance of ~ 123 Mpc (for H_o = 50 km/sec Mpc). MCG 8-11-11 is known to be variable both in the optical and X-ray band on timescales ranging from a few days to months (Miller 1979, Mushotzky & Marshall 1980 and Tennant & Mushotzky 1983). In particular the 2-10 keV luminosity changes from $5 \ 10^{43}$ to $1.2 \ 10^{44}$ erg/sec. The HEAO 1/A2 and Ariel VI satellites have measured the emission spectrum of this object up to ~ 30 keV (Mushotzky et al. 1981; Hall et al. 1981) and have shown evidence of spectral variability. While the HEAO 1/A2 spectrum is compatible with a power law spectrum of index α = 1.7, the Ariel VI results are better fitted by a power law of index $\alpha = 2.1$ together with an iron line feature at 6.2 keV. Hard X-ray emission above 20 keV from the region of sky containing MCG 8-11-11 has been reported by Frontera et al. (1979).

The recent low energy gamma-ray measurement of MCG 8-11-11 (Perotti et al. 1981b) exhibits similar spectral characteristics of those observed by the MISO telescope from NGC 4151 in 1977-1979. A power law representation of the spectral data requires a break between 2 and 4 MeV, followed by a steeper spectrum at higher energies. At energies immediately below the break the best fit photon spectral index gives a value close to unity. If it is assumed that for energies above the break, the spectrum must pass through the upper limits set by the SAS-2 instrument (Bignami et al. 1979), an index of $\alpha \sim 4$ is required. The upper limits set by Graser & Schönfelder (1982) in the 1.1-10 MeV range are not in severe contradiction with the positive MISO measurement. Figure 6 shows the photon spectrum above 1 keV from the region of the sky containing this active galaxy. The iron emission observed by Ariel VI may be due to fluorescent excitation of the gas in the broad line region surrounding the X-ray emitting nucleus (Hall et al. 1981). It is interesting to note that one requirement of this model, is that the flux several weeks before the observation be a factor of 3 greater (to allow for light travel time and force a decrease in the required solid angle of the target as seen from the source). This is just consistent with the flux measurement obtained by MISO 8 weeks before the Ariel VI observation. Therefore, although no other gamma-ray data exist on this source, it is very likely that in this case, as for NGC4151, the gamma-emission is variable.

The Peculiar Galaxy NGC 1275

While NGC 1275 was originally classified as a Seyfert Galaxy (Seyfert 1943), its classification as a true Seyfert must be considered notional. The existence of strong radio emission has led this object to be described as a 'peculiar elliptical radio galaxy' by Adams (1977) and Veron (1978) has recently suggested that it may well be a BL Lac object. The main characteristic of NGC 1275 is probably its particular location within the Perseus Cluster (Branduardi-Raymont et al. 1981). X-ray observations of this cluster of galaxies using detectors on HEAO-1 (Primini et al. 1981 and Rothschild et al. 1983) and OSO-7 (Rothschild et al. 1981) have measured a hard non thermal component in addition to a thermal bremsstrahlung spectrum (KT ~ 6 KeV). In both cases the authors attribute the non thermal source to the compact nucleus of NGC 1275. As determined by the grazing incidence X-ray telescope on Copernicus (Fabian et al. 1974), the best

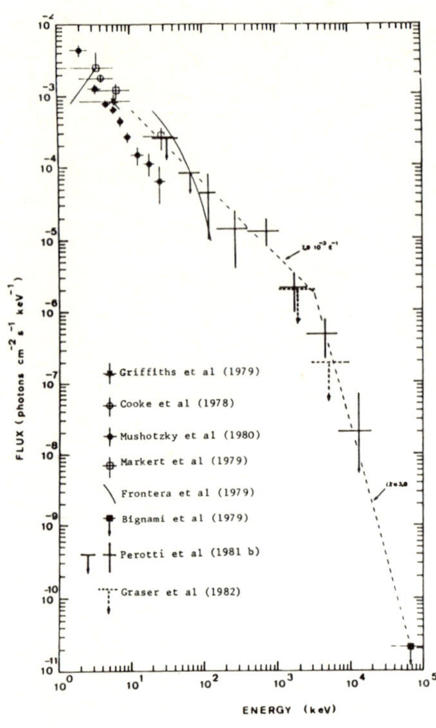

FIG. 6.   Photon spectrum of MCG
8-11-11 above 1 keV (Perotti
et al. 1981b)

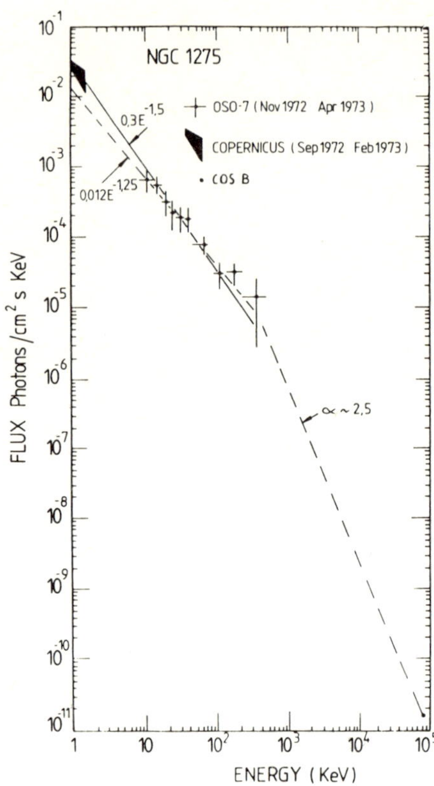

FIG. 7.   X/γ -ray spectrum of
NGC 1275 (adopted from Rothschild et al. 1981)

estimate of the spectral shape of the emission from NGC 1275, passes
through the non-thermal spectral component as deduced by OSO-7 with a
photon spectral index of α = 1.5. If, however, the UCSD data are
considered alone, a good fit is obtained using a photon spectral index
of α close to unity (Rothschild et al. 1981). Consequently a flattening of the emission spectrum above about 100 keV is indicated in a
similar way to that observed by the MISO telescope from Seyfert
galaxies. Recently Strong and Bignami (1983) have noticed that the
COS-B data show a high energy gamma-ray excess (> 70 MeV) from the
general direction of this galaxy, although some additional emission
extending several degrees to the south east must be explained in other
ways. The data are consistent with the presence of a point source of
flux I_γ (> 70 MeV) = 0.83 10^{-6} ph cm^{-2} sec^{-1}. The fit to the expected
point source profile is striking in the 70-300 MeV band, less so in
the 70-5000 MeV band, suggesting a soft source. The residual inferred
incident X/γ ray spectrum of the Perseus Cluster after subtraction of
the best estimate of the thermal bremsstrahlung contribution is shown
in figure 7, where we have added the recently measured COS-B data
point at 70 MeV.

Even though the COS-B & OSO 7/HEAO 1/A2 observations were not contemporaneous, there is an indication in the spectral data points that
the hard X-ray power law(α = 1.5) must have a break or significant

steepening somewhere below 70 MeV in a similar way as observed in NGC
4151 and MCG 8-11-11.

A comparison between observations of this source over a period of
several years shows that the non thermal hard X-ray source is variable
on timescales of years (Rothschild et al. 1981) and implies an emission region of less than one parsec, which is a value consistent with
the VLBI size of the compact source at the galaxy center (Preuss et
al. 1979).

The Radio Galaxy Centaurus A

Centaurus A (NGC 5128), generally believed to be the closest radio
galaxy ($\sim$ 5 Mpc for H_o = 50 km/sec Mpc), is also the only one seen so
far in gamma-rays. It has now been observed over a wide range of
gamma-ray frequency bands from low energy (up to a few MeV) to very
high energy ($\sim 10^{12}$ eV). The source has been observed 3 times in the
soft gamma-ray energy range (Hall et al. 1976, Baity et al. 1981,
Gehrels et al. 1984) and while only upper limits to the flux exist in
the range 30-200 MeV (Bignami et al. 1979) there is an indication of
detection at 300 GeV (Grindlay et al. 1975). Figure 8 shows the
combined X, gamma-ray photon spectrum of Cen-A as obtained by different experiments. Hall et al. (1976) indicate that a good fit to the
continuum radiation at the time of their observation is of the form
$KE^{-\alpha}$ with $\alpha = 1.9$, in the energy range 0.03-12 MeV. More recently this
object has been studied in the range 80-2300 keV, using the A4 instrument on HEAO-1 (Baity et al. 1981).

Figure 8 shows the combined X, gamma ray photon spectrum of Cen-A
as obtained by different experiments. In particular the HEAO A2 and A4
detector observations indicate that a power law photon spectrum with
index α = 1.60 breaking to $\alpha \sim 2$ at 140 keV best describe both the 1978
January and June data. However, the data are also consistent with a
single power law spectrum having index α = 1.65. A recent balloon
observation of this object, with a low energy gamma-ray spectrometer
(LEGS), has measured a similar power law spectrum (α = 1.59) in the
range 70-500 keV, with no evidence for a break at 100 keV (Gehrels et
al. 1984). This unbroken spectrum, when extrapolated, lies above the
2 σ upper limits obtained from the SAS-2 observation of this galaxy in
the energy range 35-200 MeV and is three decades above the reported
flux at energy 300 GeV. Even taking into account the fact that these
observations were not contemporaneous, they give an indication that
the power law must significantly steepen somewhere below 30 MeV.

During the period of each HEAO 1/A4 observation there was no significant indication of variability on a timescale of days or less but
there was a significant ($\sim$ 50%) decrease in the intensity between the
two observations (Δ t = 6 months). Upper limits can, however, be set
on short term variations of the order of days (35%) and hour (25%)
(Baity et al. 1981, Gehrels et al. 1984). The average luminosity in
the low gamma-ray energy range is 2-4 10^{43} erg/sec, exceeding that
observed in any other part of the spectrum. It is interesting to note
that the detection at 0.3 TeV, includes evidence for intensity variability as well: the flux in 1972 (April-June) was stronger than in
1973 (no significant detection) or in 1974. Comparison with previous
measurements indicates the presence of low energy variability (100
keV) on a similar (yearly) timescale (figure 9a from Baity et al.
1981). Contemporaneous spectral index measurements for the period
1968-1979 are also shown (figure 9b). It can be seen that there is no
clear correlation between the 100 keV flux intensity and the spectral
index (figure 10) suggesting a different behaviour than the one seen
in NGC 4151. In addition to the continuum radiation, the Rice group

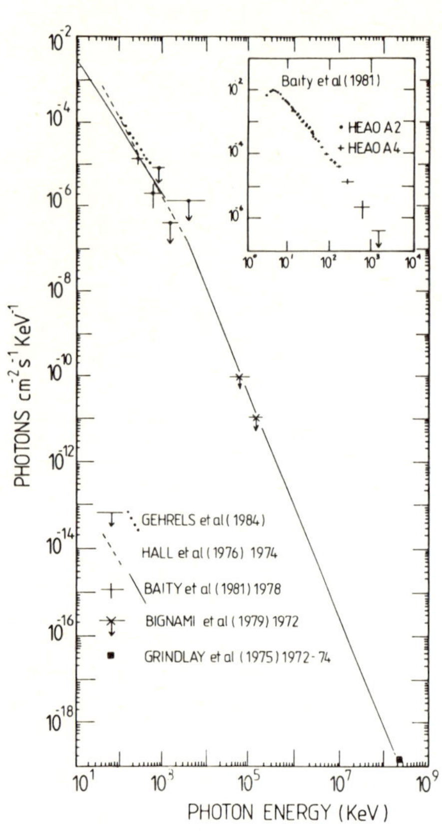

FIG. 8.   X/γ ray spectrum of
Centaurus A

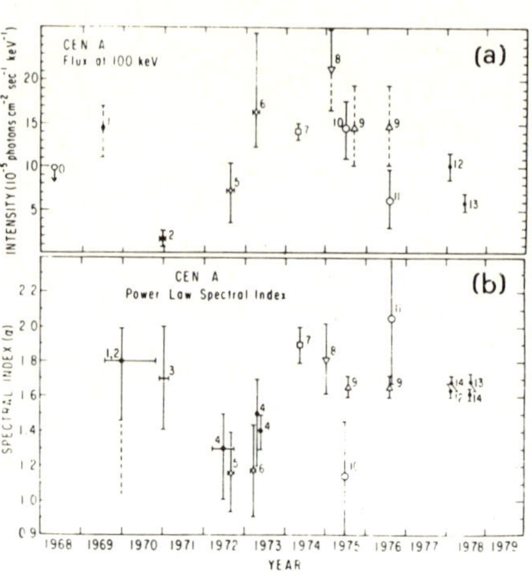

FIG. 9.   CEN A: Time history of
the 100 keV flux (a) and of
power law spectral index (b)
(Baity et al. 1981)

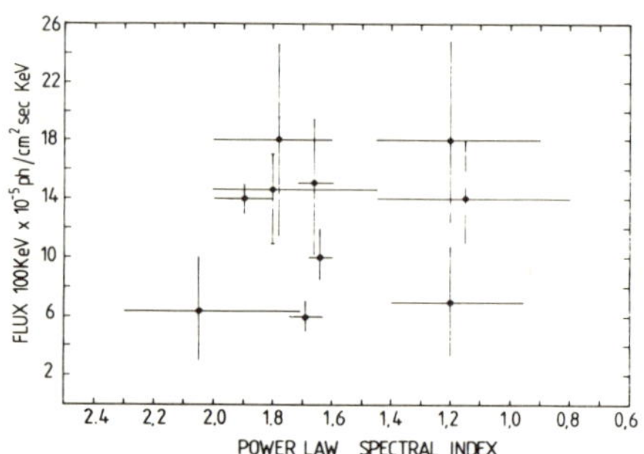

FIG. 10.   Correlation between
spectral index and 100 keV
flux for Cen A

(Hall et al. 1976) also measured in 1974 two gamma-ray lines at 1.6 and 4.5 MeV, with fluxes of $(3.4 \pm 1) \times 10^{-3}$ and $(9.9 \pm 3) \times 10^{-4}$ photons cm^{-2} s^{-1}, respectively. No evidence of line emission was found in the HEAO 1/A4 and LEGS data and upper limits were set for the lines at 511 KeV, 1.6 MeV and 4.5 MeV.

The Quasar 3C273

Quasars have been established as possible gamma-ray sources since the detection of 3C273 inside the COS B error box CG 289+64 (Bignami et al. 1981).

At an X-ray (2-10 keV) luminosity of $\sim 10^{46}$ erg/sec, 3C273 is the brightest quasar viewed from Earth and it is the only active galaxy having its spectrum clearly measured at energies greater than 100 MeV. The measured differential photon spectrum obtained by combining the data of 3 COS-B observations is well fitted by a power law of the form $kE^{-\alpha}$ with $\alpha = 2.6$ as shown in figure 11.

Contemporary X-ray measurements were available from the HEAO1 satellite for one of the COS-B observations (1978) and yielded spec-

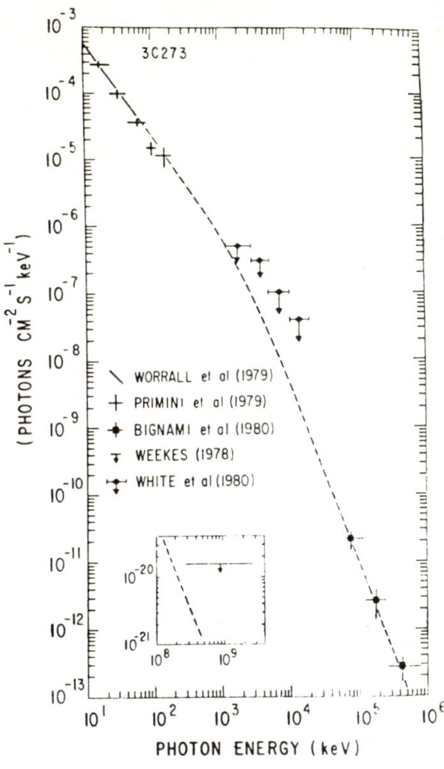

FIG. 11. X/γ ray spectrum of 3C273
(adopted from Fichtel 1982)

tral data in the 2-60 KeV range (A2: Worral et al. 1979) and in the 13-120 KeV range (A4; Primini et al. 1979). Both sets of data are well represented by power law spectra having index α = 1.4 $\pm$ 0.02 and α = 1.67 $\pm$ 0.14 respectively.

Results consistent with the A4 spectrum have been obtained at hard X-ray energies by balloon observations of the MIT/LEIDEN and AIT/MPE telescopes in the period 1978-79 (Pietsch et al. 1981). This last instrument has reobserved 3C273 in late 1981: the source was found to emit a high luminosity (L_x (20-200 keV) = 1.2 10^{47} erg/sec) and a very hard spectrum (α = 1.2) with no apparent break or cut-off up to around 200 keV (Bezler et al. 1984). However, when the X/γ -ray data are combined it becomes evident that the spectrum of 3C273 steepens strongly from the X-ray range to the high energy gamma-ray region, with the shape of the differential energy spectrum changing by $\Delta\alpha$ = 1, or more. The upper limits in the soft (White et al. 1980) and very

high gamma-ray energy interval (3 10^{11} to 3 10^{12} eV, Weeks 1978) are consistent with the steep COS-B spectrum continuing to those energies. The change in spectral shape observed in 3C273 is therefore similar to that seen in Seyfert galaxies, suggesting that this quasar should be a primary target of future observations in the 1-10 MeV energy range. Furthermore, future observations should be aimed at establishing the possible correlation between spectral index and flux indicated by the hard X-ray data and similar to that observed in NGC 4151. The COS-B instrument has observed the source on three different occasions, two years apart but no significant variation in the gamma-ray fluxes was observed (Bignami et al. 1981; Hermsen et al. 1981). However, the soft X-ray source is well known for its variability, characterized by flux variations of large factors on timescales of hours to days (Tananbaum 1979, Marshall et al. 1981).

At hard X-ray energies, the recent observation of Bezler et al. (1984) represents the first measurement of time variability in this frequency range: a factor of 3 increase in luminosity on timescale of 2 years.

The closest known quasar is QSO241+622. Although the original error box of the COS-B source CG 135+1 included this quasar, subsequent analysis has shown this positional coincidental to be marginal (Pollock et al . 1981). The MISO low energy gamma-ray telescope has detected a source emission coming from this region of the sky (Perotti et al. 1980). The MISO error box favours the alternative association with the variable radio star LSI 61°303 (Gregory et al. 1979) which is well within the new COS-B error box. We therefore regard the identification of the gamma-ray source with this QSO highly unlikely.

The detection of the QSO 630+180 inside the "Geminga" (CG 195+4) COS-B error box (Moffat et al. 1983) marks the third case of a QSO inside the field of a gamma-ray source, although there is an alternative suggestion of identification with a galactic counterpart (Bignami et al. 1983). This QSO, if proved to be the gamma-ray source, differs from 3C273 in its high ratio R of high energy gamma-ray luminosity to the luminosity measured at lower frequencies (radio, optical or X-rays). While the ratio is roughly 1 for 3C273, QSO 630+180 has R varying from 10^3 to 10^7 for L_γ (> 70 MeV) $\simeq$ 5 10^{49} erg/sec (assuming isotropic emission). Although the energy requirements seem exceptional for this source, beaming of the gamma-ray emission may ease the problem. Combining the gamma-ray luminosity with the variability timescale of 1 year seen at radio frequencies (Spoelstra & Hermsen 1984), this quasar appears to be just at the Eddington limit for an accretion disk located at 10 Schwarschild radii. Much shorter timescales will certainly require a progressively higher degree of collimation of the gamma-ray reducing the luminosity to more acceptable values. Given the exceptional characteristics of the gamma-ray source, whether galactic or extragalactic, we will not consider this object in our discussion on the observational data of gamma-ray emitting AGN.

General Characteristics of the Gamma-ray Emission from AGN

It is clear that in view of the limited number of measurements of AGN at gamma-ray energies, a discussion of any general features is necessarily tentative. However, a few specific characteristics seem to emerge from the available data. Gamma-rays have now been observed from the brightest active galaxies, representing different classification including Seyferts, QSO and radio galaxies. In the case of BL Lac objects, there are no positive detections at gamma-ray energies, and only upper limits exist on the Markarian galaxy 501 (Bassani et al. 1982a). On the base of the data we may conclude that in general active galaxies are gamma-ray emitters and future gamma-ray surveys should aim at similar targets.

Whereas at the soft end of the X-ray spectra many objects may be explained in terms of thermal spectra, it is apparent that as observations are made at higher photon energies, the emission becomes dominated by power law processes. In table 3, the power law spectral indices covering a range of photon energies are given for a number of AGN. The spectra of this limited sample of gamma-ray emitting active galaxies are similar in that all show a marked increase in the spectral index going from X-ray to high gamma-ray energies. This is particularly noticeable in the case of the Seyferts NGC 4151 and MCG 8-11-11, where a break at 1 MeV is evident in the data. Also for other Seyferts, the upper limits derived from the SAS-2 satellite, are substantially (by an order of magnitude or more) below the extrapolation of the power law X-ray spectra, suggesting that a sharp spectral change in the low gamma-ray region may be a general feature of these galaxies (Bignami et al. 1979). In several other galaxies, like in 3C273, NGC 1275 and Cen-A, a similar behaviour is indicated by comparing the existing X-ray and high energy gamma-ray data. Table 3 also contains the average X and gamma-ray luminosities for the five active galaxies under consideration.

TABLE 3. General Characteristics of the X- gamma-ray Emission Active Galaxies

Source	D (Mpc)	L, (2–10 keV)	L	X-ray variability	γ-ray variability	Approximate spectral index from X- to γ-ray energies
NGC 4151	20	7×10^{42}	$5.5 - 20 \times 10^{44}$ (0.5–5 MeV)	700 s 0.5 day	≤ 6 months	1.43 · 1.0 · $\gtrsim 3.0$
MCG 8-11-11	123	8.5×10^{43}	10^{43} (0.5–5 MeV)	30 days	No variability reported yet	1.66 · 1.0 · $\lesssim 3.8$
3C 273	948	1.1×10^{46}	4×10^{46} (50–800 MeV)	6×10^{3} s 0.5 day	No variability observed	1.67 · No data · 2.6
NGC 1275	110	2.4×10^{44}	2.8×10^{45} (0.1–0.5 MeV)	No variability reported yet	$\leq$ years	1.5 · 1.25 · $\gtrsim 2.5$
CEN A	5	2.5×10^{42}	2.6×10^{43} (0.1 – 2.3 MeV)	$(7 - 18) \times 10^{3}$ s 2 days	≤ 6 months $\gtrsim 10$ days	1.64 → $\gtrsim 2.5$

It is noticeable that the power output for these objects appears to peak in the low energy gamma-ray spectral region as shown in figure 12. This is a consequence of the spectral change discussed above and underlines the astrophysical significance that must be contained within the low energy gamma-ray data.

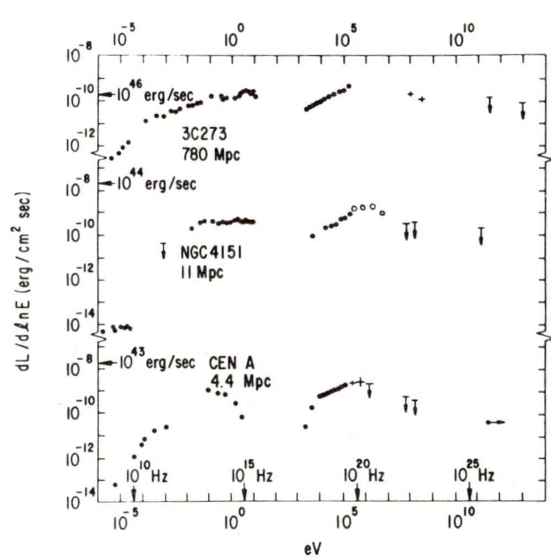

FIG. 13. Comparison between luminosities at various energies for few AGN (taken from Ramaty & Lingenfelter (1980)

When the luminosity at various photon energies over the whole electromagnetic spectrum are compared, it becomes evident that the peak luminosity occurs at low gamma-ray energies above 0.1 MeV (figure 13 from Ramaty & Lingefelter 1980).

Since a large fraction of the emitted power is contained at these wavelengths the production of gamma-ray photons has to be intimately connected to the central power source which must lurk deep within the nuclei of active galaxies. A good understanding of the observed gamma-ray spectral characteristics may, therefore, be vital to the

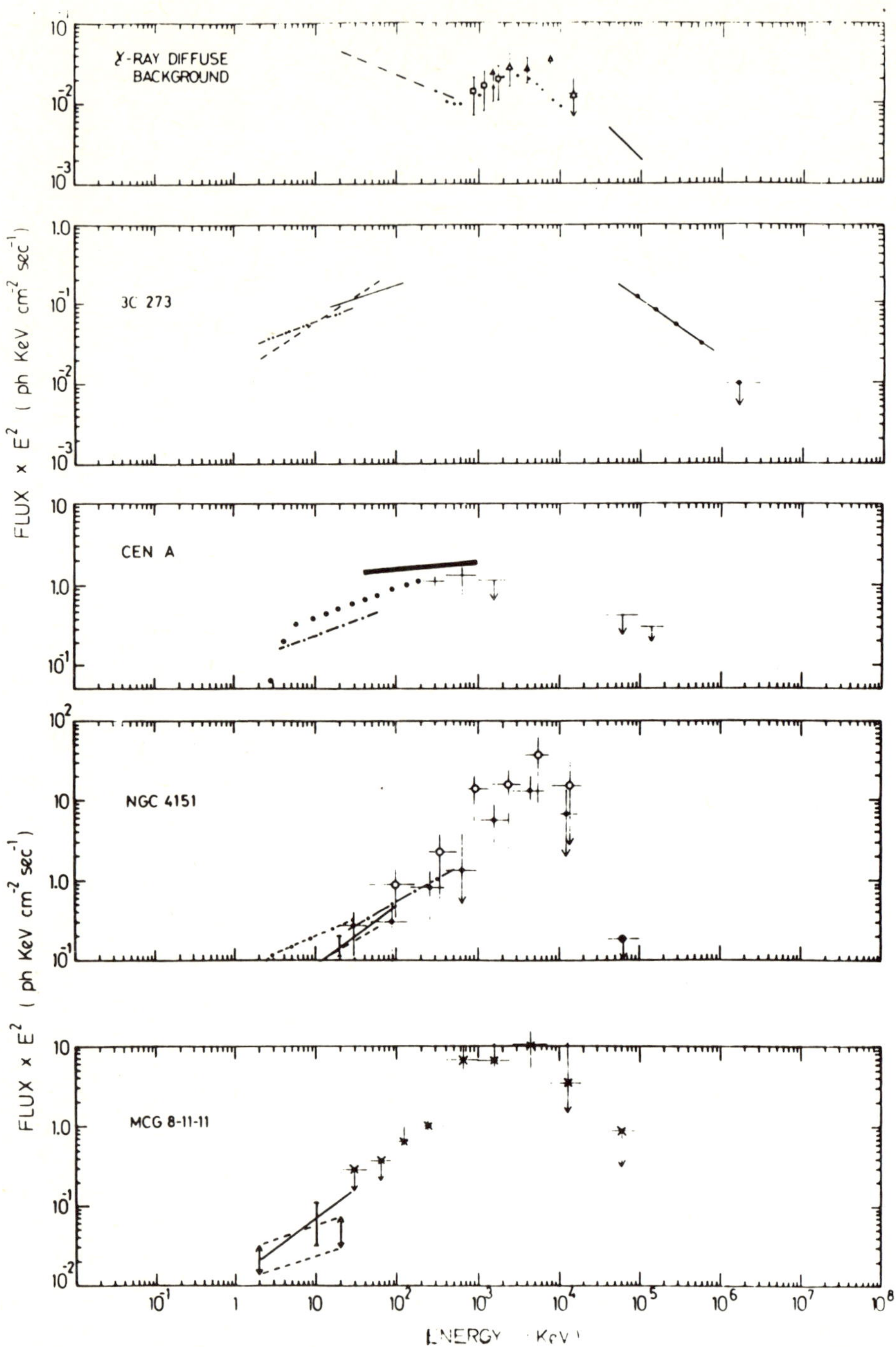

FIG. 12. Power output of a few AGN and of the cosmic diffuse background in the X/γ ray energy region.

understanding of the mechanism by which the central source of power operates. The extreme penetrating power of gamma-ray photons, such a disadvantage to the design of effective astronomical telescopes, could therefore prove invaluable in understanding these active cores.

Whereas individual active galaxies have been observed a number of times at X-ray wavelenths over the past few years and conclusive evidence revealed their time variability, very few of these objects have been viewed more than once in the gamma-ray region of the spectrum. The best studied object is NGC 4151 for which the gamma-ray measurements set an upper limit to the size of the gamma-ray emitting region of approximately 0.15 pc (i.e. $R < c \Delta t$). A similar variability timescale may characterize the soft gamma-ray emission of Cen-A probably NGC 1275, and 3C273, thus suggesting that active galactic nuclei of different classifications behave in the same way. It is clear that a number of further observations with higher sensitivity are necessary before a clear statement can be made as to the detailed characteristics of the variability in the gamma-ray emission from active galactic nuclei. If, however, gamma-ray variability is a real feature and the timescales are found to be similar to that observed for the X-ray emission, it would present a powerful argument for the case that the X- and gamma-ray emission originate from the same source region and would set severe limitations on the model scenarios. It is interesting to note that no variation has been reported on the high energy gamma-ray emission of 3C273 by the COS-B collaboration on a total of three observations. If this is confirmed for a number of objects, it may suggest that different mechanisms are involved in the production of the soft and hard gamma-rays.

INTERPRETATION OF THE GAMMA-RAY DATA

At the time of writing, despite the wealth of data on active galaxies, no clear picture has emerged as to the energy production mechanisms, the physical environment of the nuclear region or the nature of the ultimate source of power. A likely source of energy is accretion into a massive black hole, a scenario which has the advantage of being able to explain the high energies, the small sizes and the preferred orientation axes as well as jet structures.

For luminosities not exceeding the Eddington limit L_E, masses in excess of 10^4 to 10^9 $M_\odot$ are implied by the observational gamma-ray data. Furthermore, a general constraint on the efficiency with which matter is converted into energy can be obtained by combining the gamma-ray luminosity with the X-ray variability timescale (Fabian & Rees 1979). The gamma-ray data on the luminosity strongly suggest that if the gamma-ray variability timescale is found to be as short as the X-ray variability timescale, one must consider models that allow high efficiencies (Bassani & Dean 1983a).

Several radiation mechanisms could be responsible for the production of gamma-rays in AGN. At least some of the possible scenarios, originally developed to cope with the restriction imposed by the X-ray observations can be discussed in the light of gamma-ray measurements: bremsstrahlung by a hot (10^9 °K) gas, Comptonisation of cool photons, and the synchrotron self Compton (SSC) model (see Bassani & Dean 1983a for a review). In the conditions thought to exist in the vicinity of the nucleus of an active galaxy synchrotron radiation and Compton radiation could become quite important. The radiation from the synchrotron process itself can in fact create enough photons for the Compton process to become important between the parent electron and the secondary photons (Baity et al. 1981, Baity et al. 1983, Bassani

1981). These Compton synchrotron models generally predict a break in the energy spectrum between the X-ray and the high energy gamma-ray ranges, based on the observation of a break in the synchrotron spectrum between the radio and optical UV energy ranges for many active galaxies (see for example the case of NGC 4151, in figure 14, Bassani 1981). Alternatively, gamma-rays may arise from the Comptonisation of seed photons by a hot thermal electron gas (Shapiro et al. 1976, Katz 1976). The repeated Compton scatterings give a non-thermal appearance (power law) to the observed spectrum. The great advantage of this mechanism is that it is directly tied to the nature of the central source and it naturally explains the power law appearance in the X-gamma-ray energy range. Bremsstrahlung radiation by thermal electrons at different temperatures has been recently discussed by Meszaros (1983). The advantage of this model lies in its property of mantaining a constant spectral shape during changes in luminosity as observed in the X-ray band. Also to fit a number of objects, the variation of a simple parameter, the temperature, may be sufficient and this parameter can be reasonably estimated from theory, instead of being arbitrary. However, some conceptual difficulties still remain in each model. For instance, the short lifetime of the electrons against synchrotron losses poses a serious problem to the SSC model, unless continuous electron reacceleration takes place within the emission region itself. On the other hand failure to reproduce the observed spectral shape at MeV energies, although very good fits are obtained in the X-ray range, suggest that models involving Comptonisation are not viable gamma ray emission scenarios. As for bremsstrahlung models, at the hot temperatures required to produce gamma-rays in the compact nuclear region af AGN, other mechanisms such as Comptonisation pair production and annihilation become competitive and the balance between them is not fully understood.

It has become clear, however, the importance of $e^+ e^-$ pair production in photon-photon collision within the nuclear source and great theoretical effort has been recently put into the understanding of this process as well as its importance, particularly in the context of mildly relativistic plasmas (Svensson 1983 & references therein). For example, pair production in a hot plasma, acts as an efficient thermostat to limit temperatures to below a few 10^9 K (Fabian, these proceedings) thereby naturally giving a break in the spectrum at around one MeV.

On the other hand, the prescence of an MeV 'bump' in the spectra of active galaxies has prompted the production of a number of ad hoc model to fit the gamma-ray data. Without going into details, the main difference between these various type of models lies in the nature and location of the MeV energy cut-off.

Low energy gamma-rays could be due to Penrose Compton scattering (Leiter 1980) which takes place in the ergosphere of a rapidly rotating black hole, where blue shifted X-ray photons from an accretion disk interact with transient matter. If a Compton scattered electron is knocked out into the hole's event horizon, the photon picks up rotational energy from the hole and emerges into free space as a gamma-ray. This model predicts bursts of gamma-rays characterized by a specific spectral cut-off at E $\leqslant 3$ MeV. Although the NGC4151 MISO (1977-1979) data are consistent with such a model (figure 15) the lack of detection in few occasions is against frequent (less than one week) Penrose Compton Scattering bursts. These bursts should have a duration Δt greater than 2.2 hours for (M $= 10^8$ M$_\odot$) separated by off periods of the order of days.

A specific prediction of a broad peak around 0.5-1.5 MeV characte-

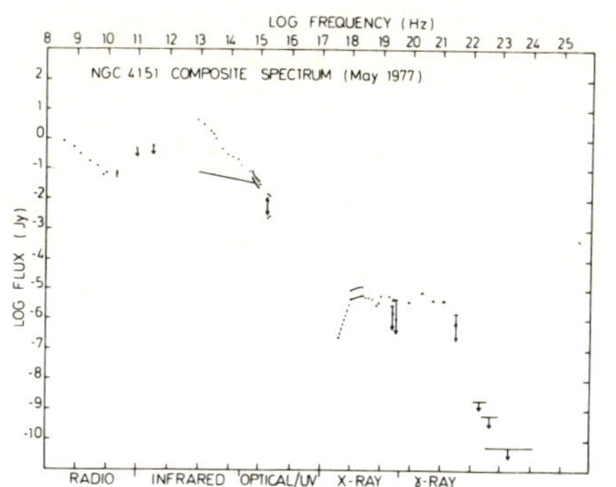

FIG. 14.   Composite spectrum of NGC 4151 (May 1977) (Bassani (1981)

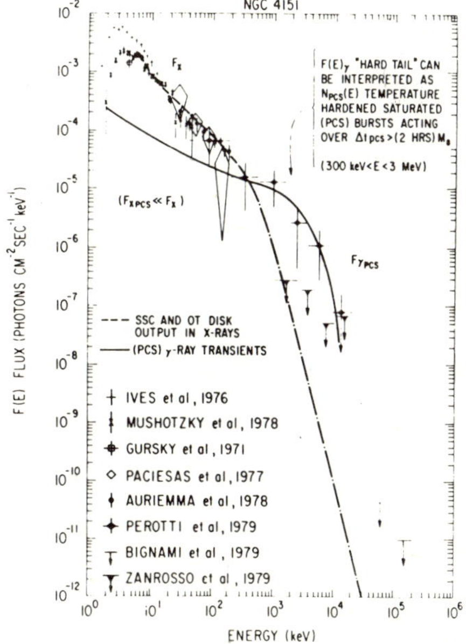

FIG. 15. Penrose Compton Scattering process in the case of NGC 4151 (Leiter 1980)

rises also the model recently proposed by Meszaros & Ostriker (1983), based on shocks in spherically accreting black holes. This type of well defined energy cut-off distinguishes the two previous processes as well as the pair production process discussed above from other possible mechanisms characterised by a spectral break which is usually taken as a free parameter used to fit the observational data (Schlikeiser 1980a, b; Pinkau 1980).

The high energy gamma-rays> (E $\geqslant$ 100 MeV) observed in 3C273 could result from meson decay. However, the observed photon energy spectrum of this QSO is much steeper than that predicted from meson decay produced by galactic cosmic ray interactions. This difference could indicate a different energetic particle spectrum in active galaxies from that in our own galaxy, or could be caused by photon-photon absorption in the more compact source region of an active galaxy. Π meson decay has also been invoked to explain the gamma-ray emission of "Geminga" if the identification with the Quasar QSO 0630 + 180 is correct (Moffat et al. 1983). The term "proton Quasar" (Schlikeiser 1984) has been coined for such objects which show an excess gamma-ray luminosity at E $\geqslant$ 100 MeV. Without further experimental evidence on the spectral shape, time variations, possible correlation with jets (see next section) and other considerations, it is not yet possible to select among the various competing models. However it is already clear that the few existing gamma

-ray data have pointed out very intriguing astrophysical problems which deserve to be studied in more detail.

Gamma-ray Emission and Jet Models

In models involving accretion into a massive black hole the maximum luminosity of an active galactic nucleus emitting isotropically is given by the Eddington luminosity, L_E, and the minimum timescale for variability is given by Δt = Schwarzchild radius/light velocity. As a consequence, the observed flux variation timescales and luminosities must obey:

$$\mathrm{Log}\ L_E \leqslant 43.1 + \log\ \Delta t \qquad \text{(Elliot \& Shapiro 1974)} \qquad (1)$$

Recently Abramowicz & Nobili (1982) considered the effects of beaming in a thick accretion disk configuration and obtained the following corrected expression for the Eddington luminosity:

$$\mathrm{Log}\ L_E \leqslant 44.3 + \log\ \Delta t \qquad \text{(Abramowicz \& Nobili 1982)} \qquad (2)$$

Figure 16 shows the location in the log L - Log Δt space diagram of a selection of about 60 active galaxies including Seyfert galaxies, quasars and BL Lac objects for which appropriate data are available. A more detailed description and complete listing of the data can be found in Bassani et al. (1983). It is noticeable from figure 16 that a few sources lie in the forbidden region defined by unequality (2) thus indicating that the observed variability timescales are inconsistent with beamed black hole models unless a higher value of the Hubble constant H_o is assumed. A choice of H_o = 100 km/sec Mpc would, in fact, reconcile the observed values of L_{bol} & Δt with those expected from the Abramowicz & Nobili formula. However, inclusion of the gamma-ray emission in the estimates of L_{bol} would prove crucial in seeking the Eddington luminosity limit for these extreme objects and as a consequence the gamma-ray channel is likely to provide a test for the existence of black holes in the nuclei of active galaxies. On the

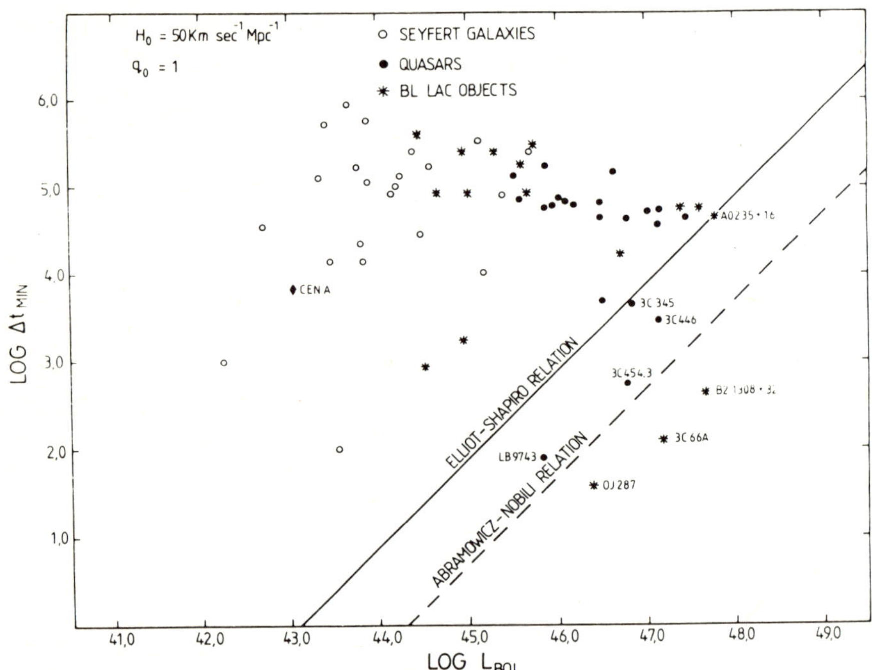

FIG. 16. The minimum variability timescale plotted against the bolometric luminosity for a selection of AGN.

other hand it is evident from figure 16, that even for high values of
H_o, a significant number of sources mainly QSOs & BL Lacs show flux
variations much shorter than those expected from the classical Elliot-
Shapiro relationship. It is therefore reasonable to conclude that the
radiation from these sources is highly beamed. This characteristic
becomes more evident when we consider that our values of L_{bol} underestimate the true bolometric luminosity since the gamma-ray emission,
which could represent a major fraction of the total output, has not
been included. However, this increase in the observed luminosity is
compensated, partially, by an increase in the Eddington limit at high
photon energies ($h\nu > m_e c^2$) due to the decrease in the photon interaction cross-section as calculated by the Klein-Nishina formulation.
In figure 17 (a) this enhanced luminosity limit is plotted in units of
the classical Eddington limit, L_E, as a function of the power law
spectral index of the emergent radiation. Figure 17 (b) shows the
location of the few known gamma-ray emitters in the Log $L_{x-\gamma}$ -
Log Δt space diagram. It is clear from the figure that the observa-

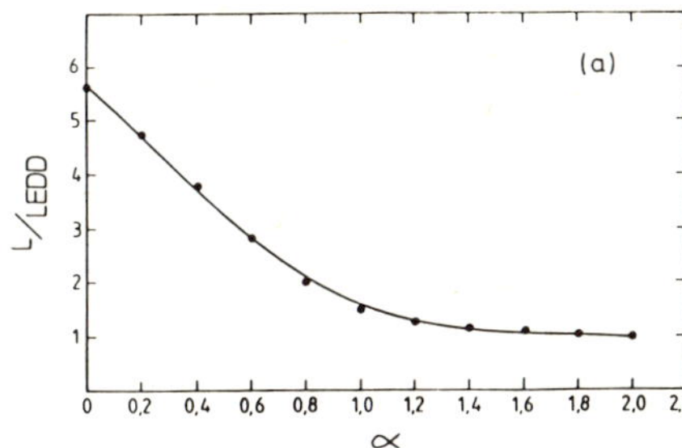

FIG. 17.(a) The minimum variability timescale against the X/γ
   ray luminosity for few AGN

FIG. 17.(b) The X/γ ray luminosity limit, obtained by using
   the Klein-Nishina cross section, is plotted against the
   power law spectral index α of
   the emergent radiation

tions of these particular gamma-ray objects are still consistent with
black hole models where no beaming is required. An independent line of
evidence for the presence of high energy emitting beams in AGN comes
from the compactness parameter L_x/R, which estimates the degree of
absorption of gamma-rays as a result of pair production in photon-
photon collision with local X-rays (Bassani & Dean 1981). A photon of
energy E_γ may collide with a photon of energy E_x and produce an $e^+ e^-$
pair provided that

$$E_x E_\gamma \geqslant \frac{2(m_e c^2)^2}{(1 - \cos \phi)}$$

where ϕ is the angle between the directions of motion of the two
photons. The optical depth for this process can be written as follows:

$$\tau_{\gamma\gamma} = R \int \sigma(E_x) N_x(E_x) dE_x$$

where the cross section and the density of X-ray photons per unit energy N_x, are assumed constant throughout the source region of size R (cm).

If, for simplicity, ϕ is set equal $\pi/2$ (i.e. isotropic emission), N_x is estimated from the 2-10 keV X-ray luminosity, L_x, (erg s^{-1}) and $\sigma(E_x)$ is approximated by a rectangular function we obtain

$$\tau_{\gamma\gamma} = 3.5 \; 10^{-28} \; f(\alpha) \; \frac{L_x}{R} \; (2E_x)^{-\alpha}$$

where $f(\alpha)$ is a slowly varying function of α and E_γ is measured in keV (is the source energy spectral index at X-ray energies).

In table 4 we list the values of L and R $\sim$ c Δt together with the optical depths calculated from the above equation for different gamma-ray energies. It is evident that whilst Seyfert galaxies are transparent to their gamma-radiation up to energies of about 1 GeV, QSO and BL Lac objects tend to be opaque to gamma-rays of few MeV or more.

TABLE 4. Absorption of γ rays in AGN

SOURCE	TYPE	L_x (2-10 keV) erg/sec	R (cm)	PHOTON - PHOTON OPTICAL DEPTH AS A FUNCTION OF GAMMA-RAY ENERGY			
				1 MeV	10 MeV	100 MeV	1000 MeV
NGC 7469	S1	$5.4 \; 10^{43}$ (6)	$5.2 \; 10^{15}$ (6)	0.02	0.08	0.30	1.20
NGC 4151	S1	$6.2 \; 10^{42}$ (22)	$7.0 \; 10^{14}$a (22)	0.02	0.06	0.26	1.03
NGC 6814	S1	$5.4 \; 10^{42}$ (6)	$6.0 \; 10^{14}$ (23)	0.02	0.07	0.26	1.04
NGC 3783	S1	$2.3 \; 10^{43}$ (6)	$5.2 \; 10^{15}$ (11)	0.01	0.03	0.13	0.51
MKN 509	S1	$2.5 \; 10^{44}$ (24)	$3.1 \; 10^{16}$ (24)	0.01	0.06	0.23	0.94
MCG 8-11-11	S1	$8.3 \; 10^{43}$ (25)	$7.8 \; 10^{16}$ (25)	0.002	0.008	0.03	0.12
NGC 2992	S2	$1.8 \; 10^{43}$ (6)	$7.8 \; 10^{15}$ (6)	0.004	0.02	0.07	0.27
NGC 5506	S2	$1.3 \; 10^{43}$ (26)	$7.8 \; 10^{15}$ (26)	0.003	0.01	0.05	0.19
NGC 526A	S2	$7.4 \; 10^{43}$ (6)	$7.8 \; 10^{15}$ (6)	0.02	0.07	0.28	1.10
NGC 7582	S2	$5.3 \; 10^{42}$ (6)	$5.7 \; 10^{16}$ (27)	$1.7 \; 10^{-4}$	$6.8 \; 10^{-4}$	0.027	0.01
MCG 5-23-16	S2	$2.2 \; 10^{43}$ (28)	$1.0 \; 10^{16}$ (28)	0.004	0.02	0.06	0.26
3C273	QSO	$1.1 \; 10^{46}$ (6)	$1.3 \; 10^{15}$b (6)	15.6	61.9	246.5	981.2
2SO 241-622	QSO	$1.0 \; 10^{45}$ (12)	$5.2 \; 10^{15}$ (6)	0.35	1.41	5.6	22.30
OX 169	QSO	$1.1 \; 10^{44}$c(23)	$1.8 \; 10^{14}$ (23)	1.24	4.93	19.62	78.12
MKN 421	BL LAC	$2.2 \; 10^{44}$ (6)	$2.6 \; 10^{13}$ (6)	0.16	0.62	2.46	9.81
PKS 2155-304	BL LAC	$6.4 \; 10^{43}$d(6)	$6.0 \; 10^{14}$ (29)	19.6	78.1	310.7	1236.9
3C 66A	BL LAC	$1.8 \; 10^{46}$e(30)	$5.7 \; 10^{15}$ (30)	6.07	24.16	96.17	382.9

a) X-ray variability also reported with ΔT = 700 sec (23) but of small statistical significance
b) 10% X-ray variability also observed on ΔT = 610^3 sec (23)
c) L_x in the range 0.5 - 4.5 keV
d) Redshift measurement doubtful
e) L_x in the range 0.2 - 3.5 keV.

If, therefore, such sources are found to be gamma-ray emitting objects, as in the case of 3C273, then one must conclude that on the basis of the above argument the X/gamma-radiation must be highly beamed in our direction. For the case of a diverging beam with opening angle ϕ_o, the limiting photon energy which is absorbed by photon-photon pair production is $E_\gamma \simeq \sqrt{2} \; m_e c^2 \; (1-\cos\phi_o)^{-\frac{1}{2}}$. In the particular case of 3C273, one would require a beaming angle of few degrees or less in order to observe gamma-rays up to 500 keV. Therefore in a simplified jet model, consisting of a stream of relatistic particles, as ϕ_o approaches zero E_γ tends to ∞ and the absorption process of gamma-rays become negligible for higher and higher energies. Both the

Eddington luminosity limit and the photon-photon absorption limit
indicate the existence of beam structures in active galaxies. Furthermore, the available data suggest that X/gamma jets must be characteristics of QSO and BL Lac objects. This is evident in figure 18, a, b,
where we have plotted the number of sources against estimated values
of L/L_E and $\tau_{\gamma\gamma}$ for each classification of AGN. It is noticeable
that while Seyfert galaxies have both L/L_E and $\tau_{\gamma\gamma}$ much less unity,
QSO and BL Lac objects tend to approach and exceed this values,
indicating once again the relevance of gamma-ray obserations of AGN to
the study of jet and beam structures.

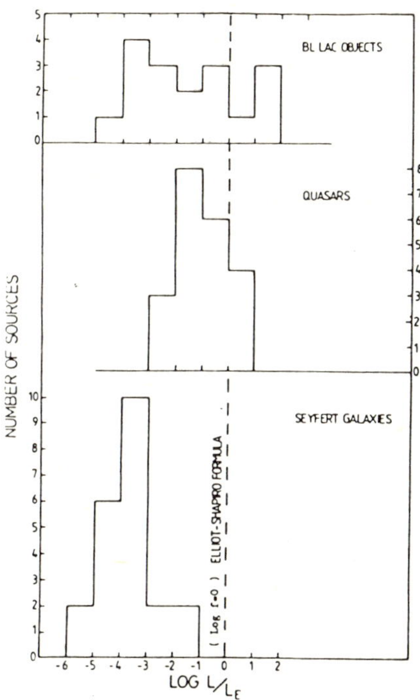

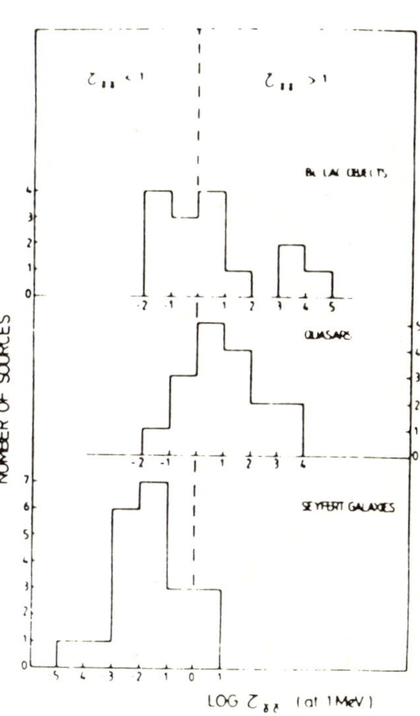

FIG. 18.(a) Distribution of L/L_E
values for different types of
AGN

FIG. 18.(b) Distribution of $\tau_{\gamma\gamma}$
for different types of AGN

Furthermore, detailed gamma-ray spectral data of active galaxies
may enlighten some of the problems related to relativistic motion in
extragalactic jets (Bassani & Dean 1984). Due to relativistic effects,
the physical parameters observed from jet structures are quite different from their corresponding rest frame values. Consequently any
discussion involving the physics of sources relating to jet phenomena
must take parameter alteration into account. The above equation for
the gamma-ray absorption optical depth is no exception. We have
therefore calculated the absorption optical depth in the rest frame of
the flow (where the emission is assumed to be isotropic) and apply
corrections for the combined effects of Doppler blue shifting of the
gamma-ray photons and for relativistic beaming which is responsible
for increasing the apparent source luminosity and decreasing the

source size. The overall effect is to reduce the gamma-ray absorption within emitters of beamed high energy photons by an appropriate factor which has a value much greater than unity in the case of relativistic motion. The equation for $\tau_{\gamma\gamma}$ may be rewritten as:

$$\tau_{\gamma\gamma}(E_{\gamma\,obs}) = 3.5\ 10^{-28} f_{obs}(\alpha)\ \frac{I_{xobs}}{R_{obs}}\ \left(\frac{2m_e^2 c^4}{E_{\gamma\,obs}}\right)^{-\alpha}\ \left(\frac{1+z}{\delta}\right)^{5+\alpha}$$

where the subscript obs refers to the observable quantities as measured at Earth, $\delta = \Gamma^{-1}(1-\beta\cos\theta)^{-1}$ is the kinematic Doppler factor of the jet, and z the source redshift. The source of radiation is assumed to move with constant velocity $\beta = (1-\Gamma^{-2})^{\frac{1}{2}}$ at an angle θ with respect to the line of sight. Again L_{xobs} is measured in erg s^{-1} over a given energy interval, R_{obs} in cm and $E_{\gamma\,obs}$ in keV. Taking I_{xobs} over the Einstein energy band (0.5-4.5 KeV) and assuming that $\alpha = 0.5$ as a reasonable description for the observed X-ray spectra of AGN one may obtain

$$\left(\frac{\delta}{1+z}\right) = \left(\frac{1.7\times10^{-31}}{\tau_{\gamma\gamma}}\ \frac{L_{xobs}}{R_{obs}}\ \sqrt{E_{\gamma\,obs}}\right)^{1/5.5}$$

If a cut-off is observed in the gamma-ray spectrum of an active galaxy at a photon energy $E_{\gamma obs}$ (> 1 MeV) due to photon-photon absorption, $\tau_{\gamma\gamma}$ can be set to unity and a direct estimate of the Doppler factor can be made. Table 5 lists values of δ calculated using

TABLE 5. <u>Values of the Doppler Factor δ for a number of AGN</u>

Source	Type	Red Shift	L_{Xobs} (0.5-4.5keV) (erg/sec)	R_{obs} (cm) x 10^{14}	δ
NGC 4151	Seyfert	0.0033	$5.7\ 10^{42}$	7.0	0.60
MCG 8-11-11	Seyfert	0.020	$6.9\ 10^{43}$	52	0.67
Cen-A	Radio G	0.0008	$(0.15-1.5)10^{42}$ *	2.2	0.77
NGC 1275	Seyfert/BL Lac	0.018	$1.2\ 10^{44}$ **	26	1.22
3C273	QSO	0.158	$1.7\ 10^{46}$	(13.0-2.0)	5.83
3C446	BL Lac/QSO	1.404	$9.6\ 10^{46}$	2.18	>11.4
3C345	QSO	0.595	$9.1\ 10^{45}$	2.14	> 4.13
3C454.3	BL Lac/QSO	0.859	$1.3\ 10^{46}$	0.299	> 8.77
CTA102	QSO	1.037	$2.1\ 10^{46}$	26.1	> 4.33
OJ 287	BL Lac	0.306	$4.3\ 10^{45}$	0.015	> 8.98
3C 66A	BL Lac	0.434	$1.2\ 10^{45}$	0.054	> 5.92
B2 1308+26	BL Lac	0.996	$1.1\ 10^{46}$	0.267	> 9.32
LB9743	QSO	0.253	$1.6\ 10^{44}$ ***	0.03	> 3.97
3C 263	QSO	0.652	$9.6\ 10^{45}$	26.0	> 3.68
PkS0537+44	BL Lac	0.894	$4.6\ 10^{45}$	26.0	> 2.97
PkS0420-01	QSO	0.915	$1.3\ 10^{46}$	25.9	> 3.71
PkS2155-30	Bl Lac	0.17	$1.0\ 10^{46}$	5.96	> 2.88
A0235+16	BL Lac	0.852	$3.0\ 10^{45}$	25.9	> 2.67

* 2-6 keV luminosity, f(α) = 0.48

** 0.5-3.0 keV luminosity, f(α) = 0.49

*** 0.1-2.6 keV, f(α) = 0.39

H_0 = 50 km sec Mpc

the above equation for a number of known gamma-ray emitting AGN as well as a number of other sources. For the latter objects an estimate of the lower limit on the Doppler factor has been made by assuming that the soft X-ray emission extends into the MeV spectral region (E_γ > 1 MeV). All known gamma-ray emitting sources with the exception of 3C273, have δ < 1 and therefore do not require relativistic beaming. In the particular case of the Quasar 3C273, we obtain 3 < δ < 6; combined with an observed velocity for the moving material of β_{obs} = 10.6 (Pearson et al. 1981) we estimate a Lorentz factor of 12 < Γ < 20 and an emission angle with respect to the Earth of 8° < θ < 10°.

<u>Gamma-ray Line Emission from Active Galaxies</u>

The Rice group has reported the detection of two features at 1.6 MeV and 4.5 MeV in the 1974 gamma-ray emission spectrum of Cen-A (Hall et al. 1976). This observation may constitute the first measurement of nuclear gamma-ray line emission from an extragalactic object. Observations of the same object at a later date by the HEAO 1/A4 and LEGS instruments set upper limits on the 1.6 MeV line which was about a factor of 8-10 below the positive measurement of the Rice group and indicates time variability in the emission if indeed the 1974 feature was real. No other line emissions have been reported from active galaxies although both the HEAO-1 (Baity et al. 1983) and MISO observations of NGC 4151 (Della Ventura et al. 1979) searched for discrete line emission from this Seyfert. Upper limits to a narrow emission line at 511 keV from NGC 4151 can be set at $\sim 8.10^{-4}$ ph/cm^2 sec (Baity et al. 1983).

We have seen that e $\pm$ pair production in photon-photon collissions can be an important process in the nuclei of active galaxies. If these pairs annihilate in an optically thin region the annihilation radiation should be observable. If the electrons and positrons cool before they annihilate then one might expect to see a narrow line emission peaked at 511 keV. This may well be the case of our galactic centre whose gamma-ray emission spectrum is characterised by an intense and variable 511 keV feature (Riegler et al. 1981). The observational data relating to our galaxy requires the production of at least 2×10^{43} positrons s^{-1}. We may therefore expect a similar or even greater production rate from AGN. In an active galaxy, however, the emitting region is expected to be much hotter (T $\simeq 10^9$ k), in which case the line will be both broadened and blue shifted (Ramaty and Meszaros 1981, Zd'iarski 1981). This broadening effect may explain the absence of a narrow 511 keV line from Cen-A (Hall et al. 1976, Baity et al. 1981, Gehrels et al. 1984) and cause the MeV 'bump' observed in the Seyfert galaxies NGC 4151 and MCG 8-11-11 (Bassani & Dean 1983b). Svensson (1984) in a study of mildly relativistic plasmas notes that this interpretation may be inconsistent with the assumption of pair balance. For temperatures corresponding to $kT_e/m_e c^2$ > 3 bremsstrahlung emissitvity dominates the pair annihilation emissivity whilst in a steady pair dominated plasma at $kT_e/m_e c^2$ < 5 Comptonisation, pair annihilation and photon-photon absorption conspire to generate a somewhat distorted and redshifted Wein peak. Therefore according to Svensson, no signature of an optically thin pair annihilation line is expected from plasmas in pair balance. Only after detailed observations of AGN in the MeV range we will be able to make definite statements about the existence of line emissions.

ACTIVE GALACTIC NUCLEI AND THE COSMIC DIFFUSE BACKGROUND

The existence of a cosmic diffuse gamma-ray background emission (CDB) has been evident from the period of the earliest gamma-ray observations.

Subsequent measurements, made with a wide variety of experimental techniques have further confirmed and quantified this isotropic flux which extends from X-ray wavelengths to photon energies in excess of 1 GeV. A variety of theories have been put forward to account for the phenomenon ranging from those which attribute it to the truly diffuse mechanisms of particle interactions in the intergalactic space to those which consider the emission to be derived from the summation of a large number of unresolved point sources. Above we have shown that some active galaxies are luminous gamma-ray sources, it is therefore appropriate to discuss the hypothesis that the cosmic diffuse gamma-radiation results from the superposition of many galaxies.

The spectrum of the diffuse background from a few keV to about 100 MeV is shown in figure 19 (references to different experimental data set can be found in Bassani et al. 1982b). Some structure is immediately apparent, in particular the observed spectrum between 5 MeV and 200 MeV gives way to a noticeable flattening close to 1 MeV (see Lockwood et al. 1981). This 'bump', which is observed in all experimental measurements made over the range 500 keV to 5 MeV, protrudes above the low energy emission which may be connected to the high energy flux (E > 40 MeV) by a simple power law (see figure 19). Boldt (1981) has shown that the X-ray background below $\simeq$ 50 keV can be well fitted by a thermal bremsstrahlung spectrum with kT $\simeq$ 40 keV. Above 50 KeV, the contribution of active galaxies becomes significant.

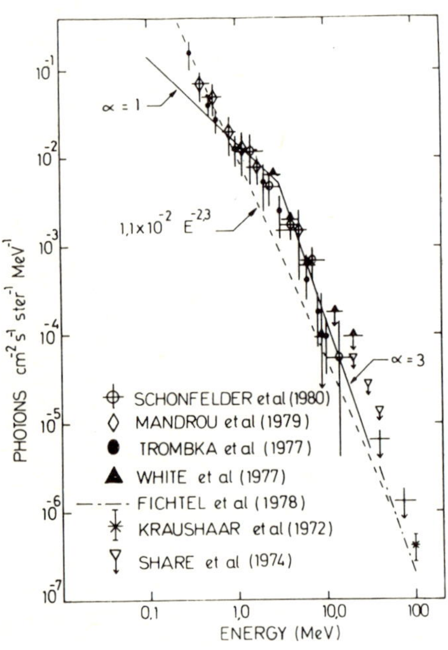

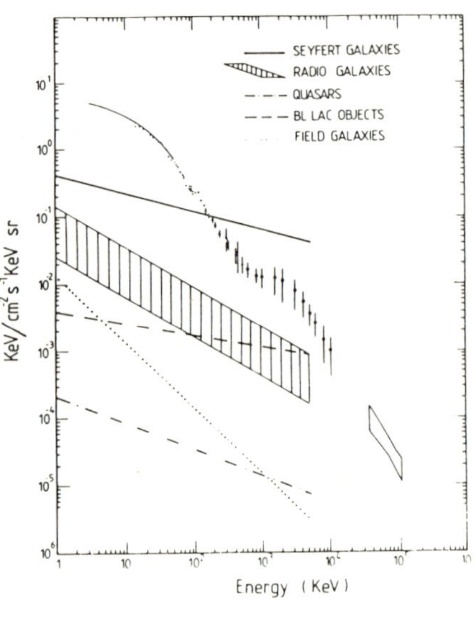

FIG. 19. Spectrum of the cosmic diffuse γ ray background (References to the figure in Bassani et al. 1982b)

FIG. 20. Contribution of different classes of AGN to the CDB at γ ray energies

The mean intensity produced by the superposition of a particular species of galaxies can be approximated by

$$I(E) = 2.5 \times 10^{-47} B_i(E) \eta_i \quad (ph\ cm^{-2} s^{-1} sr^{-1} keV^{-1})$$

where $B_i(E)$ represents the emissivity at a given energy E of a class of objects and η_i their space density.

In the following discussion the existing gamma-ray data has been used whenever possible, in order to estimate the contribution of AGN to the cosmic diffuse gamma-ray background. Otherwise it has been necessary to extrapolate X-ray measurements. Except in the case of radio galaxies, X-ray luminosity functions have been adopted as the best indicators of the space distribution of different types of gamma-ray emitters. Table 6 lists the values of B(E) and η_i used in the calculation together with some key references. The relative contributions of different types of AGN to the diffuse background is shown in figure 20.

TABLE 6. <u>Emissivity and space density values for different classes of AGN</u>

Class of sources	Prototype	$B_i(E)$ (ph/sec keV)	Reference	η_i (Mpc^{-3})	Reference
Seyfert Galaxies	NGC 4151	$9.7\ 10^{50}E^{-1.3}$	Perotti et al (1981a)	$1.4\ 10^{-5}$	Piccinotti et al (1982)
	MCG 8-11-11	$1.4\ 10^{52}E^{-1.0}$	Perotti et al (1981b)	$5.6\ 10^{-8}$	
Quasars	3C 273	$2\ 10^{54}\ E^{-1.41}$	Worrall et al (1979)	$4\ 10^{-12}$	Schwartz & Ku (1983)
Radio Galaxies	CEN A	$5.4\ 10^{50}E^{-1.64}$	Baity et al (1981)	$>2\ 10^{-6}$ $<10^{-5}$	Schmidt (1978)
BL Lac objects	MkN 501	$9.7\ 10^{51}E^{-1.2}$ $1.6\ 10^{53}E^{-2.5}$	Worrall et al (1981)	$1.6\ 10^{-8}$	Schwartz & Ku (1983)

Estimates of the Seyfert contributions to the background at hard X-ray wavelengths indicate that these galaxies account tor typically $\sim$ 20% of the diffuse flux at energies close to 50 keV (Rothschild et al. 1983). Leiter and Boldt (1982) have shown that, if at earlier stages of their evolution Seyfert galaxies were more luminous than at the present epoch, these objects could also account for the X-ray background. However, at low energy gamma-ray wavelengths estimates based on observations of 10 active galaxies between 21-165 keV (Rothschild et al. 1983) and on the MISO observations of NGC 4151 and MCG 8-11-11 (Perotti et al. 1979, 1981.a, b) show that Seyferts can explain the entire observed flux. Indeed the MISO data indicates that the summed contribution of Seyfert galaxies can exceed the diffuse emission as much as a factor of $\sim$ 5 at MeV energies with no evolution assumed for this class of AGN (Bassani et al. 1982b). In fact we are faced with the problem of limiting their contribution (even after due allowance for short timescale variability) either by postulating that for only a portion of the X-ray active lifetime a typical Seyfert galaxy is also a gamma-ray emitter or by assuming a different geometry for the X-ray and gamma-ray emitting regions, i.e. collimation of the gamma-radiation in jets. In the first case, we require a gamma-ray Seyfert phase 10-12 times shorter than the X-ray phase, while in the second case, jet opening angles as small as 5° must be postulated. No

evidence for collimation exists so far in all available data on
Seyferts (Bassani & Dean 1981). Although no high energy (E > 30
MeV) gamma-ray data exists on individual Seyfert galaxies, it is
possible that these objects contribute to produce the high energy
diffuse emission (Bignami et al. 1979). If Seyferts are the cause of
the high energy gamma-ray background, it follows that the similarity
in the measured CDB spectral index (α = 2.7 $\pm$ 0.4) close to 100 MeV
(Fichtel et al. 1978) and the steepening of the emission spectrum in
NGC 4151 and MCG 8-11-11 naturally explains the form of the high
energy diffuse emission.

If the above estimate of the integrated Seyfert intensity is
correct, then the question of the contribution of other types of AGN
to the CDB becomes critical. In the case of Quasars, low energy
gamma-ray measurements do not exist, although in the case of 3C273 the
extrapolation of the hard X-ray and high energy gamma-ray data indicates that a maximum in the emission lies in the few MeV range . Assuming 3C 273 to be typical of this sample, the quasar contribution can
account for approximately 1% of the isotropic flux, this value may be
increased if evolutionary effects, a strong possibility for these
objects, are taken into account. The same fractional contribution may
be assumed to extend to higher photon energies since the spectral
index of 3C273 (E $\geqslant$ 100 MeV) as measured by COS-B (2.5 $\pm$ 0.6) is
similar to that of the CDB.

An estimates of B(E) for radio galaxies can be made on the basis of
recent observations of Cen A at MeV energies. The space density of
radio galaxies having a radio luminosity equal to Cen-a ($L_R \sim 10^{41}$ erg
s^{-1}) is rather uncertain and at best can be set between the limits
given in table 6. For these values of B(E) and η , the summed intensity of radio galaxies is expected to lie between 2-10% of the total
CDB.

Apart from the possible identification of the BL Lac candidate NGC
1275 with a COS-B excess (Strong & Bignami 1983), no BL Lac objects
have been detected as gamma-ray emitters, and only upper limits exist
in the case of MkN 501 at MeV energies (Bassani et al. 1982a). Here,
we use the softest and hardest X-ray spectra reported to date together
with the recently estimated local volume density of X-ray BL Lac
objects to evaluate I(E) for this class. The BL Lac contribution to
the CDB ranges approximately between 0.2% to 5% at MeV energies. Finally, cluster of galaxies, like the Perseus cluster, may produce an
important contribution to the diffuse extragalactic gamma-ray flux
(Houston & Wolfendale 1984).

Although there are large uncertainties in both the luminosities
function of AGN as well as in their space densities the indications
are, that AGN can not only account for all the diffuse gamma-radiation, but even overproduce it. Evolutionary consideration would
aggravate the situation further. In fact the problem revolves around
methods of reducing the AGN contribution. The answer to this situation
may be related to the variability of the γ ray emission and to possible collimation into jets (Bassani & Dean 1981). Study of background
fluctuations due to 'empty' regions of the sky may provide a complementary method to test the hypothesis that AGN are main contributors
to the CDB. On the hypothesis of a uniform distribution of N $\sim 10^8$
faint and distant sources of constant absolute luminosity, fluctuations of $\sim$ 5% may be expected for angular resolution of about 10^{-3}
steradiant. Gamma-ray telescopes at present under development should
be able to detect these fluctuations in the background with a suitable
long observation period.

FUTURE PROSPECTS IN GAMMA-RAY ASTRONOMY FOR AGN

Although, to date, a limited number of gamma-ray observations have been made of active galaxies, the measurements indicate that a major fraction of their emitted power lies in the gamma-ray band. It follows, therefore, that much is to be learnt from a more comprehensive study of AGN at gamma-ray wavelengths. The next observational phase of active galactic astronomy augers well for an advance in this branch of astronomy since the design of gamma-ray telescopes has improved in terms of sensitivity, imaging capacity and spectral resolution.

Increases in sensitivity mean that the number of AGN which are detectable at gamma-ray wavelengths will increase significantly. If a log N/log S distribution of 3/2 is valid, then we may expect that the number of visible AGN will increase to more than one hundred for a 10 fold improvement in sensitivity over and above the past generation of instrumentation. This number of sources is sufficient to enable a gamma-ray luminosity function to be constructed for the various types of AGN. Morphological characteristics, if any, between the various types (i.e. BL Lac, Seyferts, QSO, Radio Galaxy) may be studied and possible clues obtained on the intrinsic differences in the individual gamma-ray production processes.

The improved sensitivity has other implications for extragalactic gamma-ray astronomy. An extensive study of the brighter objects will enable the spectral and time profiles of the gamma-ray emission to be recorded over a wide energy band yielding improved estimates of the parameters which characterize the gamma-ray emitting region. For example the next generation of instrumentation will be capable of detecting significant changes in the gamma-ray emission on timescales associated with massive black holes of 10^6-10^{10} $M_\odot$ (tens of sec to hours). Thus the combined spectral and time profile of the high energy emission will not only identify the dimension of the gamma-ray region, but may well distinguish between such phenomenon as Penrose Compton bursts, short time fluctuations associated with hot accretion disks, synchrotron self Compton flares etc.

One of the most obvious advantages of the next generation of gamma-ray telescope will lie in their improved angular response, made possible by recent technical developments. Future gamma-ray detectors will have a point source location capacity of typically 1 arc minute and should lead to unambiguous identification in most cases. Furthermore, the introduction of the coded aperture mask allows the gamma-ray images of the sky to be made, thus placing the gamma-ray work on a more equal footing with other branches of astronomy. Galaxy clusters are promising targets for gamma-ray imaging. For example, the Perseus cluster contains a source of hard non-thermal photons measured by HEAO-1 (Primini et al. 1981) and by OSO-7 (Rothschild et al. 1981). The source is located towards the core of the cluster but it is not clear whether it is associated with the galaxy NGC 1275, the intergalactic gas or some other cluster member.

In principle, the coded aperture mask technique may be improved and extended to very fine limits of angular resolution in future. At this point it should be feasible to perform gamma-ray imaging studies of the relativistic jet structures associated with AGN and this should lead to an improved understanding of these phenomena beyond the knowledge gained via the spectral time analysis. The gamma-ray emitting radio galaxy Cen-A, for example, has well defined jet structures at other wavelengths (optical, radio, X) and being both close and bright should prove an excellent prototype for the study of any gamma-ray jet structures in AGN.

The fine angular imaging quality and sensitivity now technically possible for gamma-ray devices will also improve our understanding of the origins of the cosmic diffuse background. Indeed we anticipate that in its exploratory phase the study of active galaxies at both high and low gamma-ray energies will become one of the most fascinating and challenging fields of astronomy.

REFERENCES

Adam TF 1977 Ap J Suppl 33:19-34

Abramowicz MA, Nobili L 1982 Nature 300:506-507

Baity WA, Mushotzky RF, Worral DM, Rothschild RF, Tennant AF, Primini FA 1983, Preprint

Baity WA, Rothschild RE, Lingenfelter RF, Stein WA, Nolan PL, Gruber DA, Knight FK, Matteson JL, Peterson LE, Primini FA, Levine AM, Lewin WHG, Mushotzky RF, Tennant AF 1981 Ap J 244: 429-435.

Bassani L 1981 Ap Spa Sci 79: 469-481.

Bassani L, Dean AJ 1981 Nature 294:332-333

Bassani L, Butler RC, Dean AJ, Di Cocco G, Della Ventura A, Perotti F, Villa G 1982a Ap Spa Sci 82: 311-316

Bassani L, Butler RC, Dean AJ, Di Cocco G, Perotti F, Villa G 1982b Ap Spa Sci 82: 199-207

Bassani L, Dean AJ 1983a Spa Sci Rev 35:367-398

Bassani L, Dean AJ 1983b Astr Ap 122:83-87

Bassani L, Dean AJ Sembay S 1983 Astr Ap 125:52-58

Bassani L, Dean AJ 1984 in preparation

Bezler M, Kendziorra E, Staubert R, Hainger G., Pietsch W, Reppin C, Trumper J, Voges W 1984, Astr Ap: in press

Bignami GF, Fichtel CE, Hartman RC, Thompson DJ 1979 Ap J 231:649-658

Bignami GF, Bennett K, Buccheri R, Caraveo PA, Hermsen W, Kanbach G, Lichti GG, Masnou JL, Mayer-Hasselwander HH, Paul JA, Sacco B, Scarsi L, Swanenburg BN, Wills RD 1981 Astr Ap 93:71-75

Bignami GF, Garaveo PA, Lamb RC 1983 Ap J 272:L9-L14

Boldt EA 1981 Comments Ap 9:97-115

Bower CS, Lampton M, Mack J 1970 Ap J 161:L1-L7

Branduardi-Raymont G, Fabricant D, Feigelson E, Gorenstein P, Grindlay J, Soltan A, Zamorani G. 1981 Ap J 248:55-60

Byram ET, Chubb TA, Friedman H 1971 Nature 229:544-546

Coe MJ, Bassani L, Engel AR, Quenby JJ 1981 MNRAS 195:241-244

Della Ventura A, Perotti F, Villa G, Baker RE, Butler RC, Dean AJ, Martin SJ, Ramsden D, Di Cocco G 1979 in:
16th Int Cosmic Ray Conf OG4-4 113-118, Conf. Proceedings

Elliot JL, Shapiro SL 1974 Ap J 192:L3-L6

Fabian AC, Zarnecki JC, Culhane JL, Hawkins FJ, Peacock A, Pounds KA, Parkinson JH 1974 Ap J 189:L59-L61

Fabian AC, Rees MJ 1979 in:
Baity & Peterson (Eds) Cospar X Ray Astronomy 381-398, Pergamon Press, Oxford

Fichtel CE, Simpson GA, Thompson DJ 1978 Ap J 222, 833

Fichtel CE 1982 NASA Tch Mem 83986

Fritz G, Davidsen A, Meekins JF, Friedman H 1971 Ap J 164:L81-L85

Frontera F, Fuligni F, Morelli E, Ventura G 1979 Ap J 234:477-480

Gehrels N, Cline TL, Teegarden BJ, Paciesas WS, Tueller J, Durouchoux P, Hameury JM 1984 Ap J: in press

Graser U, Schönfelder V 1982 Ap J 263:677-689

Gregory PC, Taylor AR, Cramton D, Hutchings JB, Hjellming RM, Hogg D, Huatum H, Gottlieb EW, Feldman PA, Kwok S 1979 Astr Ap 84:1030-1036

Grindlay JE, Helmken JF, Brown RH, Davis J, Allen LR 1975 Ap J 197:
    L9-L12
Gyrsky H, Kellogg EM, Leong C, Tananbaum H, Giacconi R 1971 Ap J
    165:L43-L48
Hall RD, Meegan CA, Walraven GD, Rjuth FK, Haymes RC 1976 Ap J 210:
    631-641
Hall R, Ricketts MJ, Page CG, Pounds KA 1981 Spa Sci Rev 30:47-54
Hermsen W, Swanenburg BN, Bignami GF, Boella G, Buccheri R, Scarsi L,
    Kanbach G, Mayer Hasselwander HA, Masnou JL, Paul JA, Bennett K,
    Higdon JC, Lichti GG, Taylor BG, Will RD 1977 Nature 269:494-494
Houston BP, Wolfendale AW 1984: preprint
Katz JI 1976 Ap J 206:910-916
Kellogg E, Gursky H, Leong G, Schreier E, Tananbaum H, Giacconi R 1971
    Ap J 165:L49-L54
Lawrence A 1980 MNRAS 192:83-94
Leiter D 1980 Astr Ap 89:370-376
Leiter D, Boldt E 1982 Ap J 260:1-19
Loockwood JA, Webber WR, Friling LA, Macri J, Hsieh L 1981 Ap J
    248:1194-1201
Lyuty VM 1978 Sov Astr 21:655-664
Marshall N, Warwick RS, Pounds KA 1981 MNRAS 194: 987-1002
Meegan CA, Haymes RC 1979 Ap J 233:510-513
Meszaros P, Ostriker JP 1983 Ap J 273:L59-L69
Meszaros P 1983 Ap J 274:L13-L17
Miller HR 1979 Ap J 227:52-53
Moffat AFJ, Schlickeiser R, Shara MM, Sieber W, Tuffs R, Kuhr H 1983
    Ap J 271:L45-L48
Mushotzky RF, Marshall FE 1980 Ap J 239:L5-L10
Mushotzky RF, Marshall FE, Boldt EA, Holt SS, Serlemitsos PJ 1981 Ap J
    235:377-385
Pearson TJ, Unwin SC, Cohen MH, Linfield RP, Readhead ACS, Seilstad
    GA, Simon RS, Walker RC 1981 Nature 290:365-368
Perola GC, Boksenberg A, Bromage GE, Clave J, Elvis M, Elvius A,
Gondhalekar PM, Lind J, Lloyd C, Penston MV, Pettini M, Snijders MAJ,
    Tanzi EG, Tarenghi M, Ulrich MH, Warwick RS 1982 MNRAS 200:293-312
Perotti F, Della Ventura A, Sechi G, Villa G, Di Cocco G, Baker RE,
    Butler RC, Dean AJ, Martin SJ, Ramsden D 1979 Nature 282:484-486
Perotti F, Della Ventura A, Villa G, Di Cocco G, Butler RC, Dean AJ,
    Hayles RI 1980 Ap J 239-L49-L52
Perotti F, Della Ventura A, Villa C, Di Cocco G, Bassani L, Butler RC,
    Carter JN, Dean AJ, 1981a Ap J 247:L63-L66
Perotti F, Della Ventura A, Villa G, Di Cocco G, Butler RC, Carter JN,
    Dean AJ, 1981b Nature 292:133-135
Perotti F, Della Ventura A, Villa G, Butler RC, Di Cocco G, Baker RE,
    Carter JN, Dean AJ, Hayles RI 1983 Adv Spa Res 3:117-119
Piccinotti G, Mushotzky RF, Boldt EA, Holt SS, Marshall FE, Serlemit-
    sos PJ, Shafer RA 1982 Ap J 253:485-503
Pietsch W, Reppin C, Trumper J, Voges W, Lewin W, Kendziorra E,
    Staubert R 1981 Astr Ap 94:234-237
Pinkau K 1980 Astr Ap 87:192-195
Pollock AMT, Bignami GF, Hermsen W, Kandback G, Lichti GG, Masnou JL,
    Swanenburg BN, Wills RD 1981 Astr Ap 94:116-120
Preuss E, Kellermann KI, Pauliny-Toth IIK, Witzel A, Schaffer DB 1979
    Astr Ap 79:268-280
Primini FA, Cooke BA, Dobson CA, Howe SK, Scheepmaker A, Wheaton WA,
    Lewin WHG 1979 Nature 278:234-236
Primini FA, Basinska E, Howe SK, Lang F, Levine AM, Lewin WHG, Roth-
    schild R, Baity WA, Gruber DE, Knight FK, Matteson JL, Lea SM,

Reichert GA 1981 Ap J 243:L13-L17
Ramaty R, Meszaros P 1981 Ap J 250:384-38.
Ramaty R, Lingefelter RE 1982 Ann Rev Nucl & Part Sci 32:235-269
Riegler GR, Ling JC, Mahoney WA, Wheaton WA, Willett JB, Jacobs AS,
    Prince TA 1981 Ap J 248:L13-L16
Rieke GH, Lebofsky MS 1981 Ap J 250-87-97
Rothschild RE, Baity WA, Marscher AP, Wheaton WA 1981 Ap J 243:L9-L12
Rothschild RE, Muszotzky RF, Baity WA, Gruber DE, Matteson JL, Peter-
    son LE 1983 Ap J 269:423-437
Tananbaum H 1979 in:
    Giacconi & Setti (Eds) X ray Astronomy 291-310 NATO Adv. Study Conf.
    Proceedings.
Tennant AF, Mushotzky RF 1983 Ap J 264:92-104
Schlickeiser R 1980a Ap J 236:945-950
Schlickeiser R 1980b Ap J 240:636-641
Schlickeiser R 1984 Ap J 277:485-486
Schmidt M 1978 Phys Scripta 17:135-136
Schonfelder V 1980 in:
    Coswik & Will (Eds) Cospar non Solar Gamma Rays 3-5, Pergamon,
    Oxford.
Schwartz DA, Ku WHM 1983 Ap J 266:459-465
Seyfert CK 1943 Ap J 97:28-40
Shapiro SL, Lightman AP, Eardley DM 1976 Ap J 204:187-199
Strong AW, Bignami GF 1983 Ap J 274:549-557
Svensson R 1983 Ap J 270:300-304
Spoelstra TA, Hermsen W 1984 preprint
Veron P 1978 Nature 272:430-431
Weeks TC 1978 in:
    Robert (Ed) Proceedings of the 1978 Dumand Summer Workshop 313-318
    (la Jolla: Scripps Institution of Oceanography)
White RS, Dayton B, Gibbons R, Long JL, Zanrosso EM, Zych AO 1980
    Nature 284:608-610
Worrall DM, Mushotzky RF, Boldt EA, Holt SS, Serlemitsos PJ 1979 Ap J
    232:683-688
Worrall DM, Boldt EA, Holt SS, Mushotzky RF, Serlemitsos PJ 1981 Ap J
    243: 53-59
Zdiarski AA 1981 Acta Astr 30:371-391

Theoretical Models of Active Galactic Nuclei

R.D. Blandford

Theoretical Astrophysics 130-33
California Institute of Technology
Pasadena, CA 91125   U.S.A.

SUMMARY

The powerhouses in the nuclei of active galaxies are interpreted as
accreting, massive black holes. It is argued that the ratio of relativistic to thermal particle production decreases with increasing mass
accretion rate relative to the mass of the hole and that this is
reflected in the different spectra observed in the different classes
of object. The possible influence of beaming on the appearance of
BL Lac objects and compact radio sources is examined. Recent work on
the structure and stability of radiation tori is described. It is
suggested that the broad emission line region might be associated with
shocked, radiatively-driven winds.

INTRODUCTION

Just as the focus of basic physics has progressed from atoms to
nuclei to the "fundamental" particles and fields, so has the attention
of extragalactic astronomers been drawn from galaxies to their
sporadically active nuclei and finally to the internal sources of
this activity. These "prime movers", which are commonly identified
with massive, spinning, black holes, are responsible for phenomena
ranging in lengthscale from the giant radio sources ($\sim 10^{22}$–10^{25} cm)
through narrow emission line regions ($\sim 10^{20}$–10^{22} cm), compact radio
sources ($\sim 10^{18}$–10^{21} cm), broad emission line regions ($\sim 10^{18}$–10^{19} cm),
"thermal" bumps and non-thermal continua ($\sim 10^{15}$–10^{17} cm) down to the
shortest scales associated with occasional rapid X-ray variability
($\sim 10^{13}$ cm). If we look along the frequency axis we see that active
nuclei are responsible for emission all the way from the lowest radio
frequencies to the highest γ-ray energies.

Active galactic nuclei constitute a heterogeneous class of objects.
The brightest have bolometric luminosities that can exceed 10^{47} ergs^{-1}
and at the other end of the scale it appears that the central regions
of most galaxies (including our own) display evidence for some form of
non-stellar emission (e.g. Keel 1983). We shall not discuss further
the lower power objects such as "liners" and "starburst" nuclei save to
emphasize that their frequency indicates that the majority of galactic
nuclei have (if the pun can be excused) the potential to become active.

It is doubtful whether active galaxies will ever be classified as
well as stars have been. Nevertheless, if we ignore much of the
well-documented spectral detail and gloss over some inconsistencies,
then it is possible to organize the more powerful objects on the basis
of their optical and radio polarization properties. At one extreme,

there are the radio-quiet quasars (and what appear to be their low
luminosity equivalents, the Seyfert 1 galaxies) which display very
little polarization ($\lesssim$1 percent); at the other end are the "blazars"
which are substantially polarized ($\gtrsim$10 percent) from radio to optical
wavelengths (Martin, this volume) and the radio galaxies (Porcas, this
volume). The non-blazar radio-loud quasars combine low polarization
optical emission with polarized radio emission.

We display some schematic total spectra of "typical" members of the
different classes of active galactic nucleus in Fig. 1. The most
striking feature of these spectra is their gaps! We still do not know
in what region of the spectrum different objects radiate most of
their power. In the case of 3C273 (Robson, these proceedings), most
power seems to emerge in the near UV. However, there seems to be
almost as much power in the far IR/sub mm and in the γ-ray regions.

Let us, as is conventional, associate the polarized emission with
synchrotron radiation by ultrarelativistic electrons and the unpolarized emission with bremsstrahlung and Comptonization by sub- (or
possibly mildly) relativistic electrons (and possibly positrons, see
Fabian, these proceedings). The sequence from radio quiet quasar to
radio galaxy can then be interpreted as an increasing dominance of
relativistic particle production in the nucleus. In this article, we
describe one general explanation for these trends and outline some
additional interpretations of the observations in this context.

The theoretical literature on the subject of active galactic nuclei
is too extensive to be properly reviewed and only representative
papers concerning a particular class of model favored by the author
will be included. More comprehensive reviews of some of the material
contained below together with a variety of competitive explanations
of active galactic nuclei are to be found in the conference proceedings
edited by Wolfe (1978), Ulfbeck (1978), Hazard and Mitton (1979),
Heeschen and Wade (1982), Riegler and Blandford (1983), Evans (1984),
Kellerman and Setti (1984) and the article by Begelman, Blandford,
and Rees (1984).

MASSIVE BLACK HOLES IN GALACTIC NUCLEI

The notion that active galaxies are powered by accretion onto
massive black holes is almost as old as the discovery of quasars (e.g.
Zel'dovich 1964, Salpeter 1964, Lynden-Bell 1969). As is well known,
the observational evidence in favor of this view is circumstantial
and not even by the laxest standards of scientific proof can we claim
to have <u>demonstrated</u> the existence of a black hole within the nucleus
of any galaxy (or indeed anywhere else). The most direct argument
probably follows from the discovery of $\sim$100–1000 s X-ray variability
in a minority of quasars and Seyferts (Tennant <u>et al.</u> 1981). Central
light cusps (Sargent <u>et al.</u> 1979), the apparent stability of radio
source axes (e.g. Kellermann and Pauliny-Toth 1981), superluminal
expansion (e.g. Cohen 1984) and observations of the Galactic Center
(e.g. Lo and Claussen 1983) provide further indirect evidence for
spinning, relativistically deep potential wells. The most
persuasive arguments for the existence of massive black holes are
essentially theoretical. They include the requirement of a high
radiative efficiency and the apparent inevitability of black hole
formation during the evolution of a nucleus (e.g. Begelman and Rees
1978).

In view of the unsatisfactory status of this hypothesis, it is
perhaps of interest to suggest possible observations which would cast

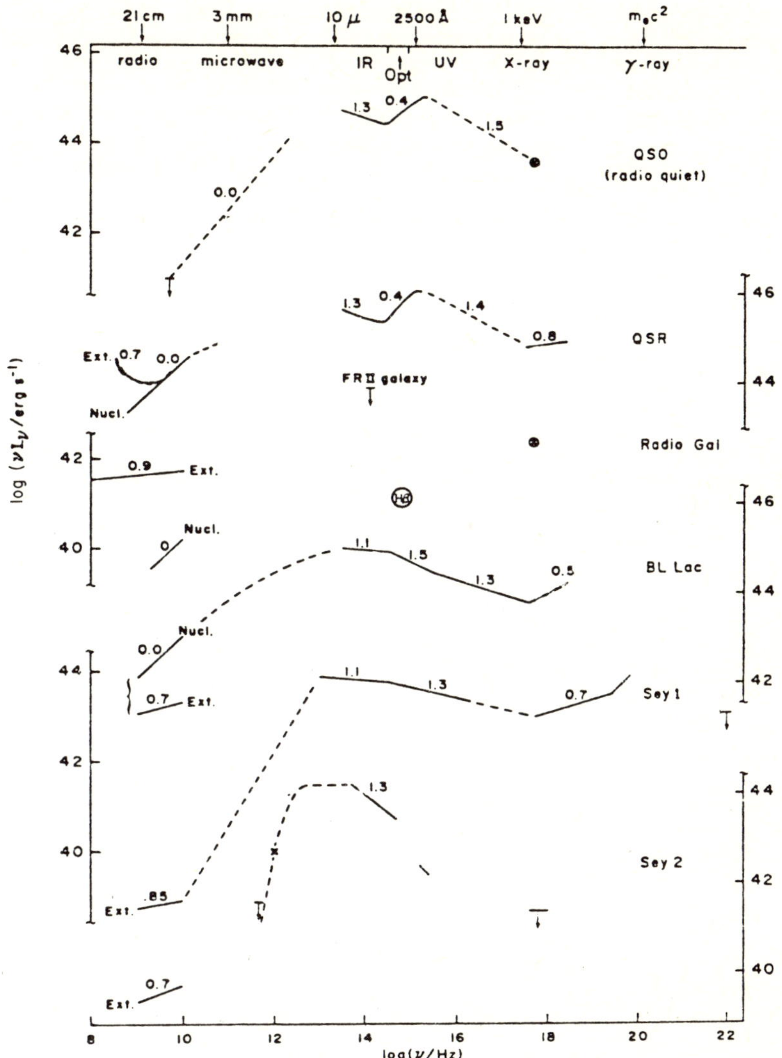

FIG. 1. Schematic spectra (plotted as power per logarithmic frequency bandwidth) for "typical" members of the different classes of active galaxy. The diagram is adapted from a similar one in Phinney (1983). The various portions of the spectrum are labeled by their spectral indices. The variation in these spectra within an individual class is substantial.

considerable doubt upon the notion that active extragalactic objects are powered by massive black holes residing in galactic nuclei. Such a list might include the following:
(i)   High-Q pulsation on timescales $\lesssim 1$ hr. By the "no-hair" theorems we conclude that there should be no fiducial marks on the surface

of a black hole and therefore there should be no well-defined
rotation period. In fact, any large amplitude aperiodic rapid
variability on timescales much shorter than the crossing time of
a massive black hole would be rather surprising. Regular
precession or orbital motion on much longer timescales is
conceivable, however.

(ii)   The discovery of X-ray or γ-ray (e.g. electron-positron annihila-
       tion) lines redshifted to the surface of a neutron star. This
       would provide strong evidence for an alternative class of quasar
       models in which the power is derived from a compact cluster of
       spinning or accreting neutron stars (e.g. Arons, Kulsrud, and
       Ostriker 1975).

(iii) Proof that a large class of quasars are genuinely quasi-stellar
       and not surrounded by galaxies or protogalaxies. We have tacitly
       been assuming in this article that all of the ultraluminous,
       compact extragalactic objects such as quasars, Seyferts, and
       BL Lacs, are indeed surrounded by galaxies as, at present, there
       is every indication is the case (e.g. Hutchings 1984). Although
       a demonstration that galaxies were absent wouldn't necessarily
       rule out massive black holes, it would raise some hard questions
       about their genesis and fueling.

(iv)   Determination that extended radio sources are contracting. It
       has generally been assumed, with no direct evidence other than
       the suggestive morphology of the radio structures, that extended
       radio sources are expanding. It may be possible to confirm this
       "galactometrically" in M87 and 3C273 over the next few years.

(v)   Dynamical determination that the central mass in Seyferts is
       less than $\sim 10^6 M_\odot$ or in energetic radio sources less than $\sim 10^8 M_\odot$.
       These are the minimum masses which must be converted into
       energy (with $\lesssim 10$ percent efficiency) to account for the require-
       ments of these objects. The central masses would have to be
       even larger if, for instance, it could be demonstrated that only
       the few percent of spirals that are now Seyferts were ever
       Seyferts. In other words, the _average_ central mass in a spiral
       galaxy should be $\sim 10^6 M_\odot$ (cf. Soltan 1982).

MODES OF ACCRETION

Much theoretical work has been devoted to spherically symmetric
accretion flows (e.g. Krolik and London 1983). Although this remains
a problem of genuine technical challenge, it is probably of limited
relevance to the observations simply because it is hard to see how
infalling gas can have so little angular momentum to move radially
near the event horizon of the hole; a view supported by the observation that all classes of active nuclei exhibit strongly aspherical
radio structure. For this reason, we confine attention to accretion
with angular momentum through a disk or torus.

Four modes of disk accretion have been distinguished and the choice
of mode depends on the ratio of the rate of mass supply to the
Eddington rate, $\dot{M}_{Edd} = 4\pi GM/\kappa c$ where M is the mass of the hole and
κ is the relevant opacity, Thomson in this case. Numerically,
$\dot{M}_{Edd} = 0.2 M_8$ $M_\odot$ yr^{-1}, measuring the hole mass in units of $10^8 M_\odot$ (see
Fig. 2).

Ion Torus ($\dot{M}/\dot{M}_{Edd} \lesssim 0.1$)
This case arises when the accreting gas is unable to cool efficiently on the infall timescale (Rees _et al._ 1982). The ions therefore

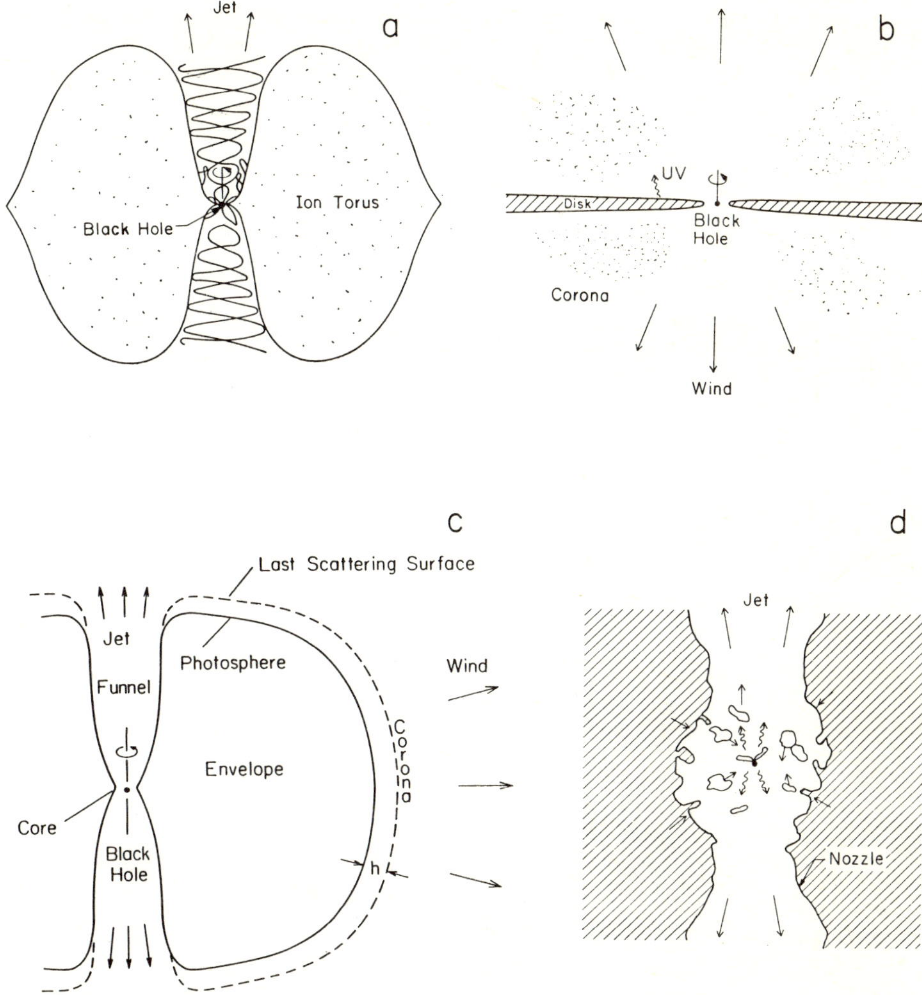

FIG. 2. Four modes of disk accretion: (a) Ion torus, (b) Disk and
corona, (c) Radiation torus, and (d) Super-critical accretion.

acquire a temperature comparable with the virial temperature ($\sim$100 MeV)
and form a thick ring in orbit around the hole. If the electrons were
to be maintained at the same temperature they would be ultrarelativistic and the torus would cool and deflate very rapidly. It is therefore necessary that the electrons be substantially cooler than the

ions. The conditions for this mode of accretion and indeed its very
existence are highly uncertain in view of our inability to estimate the
viscous stress (assumed here to be a substantial fraction of the gas
pressure) and the electron-ion coupling rate which determines the
cooling. An ion torus would probably be a copious source of pair-
producing gamma rays.

If the black hole is threaded by a uniform magnetic field, then
electro- or hydromagnetic torques can release a substantial fraction
of its spin energy (e.g. Macdonald and Thorne 1982; Phinney 1984). The
spin kinetic energy residing in the hole can amount to as much as 29
percent of the rest mass and up to half of this may be extracted.
In fact, the rate at which energy is extracted from the hole may exceed
the rate at which energy is released by the accreting gas. The
efficiency of extraction of spin energy for a realistic hydromagnetic
model has recently been estimated to be as high as $\sim$30 percent (Phinney
1984). This means that $\sim 10^{61} M_8$ ergs can be released as the hole spins
down, probably in the form of an electromagnetic Poynting flux combined
with a relativistic electron-positron wind. This is adequate to supply
the energy requirements of the most energetic extended radio sources.

Disk Accretion ($0.1 \lesssim \dot{M}/\dot{M}_{Edd} \lesssim 10$)
If the mass accretion rate is increased, then the gas is generally
expected to be accreted onto the hole through a thin disk. The disks
are similar to those invoked and extensively analyzed in the context
of X-ray binary stars and cataclysmic variables. The innermost parts
of the disk are supported by radiation pressure and the dominant
opacity is due to Thomson scattering. At larger radii, gas pressure
starts to dominate and if the disk extends out to very large radii it
will become sufficiently cool that free-free opacity will dominate
(Shakura and Sunyaev 1973). If a disk were to radiate the locally
released binding energy (actually three times this if we allow for
energy transport from the inner parts of the disk) at the black body
temperature, then most of the power would be radiated from the innermost regions of the disk at ultraviolet wavelengths. The visible
spectrum would then be a power-law with spectral index
$\alpha = -d \ln I_\nu/d \ln \nu \sim -0.3$, nothing like what is observed. However,
the outer parts of a concave disk may be irradiated by the inner parts
and the reprocessed radiation may have a quite different spectrum.
Furthermore, if hard X-rays originate from the inner disk, then they
can Compton heat gas in the outer disk until the temperature is large
enough to drive a powerful wind (Begelman, McKee, and Shields 1983).

In a steady state, the disk must adjust so that its height-integrat-
ed coefficient of dynamical viscosity equals $\sim \dot{M}/3\pi$. This may not be
achievable in practice and real disks may be prone to a variety of
thermal and secular instabilities (e.g. Pringle 1981). The net result
may be that accretion proceeds at a rate determined self-consistently
by the angular momentum transport associated with the non-linear
development of these unstable modes. Accretion disks will probably be
sandwiched by active coronae and it is certainly possible that a
substantial fraction of the binding energy released by the accreting
matter be dissipated in or at least be reprocessed by an optically thin
hot plasma. This is an ideal location for Comptonization regions in
which soft photons are "Fermi-accelerated" to generate power law X-ray
spectra. If the Thomson optical depth of the source region is τ ,
then the spectral index is given by α where

$$(3 + \alpha)\alpha = \frac{(m_e c^2)}{\tau^2 kT} \tag{1}$$

(e.g. Katz 1976; Takahra, Tsuruta and Ichimaru 1981).

Magnetic fields are highly likely to be amplified by dynamo processes in a differentially rotating accretion disk (e.g. Pudritz 1981). In
fact the field strength may saturate through reconnection at an energy
density comparable with the pressure. (N.b. It has been argued that
it is appropriate to use just the gas pressure in the radiation-supported inner regions of a disk in which case, the magnetic energy
density would be somewhat smaller than this (Sakimoto and Coroniti
1981).) If accretion disks were formed out of this "low beta" plasma,
then the dominant viscosity would probably be magnetic rather than
turbulent.

These dynamo processes may also be responsible for generating large
scale poloidal magnetic fields stretching out of the disk. These
fields may become twisted around the rotation axis a long way above
the disk and thereby collimate a pair of anti-parallel jets through
magnetic pinching forces (e.g. Lovelace 1976; Blandford and Payne 1982).
A simple way to see that gas can also flow along the magnetic field
lines is to imagine a wire poking out of the surface of the disk and
being carried around by it at the Keplerian speed. Now put a bead on
this wire. If the wire extends horizontally in the radial direction,
then the bead will be flung outwards. If the wire is vertical, then
the bead can only execute stable vertical epicyclic motion. It turns
out that the wire must make an angle of more than 60° with the radial
direction for the bead not be flung out. In just the same way, the
magnetic field is likely to drive a centrifugal wind away from the
surface of the disk. This constitutes a quite attractive mechanism
for creating radio jets and thus accounting for the origin of double
radio sources. It has some observational support in the discovery that
several jets appear to have minimum internal equipartition pressures
that exceed the maximum external gas pressures constrained by X-ray
observations (e.g. Potash and Wardle 1979). This has been taken to
imply that large scale jets have to be confined by toroidal field.

This mechanism may also solve a quite different problem. As gas
accretes onto a black hole, it must lose angular momentum. Conventionally, the angular momentum is transported outwards through the disk
by viscous stresses and must accumulate at large radius. This
accumulation presents no problem in a close binary system where
non-axisymmetric motion in the outer disk allows a torque to be exerted
on the orbiting companion star (Paczynski 1977). However, there is no
analogous repository of angular momentum in the case of a galactic
nucleus. One possible solution to the angular momentum problem is to
postulate that it is transferred to a co-extensive non-rotating star
cluster coupled to the disk through dynamical friction (Ostriker 1983)
or similarly to a gas cloud coupled through Kelvin-Helmholtz instability (Gunn 1979). An alternative possibility, which we advocate here,
is that magnetic torques extract most of the angular momentum in an
analogous fashion to their action on the sun which has presumably been
decelerated over its lifetime by hydromagnetic stresses in the solar
wind.

Radiation Torus $(10 \lesssim \dot{M}/\dot{M}_{Edd} \lesssim 100)$

At higher values of the mass accretion rate, the inner parts of the
disk will thicken to form a radiation-supported torus. Radiation tori

have been actively investigated over the past five years and clearly
have a lot of desirable features when it comes to modeling active
galactic nuclei. They can be thought of as toroidal stars which either
approach or overflow their Roche lobes. We discuss this case further
below.

<u>Super-critical Accretion</u> (100 $\lesssim$ M/M$_{Edd}$)
   Finally, at the highest mass accretion rates, most of the infalling
gas must be expelled with kinetic energy derived from that small
fraction of the gas which is accreted. The outflowing gas may either
take the form of a wind as described by Meier (1982abc) or a pair of
radiation-dominated jets as asserted by Begelman and Rees (1984) in
their "cauldron" model.

RADIATION TORI

If the accretion rate is large, the inner portions of the disk will
swell under the pressure of the trapped radiation. The radiation will
then emerge from a quasi-spherical structure rather than a planar
surface. We now consider some of the implications of this.

<u>Structure</u>
   It is helpful to regard radiation tori in an analogous fashion to OB
stars. Although they are geometrically more complicated, tori are
microphysically simpler in that the dominant opacity is just that due
to Thomson scattering and because radiation pressure dominates gas
pressure. The structure of a torus can be computed if we specify the
angular momentum and entropy distributions which must in turn be
determined by the internal heat transport, the angular momentum
transport and the history. This is far too hard a problem to solve
self-consistently and so in most existing work these distributions
have been specified by fiat. The resulting structures have been
computed in the full Kerr geometry (e.g. Jaroszynski, Abramowicz, and
Paczynski 1980) and in an instructive Newtonian approximation to the
Schwarzschild geometry (e.g. Paczynski and Wiita 1980). The isobaric
surfaces are limited by a critical surface that includes a cusp in
the equatorial plane at its innermost radius. If the gas expands to
fill this critical surface then it can flow through the cusp onto the
hole. Note that as we are ignoring gas pressure and using Thomson
opacity, there is no defined mass scale in the problem and structures
with different central gas densities are homologous, subject only to
the requirement that there be sufficient mass in the torus to ensure
that it is optically thick, and insufficient to make its self gravitation important.
   In a steady state, most of the energy will be released near the
pressure maximum (the core) located within a few Schwarzschild radii
of the horizon. The heat will leak outwards through the surrounding
envelope by a combination of radiative conduction and convection. The
structure must everywhere adjust so that the local acceleration
relative to a freely falling frame balances the radiative heat flux
divided by κ_T . If the heat flux that has to be transported exceeds
this value then the excess may be convected. If the dissipation within
the torus does not balance the radiative heat loss then the torus will
deflate to form a thin disk. Again, in a steady state, the mechanical
energy dissipated in the envelope where the binding energy is smaller
is a minor contributor to the total heat production.
   Although Thomson scattering dominates, we must rely upon free-free

emission to create and thermalize the photons. It turns out that an optical depth $\tau_T \gtrsim 1000$ is adequate to ensure that the radiation field is locally Planckian (Blandford 1984).

<u>Stability</u>
    Radiation tori are similar in many respects to the massive objects originally postulated by Hoyle and Fowler (1963) and, ironically, may suffer a similar fate by being prone to dynamical instability. There are possible axisymmetric local instabilities caused by unfavorable entropy and angular momentum gradients (e.g. Seguin 1975). (The formal instability criterion is a linear combination of the better known Schwarzschild and Rayleigh criteria.) Unstable regions presumably evolve to form marginally stable convection zones just as in a star. The circulatory motions will transport entropy and angular momentum and it seems a reasonable speculation that the fluid be "gyrotropic", that is to say that entropy be a unique function of the specific angular momentum, within these zones (Bardeen 1973, Paczynski and Abramowicz 1982).
    More threatening are non-axisymmetric instabilities. In recent important work, Papaloizou and Pringle (1984), have demonstrated that a toroidal configuration known to be marginally stable to axisymmetric disturbances possesses global non-axisymmetric dynamical instabilities. It will apparently destroy itself in a few orbital periods unless non-linear effects saturate the instability at a small amplitude. Analogous modes can also be found in models of thin disks that are quite stable locally (Narayan 1983). It appears that modes of a given frequency can be trapped between the inner edge of the disk and the inner Lindblad resonance with negative energy density. These modes are evanescent in the region between the inner and outer Lindblad resonance (and are not seriously affected by the corotation resonance). They carry away positive energy outside the outer resonance which is compensated by the growth of negative energy within the inner resonance. Necessary conditions for the growth of these modes clearly include mode trapping by the resonances and good reflection at the inner edge of the disk. Perhaps the instabilities will develop so that the inward flow from the inner edge is just large enough to limit the amplitude to a value that gives sufficient angular momentum transport to maintain the flow. Alternatively, the cores of thick disks may become supported as much by turbulent motions as by centrifugal forces. This research offers the exciting prospect of furnishing a believable prescription for the viscosity of accretion disks in general. Unfortunately, it is not yet clear if dynamical instabilities will always be present. Numerical calculations along the lines of those already carried out by Hawley and Smarr (1984) may be necessary to settle the issue.
    If there are regions that are stable to all of these dynamical instabilities, then the radiative zones will probably be subject to Goldreich-Schubert instability (e.g. Seguin 1975). A simple estimate (Blandford 1984) indicates that these modes are capable of transporting angular momentum and keeping the fluid marginally stable--i.e. barytropic. (This means that the entropy is now a fixed function of the pressure.) Meridional circulation, on a thermal timescale, will also be important in a radiative zone.

<u>Photosphere</u>
    Just as is the case with a star, the spectrum of the radiation emerging from the surface of a torus is determined by the thermal structure of the atmosphere. (In fact, this discussion is applicable

to any quasi-spherical gas distribution.) As electron scattering
still dominates free-free opacity, the photosphere, where the emergent
photons are emitted will lie well below the last scattering surface
(e.g. Wiita 1982). Even if the envelope is fully convective, convection will be decreasingly efficient below an optical depth $\tau_T \sim 1000$
and will be ignorable below $\tau_T \sim 30$. The upper atmosphere is then
radiative and the gas must be essentially "buoyant" with the upthrust
caused by the radiation pressure counteracting gravity. We consider
two cases.

If convective transport is unimportant and the scale height of the
atmosphere is small compared with the radius (or the geometry accurately spherical) then the opacity must be constant in order that the
atmosphere remain balanced. The Thomson opacity depends upon the mean
molecular weight per electron and when the gas temperature falls below
$\sim 30{,}000$ K , helium will start to recombine giving a 9 percent decrease
in the opacity. This suggests that $T_{eff} \sim 30{,}000$ K is a sort of
limiting effective temperature for a radiation-dominated photosphere.
If the torus attempts to expand so that the effective temperature is
reduced below this value, then the surface layers will sink. Now,
quasar spectra often exhibit "ultraviolet excesses" or "30,000 bumps"
which have been interpreted in terms of the superposition of a
 30,000 K black body on a power-law extending into the far infrared
(e.g. Grewing and Limla 1968; Malkan 1983; Fig. 1). The change of
opacity at this temperature provides one fairly general explanation for
these features.

Alternatively, if the convective transport well below the photosphere is significant then the radiative flux in the upper atmosphere
will exceed the Eddington limit and so the outer layers must be
continuously blown off. The resulting wind will have a momentum flux
that can be as large as the super-Eddington momentum flux in the radiation field. It will presumably also have a terminal velocity several
times the escape velocity from the surface of the torus. Numerically,

$$V_{\infty} \sim 50{,}000 \ M_8^{1/4} \ (T_{eff}/30{,}000 \ K) \ km \ s^{-1} \tag{2}$$

$$\dot{M}_W \sim [(L - L_{Edd})/L_{Edd}] \ M_8^{3/4} \ (T_{eff}/30{,}000 \ K)^{-1} \ M_{\odot} \ yr^{-1} \tag{3}$$

Radiation tori may also contain active optically thin coronae in
which most of the power-law infrared and optical radiation is generated
through Comptonization. If, as appears to be true observationally,
the extra luminosity is a significant fraction of the "thermal" power,
then this may also be able to drive a strong wind. Coronae may be
energized and the non-thermal particle distribution be maintained by
the non-radiative heat flux from below the photosphere (e.g. shocks,
Alfvén waves). A particularly promising means of energization of this
corona is via the "funnels" that bore down to the black hole (see Fig.
2). (It was thought that these funnels might channel a large super-
Eddington radiative luminosity which would be capable of accelerating
an optically thin plasma to ultrarelativistic speeds and thus account
for the superluminally expanding compact radio sources. However, it
is now believed that the luminosity is mostly carried by matter
accelerated by shear stresses along the funnel walls to sub- or mildly
relativistic speeds; e.g. Narayan, Nityananda, and Wiita 1983).

The radiation that emerges from the photosphere will be subject to
substantial reprocessing. The electrons in the optically thick layer

above the photosphere will probably be maintained at a temperature
dictated by the radiation field, T = 1/4 h$\bar{\nu}$/k , where $\bar{\nu}$ is the
intensity-weighted frequency in the spectrum. However, the scattering
by these electrons will probably cause some Comptonization of the
spectrum towards a Bose-Einstein distribution at the same photon
number and energy density. Higher energy electrons accelerated in the
corona may have an even more pronounced effect as may radiative transfer through a wind (cf. Begelman <u>et al.</u> 1983). These effects should
be borne in mind when attempting to fit continuum fluxes to specific
models of the accreting gas.

<u>Tori vs. Disks</u>
    In much recent work, it has been hypothesized that quasars contain
thin accretion disks and many observations have been interpreted in
these terms. However, there may be some difficulties associated with
disk models that are alleviated if the accreting gas is distributed
more spherically at least out to a few hundred Schwarzschild radii.
    The major difference between a disk and a sphere is that in the
case of a disk, most of the power will emerge from the innermost radii,
presumably at the local effective temperature which is in the ultraviolet, whereas in the sphere the photons will be reprocessed to have
a lower effective temperature corresponding to a larger emitting area.
This difference is not as great as might at first be imagined because
a substantial amount of energy is transported from the innermost radii
to larger radii by viscous stresses and also because the photons
created near the hole will be somewhat redshifted (e.g. Novikov and
Thorne 1973). The most detailed comparison has been carried out by
Malkan (1983) who has compared his extended spectra of quasars and
Seyferts with the prediction of a model in which the surface of a thin
stationary disk in orbit about a Kerr black hole radiates the energy
dissipated locally with a black body spectrum. We replot his data
for three quasars in Fig. 3 in terms of the power per logarithmic
bandwidth, νL_ν. This plot has the virtue that the underlying power-law
continuum is a nearly horizontal line which is easy to remove by eye.
The observations do seem to indicate that there is very little radiation emitted with an effective temperature in excess of 27,500 K. This
is consistent with radiation from the surface of a torus but not from
a thin disk. (It may also be attributable to Ly α absorption.)
    A second problem with the disk interpretation is that the model fits
give luminosity $\gtrsim L_{Edd}$. The disks must therefore be thickened by the
pressure of the escaping radiation. A straightforward calculation
shows that the half thickness of the disk, h , at radius r satisfies

$$(h/r)_{max} = K \ (\dot{M}/\dot{M}_{Edd}) \tag{4}$$

where the constant K increases monotonically from 0.022 to 0.11 as
the specific angular momentum of the hole, a , increases from 0
(Schwarzschild) to m (extreme Kerr). Comparison with the models
confirms that the disks are not self-consistently thin. This is not a
very strong argument however, because the black body approximation on
which the models are based must be suspect.
    A final difficulty with thin disk models is that most optical
quasars have very small degrees of linear polarization ($\lesssim 1\%$; Martin,
this volume). It might be expected that radiative transfer through a
geometrically flattened electron scattering atmosphere would produce a
larger degree of polarization (e.g. Angel 1969).

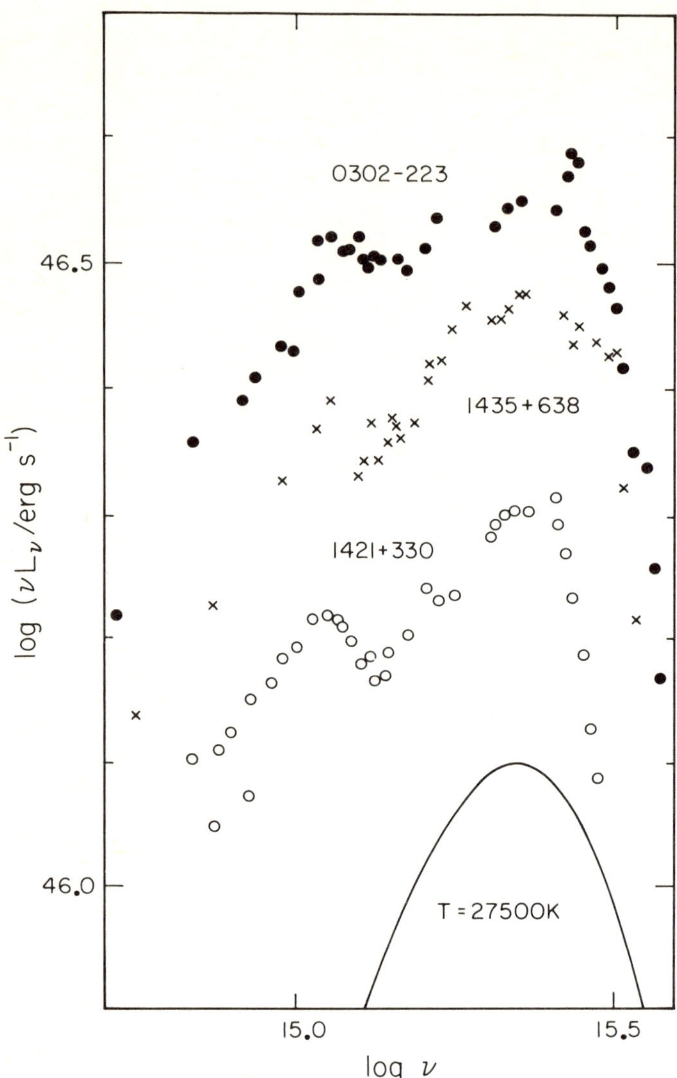

FIG. 3.   Three quasar spectra from the data presented by Malkan (1983)
   and replotted in terms of luminosity per logarithmic frequency
   interval.   For comparison a black body spectrum with temperature
   27,500 K is also exhibited.   (I am indebted to R. Narayan for
   preparing this figure.)

Broad Emission Line Region
   We have argued that a radiation torus is likely to power a fast wind
that carries as much momentum as the radiation field.   The momentum
flux in this wind will eventually fall below the ambient nuclear
pressure and so the wind will be decelerated by passing through a
strong shock wave.   The broad emission lines that distinguish quasars
and Seyfert 1 galaxies may be located behind this shock front (e.g.
Camerzind and Courvoisier 1983).

One attraction of this idea is that it can account quantitatively
for several puzzling features of broad emission line regions. The
pressure in the post shock region (radius R_S) is essentially the
momentum flux in the wind, $M_W V_W / (4\pi R_S^2)$, where V_W is the wind speed.
Hence, the ionization parameter, the ratio of the density of ionizing
photons to the electron density is roughly

$$\Gamma \stackrel{\sim}{\scriptstyle\sim} (L_{ion}/L_{bol}) \; [\dot{M}_W V_W / (L_{bol}/c)]^{-1} \; (kT_e/h\nu_{ion}) \tag{5}$$

where L_{ion} is the ionizing luminosity and L_{bol} is the bolometric
luminosity. We observe that $L_{ion} \sim 0.1 \, L_{bol}$ and the emission line
gas should cool to a temperature $T_e \sim 0.1 \, h \, \nu_{ion}/k$ (e.g. Davidson and
Netzer 1979). Hence if the wind and the radiation carry comparable
momenta, $\Gamma \stackrel{\sim}{\scriptstyle\sim} 10^{-2}$ as appears to be generally true.
Gas in the outflowing wind will be cooled by the adiabatic expansion.
It may also be heated by the X-ray photons if these are present.
Beyond some radius, which a conservative estimate places within
$\sim 10^{17}$ cm, the cooling will dominate and the gas may become cold enough
to cool radiatively. It may then become thermally instable, creating
dense filaments (cf. Beltrametti 1981). It is these filaments which
create the emission line clouds after they have been compressed in
the post-shock flow. (The viability of this model depends upon a
detailed comparison of the cooling times with the flow times along the
lines of the analysis presented here by Dyson and Perry.)
The shock may be located at the radius where the shape of the
gravitational potential changes from being dominated by the black hole
to being dominated by the central star cluster, i.e. $R_S \sim GM/\sigma_c^2$,
where σ_c is the central velocity dispersion. For Seyferts,
$R_S \sim 0.1\text{-}1$ pc and for quasars, $R_S \sim 1\text{-}10$ pc , consistent with the
indications from emission line variability data (e.g. Penston, these
proceedings). These radii guarantee that the gas density in the broad
line regions is $\stackrel{\sim}{\scriptstyle\sim} 10^9\text{-}10^{10}$ cm^{-3} as the detection of semi-forbidden
CIII] and the anomalous $L_{y}\alpha/H\beta$ ratios together seem to indicate (e.g.
Carswell, these proceedings). Numerically,

$$n_e \sim 10^8 \; \sigma_{300}^4 \; M_8^{-1} \; [\; MV_W / (L_{bol}/c)] \tag{6}$$

The shock itself is likely to be quite unsteady and to move with a
speed comparable with V_W . The cooling gas clouds will, to a large
extent, track the shock, and so the cloud velocities will have a radial
outflow superposed on a large random dispersion. Note that the velo-
city width $\stackrel{\sim}{\scriptstyle\sim} V_{esc} \stackrel{\sim}{\scriptstyle\sim} 10{,}000\text{-}30{,}000$ km s^{-1} is generally much larger
than the virial velocity. Conversely, the mass of the central black
hole is much smaller than the value $\sim 7 \times 10^8$ (V/10,000 km s^{-1})^2
(R/10^{17} cm) (G/7$\times 10^{-8}$ g^{-1} cm^3 s^{-2})$^{-1}$ M$_\odot$ inferred from assuming that
the gas is in orbital motion about the hole (Ulrich et al. 1984). The
column density in the broad line clouds can be estimated by assuming
that the clouds remain in the post-shock region for a dynamical time
$\stackrel{\sim}{\scriptstyle\sim} 30 \, R_S/V_W$. The column density is then $\stackrel{\sim}{\scriptstyle\sim} 30 \, (\sigma/V_W)^2 \, \kappa_T^{-1} \, C^{-1} \stackrel{\sim}{\scriptstyle\sim}$
10^{-2} g cm^{-2} where $C \stackrel{\sim}{\scriptstyle\sim} 0.1$ is the covering factor. Again, this is
consistent with the observations.

BEAMING

The observation of superluminal expansion, in a large fraction of

the brightest compact radio sources and the prevalence of one-sided
jets in radio-loud quasars and Class 2 radio galaxies strongly suggests
that radio jets are created with relativistic speeds. As has been
widely discussed, this in turn implies that our view of a particular
source will be strongly influenced by our orientation and that flux-
limited samples of radio sources will be strongly biased to include
those objects that contain jets that are pointed in our direction.

The OVV quasars and the BL Lac objects, with their distinctive
spectral and polarization properties have been respectively associated
with beamed radio-loud quasars and radio galaxies of intermediate
intrinsic strength (e.g. Blandford and Rees 1978; Browne 1983). The
steep, polarized, featureless power-law continua in the infrared and
optical are Doppler-boosted synchrotron radiation by relativistic
electrons accelerated in the jet. This outshines any emission lines
associated with the nucleus. The X-ray emission may also arise from
the jet and be due to the inverse Compton mechanism (Königl 1981).

Some evidence for this model comes from the observations of
Antonucci, reported here (and other references therein) of extended
radio emission surrounding the compact radio components. The morphology of these radio sources is generally consistent with that of an
extended double source foreshortened by projection. One objection
that has been raised to this beaming model is that the OVV quasars do
not show narrower equivalent widths in their permitted lines than the
QSRs as might be expected. However, if the brightest BL Lac objects,
like AO 0235+164 are also presumed to be beamed quasars then the
distribution of equivalent widths is more similar to that found in the
low luminosity BL Lacs.

The compact radio sources (which are mostly quasars) have also been
postulated to be strongly beamed. There are two candidates for the
unbeamed sources: the radio-quiet quasars (Scheuer and Readhead 1979)
and the extended radio-loud quasars (Orr and Browne 1982). However,
the former possibility now seems to be unlikely following the discovery
of extended (and presumably unbeamed) radio emission around the compact
sources and absent from the radio-quiet objects. The latter hypothesis
is in similar trouble if, as seems to be the case, there are too few
extended quasars per compact quasar compared to the same optical
emission line strength. Fortunately, two recent trends in VLBI offer
a possible resolution of these problems. The first is that, as
expected, compact radio sources are proving to be more complicated
than simple core-jet sources when mapped with high dynamic range
(e.g. Wilkinson 1984). From a theoretical perspective, compact radio
sources are indeed likely to be far more complex than the simple
kinematic models favored by observers. If the radio emission is
produced in an underlying jet, then the emissivity of an optically
thin fluid element will decrease rapidly as the element expands. The
brightest features should occur at places where fresh particles are
being accelerated and magnetic field is being amplified--in other
words, where the flow is most dissipative. Whatever the nature of
this dissipation, be it shock fronts (e.g. Lind and Blandford 1984)
or non-linear waves (e.g. Eilek and Henriksen 1984) it is probably
true that the velocity of the emitting fluid differs from the pattern
velocity of the emitting region. Furthermore, it seems extremely
unlikely that there only be one speed that characterizes the Doppler
boosting of a jet. There is likely to be a range of Lorentz factors
perhaps increasing from $\gamma \sim 1$ in a boundary layer to a much larger value
in the center of a jet. An observer oriented at an angle θ to the
jet axis is likely to see that part of the source moving with $\gamma \sim \theta^{-1}$.

These complications have the effect of making the fraction of compact
radio sources that are seen by us much larger than the usual estimate
$\sim (V_{SL}/c)^{-2}$ where $V_{SL} \approx 10\,c$ is the typical speed of a strong
superluminal source.

The second trend is that superluminal expansion appears to be common
only among the brightest sources (e.g. Readhead 1984). It is possible
to account for this if the intrinsic luminosity function of the sources
steepens at the bright end. In this case a small fraction f of intermediate power sources boosted to large observed flux will constitute
a larger proportion of the brightest sources than the proportion of
intermediate flux sources associated with the same fraction f of
boosted lower power objects.

UNIFIED SCHEME

We have described four different modes of accretion that have been
distinguished in the literature and argued in general terms that
the progression of increasing mass accretion rate relative to the
Eddington rate corresponds to a decreasing ratio of relativistic to
thermal particle production. It is this fact that can be used to map
the observational classification onto the theoretical models (Fig. 4).

We identify the radio-quiet quasars with radiation tori radiating
at just over the Eddington limit. (The Broad Absorption line quasars
may be accreting supercritically and thus driving a powerful wind as
we described above.) The associated black hole masses are
$\sim 10^7$-$10^9 M_\odot$. Similarly, the Seyfert 1 galaxies are identified with
critical accretion onto $\sim 10^6$-$10^7 M_\odot$ holes. This has an immediate
implication that the Seyfert stage cannot have lasted much more than
10^8 yr. As several percent of spiral galaxies are Seyferts, this in
turn implies that most spirals should be dormant Seyferts and contain
$\sim 10^6 M_\odot$ black holes in their nuclei (cf. Bailey and Clube 1978). A
small minority will be old quasars which presumably harbor heavier
holes.

Radio galaxies are apparently highly non-thermal objects and on
the above scheme derive their power from the spin energy of massive
black holes spun up during an earlier phase of rapid accretion. (It
has been alternatively suggested that the radio jets are formed in
the funnels of the radiation tori. It is very hard to see how this
could be the case as, except for the QSRs, we do not see a nuclear
Eddington-limited flux from a black hole heavy enough to have produced
the minimum energy of the extended radio components.)

The radio loud quasars are an intermediate class displaying both the
unpolarized optical-IR continua of the quasars and the non-thermal
energy production of the radio galaxies. They can be associated with
intermediate values of $\dot{M}/\dot{M}_{Edd}$. It is then natural to associate
Seyfert 2 galaxies with their larger radio to optical flux ratios with
lower-luminosity versions of the QSRs.

The association of radio galaxies with ellipticals and Seyferts with
spirals and the predictions for the quasars are explicable if ellipticals grow larger black holes in their nuclei than spirals. This may
be because it is harder for gas to accrete into the nucleus from the
outer parts of a galaxy when it has a large specific angular momentum
as in a spiral.

OBSERVATIONAL PROSPECTS

There have been dramatic observational advances in the study of

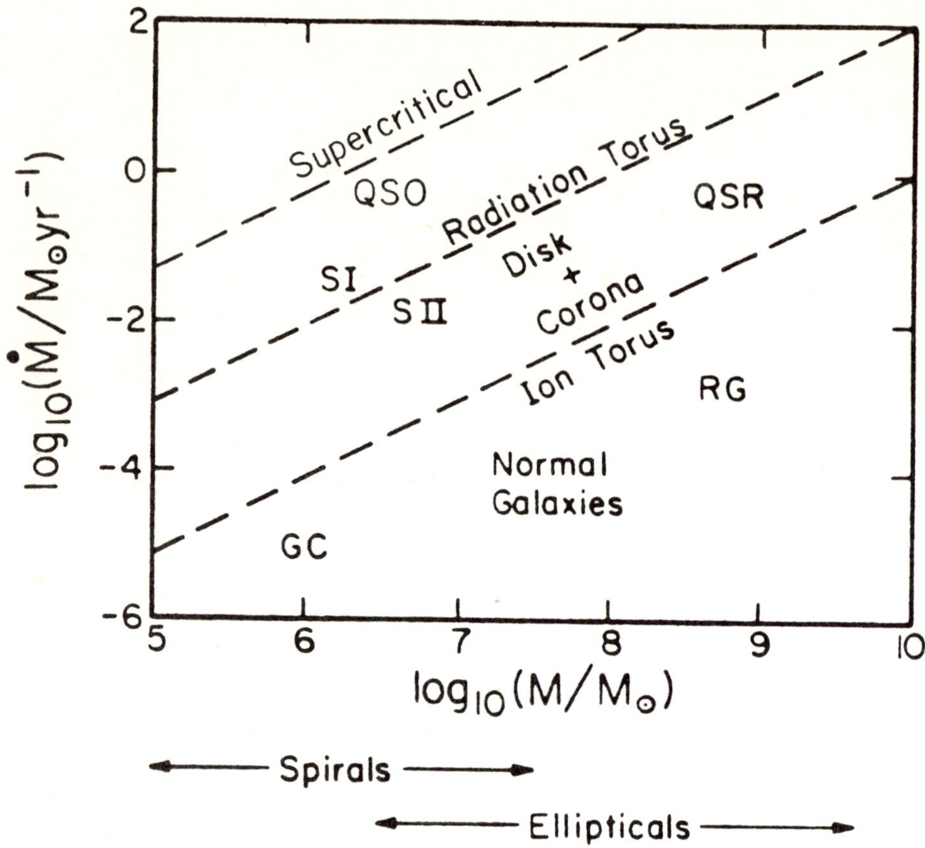

FIG. 4. Simplified unified scheme for different types of extragalactic
   object. In this scheme, the observed properties are believed to be
   controlled by the ratio of the mass accretion rate to the mass of
   the hole and the viewing angle. QSO denotes a radio quiet quasar,
   QSR a radio loud quasar, SI and SII are types I and II Seyfert
   galaxies respectively. RG is a radio galaxy and GC our galactic
   center which is supposed typical of a dormant nucleus that may once
   have been a low luminosity Seyfert. OVV quasars are identified with
   intermediate power QSRs beamed in our direction and BL Lac objects
   with intermediate power radio galaxies likewise beamed towards us.

active galactic nuclei over the past ten years. It is of obvious
interest to ask what questions we can hope to settle over the next
ten years. As there should be several new instruments deployed, there
are good grounds for optimism.

<u>VLBI</u>
   The quality of VLBI maps is steadily improving, especially with the
increased availability of the Mark III recording system. The next
major technical advance should come with the dedicated arrays proposed
in the United States, Canada, and Europe. These should produce higher
dynamic range maps and more frequent monitoring of the superluminally
expanding sources and thus help unravel their kinematic properties.

It will be of particular interest to see if the moving features can be interpreted as shock fronts.

There is some hope that these observations can also be used to quantify the importance of beaming in general, though, as was discussed above, the source geometries are expected to be too complex to do this in any simple way.

In the more distant future, orbiting VLBI will produce an improvement in the angular resolution and should be able to resolve features $\sim 10^{16}$ cm in size at the distance of Centaurus A. This is only 30 Schwarzschild radii for a $10^9 M_\odot$ black hole.

Sub mm Astronomy

As exhibited in Fig. 1, a significant fraction of the total bolometric power of a nucleus may emerge in the sub mm region. Direct observations should determine the position of the spectral breaks expected from the IR and radio slopes.

Space Telescope

The space telescope should have an angular resolving power of $\sim .05$ arcsec and so should be able to detect galaxies surrounding distant quasars. In particular it should be able to determine what type of galaxy surrounds different classes of objects and to see whether or not radio quiet quasars are mainly located in spirals with radio loud quasars and BL Lac objects being preferentially in ellipticals as asserted above. Existing observations (e.g. by Boroson, Oke, and Green 1982) have indicated that most quasars are in spirals. (Note that the issue may be confused if the onset of a quasar phase is triggered by a galaxy merger or close encounter.)

The space telescope should also be able to probe the nuclei of nearby Seyferts and normal galaxies repeating the work on the nucleus of M87 on a smaller angular scale. It should be capable of detecting $10^8 M_\odot$ central masses as far away as the Virgo cluster. (Recent work by Dressler (1984) on the nucleus of M31 is very encouraging.) These observations should help see if a large fraction of spirals have been Seyferts (as argued above) and perhaps find some dead or dormant quasars. They should also be able to detect large black holes in powerful radio galaxies (cf. Young et al. 1979).

The ultraviolet capability should help define the shape of the "black body" humps and perhaps find some correlation between effective temperature and total luminosity analogous to that which exists for stars. The extension of polarization observations into the UV would also be of considerable interest as it is predicted here that it should be reduced as the proportion of "black body" emission increases.

Finally, the space telescope may be capable of detecting optical emission from distant radio jets on similar angular scales to those probed by the Merlin radio telescope.

X-rays

The launch of AXAF, currently projected for the summer of 1991, will herald a new era in X-ray astronomy. It should be able to determine the X-ray fraction of the total bolometric luminosity of different types of active galaxy and monitor its variability. The X-ray timing explorer should increase the number of hard X-ray spectra from active nuclei.

γ-rays

The Gamma Ray Observatory should be launched in 1988. It should

measure $\sim$100 keV – 1 MeV γ-ray fluxes from the brighter active
galaxies. In the context of the ideas discussed above, it ought to
be able to see how much power is radiated at γ-ray energies, and, by
extension determine the importance of electron-positron plasmas in
different classes of objects. It may also be able to detect evidence
for positrons from nearby radio galaxies.

<u>Acknowledgements</u>
    I thank Drs. Dyson and Kahn for the opportunity to attend this
meeting and several of the other participants for helpful discussions
on active galactic nuclei. I also thank my collaborators, M. Begelman,
A. Fabian, M. Jaroszynski, S. Kumar, K. Lind, R. Narayan, D. Payne,
S. Phinney, and M. Rees for their contributions. Financial support
by the National Science Foundation under grant AST82-13001 and the
Alfred P. Sloan Foundation is gratefully acknowledged.

REFERENCES

Angel JRP 1969 Ap J 158:219
Arons J, Kulsrud RM, Ostriker JP 1975 Ap J 198:687
Bailey J, Clube SVM 1978 Nature 275:278
Bardeen JM 1973 In: DeWitt C, DeWitt B (eds) Black Holes. Gordon and
    Breach, New York
Begelman MC, Blandford RD, Rees MJ 1984 Rev Mod Phys 56:255
Begelman MC, McKee CF, Sheilds G 1983 Ap J 271:70
Begelman MC, Rees MJ 1978 MNRAS 188:847
Begelman MC, Rees MJ 1984 MNRAS 206:209
Beltrametti M 1981 Ap J 250:18
Blandford RD 1984 In: LeBlanc (ed) Numerical Astrophysics, in press
Blandford RD, Payne DG 1982 MNRAS 199:883
Blandford RD, Rees MJ 1978 In: Wolfe (ed) Proceedings of Pittsburgh
    Conference on BL Lac Objects, Pittsburgh
Boroson T, Oke JB, Green RF 1982 Ap J 263:32
Browne IWA 1983 MNRAS 204:23
Camerzind M, Courvoisier TJC 1983 Ap J Lett 266:L83
Cohen MH 1984 In: Kellermann, Setti (eds) Proceedings of IAU Symposium
    No. 110. Reidel, Dordrecht, Holland, in press
Davidson K, Netzer H 1979 Rev Mod Phys 51:715
Evans DS 1984 (ed) Eleventh Texas Symposium on Relativistic Astrophys-
    ics, Annals New York Acad Sci 422
Grewing M, Limla E 1968 Z Astrophys 68:473
Gunn JE 1979 In: Hazard, Mitton (eds) Active Galactic Nuclei.
    Cambridge University Press, Cambridge
Hawley J, Smarr LL 1984 In: Proceedings of the Toulouse Conference on
    Numerical General Relativity. Reidel, Dordrecht, Holland
Hazard C, Mitton SA 1979 (eds) Active Galactic Nuclei. Cambridge
    University Press, Cambridge
Heeschen DS, Wade CM 1982 (eds) Extragalactic Radio Sources (IAU
    Symposium No. 97) Reidel, Dordrecht, Holland
Hoyle F, Fowler WA 1963 MNRAS 125:169
Hutchings JB, Crampton D, Campbell B 1984 Ap J 280:41
Jaroszynski M, Abramowicz MA, Paczynski B 1980 Acta Astron 30:1
Katz JI 1976 Ap J 206:910
Keel WC 1983 Ap J 269:466
Kellermann KI, Pauliny-Toth IIK 1981 Ann Rev Astr Ap 19:373
Kellermann KI, Setti G 1984 (eds) Proceedings of IAU Symposium No. 110.
    Reidel, Dordrecht, Holland, in press

Königl A 1981 Ap J 243:700
Krolik JH, London R 1983 Ap J 267:18
Lind K, Blandford RD 1984 Ap J, submitted
Lo KY, Claussen MJ 1983 Nature 306:647
Lovelace RVE 1976 Nature 262:649
Lynden-Bell D 1969 Nature 223:690
Malkan M 1983 Ap J 268:582
Meier D 1982a Ap J 256:386
Meier D 1982b Ap J 256:681
Meier D 1982c Ap J 256:693
Macdonald D, Thorne KS 1982 MNRAS 205:1103
Narayan R 1983 unpublished work
Narayan R, Nityananda R, Wiita PJ 1983 MNRAS 205:1103
Novikov ID, Thorne KS 1973 In: DeWitt C, DeWitt B (eds) Black Holes.
    Gordon and Breach, New York
Orr MJL, Browne IWA 1982 MNRAS 200:1067
Ostriker JP 1983 Ap J 273:99
Paczynski B 1977 Acta Astron 28:91
Paczynski B, Abramowicz MA 1982 Ap J 253:897
Paczynski B, Wiita PJ 1980 Astron Astrophys 88:23
Papaloizou JCB, Pringle JE 1984 MNRAS, in press
Phinney ES 1983 unpublished thesis, University of Cambridge
Phinney ES 1984 MNRAS, in press
Potash RI, Wardle JFC 1979 Astron J 84:707
Pringle JE 1981 Ann Rev Astron Astrophys 19:137
Pudritz RE 1981 MNRAS 195:881
Readhead ACS 1984 In: Kellermann, Setti (eds) Proceedings of IAU
    Symposium No. 110. Reidel, Dordrecht, Holland
Rees MJ, Begelman MC, Blandford RD, Phinney ES 1982 Nature 295:17
Riegler C, Blandford RD 1983 (eds) The Galactic Center. AIP, New York
Sakimoto P, Coroniti FV 1981 Ap J 247:19
Salpeter EE 1964 Ap J 140:796
Sargent WLW, Young PJ, Boksenberg A, Shortridge K, Lynds CR, Hartwick
    FDA 1979 Ap J 221:731
Scheuer PAG, Readhead ACS 1979 Nature 277:182
Seguin F 1975 Ap J 197:745
Shakura NI, Sunyaev RA 1973 Astron Astrophys 24:337
Soltan A 1982 MNRAS 200:115
Takahara F, Tsuruta S, Ichimaru S 1981 Ap J 25:26
Tennant AF, Mushotzky RF, Boldt EA, Swank JH 1981 Ap J 251:15
Ulfbeck O 1978 Physica Script 17:135
Ulrich MH, Boksenberg A, Bromage GE, Clavel J, Elvius A, Penston MV,
    Perola GC, Pettini M, Snijders MAJ, Tanzi EG, Tarenghi M 1984
    MNRAS 206:221
Wiita PJ 1982 Ap J 256:666
Wilkinson PN 1984 In: Kellermann, Setti (eds) Proceedings of IAU
    Symposium No. 110. Reidel, Dordrecht, Holland
Wolfe A 1978 (ed) Proceedings of Pittsburgh Conference on BL Lac
    Objects, Pittsburgh
Young PJ, Sargent WLW, Kristian J, Westphal JA 1979 Ap J 234:76.
Zel'dovich YaB 1964 Soviet Phys Doklady 9:195

BL Lac Nuclei in Elliptical Galaxies

Ismael Pérez-Fournon

Max-Planck-Institut für Radioastronomie, Auf dem Hügel 69, 5300 Bonn 1,
Federal Republic of Germany

SUMMARY

Orientation effects are believed to be important in sources which harbour relativistic jets in their nuclei. BL Lac type objects are interpreted as a type of this class of objects whose galaxy counterpart is a bright elliptical galaxy. This interpretation requires every bright elliptical galaxy to have a BL Lac type nucleus beamed usually away from the observer. We discuss the possibility of identifying M87 as one of those misaligned BL Lacs.

INTRODUCTION

Relativistic beaming of the emission from a compact jet is the most successful explanation for the extreme properties of BL Lac objects at all frequencies. In that picture we only identify BL Lacs when the axis of their relativistic jet makes a small angle to the line of sight. The morphology of the extended radio emission, if any can be detected, should resemble a halo around the compact nucleus rather than the typical double radio structure seen in radio galaxies and steep spectrum quasars. This has been proven by recent VLA observations (Ulvestad and Johnston 1984) which show that only objects with weak BL Lac properties, or misclassified objects, have large scale extended radio structure. As far as the galaxy counterparts of BL Lacs are concerned, all galaxies identified with BL Lacs are bright ellipticals. In a previous work (Pérez-Fournon and Biermann 1984) we used this evidence and the existence of a first complete X-ray selected sample of BL Lacs with known redshifts (Schwartz and Ku 1983) to derive the space density of the misaligned BL Lacs. From the comparison of the space densities of BL Lacs and elliptical galaxies we came to the conclusion that relativistic beaming in BL Lacs strongly suggests that all bright elliptical galaxies contain a BL Lac type nucleus usually pointed elsewhere. Therefore,we asked ourselves if there is any evidence for these misaligned BL Lacs among the few hundred bright ($M_B < -21$ for $H_O = 50$ km s^{-1} Mpc^{-1}), nearby ($z < 0.02$) elliptical galaxies. I shall report here on an interpretation of observations providing evidence for relativistic beaming in objects in which the jet is not pointing at us.

M87

One of the candidates for being a counterpart of BL Lac objects is the well-known active galaxy M87 near the center of the Virgo cluster. The system of optical knots was already known (Curtis 1918) before external galaxies were distinguished from galactic nebulae. The structure and spectrum of the jet are now well known. The radio structure (Owen et al. 1980, Charlesworth and Spencer 1982, Biretta et al. 1983) correlates extremely well with the optical one (de Vaucouleurs and Nieto 1979, Nieto and Lelièvre 1982) and is also morphologically consistent with X-ray and infrared observations (Schreier et al. 1982, Stocke et al. 1981). From X-ray observations (Schreier et al. 1982, Lea et al. 1982) there is evidence for an extended hot intragalactic gas that may be responsible for confinement and recollimation of the jet (see, e.g., Sanders 1983). Since the jet emission is polarized at optical and radio frequencies that emission is best interpreted to be of synchrotron origin. However, the X-ray emission from the jet cannot be easily explained by synchrotron self Compton models (SSC).

<u>The spectrum of the knots</u>

For knots A and B the spectrum is a power law of spectral index $\alpha = 0.6$ ($S \sim \nu^{-\alpha}$) from radio frequencies to the red and shows a break at frequencies near 3×10^{14} Hz (Stocke et al. 1981). There is evidence for similar cutoffs in the Cen A inner jet (Brodie et al. 1983, Lépine et al. 1984). The extrapolation of the steep optical spectrum to the X-rays underestimates the observed emission. This suggests that the X-ray emission is not synchrotron in origin. Furthermore, at the equipartition magnetic field and number density of relativistic particles required to explain the synchrotron luminosity, reacceleration of relativistic particles must occur in the knots. The best way to achieve this is by reacceleration of relativistic particles in shock waves (Axford et al. 1977, Krymsky 1977, Bell 1978a,b, Blandford and Ostriker 1978). Actually, the M87 high resolution VLA observations of Biretta et al. (1983) suggest a multiple shock structure (Sanders 1983, Norman et al. 1982).

Reacceleration together with radiative losses produces a power law distribution of the relativistic particles with a cutoff at the energy $\gamma_c mc^2$ at which acceleration compensates the energy losses (Webb et al. 1984, Biermann et al. 1984, Schlickeiser 1984). The resulting spectrum is a power law with an exponential cutoff at the frequency $\nu_c \simeq 4.2 \times 10^6$ γ_c^2 B(Gauss) Hz. An exponential cutoff spectrum fitted to the observed bend frequency of 3×10^{14} Hz lies far below the observed X-ray flux. Therefore, synchrotron radiation cannot be the correct explanation for the X-rays in the context of this simple model and other models must be explored.

INVERSE COMPTON MODELS FOR THE X-RAY EMISSION

Schreier et al. (1982) discarded thermal models as well as inverse Compton emission due to scattering of relativistic electrons off the microwave background or stellar photon field. They also found implausible a synchrotron self Compton origin for the X-rays. We show with stronger arguments that the SSC mechanism actually does not work in the M87 knots for a large range of models.

Assuming a distance of 16 Mpc to M87, the total synchrotron luminosity between 10^8 Hz and 3×10^{14} Hz in Knot A is 4.5×10^{41} erg s^{-1}. We estimate an emitting volume of 1.6×10^{61} cm^3 from the radio observations (Biretta et al. 1983). For spherical symmetry and filling factor one, this means that the photon energy density in synchrotron emission at the center of the knot is 1.5×10^{-10} erg cm^{-3}. Minimum energy calculations give a value of the magnetic field and particle number density of B = 4×10^{-4} G and k = 3×10^{-3} cm^{-3}, where k is the normalization constant of the electron energy distribution dn = k $\gamma^{-(2\alpha+1)}$ dγ. We have also assumed the same energy in heavy particles as in electrons. At minimum energy the pressure is 7×10^{-9} dynes cm^{-2}, already somewhat above the pressure of the external hot plasma. As shown by Sanders (1983) this is still compatible with the transition from a free inner jet to a confined one at knot A.

From standard inverse Compton (IC) calculations for a homogeneous spherical source (Gould 1979) we obtain a value of the monochromatic X-ray luminosity at 4×10^{17} Hz of 1.4×10^{19} erg s^{-1} Hz^{-1}, 4300 times less than the observed value we obtain from the 0.2 μJy that Schreier et al. assign to knot A at that frequency. To match both the integrated synchrotron luminosity L_S and the monochromatic X-ray luminosity at 4×10^{17} Hz, $L_{\nu x}$, we discuss four different ways (in the following a subscript 'x' denotes parameter values which explain L_S and $L_{\nu x}$):

1) No equipartition and filling factor $\Phi = 1$

We obtain then $B_x = 2$ μG, $k_x = 14$ cm^{-3} and a pressure of $P_x = 6 \times 10^{-6}$ dynes cm^{-2}. This high pressure in the knot (dominated by the relativistic particles) is difficult to understand if the knot is confined by the external hot gas. The external pressure is four orders of magnitude less than the pressure in the knot. Furthermore, the magnetic field energy density is eight and three orders of magnitude below the energy densities of the relativistic particles and of the synchrotron photon field respectively. The latter requires catastrophic inverse Compton cooling.

2) No equipartition and filling factor less than one.

Since the knots are resolved in the optical and in the radio, we assume that the emission comes from N clouds of radius R_C inside the volume defined by the observed radius R. R_C and N are constrained to give an emitting volume smaller than the observed one: $N\,R_C^3 < R^3$.

As a first approximation, the synchrotron energy density, and hence the IC emissivity, is dominated by the integrated emission of all clouds if $R_C > R\,N^{-1/2}$ and by the emission of each cloud if $R_C < R\,N^{-1/2}$. We denominate these regimes as non local and local respectively.

a) <u>Non local</u>

For given k and B, L_S and $L_{\nu x}$ as a function of L_S and $L_{\nu x}$ for filling factor one (distinguished by the subscript '1') L_{S1}, $L_{\nu x1}$ are

$$L_S = L_{S1} \quad \Phi \sim \Phi \quad k \quad B^{1+\alpha}; \quad L_{\nu x} = L_{\nu x1} \quad \Phi^2 \sim \Phi^2 \quad k^2 \quad B^{1+\alpha}$$

Then, for all values of R_C and N in the non local region to account for the observed L_S and $L_{\nu x}$ we obtain the relations:

$$B_x = B_{x1}; \quad k_x = k_{x1} \quad \Phi^{-1}$$

Since $P \sim k$ (the pressure is dominated by the relativistic particles), then $P_x = P_{x1}\,\Phi^{-1}$. The internal pressure is inversely proportional to the filling factor, (larger than the already large value we obtain for $\Phi = 1$).

b) <u>Local</u>

L_S and $L_{\nu x}$ are now given by

$$L_S = L_{S1}\, N\left(\frac{R_C}{R}\right)^3 \sim N \quad R_C^3\, k\, B^{1+\alpha}; \quad L_{\nu x} = L_{\nu x1}\, N\left(\frac{R_C}{R}\right)^4 \sim N R_C^4\, k^2\, B^{1+\alpha}$$

and B_x and k_x as a function of B_{x1}, k_{x1}, R_C, and N are

$$k_x = k_{x1} \quad \left(\frac{R_C}{R}\right)^{-1}; \quad B_x = B_{x1} \quad \left(N\left(\frac{R_C}{R}\right)^2\right)^{-\frac{1}{1+\alpha}}$$

The internal pressure of the small clouds increases with decreasing radius and, as before, the external gas pressure is not high enough to confine these clouds.

3) Relativistic bulk motion of the emitting region of the knot

The problem we want to solve is just the opposite to that present in compact sources. In these we observe less flux than expected from the SSC theory. In M87 knots we observe more than what is expected. Therefore, we try to solve it by assuming relativistic bulk motion of the knots, of Lorentz factor γ_j, but in the plane of the sky ($\theta = 90°$). This increases the synchrotron energy density in the comoving frame (subscript 'o') relative to the apparent one obtained from the observed flux (subscript 'ap').

$$L_{So} = L_{Sap}\ D^{-4} \sim k_o\ B_o^{1+\alpha} \sim k_{ap}\ B_{ap}^{1+\alpha}\ D^{-4}$$

$$L_{\nu xo} = L_{\nu xap}\ D^{-(3+\alpha)} \sim k_o^2\ B_o^{1+\alpha} \sim k_{ap}^2\ B_{ap}^{1+\alpha}\ D^{-(3+\alpha)}$$

where $D = \dfrac{1}{\gamma_j(1-\beta\cos\theta)}$. Since at $\theta = 90°$ $D = \gamma_j^{-1}$, we obtain for $\gamma_j = 5$

$$k_o = k_{ap}\ \gamma_j^{-(1-\alpha)} = 7\ cm^{-3}; \quad B_o = B_{ap}\ \gamma_j^{\frac{5-\alpha}{1+\alpha}} = 170\ \mu G$$

$$P_o = P_{ap}\ \gamma_j^{-(1-\alpha)} = 3 \times 10^{-6}\ dynes\ cm^{-2}$$

Relativistic motion of the knots alleviates the problem. k and P decrease by a factor of $\simeq 2$ and B increases by a factor of nearly 100; but the internal pressure is still much higher than the external one. Furthermore, the observed luminosity of the knots is too large for a relativistic kpc-scale jet viewed sideways.

4) Synchrotron external inverse Compton emission (SEC)

We then come to one possible solution that allows a low pressure for the knots. To increase the X-ray IC emissivity for the minimum energy k and B obtained from the observed L_S we assume that the knots are 'seeing' a radiation field of higher energy density than their own synchrotron one. The best source for that low frequency photon field is the VLBI nucleus itself. In this case the photon field incident on the knot is nearly monodirectional. For an isotropic distribution of relativistic electrons in the knot the IC emission is viewing angle, θ, dependent. As a first approximation $L_{\nu x} \sim (1-\cos\theta)^{1+\alpha}$. $L_{\nu x}$ is maximum in the direction to the photon source (head-on collisions). From the calculations of Reynolds (1982) and assuming for the central source a flat spectrum up to 5×10^{11} Hz we need a photon energy density at knot A of 4×10^{-7} erg cm^{-3}. At a distance of $\simeq 1$ kpc from the center this requires an apparent luminosity for the central source of $\simeq 10^{48}$ erg s^{-1}. Although this value is quite high it is consistent with the apparent luminosity of other sources believed to be beamed towards us (e.g., S5 0014+81, Kühr et al. 1983).

The apparent luminosity of that compact relativistic jet when viewed sideways is $10^{48}\ (2\gamma_j^2)^{-4} \simeq 10^{41}$ erg s^{-1} and is again consistent with the weak observed luminosity of the central source in M87 (Reid et al. 1982, Young et al. 1978, Duncan and Wheeler 1980).

CONCLUSIONS

Relativistic beaming of the emission of the M87 compact jet (the VLBI core-jet) with its velocity vector in the plane of the sky can explain the X-ray emission of the knots. We interpret it as inverse Compton scattering of the relativistic electrons in the knots off the beamed photon field from the nucleus. Furthermore, boosting of the compact jet emission away from the observer is consistent with the otherwise remarkable weakness of the central source given the high accretion rate deduced from X-ray spectroscopy.

Various other models will be considered elsewhere in competition with

the model presented here. We note that the luminosity function of misaligned BL Lacs leads us to expect about one object with 10^{48} erg s^{-1}
beamed luminosity per volume of 150 Mpc scale. Were we to find several
objects similar to M87 close to us in space, the above interpretation
may become unacceptable.

A possible consequence of having a beamed source at the center of a
galaxy is the efficiency of the strong beamed photon field as a photoionization mechanism. It is well known that in some cases photoionization
models do not work if the observed weak ionizing continuum is isotropic
(see, e.g.,Ford and Butcher 1979, for the optical filaments in M87).

The interpretation presented here that M87 may have a misaligned BL
Lac type nucleus can be tested. Crucial observations would be: 1) the
monitoring of the VLBI jet, 2) a better determination of the optical to
X-ray spectrum, 3) the determination of the X-ray spectrum of the knots,
and 4) spectroscopy to use photoionization models as a probe of an anisotropic radiation field. This last test may be the most critical one
with present technical possibilities.

REFERENCES

Axford WI, Leer E, Skadron G 1977 Proc 15th Int Cosmic Ray Conf (Plovdiv
    Bulgaria) 11:132-137
Bell AR 1978a M N R A S 182:147-156; 1978b M N R A S 182:443-455
Biermann P, Strittmatter PA, Drury LO'C, Webb GM, Schlickeiser R 1984
    (in preparation)
Biretta JA, Owen FN, Hardee PE 1983 Ap J 274:L27-L30
Blandford RD, Ostriker JP 1978 Ap J 221:L29-L32
Brodie J, Königl A, Bowyer S 1983 Ap J 273:154-166
Charlesworth M, Spencer RE 1982 M N R A S 200:953-960
Curtis HD 1918 Publ Lick Obs 13:11-44
de Vaucouleurs G, Nieto J-L 1979 Ap J 231:364-371
Duncan MJ, Wheeler JC 1980 Ap J 237:L27-L31
Ford HC, Butcher H 1979 Ap J Supp Ser 41:147-172
Gould RJ 1979 A A 76:306-311
Krymsky GF 1977 Dokl Akad Nauk SSSR 234:1306-1308 (Engl transl Sov Phys
    Dokl 22:327-328)
Kühr H, Liebert JW, Strittmatter PA, Schmidt GD, Mackay C 1983 Ap J 275:
    L33-L37
Lea SM, Mushotzky RF, Holt SS 1982 Ap J 262:24-32
Lépine JRD, Braz MA, Epchtein N 1984 A A 131:72-76
Nieto J-L, Lelièvre G 1982 A A 109:95-100
Norman ML, Smarr L, Winkler K-HA, Smith MD 1982 A A 113:285-302
Owen FN, Hardee PE, Bignell RC 1980 Ap J 239:L11-L15
Pérez-Fournon I, Biermann P 1984 A A 130: L13-L15
Reid MJ, Schmitt JHMM, Owen FN, Booth RS, Wilkinson PN, Shaffer, DB,
    Johnston KJ, Hardee PE 1982 Ap J 263:615-623
Reynolds SP 1982 Ap J 256:38-53
Sanders RH 1983 Ap J 266:73-81
Schlickeiser R 1984 (this volume)
Schreier EJ, Gorenstein P, Feigelson ED 1982 Ap J 261:42-50
Schwartz DA, Ku WH-M 1983 Ap J 266:459-465
Stocke JT, Rieke GH, Lebofsky MJ 1981 Nature 294:319-322
Ulvestad JS, Johnston KJ 1984 A J 89:189-194
Webb GM, Drury LO'C, Biermann P 1984 A A (in press)
Young PJ, Westphal JA, Kristian J, Wilson CP, Landauer FP 1978 Ap J 221:
    721-730

Radiatively-driven shock waves in quasar envelopes

L. Mestel

Astronomy Centre, University of Sussex, Falmer, Brighton BN1 9QH,
England.

SUMMARY

The response of a quasar envelope to periodic disturbances at its
base is studied, using a simple form for the density-dependent
radiative acceleration. Quasi-steady winds flow, with strong shock
waves propagating outwards as the gas absorbs momentum from the quasar
radiation field. The compressed gas behind a shock has high speed in
the quasar frame and so can yield broad emission lines. A short-wave
approximation method yields self-consistent solutions in which strong
shocks forming near the quasar surface decelerate outwards and steadily
weaken, but does not describe rigorously solutions with shock-trains
that simultaneously accelerate and strengthen.

INTRODUCTION

The work in this report (discussed in detail in Mestel & Moore
(1984)) continues the study of the dynamical effect of the intense non-
thermal radiation field emitted by quasars and other active galactic
nuclei. When photo-ionization followed by radiative recombination is
the dominant process of energy input, the associated momentum input
corresponds to a radiative body force per gram of the approximate form

$$g_r = k\rho \qquad (1)$$

where ρ is the density and k is a quantity depending primarily on
fundamental physical constants and varying only weakly with gas
temperature and the quasar luminosity (Kippenhahn et al. 1974). When
hydrogen continuum absorption is the most important process,
$k \simeq 4\chi\alpha/3cm_H^2 \simeq 6\times10^{13} T_4^{-1/2}$, where χ is the ionization energy of
hydrogen, α the electron-proton recombination coefficient, and T_4 the
temperature in units of 10^4 ^{0}K.
To maintain the link with earlier papers in the series (Kippenhahn
et al. 1974, 1975; Mestel et al. 1976; Kippenhahn 1977), and because
the principal aim is to explore further the consequences of the law (1),
we assume that this law is acceptable all the way to the quasar
"surface" (defined operationally as the radius within which virtually
all the radiation is generated), even though this may very well be
unrealistic. We note however that other recent studies (Dyson et al.
1980; Falle et al. 1981) find that a limited "kρ domain" develops
spontaneously as a consequence of the interaction between a hot quasar
wind and an ambient galactic medium, and that radiatively driven shock
waves are again to be expected.
In the absence of the radiation force (1), an isothermal atmosphere
supported against gravity by the pressure $p = \rho a^2 = 2nkT$ would have a
number density n that exponentiates according to

$$n = n(r_Q) \exp \left[-\eta_1 (1 - \frac{r_Q}{r}) \right] , \qquad (2)$$

where r_Q is the radius at the base of the atmosphere, and the parameter η_1 is given by

$$\eta_1 = (\frac{GM}{r_Q^2}) \frac{r_Q}{a^2} = \frac{2.7 \times 10^3 M_8}{(r_Q)_{pc} T_4} , \qquad (3)$$

with the quasar mass measured in units of $10^8 M_\odot$ and $(r_Q)_{pc}$ in parsecs. With $T_4 \simeq 1$ (low enough for the law (1) to hold), $\eta_1 >> 1$ the thermal scale height is much less than the radius, and (2) would predict a very rapid decline; however, support by the radiation force yields the slowly declining

$$n \simeq \frac{10^7 M_8}{r_{pc}^2} . \qquad (4)$$

Of greater interest are solutions in which the gas flows out as a steady wind (Kippenhahn et al. 1975). These flows are necessarily constrained by the continuity equation: in spherically symmetric outflow with steady mass loss $-\dot{M}$ and local velocity v

$$4\pi \rho v r^2 = -\dot{M} \qquad (5)$$

so that the driving radiation force $k\rho \propto 1/vr^2$ and so is self-limiting. These steady flows do not therefore yield very high asymptotic speeds; also, they have a built-in anti-correlation between high density and high velocity, so that predicted emission lines would not have even the breadth associated with the modest asymptotic speeds. However, these flows are violently unstable (Mestel et al. 1976). A linear WKBJ treatment shows that a periodic density disturbance of frequency ω and of amplitude $A(r_Q)$ at r_Q developes according to

$$\tilde{\rho}(r,t) = A(r_Q) \exp \left[\frac{\eta_1}{2} (1 - \frac{r_Q}{r}) \right] \exp \left[i\omega(t - \frac{r}{a}) \right] ; \qquad (6)$$

the input of momentum from the radiation field yields amplifying sound-waves, with e-folding distance $\simeq (2/\eta_1)r_Q << r_Q$. Thus the linear treatment breaks down very quickly, and the present work is an attempt to construct an approximate representation of the asymptotic non-linear response of the atmosphere to a periodic disturbance at the base. As with familiar laboratory gas dynamics, we expect the waves to steepen into shocks, which will move supersonically through the expanding quasar atmosphere. The compressed gas behind the shock is subsonic in the shock frame and so is hypersonic in the quasar frame. In contrast to the smooth steady winds, high density is now correlated with high speed with respect to the quasar: flow over a large solid angle should therefore yield broad emission lines, as well as a narrow absorption line in the line-of-sight with a blue-shift to be subtracted from the basic cosmological red-shift.

SHOCK-TRAINS IN A HOMOGENEOUS MEDIUM

The method of attack on the temporally-periodic, spatially inhomogeneous problem as formulated is suggested by the rigorous WKBJ theory for linear problems. We study first a homogeneous medium with a flow containing a spatially periodic system of shocks. The system is supposed steady when viewed from a frame with a uniform acceleration (positive or negative) with respect to an inertial frame. The problem is a generalization of a study by Kippenhahn (1977) of a single shock moving with constant speed. From the gas dynamics of the smooth flow between shocks together with the jump conditions for isothermal shocks, one finds

$$\lambda(1+f) = \frac{1}{2} (V^2 - 1/V^2) + (V - 1/V) \equiv F(V) \qquad (7)$$

where

f = acceleration of frame in units of the local gravity g,
λ = wave-length (distance between shocks) in units of a^2/g,
V = velocity just ahead of a shock in units of a.

To complete the solution we need to prescribe the mass between shocks, with the density ρ measured in units of ρ_0 = g/k, and ε a local constant, as yet undetermined:

$$\varepsilon\lambda = \int_\lambda \rho dx = (V - 1/V) + 2\log V . \qquad (8)$$

With ε and λ given, equation (8) fixes V in terms of $(\varepsilon\lambda)$, whence $F(V)$ defined by (7) is equivalent to a function $\psi(\varepsilon\lambda)$. Note that a longer λ implies by (8) a larger V and so a stronger compression. For very strong shocks ($\varepsilon\lambda \gg 1$) and with ε not too small, $V \simeq \varepsilon\lambda$, $f \simeq \varepsilon^2 \lambda/2$ and the densities just ahead of and just behind a shock are respectively $\varepsilon/2$ and $\varepsilon^3\lambda^2/2$. The particular case ε = 1 corresponds by (8) to the mean density over a wavelength being equal to ρ_0 = g/k, for which radiation force balances gravity. Since $(V^2 - 1/V^2)/2 > 2\log V$, when ε = 1 it follows from (7) and (8) that $f > 0$, a special case of a general result (cf. Mestel and Moore 1984). In order to have a steady non-accelerating system, ε must have a particular value (<1); and for ε still smaller, the mean input of momentum from the radiation field is insufficient to off-set gravity, and the system decelerates.

APPLICATION TO A PERIODICALLY DISTURBED ATMOSPHERE

The theory just summarised is applicable to a non-homogeneous medium if the wavelength λ is small compared with the scale of variation of all the macroscopic quantities, including V, ε and f. We now attempt to apply it to the problem as formulated — the response of the quasar atmosphere to periodic disturbances at the surface. Since the e-folding distance of the linear solution (6) is so short, it is reasonable to postulate that within a short distance the waves have steepened into a train of shocks. Thus we suppose that at radius r there is a shock moving with speed u_s in the inertial frame at rest in the quasar, and that in the frame with outward speed u_s the local flow field is plane-parallel and steady. However, this frame will have an outward acceleration $u_s du_s/dr$, so that we write

$$u_s \frac{du_s}{dr} = g(r) \, f(\varepsilon, \lambda) \qquad (9)$$

where $g(r)$ is the local gravity, f is the non-dimensional acceleration appearing in (7), ε is the local mass parameter defined by (8), and $\lambda = \Lambda g(r)/a^2$ with Λ the local distance between shocks. In order that the short wavelength theory be applicable, we require that the change in u_s be small over a wavelength, i.e.

$$\frac{\Lambda}{u_s} \left| \frac{du_s}{dr} \right| \ll 1 \, . \qquad (10)$$

With this satisfied, then the imposed condition of strict temporal periodicity P is well approximated by

$$\Lambda = Pu_s \, , \qquad (11)$$

and (9) becomes

$$u_s \frac{du_s}{dr} = g(r) \, f\left[\varepsilon(r), \, Pu_s g(r)a^{-2}\right] \, . \qquad (12)$$

The mass parameter ε — now a slowly varying function of r — is determined by the condition that averaged over a period the mass flux across any radius r is steady, yielding

$$\varepsilon \left(\frac{u_s}{a}\right) - (1 + f) = m \, , \qquad (13)$$

where

$$-\dot{M} = 4\pi r^2 \rho_o(r) am = \left(\frac{4\pi GMa}{k}\right) m \, . \qquad (14)$$

If we now scale by writing

$$u_s = a\omega, \qquad s = r/r_Q \, , \qquad (15)$$

then (11) becomes

$$\lambda = \frac{\eta_2 \omega}{s^2} \, , \qquad (16)$$

where the second parameter of the problem η_2 is given by

$$\eta_2 = \frac{P \, g(r_Q)}{a} = \frac{3 \times 10^{-2} \, P(\text{years}) \, M_8}{(r_Q)^2_{pc} \, T_4^{1/2}} \qquad (17)$$

and is typically of order unity for oscillation periods of a few months and for $(r_Q)_{pc} \sim 10^{-1}$. Equation (13) combines with (16) to yield

$$\omega(\varepsilon\lambda - \frac{m\eta_2}{s^2}) = \psi(\varepsilon\lambda) , \tag{18}$$

where again $\psi(\varepsilon\lambda) \equiv F[V(\varepsilon\lambda)]$ from equations (7) and (8). Equation (12) now becomes

$$\omega \frac{d\omega}{ds} = \frac{\eta_1}{s^2} f = \frac{\eta_1}{s^2} (\varepsilon\omega - (m+1)) = \frac{\eta_1}{s^2} (\frac{\psi}{\lambda} - 1) = \frac{\eta_1}{\eta_2} (\varepsilon\lambda - \frac{\eta_2(m+1)}{s^2}). \tag{19}$$

The criterion (10) for the validity of the short wavelength approximation becomes

$$\frac{\eta_2}{\eta_1} \left| \frac{d\omega}{ds} \right| = \frac{\eta_2|f|}{\omega s^2} = \frac{1}{\omega} \left| \varepsilon\lambda - \frac{(m+1)\eta_2}{s^2} \right| << 1 . \tag{20}$$

Equations (19) and (18) have two classes of solution of interest, of which however only the first is rigorous in the sense of satisfying (20). This class consists of shock-trains which simultaneously <u>decelerate</u> outwards and decrease in strength, with

$$\omega \simeq \frac{1}{2} \frac{(m+1)^2\eta_2}{s^2} >> 1 , \quad \varepsilon \simeq \frac{2s^2}{(m+1)\eta_2} << 1 , \tag{21}$$

$$f \simeq - \frac{(m+1)^4\eta_2^2}{2\eta_1 s^3} .$$

Thus the density parameter has adjusted itself so as to yield a deceleration $f = O(1/\eta_1)$, so that since $\eta_2/\eta_1 << 1$ the condition (20) is satisfied. The correcting terms to this approximate solution are small provided m < 35, for typical values of η_1 and η_2. In fact, physical constraints — e.g. the condition for validity of the isothermal shock approximation — require m to be less than 19. With this value, the post-shock particle density is $\simeq 3\times10^{10} M_8$, and the corresponding velocities in the quasar frame are $\simeq c/150$.

The equations apparently allow also a class of solution with outwardly accelerating shock-trains. A necessary (though not a sufficient) condition is that $(\varepsilon\lambda)_{r_Q} > (m+1)\eta_2$. Asymptotically these solutions have $d\omega/ds \simeq 2\eta_1/\eta_2$, $\varepsilon \simeq 2s^2/\eta_2$, $\lambda \simeq 2\omega/\varepsilon$. They cannot be rejected simply because they yield ω large at infinity: if acceptable on other grounds they would demonstrate that the assumed unlimited supply of momentum from the radiation field can sometimes accelerate the system of shocks until relativistic corrections to inertia can no longer be ignored. The objection comes from finding that the short wavelength consistency condition (20) is normally not satisfied. Intuitively it seems clear that accelerating solutions should exist. Our results do not cast doubt on this, but rather show that the input of momentum can sometimes be embarrassingly efficient: because of the presence in equation (19) of the large parameter η_1, the shock speed in an accelerating solution increases over a wavelength by too large a factor for the short wavelength theory to be applicable.

An inevitable shortcoming of a purely atmospheric theory is that it involves as free parameters in addition to η_1 and η_2 the mass flux m and the base-value $\omega(1)$ of ω. A complete theory would link the envelope with the quasar proper, showing how m and $\omega(1)$ are correlated with each other and with η_1 and η_2; as it is, we have to take them as independent — within the constraints, both mathematical and physical, imposed on m by the atmosphere theory — and then to relate them to the different classes of solution (e.g. through the condition quoted for an accelerating solution to be possible). Nevertheless, it is gratifying that one can construct at least one class of satisfactory approximate solutions to this time-dependent, non-linear gas dynamical problem, which predict high-density domains moving with high speed in the quasar frame.

REFERENCES

Dyson JE, Falle SAEG, Perry JJ 1980 Mon Not R astr Soc 191:785-819

Falle SAEG, Perry JJ, Dyson JE 1981 Mon Not R astr Soc 195:397-427

Kippenhahn R 1977 Astron & Astrophys 55:125-133

Kippenhahn R, Perry JJ, Roeser H-J 1974 Astron & Astrophys 34:211-224

Kippenhahn R, Mestel L, Perry JJ 1975 Astron & Astrophys 44:123-138

Mestel L, Moore DW 1984 to be submitted

Mestel L, Moore DW, Perry JJ 1976 Astron & Astrophys 52:203-212

SHOCK FORMATION OF THE BLR IN QSOs AND AGNs

Judith J. Perry * and John E. Dyson +

* Institute of Astronomy, University of Cambridge
+ Department of Astronomy, University of Manchester

SUMMARY

We present a model for the formation of the BLR of QSOs and AGNs which accounts simultaneously for the formation, structure and velocities of the clouds. These emerge as a natural consequence of the interaction of optically thin supersonic flows with a dense stellar cluster in the immediate circumquasar environment. Low density gas near a QSO or AGN must be hot ($T_e \approx 10^6 - 10^8 K$); if it flows supersonically stand-off shocks inevitably form around all obstacles to the flow. The high pressure, superheated shocked gas cools rapidly by inverse-Compton cooling off the radiation field. If the flow time around the obstacles is longer than the cooling time, cold condensations – which we identify with the BLR clouds – form in the shocked gas. Both supernovae shells and groups of stars with strong stellar winds are large enough to act as cooling obstacles. The clouds are continuously generated and need no external confinement mechanism. They are accelerated to the local flow speed by pressure gradients in the shock and thus the line widths reflect the flow speeds and no other acceleration mechanism is required. Their maximum column density is of the order of $10^{22} cm^{-2}$. The total mechanical energy input required to create the obstacle-shock system is modest compared to the output in continuum radiation, and the system is shown to be self-generating and self-consistent.

INTRODUCTION

The properties of the broad emission line region (the BLR) as they can be inferred from the observations have been reviewed most excellently by Carswell (1984) and we refer the reader to that paper for a detailed discussion of the density, temperature, column densities, velocities, location and filling factors of the BLR clouds. No simple radial flow appears to be compatible with the observations, and the kinematic properties and ionization of the clouds appear to be linked.

It is now generally thought that QSOs are embedded in galaxies. Present dogma suggests that the QSO phenomenon arises because of the presence of a massive black hole ($M > 10^8 M_\odot$) surrounded by a dense stellar cluster in the galactic centre. The BH must accrete gas and a thick accretion disc may form. The continuum radiation may come from the surface of such a disc as well as from the immediate circumhole plasma. For a review of current QSO models we refer the reader to the interesting review by Blandford (1984).

The circumquasar region must be surrounded by optically thin, complex, high speed gas flows. This gas is illuminated by the continuum radiation and subjected to a powerful radiation acceleration. The radiation maintains the gas at temperatures of $O(10^7 - 10^8 \,^\circ K)$; furthermore, it contains enough energy to accelerate winds of $O(10-100) M_\odot yr^{-1}$ with velocities up to 0.1 c by converting no more than one percent of its luminous energy into mechanical energy of the wind.

Our model for the creation of the BLR depends only upon the existence of optically thin, supersonic, flows and is _not_ dependent upon that flow being an outflow. The total mass of this gas within the BLR is only of $O(10^4)M_\odot$ as compared to an assumed stellar population of 10^8 to 10^9 stars in the dense circumquasar stellar cluster. Stand-off shocks form around any obstacles in such flows. The superheated shocked gas is efficiently cooled by inverse-Compton and radiative line cooling. When the obstacles to the flow are large enough so that the flow time in the shocked layer is longer than the cooling time cool condensations are formed, which we identify with the BLR clouds. Both supernovae and winds from stellar associations form effective, cooling, obstacles. The number of obstacles which are required is determined from the condition that the system process the QSO flow gas into as much cold matter as is observed. The expected column densities of the clouds, their velocity structure and the line profiles are all in reasonable agreement with those observed.

GAS FLOWS AROUND QSOs

Whatever the details of the velocity field, any gas flow exterior to the continuum source must be globally optically thin to electron scattering. The observed optical variability of the central continuum source would be suppressed were the optical depth, τ_e, between the emitting surface and the observer greater than unity. Furthermore, the low density gas is hot, $T_e \approx 10^7 - 10^8$ K, and if the optical depth were much greater than unity spectral redistribution would wipe out most of the features of the observed spectrum. This constraint limits the density and mass injection rates within the BLR, which in turn constrains possible models for the formation of the BLR.

The details of the radial variation of the density are inexorably linked with the velocity field and the injection of material into the region. The flow is unlikely to be strictly radial. There may be outflow over a substantial solid angle (perhaps about the rotation axis) and inflow in the remaining volume (parallel to the plane of the nuclear accretion). Injection of material throughout an extended region, e.g. through ablation off the surface of stars, through strong stellar winds or supernovae, or through stellar collisions, may well result in $n(r)$ being roughly constant there. τ_e is sensitive to that region's outer radius, R_l $(\equiv 1 \, r_\ell \, \text{pc})$*. Above R_l, the gas density must fall off rapidly with radius, and does not contribute appreciably to τ_e. For a discussion of general density (and velocity) laws we refer the reader to Perry and Dyson (1984a, henceforth PD1). Here we limit our discussion to constant density flows.

The condition that the gas be optically thin to electron scattering,

$$1 \geqslant \tau_e = \int_{R_Q}^{\infty} n_e(r) \, \kappa_e \, dr \quad , \tag{2-1}$$

yields for the total (ambient background) gas density at r_{pc}, $n(r_{pc})$,

$$n(r_{pc}) = 5 \times 10^5 \, \tau_e / r_\ell \; ; \qquad r_{pc} < r_\ell \quad . \tag{2-2}$$

---

*In general, we express all variables in terms of parameters of order unity; e.g. the radial variable is written in units of parsecs, $(r = 3.08 \times 10^{18} \, r_{pc} \, \text{cm})$, the velocity in units of 0.01c $(V = 3 \times 10^8 \, \omega \, \text{cm/s})$ and the radius of the continuum emitting region, R_Q, in units of 0.01 pc $(R_Q = 0.01 \, R_o \, \text{pc})$. The BLR characteristically occurs at $r_{pc} \approx R_o \approx 1$. Temperatures are written as $T = 10^n \, T_n$ K. The bolometric luminosity is written $L_{bol} = 10^{47} \, L_{47}$ erg/s).

The electron density, $n_e(r_{pc})$, has been set equal to the total gas density. κ_e, the electron scattering cross-section $= 6.6 \times 10^{-25}\, cm^{-2}$. X-ray absorption studies indicate that τ_e is of $\mathcal{O}(0.1)$. The density contrast between emission line clouds at $r_{pc} \approx 1$ and the surrounding flow is of $\mathcal{O}(10^4 - 10^5)$ if $R_l > 1$ pc.

The total mass of gas interior to the observed BLR is $\approx 5 \times 10^4\, (\tau_e/r_\ell)\, r_{pc}^3\, M_\odot$. If the typical local velocity at the BLR is V_w, then the gas will remain in the BLR for a time roughly given by $t_f \approx \sqrt{2}\, R/V_w \approx 1.5 \times 10^{10}\, (r_{pc}/\omega)$ sec. It must therefore be resupplied from throughout the volume at a total rate of

$$\dot{M} \approx 1.1 \times 10^2\, (\tau_e/r_\ell)\, r_{pc}^2\, \omega \qquad M_\odot\, yr^{-1}\; . \tag{2-3}$$

This flow has a mechanical luminosity (kinetic energy transport past r_{pc}) of $L_w \approx 1.6 \times 10^{44}\, (\tau_e/r_\ell)\, \omega^3\, r_{pc}^2$ erg/s which is a small fraction, $\eta \approx 10^{-3}\, (\tau_e/r_\ell)\, \omega^3\, (r_{pc}^2/L_{47})$, of the the bolometric luminosity, L_{bol}. These mass-flux rates should be compared to the minimum accretion rates required to generate the bolometric luminosity of the QSO,

$$\dot{M}_a = L_{bol}/\epsilon c^2 = 1.75\, \epsilon^{-1}\, L_{47} \qquad M_\odot\, yr^{-1}\; , \tag{2-4}$$

where ϵ, the efficiency of mass conversion, is usually taken to be $\approx 0.1-0.4$.

The velocity, ω, is as yet a free parameter. Line widths, which in this model are a reflection of the local flow speed, imply that characteristically $\omega \approx 1.7$, with a normal range of $\approx 1 - 4$.

The thermal and ionisation structure of the gas can be expressed most conveniently in terms of the constant pressure ionisation parameter, Ξ, (Krolik, McKee and Tarter, 1981; henceforth KMT), defined as

$$\Xi = \frac{F_{ion}}{n\,kT\,c} = 2 \times 10^6\, \frac{L_{ion}}{L_{bol}}\, \frac{L_{47}}{n_e\, r_{pc}^2\, T_8}\; . \tag{2-5}$$

F_{ion} and L_{ion} are the ionizing flux and luminosity. Ξ is proportional to the ratio of the radiation pressure in the ionising spectrum to the gas pressure ($\Xi = 2.3\, p_r/p_G$). KMT give constant pressure cooling and two-phase equilibrium temperature curves for several assumed QSO spectra. Throughout this paper, we use their results for their "standard" spectrum. Later (Perry and Dyson, 1984b) we explore the consequences of deviations from the "standard" spectrum on the BLR models. The general conclusions reached in this paper are, however, unaffected by most changes in the continuum which are compatible with observations. KMT's standard spectrum has $L_{bol} \approx 5.4\, L_{ion}$. Therefore, for flows satisfying (2-2),

$$\Xi_w \approx 0.74\, (L_{47}/r_{pc}^2)\, (r_\ell/\tau_e)\, T_8^{-1}\; . \tag{2-6}$$

For bolometric luminosities greater than $\approx 10^{46}$ erg/s, optically thin, hot, flows are thermally stable due to the dominance of Compton cooling over bremsstrahlung (PDI). Thus thermal instability cannot be generally responsible for the formation of the BLR of QSOs. However, it may play an important role in e.g. Seyfert galaxies.

Cold dense gas clouds directly injected into such flows can be pressure confined under certain well defined conditions. Such a stable two-phase equilibrium description of the BLR has been proposed by KMT. Two phase equilibria occur only for $0.1 < \Xi < 10$. However, the observed BLR ionisation state, Ξ, is restricted to the narrow range

$\approx 0.3 - 2$. Thus such two-phase BLRs exist only for a narrow range of luminosities. These are coupled to the mass-flux rates (2-3) through τ_e. Ξ_w can be written in terms of $\dot{M}$ as $\Xi_w = 82 \, L_{47} \, \omega / \dot{M} T_8$. This constraint ($0.3 < \Xi_w < 2$) imposes the condition $270 > \dot{M} T_8 / L_{47} \, \omega > 42$. Since L_{47} can be at least as large as 10 and ω as 4, $\dot{M}$'s of the order of $10^3 \, M_\odot \, yr^{-1}$ would then be required. Such mass loss rates are difficult to reconcile with present models of galactic centres. Thus, whereas simple stable two-phase BLRs are possible low luminosities they are unlikely in high luminosity QSOs. Furthermore, in such scenarios the clouds must be injected into the flow in essentially their final configuration; they do not arise in the flow in a self-consistent fashion.

SHOCK WAVES IN THE HIGH SPEED GAS FLOW

The flow is hypersonic (with respect to its internal sound speed, ω_C) provided only that $\omega > \omega_C = 0.4 \, \sqrt{T_8}$. Since $T_8 < 1$ this criterion is satisfied even for flows whose velocities are well below that inferred for the BLR clouds. Any obstacles in such a flow will cause the wind to shock. Furthermore, the orbital velocity of a star at r_{pc} from a central mass condensation of $10^8 \, M_8 \, M_\odot$ is $\omega_{orb} \approx 0.22\sqrt{(M_8/r_{pc})}$. Thus strong shocks form even in quiescent atmospheres if $\omega_{orb} \gg \omega_C$ or $M_8 \gg 4 \, T_8 \, r_{pc}$.

These shocks heat and compress the QSO wind material which is then no longer in equilibrium with the QSO radiation field. Paradoxically, if the obstacles are large enough, and close enough to the QSO, this superheated gas can cool to form high density clouds. As shall soon become clear, in order to be effective in the formation of the BLR, obstacles must have dimensions of the order of hundredths to tenths of parsecs. Only e.g. supernovae or giant stellar complexes with strong stellar winds can form obstacles of that size. They inject mass locally into the flow at high-speed (the combination of the stellar-wind speed and the orbital velocity). This mass creates an obstacle to the wind by virtue of its momentum which deflects the ambient flow and causes it to form a stand-off shock. This ejected mass is then incorporated into the ambient flow itself (PDI). So long as the flow pattern is not strictly radial, the material from upstream obstacles then participates in the shocks giving rise to the cold downstream BLR gas.

The obstacles are not rigid and involve supersonic outflows. The configuration of the shocks depends on the details of the mass injection (PDI). Here we discuss only the classical open bow shocks which form around slow SN ($\omega_* < \omega$) or groups of stars with strong stellar winds. A full calculation of the flow-obstacle interaction including cooling is an impossible proposition particularly when the obstacle is both deformable and movable. The few numerical studies of the interaction of shock waves with compressible objects show that pronounced geometrical distortion of the object takes place (e.g. Nittman et. al. (1982)). Isotropic mass injection into a hypersonic gas flow should produce an obstacle rather more egg shaped than spherical with the blunt end pointing upstream. Two oppositely facing shock waves form. There will be streaming along the interface, and a mixing layer forms. We discuss an appropriately simplified flow in some detail in PDI and here give only the essential results.

We assume that quasi-steady flow around the obstacles is established. Since the shock waves are strong, the post-shock temperature and density are, respectively,

$$T_{sw} = 1 \times 10^8 \, \omega^2 \; K \quad ; \quad n_{sw}(r_{pc}) = 4\,n(r_{pc}) \quad , \tag{3-1}$$

where we use the subscript "sw" to denote the shocked wind. Because of the oblique nature of these shocks, the shocked gas accelerates away from the stagnation point and eventually passes through a sonic line, subsequently reentering the general flow at roughly the streaming velocity of the unshocked flow. The flow region between the stagnation point and the sonic line is effectively isobaric. Because the flow there is subsonic the shocked gas spends a relatively long time in this region.

The only important radiative process occurring in the immediate post shock gas is Compton cooling (PDI). This cooling takes place essentially isobarically at the stagnation pressure of the wind. The ratio of the value of the parameter Ξ in the shocked gas to that in the surrounding flow is

$$\frac{\Xi_{sw}}{\Xi_w} = \frac{n(r_{pc})\,T}{n_{sw}(r_{pc})\,T_{sw}} \approx T_8/4\omega^2 \quad . \tag{3-2}$$

As would have been anticipated, the effect of a shock is to strongly decrease Ξ. The reduction, even for ω of say 2, is by a factor of 16 at $T_8 = 1$, and is yet greater if $T_8 < 1$. Since the gas behind the shock cools at effectively a constant distance from the radiation source, Ξ remains $\approx$ constant during the cooling and is, from (2-6) and (3-2),

$$\Xi_{sw} = 9.8 \times 10^4 \, \frac{L_{47}}{n_w \, \omega^2 \, r_{pc}^2} \, , \tag{3-3}$$

which can be written using (2-2) and (2-3), as

$$\Xi_{sw} = 0.2 \, (L_{47}/r_{pc}^2)(r_\ell/T_e)\,\omega^{-2} = 22 \, (L_{47}/\dot{M}\omega) \quad . \tag{3-4}$$

Ξ_{sw} is less than 10 for even the most luminous QSOs, for almost all relevant flows — in agreement with the observations. Imposing the limit $2 > \Xi_{sw} > 0.3$ and replacing L_{47} by $\dot{M}_a$ from (2-4) yields

$$42 \, \epsilon/\omega > \dot{M}/\dot{M}_a > 6.3 \, \epsilon/\omega \tag{3-5}$$

We thus see immediately that *strong shocks in flows whose mass-flux rates are of the order of those required to fuel the nucleus have internal pressures capable of compressing the gas to the densities and ionisation degrees which correspond exactly to those observed in the BLR.*

Furthermore, it is noteworthy that the mass flux rates required to explain the BLR by thermal instability or by two-phase equilibrium are much greater, particularly at high velocity, than that required by the shock-production mechanism (PDI). We naturally find these low mass injection rates attractive.

Mushotsky and Ferland (1983) find that lower luminosity objects tend to have higher excitation BLRs with $\Xi \propto L^{-0.25}$. On the basis of this model, this can be expressed (3-4) as $\omega \dot{M} \propto L^{(1+x)}$ where $x \approx 0.25$. Such a relationship may be expected on two grounds: if the luminosity is a direct result of the mass flux rates then $\dot{M} \propto L$. Furthermore since the ambient gas is optically thin the radiation force can directly effect the dynamics of the gas flow and the flow velocities would then be expected to be proportional to some power of the luminosity. In the particularly simple electron scattering wind studied by Beltrametti and Perry (1980) it was found that $\omega \propto L^{0.5}$.

The time required to cool from an initial temperature, T_i, to a final temperature, T_f, is given by

$$t_c = \frac{3k}{m_H} \int_{T_f}^{T_i} \frac{dT}{\mathbb{L}(T)} = 1.1 \times 10^{-29} \frac{r_{pc}^2}{L_{47}} \int_{T_f}^{T_i} \frac{\Xi T}{\Lambda_\Xi(T)} dT \quad , \qquad (3-6)$$

where $\mathbb{L}(T)$ is the cooling rate in $\mathrm{erg\,sec^{-1}\,gm^{-1}}$. We have recast $\mathbb{L}(T)$ in terms of the isobaric cooling function $\Lambda(T) = m_H \mathbb{L}(T)/n_e$ given by KMT, replacing n_e by Ξ (2-5). In our standard units then

$$t_c = 10^9 \, I_\Xi(T_i)(r_{pc}^2/I_{47}) \quad s. \qquad (3-7)$$

The function $I_\Xi(T)$ is not sensitive to T_f for $T_f \ll T_i$. $I_\Xi(T)$ is ≈ 1 for $\Xi \approx 1$, $T \approx 10^8$; it varies approximately as ΞT for $\Xi T < 10^8$ and as $\ln T$ for larger T (PDI). The first cooling phase – due to Compton cooling – is thermally stable; *the second cooling phase – due to radiative line cooling – is thermally unstable.* During this phase, the gas will break up into clouds or filaments.

If the flow time is long enough, the gas will cool fully – i.e. to equilibrium with the radiation field. The equilibrium temperature is dependant on Ξ_{sw}. KMT show that if $\Xi_{sw} < 0.3$, $T_f \approx 10^4$ always. If $0.3 < \Xi_{sw} < 10$ there are two possible equilibrium values, $\approx 10^4$ and $\approx 10^8$. It is not possible to determine, a priori, the relative distribution of these two phases in the complex, turbulent flow behind the shock. If we assume that Ξ_{sw} is always low (< 0.3) then all the gas will cool to the lower temperature. This cool material expands as it flows around the obstacles and automatically evolves into a range of ionization parameters extending upwards of 0.3 – as is observed.

The time it takes an element of gas to reach the sonic line after it enters the shock depends upon how close to the symmetry axis (the velocity vector of the incident flow passing through the stagnation point) it enters the shock. We define the nominal size of the obstacle, d_0, to be the distance of the upstream end of the shock front from the centre of the obstacle. Because of flattening the lateral distance from the centre is perhaps $2d_0$. That element of flow which enters the shock a distance $\approx d_0/2$ from the axis requires a time

$$t_{fl} \approx 4 \times 10^{10} \, d_0/\omega \quad s \qquad (3-8)$$

to reach the sonic line. All gas which enters the shock closer to the axis spends longer in the shocked region and will certainly have cooled if that element has cooled by the time it reaches the sonic line.

The density of the cold gas is given by the condition that its thermal pressure be equal to the stagnation pressure in the flow. This gives $n_C < 4 \times 10^4 \, n_w \omega^2/T_4$. From (3-3) this is

$$n_C < 3.7 \times 10^9 \frac{I_{47}}{T_4 \, \Xi_{sw} \, r_{pc}^2} \quad . \qquad (3-9)$$

We use the inequality since as the gas flows away from the stagnation line its pressure, and hence density, falls.

GENERAL DESCRIPTION OF SHOCK PRODUCED BLRs

Size and Number of Obstacles Required

The criterion for effective cooling behind the shocks is that $t_c < t_{fl}$, (3-7) and (3-8). This in turn becomes a criterion for the minimum obstacle size. Gas entering the stand-off shocks around smaller obstacles will also cool by the Compton processes, but will have reached the sonic line well before catastrophic line cooling sets in and will expand back into the general flow at a high temperature, and probably not contribute to the BLR clouds. The condition $t_c < t_{fl}$ will be satisfied if

$$d_o > 2.4 \times 10^{-2} \, \omega \, \Xi(\omega) \, (r_{pc}^2/L_{47}) \ . \qquad (4-1)$$

In the parameter range of interest d_o varies roughly $\propto \omega^2 \, \Xi^{1/2}$ (PDI).

A natural explanation for the observation that the preponderance of BLR systems are characterised by $\langle\omega\rangle \approx 1.7$ and $\Xi \approx 0.4$ may well lie in the behaviour of the cooling times as a function of ω and Ξ : we do not often expect high ω, high Ξ systems in any scenario involving the cooling of hot gas, because flow time scales around moderately sized obstacles are too short compared to the cooling times.

In order to determine how many such obstacles must be present to produce the BLR we must first determine how much cool gas they must collectively produce. It is straightforward to show (PDI) that M_C, the total amount of cold gas in the region is $\approx 6.7 \times 10^3 \, f_C \, N_{23} \, r_{pc}^2 M_\odot$, where f_C is the fraction of sky covered by the clouds and N_{23} is their column density (in units of 10^{23}). The observations imply that $N_{23} < 0.5$ and that f_C lies roughly between 0.02 and 0.08 (PDI, Carswell, 1984). It is extremely unclear what fraction of the clouds have column densities as high as the maximum, $N_{23} \approx 0.5$, yet it is precisely these which contribute the bulk of the mass of the region. In a two, or more, component model where there is a spectrum of clouds present with optical depths ranging downward from the maximum to roughly 1, f_C for the thick component is sensitive to the mix. It is conservative to estimate that 80% of the clouds are optically thick, giving $f_C < 0.06$, which implies that

$$M_C/L_{47} < 2 \times 10^2 \, (r_{pc}^2/L_{47}) \ . \qquad (4-2)$$

The shocked gas is effective as BLR emitting gas only so long as it is cold — i.e. so long as it experiences the high pressure of the shocked region. Its lifetime, t_L, is therefore the time between cooling and flowing out of the shocked region — when it reexpands and heats up again. It is easy to show that this time is $\approx t_{fl}/2$. The rate of destruction of cold material is therefore

$$\dot{M}_d \approx M_C/t_L = 8 \times 10^{-4} \, M_C \, (\omega/d_o) \quad M_\odot \, yr^{-1} \ . \qquad (4-3)$$

The rate of conversion of flow material into cold gas by a single obstacle is

$$\dot{M}_C \approx 5.5 \times 10^{-5} d_o^2 \, n_w \, \omega \quad M_\odot \, yr^{-1} \ , \qquad (4-4)$$

where $\pi(d_o/2)^2$ is the effective cross-section.

In addition, there will be cold line-emitting gas present due to cooling of the shocked obstacle gas. The presence of this gas will reduce the number of obstacles which are required, by factors of perhaps as much as 2. However, we ignore this contribution here, noting only that by doing so we are overestimating the number of obstacles which are required.

The number of obstacles, N_O, required to maintain the BLR is found by setting $N_O \, \dot{M}_c < \dot{M}_d$, and is

$$N_O < 28 \, M_c/(n_w d_o^3) \; , \qquad\qquad (4\text{-}5)$$

where d_o is given by (4-1). We may now replace the ambient flow density, n_w, by the observed ionisation parameter of the clouds through (3-3) and M_c through (4-2). Then

$$N_O \ll 4 \times 10^2 \, \mathbb{N}_\Xi(\omega) (L_{47}/r_{pc}^2) L_{47} \; , \qquad\qquad (4\text{-}6)$$

where $\mathbb{N}_\Xi(\omega) \equiv 9 \Xi_{sw}/\omega \, I_\Xi^3(\omega)$. In the parameter range of interest $\mathbb{N}_\Xi(\omega)$ varies roughly as ω^{-2}. The observations imply that (L_{47}/r_{pc}^2) is roughly constant from one QSO to another. Therefore, for the same observed degree of ionisation, we require $N_O \propto L_{47}$. It is interesting to note that recent observations by Hutchings, Campbell and Crampton (1984) have found a correlation between the luminosity of QSOs and the mass of the galaxy in which they are situated.

There is a trade-off between the required size and number of obstacles. In a high speed flow very few, very large, obstacles are required; in a very low speed flow, a very large number of quite small obstacles are required. For example, if $\omega \approx 2$ and $(r_{pc}^2/L_{47}) \approx 1$ the number of obstacles required is $< 80 \, L_{47}$. The symmetry of the line profiles implies a high degree of symmetry of the BLR and we therefore expect that N_O should be large; we note here that the number of clouds produced by each obstacle is very large. For very low luminosity objects, e.g. Seyfert Galaxies, the number of obstacles required at any one time is very small, implying that the observed variability and irregularity of the line profiles could be associated with a patchy distribution of, and possible time variation in, the obstacles.

Global Energy Requirements

We may now use a rather simple arguement to estimate the energy required to create the system of obstacles. Each obstacle must have an internal pressure equal to that in the shocked layer and exerted on the interface between the two gas streams. This pressure is equal to the stagnation pressure in the QSO ambient flow plus any radiation pressure exerted on the cooled gas. If a cooled gas layer forms over most of the surface of the obstacle which faces the central source, and if its optical depth is large the radiation force will be at its maximum, and be $p_r \approx p_G \Xi_{sw}/2.3$. The total pressure exerted on the obstacle will then be $< \frac{3}{4} \rho_w V_w^2 (1 + \Xi_{sw}/2.3)$. Its internal pressure is $\approx \frac{2}{3} E_o/V_o$ where E_o is the energy and V_o the volume of the obstacle. Because of the enlargement due to flattening, the volume is $\approx 4\pi d_o^3$ pc^3. Therefore the energy contained in the bubble must be $E_o \approx 4.2 \times 10^{56} \rho_w V_w^2 d_o^3 (1 + \Xi_{sw}/2.3)$ erg. Substituting for d_o from (4-12) and for $\rho_w V_w^2$ from (3-3) we find that

$$E_o \approx 8.3 \times 10^{49} \, [\omega^3 \, I_\Xi^3(\omega)/\Xi_{sw}] (1 + \Xi_{sw}/2.3)(r_{pc}^2/L_{47})^2 \; \text{erg.} \qquad (4\text{-}7)$$

It is immediately clear that a supernovae of explosive energy E_* $(= E_{51} 10^{51} \text{ergs})$ can easily satisfy (4-7). The details of supernovae as obstacles are treated in PDI. Setting $E_o = E_*$ in (4-7) yields a relationship between (r_{pc}^2/L_{47}) and E_{51} for any Ξ and ω. The exact relationship depends on the details of the expansion and subsequent flow downstream of the SN shell, but is roughly equal to that found from (4-7). We find (PDI) always that $(r_{pc}^2/L_{47}) \approx 1$, in striking agreement with that inferred from the observations.

The energy contained in an obstacle "bubble" flows out of the region in a sound crossing time, t_s, and must therefore be resupplied at a rate given by $\dot{E}_o \approx E_o/t_s$. If we assume that the entire bubble has a temperature equal to that behind the inner shock we shall underestimate t_s and $\dot{E}_o$ will be an upper limit. The temperature behind the inner shock is $T_8 \approx \omega_*^2$ where ω_* is the typical mass ejection velocity in the obstacle. Since the typical size of an obstacle is $\approx 3d_o$, $t_s > 7.7 \times 10^{10} \, d_o/\omega_*$ sec. The energy input rate required is then

$$\dot{E}_o < 4.6 \times 10^{40} \, \omega_* \, \mathbb{E}_\Xi(\omega) \, (r_{pc}^2/L_{47}) \quad \text{erg/s}, \tag{4-8}$$

where $\mathbb{E}_\Xi(\omega) = [\omega^2 \, L_\Xi^2(\omega)/\Xi_{sw}](1 + \Xi_{sw}/2.3)$. The function $\mathbb{E}_\Xi(\omega)$ varies approximately as ω^5 for ω between 1 and 4 (PDI).

A stellar association of stars whose winds have an average energy of $\dot{E}_*(= \dot{E}_{36} 10^{36}$ erg/s), must therefore contain a minimum number of stars, N_S, found by equating the total energy input rate, $N_S\dot{E}_*$, to the total required, $\dot{E}_o$ (4-8). This gives

$$N_S < 4.6 \times 10^4 \, \mathbb{E}_\Xi(\omega) \, (\omega_S/\dot{E}_{36}) \, (r_{pc}^2/L_{47}) \, . \tag{4-9}$$

For O stars, $(\omega_S/\dot{E}_{36}) \approx 0.14$ and for W-R stars is $\approx 7 \times 10^{-3}$. The minimum number of stars required occurs for $\Xi \approx 2$, $\omega \approx 1$ when either 6400 O stars or 320 W-R stars are necessary. This requirement rises, by $\omega \approx 4$, $\Xi \approx 0.3$ to 4.8×10^6 O stars or 2.4×10^5 W-R stars! For $\langle\omega\rangle \approx 1.7$, associations of 1.3×10^5 O stars or 6400 W-R stars are required.

The overall energy input requirement into the BLR (that contributed by all the obstacles, $\dot{E}_T = N_O\dot{E}_o$) is, from (4-2), (4-6) and (4-8),

$$\dot{E}_T \ll 1.7 \times 10^{44} \, [\omega/L_\Xi(\omega)] \, (1 + \Xi/2.3) \, \omega_* \, L_{47} \quad \text{erg/s}. \tag{4-10}$$

Interestingly, *the total energy input requirement to maintain the BLR producing system of obstacles is essentially independent of both Ξ and ω.* At maximum $\dot{E}_T/L_{bol} < 2 \times 10^{-3} \, \omega_*$.

The overall total number of stars (with stellar winds of $\dot{E}_{36}$) which are required in the BLR to create the entire obstacle complex can be found from (4-10). It is at maximum

$$N_T < 2 \times 10^8 \, (\omega_*/\dot{E}_{36}) \, L_{47} \quad , \tag{4-11}$$

e.g. less than $2.8 \times 10^7 \, L_{47}$ O stars or $1.4 \times 10^6 \, L_{47}$ W-R stars.

Fragmentation

Once radiative (metal-line) cooling takes over in the shocked gas, thermal instability sets in. The maximum scale size of the fragments which can form is equal to the largest coherence length for pressure waves at the onset of the instability due to radiative cooling. Since it is not clear whether the fragments will be filamentary or globular it is difficult to estimate their mass and follow their subsequent development. For simplicity we assume that they are spherical and maintain their identity throughout the cooling. Then at equilibrium we can show (PDI) that the the characteristic size of the <u>largest</u> possible cool fragment for $\Xi = 0.1$ is $\approx 4 \times 10^{13}$ cm and for $\Xi = 1$ is $\approx 10^{14}$. The maximum possible total hydrogen column density of a fragment, $N_H = n_c \ell_c$, is distance _independent_, and lies between 10^{23} and 10^{24} (PDI).

Independent limits on the column density of the cold clouds follow from considerations of continuity in the shocked layer. Although the cooling flow will be turbulent and the cold material clumpy, we may estimate mean column densities by treating the flow as if it were smooth and imposing mass conservation conditions. Equating the mass flux into the shock to that crossing the sonic line yields a mean thickness of cold gas at the sonic line, Δ_s, from which the column density, $n_c\Delta_s$, follows. This limit on the column densities is also _distance independent_ and is only very weakly dependent on Ξ and ω. It lies between $\approx 1 - 4 \times 10^{2l}$ — a factor of roughly 10^2 less than that given above (PDI).

Thus a "steady-state" system of clouds all with column densities as high as that given by the coherence length arguement cannot occur. However, it may well be that the fragmentation results in a population of clouds whose maximum size approaches a few times 10^{22}. A spectrum of column densities lying below this maximum is to be expected both because of the fragmentation process itself and because as the clouds accelerate due to the pressure gradients in the shocked region they expand with a resultant drop in column density.

Once the gas has cooled radiative driving sets in. The radiative instability discussed by Mestel, Moore and Perry (1978) may then be operative and be responsible for further fragmentation. It is interesting to note here that the observations limit Ξ to less than 2 — precisely that value where the radiation pressure becomes equal to the gas pressure. The shock configuration itself may be disrupted for larger values of Ξ by radiation driven instabilities.

Acceleration Mechanisms and Line Profiles

The gas behind the shock accelerates as it moves away from the stagnation point towards the low pressure region where it escapes and reexpands. Ignoring radiative driving, the gas will reexpand into the wind with a maximum velocity $\approx \omega$. The velocity of the BLR gas therefore reflects primarily the wind velocity and _no_ acceleration mechanism — other then the existence of pressure gradients in shocked gas — is necessary. This seems to get rid of one of the most puzzling aspects of the BLR.

We defer detailed discussion of the expected line profiles to a later paper but it is hard to see how they can be other then generally symmetric if large numbers of obstacles are involved. The flow itself is expected to be complex and the velocity pattern of the condensations yet more random since they will almost certainly not be symmetrically aligned about the local flow velocity vector. The orbital velocity of a star at typical BLR radii, and the velocities of stars in bound stellar groups which must exceed ω_{orb}, can be significant compared to either the velocities of stellar winds or supernovae ejecta. Thus the obstacles will be further distorted; additionally, the shocks behind large neighboring obstacles may overlap.

The largest fragments will be optically thick. Optically thin clouds will also exist, either formed as such or produced by damage to or expansion of optically thick clouds. The pressure is lower as the clouds enter the general flow and here their velocity is higher. The optically thinner clouds may then exist preferentially at higher velocities. Thus, although the velocity dispersion of all the clouds ought to be broadly similar, the velocity dispersion may be greater in

the high ionisation lines (which we assume are produced mainly in optically thin clouds) than in the low ionisation lines.

There is some observational evidence that the full width of the
H-α lines is proportional to the x-ray and/or non-thermal continuum
luminosities (Kriss, Canizares and Richer, 1980; Steiner, 1981; Ward,
1981). This may reflect an underlying tendency of the QSO flow velocities themselves to be proportional to the luminosity as well as the
effect of radiation driving acting on the cooled gas.

DISCUSSION

The model presented here for the formation of the BLR of QSOs and
AGNs has many appealing features. Optically thin supersonic gas flows
in the circumquasar region inevitably form stand off shocks around all
the obstacles they encounter The superheated shocked gas cools rapidly
by inverse Compton cooling and provided only that the obstacles are
large enough so that the flow time around them is longer than the
cooling time, cold condensations form in the shocked gas.

The BLR clouds are continuously generated behind the shocks and
need no external confinement mechanism. It is the stagnation rather
than the thermal pressure of the ambient flow which provides the high
pressure typical of the BLR clouds. The stagnation pressure is much
greater than the thermal pressure in the flow and therefore the mass
and energy input requirements to the BLR are significantly reduced
over those required in any thermal pressure two-phase equilibrium
model for the BLR. In addition, whereas the thermal pressure of the
hot medium is critically dependent on the spectral shape of the radiation field which determines its temperature, the stagnation pressure
of the flow is not strongly dependent on the characteristics of the
radiation field. Unless the flow is radiatively driven it will not,
in fact, be dependent on the radiation field at all. If it is radiatively driven, because it is optically thin and hot it will be coupled
to the radiation field through electron scattering opacity, and thus
be dependent only upon the total bolometric luminosity and not on the
spectral details. The ionization parameter of clouds generated at any
radius, r_{pc}, depends only upon the ratio of the bolometric luminosity
to the local flow speed and mass-flux rate. These mass-flux rates
must be roughly equal to the accretion rates required to create the
bolometric luminosity itself. Thus the remarkable uniformity of the
characteristics of the BLR from object to object emerges naturally
within the framework of this model.

The clouds are generated with a spread in velocity, and are
accelerated to the local flow speed by the presure gradients in the
shock; there is an additional component of velocity which is due to
radiative acceleration. The clouds are born with maximum typical
column densities of the order of 10^{22} cm^{-2}, but since they result from
fragmentation during catastrophic radiative cooling behind the shock
they are expected to have a wide range of sizes providing a natural
explanation for the observed multicomponent BLR. Furthermore, as the
clouds accelerate behind the shock prior to reentering the general
flow (when they cease to exist) they expand due to the falling pressure. Simultaneously their total column density decreases and their
degree of ionization increases.

The characteristic size of an obstacle which is required for
effective cooling lies between a few hundredths to tenths of parsecs
at distances of the order of a parsec from the nucleus. Individual
normal stars therefore are not effective as obstacles. Supernovae (of

$\approx 10^{51}$ ergs) are, however, as are associations of typically a few thousands to a few tens of thousands of O stars. The kinetic energy input rate in supernovae explosions or stellar winds, required to maintain the system of obstacles is always less than one thousandth of the magnitude of the observed bolometric luminosity of the continuum source. The material which forms the obstacle system must be supplied continuously as it flows out of the obstacle shock system in times of the order of the sound crossing time of the obstacle. It turns out that the rate at which material must be supplied to maintain the obstacle system is of the order of the mass-flux inferred for the ambient hot flow. A satisfying self-consistent picture thus emerges of the entire flow-obstacle system.

The structure and details of the BLR is expected, in this model, to reflect the structure and evolution of the central galactic regions and we would therefore expect some correlations of BLR properties with Hubble type to emerge from further, and future, observations. It remains a matter for conjecture whether or not e.g. massive stars are plentiful in the circumquasar region, leading either to stellar winds or frequent supernovae explosions both more powerful than those associated with our, and neighbouring, galaxies. Such a population would, for instance, result in this model functioning in a yet more natural fashion than it does under the rather conservative assumptions adopted in this paper. The BLR will vanish if either the QSO wind ceases to carry sufficient mass flux or if suitable obstacles are no longer available. Thus the mechanism has a possible, and natural, evolutionary cut-off.

Acknowledgements

We are grateful to Drs. M. Whittle, R. Carswell and M. Sohnius for many very helpful discussions.

REFERENCES

Beltrametti, M., and Perry, J.J., 1980, A & A, __82__, 99 (herein BP).

Blandford, R., 1984, in "Active Galactic Nuclei, Proceedings of the 1984 Manchester Conference on AGN", ed. J. E. Dyson (Univ. of Manchester Press).

Carswell, R., 1984, in "Active Galactic Nuclei, Proceedings of the 1984 Manchester Conference on AGN", ed. J. E. Dyson (Univ. of Manchester Press).

Hutchings, J. B., Campbell, B. and Crampton, D., 1984, preprint

Kriss, G. A., Canizares, C. R., and Ricker, G. R., 1980, Ap. J., __242__, 492.

Krolik, J.H., McKee, C.F., and Tarter, C.B., 1981, Ap.J., __249__, 422 (herein KMI).

Mestel, L., Moore, D. W., and Perry, J. J., 1976, A & A, __52__, 203.

Mushotzky, R. and Ferland, G.L., 1984, Ap. J., __274__,   .

Nittmann, J., Falle, S. A. E. G., and Gaskell, P. H., 1982, MNRAS, __201__, 833.

Perry, J. J., and Dyson, J. E., 1984a, MNRAS, submitted (herein PDI)

Perry, J. J., and Dyson, J. E., 1984b, in preparation

Steiner, J. E., 1981, Ap. J., __250__, 469.

Ward, M., 1981, in "Noyaux Actifs des Galaxies", ed. E. Schatzman, (Suppl. au Journal des Astron. Francais) __13__, 1.

Are Radio-Quiet Active Nuclei the Result of Thermal Winds
Driven by Bubbles?

I.N. Evans and M.A. Dopita

Mount Stromlo and Siding Spring Observatories
Research School of Physical Sciences
The Australian National University, Canberra

SUMMARY

In constructing theoretical models of quasars, we now have available a large body of data collected in many spectral regions, which collectively impose severe restraints on allowable models. It is important to remember that over 90% of the quasars are radio-quiet (Savage et al. 1982), which implies that the processes of nozzle formation, relativistic jets and the associated non-thermal emission are not essential to a description of the quasar phenomenon. However, when surveying the published data base one is struck by the heavy observational selection towards the small minority of radio-loud objects.

In this paper we will show how the observational data tend to support the idea that ordinary thermal processes are important in determining quasar properties, that an optically thick thermal wind model appears most attractive and briefly show how this wind may be driven by supersonic radiation or plasma bubbles generated close to the central collapsed object.

OBSERVATIONAL CORRELATIONS IN QUASARS

The most striking observational correlation which has so far emerged in the study of quasars is the correlation shown in figure 1 between optical continuum luminosity, L_O, and the Hβ luminosity, $L_{H\beta}$. Yee (1980) showed that this extends over at least five orders of magnitude in L_O, with a slope of 0.97 ± 0.04; interestingly close to unity! Since the Balmer emission results from recombination after photoionisation by UV photons, this implies that there exists a self-similar structure in quasars in which the broad line region is causally related to the optical *and* UV emission, and that the geometry of the dense clouds giving rise to the broad line emission is fixed. That is to say that the volume filling factor, and geometrical covering factor remain constant, and that the spatial extent of the broad line region scales in a one-to-one fashion with the size of the quasar photosphere.

For the radio-quiet quasars and Seyfert galaxies, there is evidence that the

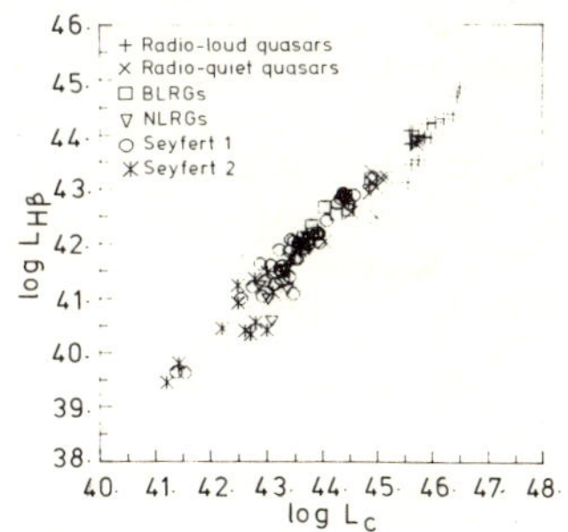

Fig.1 Correlation between Hβ line luminosity and optical continuum (3000 Å -9000 Å) luminosity. (After Yee 1980).

presumed L_o, L_{UV} correlation may extend to X-ray frequencies. Kriss et al. (1980) found, in a sample of 37 Seyfert galaxies observed with the Imaging Proportional Counter on the Einstein Observatory, that correlations exist between the 2 keV X-ray flux, F_X, the optical B band flux corrected for the underlying stellar component, F_B, and the corresponding 3.5 μm infrared flux, $F_{3.5}$. The data are compatible with $F_{3.5} \propto F_B \propto F_X$, though with some scatter. Similar correlations exist in the X-ray selected quasar sample of Grindlay et al. (1980). Thus there is some evidence that the (presumed) thermal processes giving rise to the optical and UV emission of radio-quiet AGN extend both into the IR and up to X-ray frequencies.

When radio-loud objects are considered, there are many lines of evidence to suggest that non-thermal processes may either underly, or indeed in some cases dominate this thermal emission. Ku et al. (1980) find in an X-ray selected sample that the ratio of broadband (0.5–4.5 keV) X-ray luminosity to broadband (3000–6000 Å) optical luminosity is larger in the radio-loud quasars than for the radio-quiet objects. Zamorani et al. (1981) find a similar result and they estimate that the average monochromatic luminosity ratio $\ell_X(2\ \text{keV})/\ell_o(2500\ \text{Å})$ is three times higher in the radio-loud quasars $(\alpha_{ro}>0.35)$. Blumenthal et al. (1982) obtain a loose correlation between the broadband X-ray luminosity and the optical continuum luminosity at 5000 Å, L_C, with $L_X \propto L_C^{1/2}$. However, apart from two discordant points, the radio-quiet subsample fits well to a correlation $L_X \propto L_C$.

These data lead us to a two component interpretation of the radio-loud spectra; that in addition to a thermal component which is *relatively* most important in the UV and optical frequencies, there exists a non-thermal component extending from the radio to X-ray frequencies. The non-thermal X-ray emission may arise by unsaturated Comptonisation of soft photons (Rybicki & Lightman 1979) while Compton scattering of flat-spectrum centimetric radio photons off electrons with a Lorentz factor of order 10^3 has also been suggested (Ku et al. 1980).

Further support for this two component concept comes from studies of spectral index correlations, which have covered radio-loud, radio-quiet quasars and Seyfert galaxies. All authors have found an anticorrelation between the optical-soft X-ray spectral index α_{ox} and any of the following indices: (1) α_{ro}, the radio-optical spectral index (Ku et al. 1980; Zamorani et al. 1981); (2) α_{io}, the 3.5 μm IR-optical index (Kriss et al. 1980); or (3) α_{VRI}, the visual-red-infrared index (Sitko et al. 1982). In each case the anticorrelations are compatible with slope -1 (see Fig. 2). All the very radio-loud strong X-ray sources are in the region of large $(\alpha_{ro}, \alpha_{io}, \alpha_{VRI})$ and small α_{ox} whereas the radio-quiet sources are weak X-ray sources and lie in the region of small $(\alpha_{ro}, \alpha_{io}, \alpha_{VRI})$ and large α_{ox}. Ku et al. (1980) have defined Type I (radio-quiet) quasars as those with $L_r(3.1\text{–}8.1\ \text{GHz}) <10^{41}$ erg s^{-1} and Type II quasars as those with $L_r>10^{41}$ erg s^{-1}. With these definitions, Type I have $(\alpha_{ox})\sim1.5$ (2500 Å–2 keV) and $\alpha_{ro}<0.2$ (5 GHz–2500 Å) whilst the Type II lie in the region $\alpha_{ox}<1.35$, $\alpha_{ro}>0.2$.

From these data, it is clear that the radio, infrared and X-ray regions of the spectrum may be dominated by non-thermal emission in the radio-loud objects; but in radio-quiet quasars there is appreciable X-ray emission from some other source. If this is thermal emission, it must arise in a hot corona surrounding the central source. Such a component is also required to give a confining pressure to the dense partially photoionised cloudlets in the broad line region (Krolik et al. 1981). It is likely that this hot component of the quasar is optically thick to self absorption for highly optically luminous radio-quiet

objects. We believe that the data of
Zamorani et al. (1981) for the radio-
quiet objects in their sample illustrate
this effect (see Fig. 3) since the highly
luminous objects have α_{ox}=1.58 with a
very narrow spread of order 0.2 in the
index, whereas quasars of low luminosity
show a much broader spread, $1.0 \lesssim \alpha_{ox} \lesssim 1.8$.

In a very small percentage of quasars,
non-thermal emission can become strong
enough to completely dominate over any
underlying thermal component. These
objects include the BL Lac class, the
optically violent variable (OVV) quasars
and the highly polarised quasars (HPQs).
They all show similar properties and lie
at the extreme high α_{ro}, low α_{ox} end of
the colour-colour diagram (Moore &
Stockman 1981; Sitko et al. 1982). Their
optical spectra are redder and smoother
than normal quasars and, with the exception of PHL5200, are strong radio sources
with flat radio spectra. They are not particularly overluminous when
compared with 'normal' quasars (Moore & Stockman 1981), but the non-
thermal component must contribute of order 70% of the visible continuum.
It is this non-thermal component which is strongly variable, since the
equivalent width of the emission lines
decreases during outburst whilst the
degree of polarisation tends to increase
(Miller & French 1978; Arp et al. 1979;
Netzer et al. 1979). The rapidity of
these optical variations puts stringent
limits on the scale size of the region
emitting the non-thermal component, whilst
the polarisation implies emission in a
region of low optical depth, since otherwise Faraday rotation or electron scattering would depolarise the radiation
(Blandford & Rees 1978). It is interesting that BL Lac type objects and the HPQs
have compact radio source structure, and
that the HPQs comprise 50% of the superluminally expanding sources known (Moore &
Stockman 1981).

All these properties suggest that one of
the relativistic jets which have been used
to explain the radio structures in radio-loud quasars (Blandford & Rees
1978; Smith et al. 1983) is directed almost along the line of sight in
these objects.

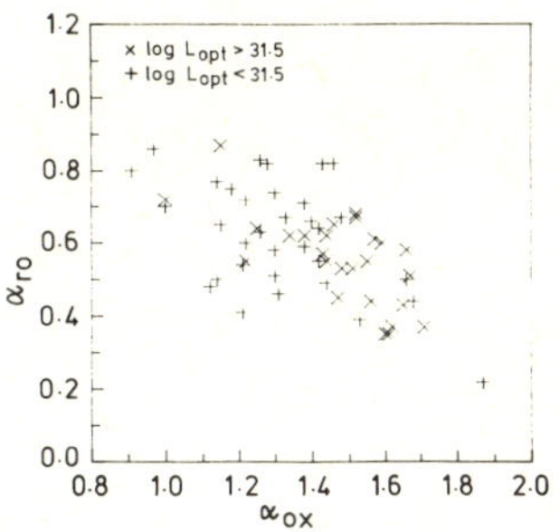

Fig.2 Correlation between α_{ro} (5 GHz–2500 Å) and α_{ox} (2500 Å–2 keV) spectral indices in the X-ray selected sample of Zamorani et al. (1981). (After Zamorani et al. 1981).

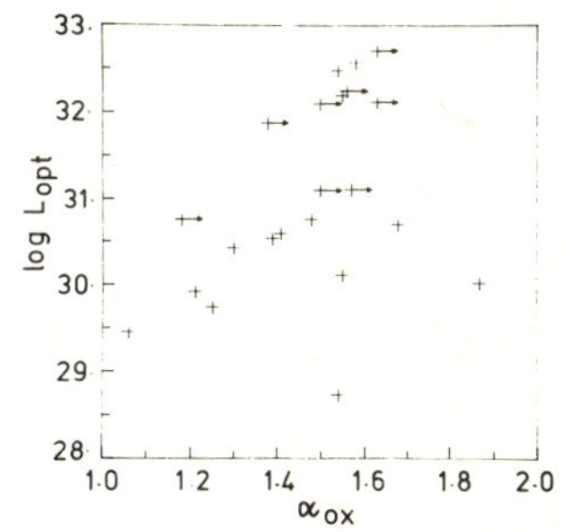

Fig.3 Optical luminosity dependence of α_{ox} for the radio-quiet quasars in the X-ray selected sample of Zamorani et al. (1981).

A THERMAL WIND MODEL

Although the observations described in the previous section may be
capable of more than one interpretation, it seems clear that a description of the radio-quiet quasars requires the following components:
(1) a hot photosphere, emitting from IR to at least far UV frequencies;
(2) a hot corona of optically thin gas which confines a dense phase of

partially ionised cloudlets responsible for the broad line emission;
(3) a scaling property such that the size of the whole structure scales
simply with luminosity; and (4) an absence of relativistic plasma jets.

We regard these properties as strongly favouring a hot thermal wind
model. Such winds are almost inevitable in the immediate vicinity of
the collapsed central object in a quasar where radiation pressure dominates in determining the dynamics of the gas (Weymann et al. 1982), and
winds driven by various mechanisms (e.g., radiation pressure, shocks
etc.) have been extensively discussed (e.g., Kippenhahn et al. 1975;
Kippenhahn 1978; Beltrametti 1981; Falle 1983). Our model differs from
these in the mechanism by which the wind is driven, which appears to
give a natural explanation for the properties of radio-quiet quasars
given above.

We assume that the energy liberated by the central object results
from supercritical accretion from a physically and optically thick disc
onto a supermassive black hole. These have been extensively studied
(Wiita 1982) and it has been shown that such discs are capable of generating luminosities up to of order 100 times the Eddington limit
(Abramowicz et al. 1980). By whatever mechanism, this super-Eddington
luminosity is largely converted to kinetic energy in a dense outflowing
wind, and this wind will have a photosphere with an opacity largely
dominated by electron scattering and with a sufficiently large radius
to obscure the central source and, possibly, much of the accretion disc
as well. It is this photosphere which gives rise to the optical
continuum of radio-quiet quasars.

In the inner region of the accretion flow, near the cusp point where
the flow changes from physically thick to thin and where it changes
direction, a turbulent region must develop. This region has a low
density photon gas and Compton heated plasma accelerating out, with
shearing motions, a much denser layer originating from the accretion
flow, and hence is subject to both Rayleigh-Taylor and Kelvin-Helmholtz
instabilities. Under these conditions it should be possible to determine the characteristic size of an instability given by the wavenumber
with maximum growth rate (Smith et al. 1983). For the Rayleigh-Taylor
instability, the asymptotic growth rates for large and small wave-
numbers k_1, k_s are given by

$$n_1 = \frac{(\rho_2 - \rho_1)\nu k_1^{-1}}{(\rho_1 + \rho_2)2g'}; \qquad n_s = \left[\frac{(\rho_2 - \rho_1)k_s}{(\rho_1 + \rho_2)g'}\right]^{1/2},$$

where g' is the effective acceleration of the dense layer (dV/dR-g), and
ν is the coefficient of kinematic viscosity for matter filled bubbles,
or is determined by the radiation diffusion length in radiation filled
bubbles. Other processes such as anisotropic Compton losses (Cheng &
O'Dell 1981) may also contribute to the early phases of bubble motion.

Once formed, bubbles will flow outward (in general supersonically
with respect to the surrounding medium) because of their buoyancy in the
already outflowing wind. Momentum transfer between the bubbles and the
surrounding gas drives an almost spherically symmetric wind. An exact
solution for the flow region has been found only for the simplest case
of a hot plasma or radiation filled bubble characterised by an equation
of state with adiabatic index γ and an internal pressure which matches
the ram pressure due to its buoyant motion. For radiation bubbles
(γ=4/3) we find that this gives rise to a wind velocity V(R) given by

$$V(R) - V_\infty = \frac{\pi^{5/12}(GM_*)^{3/8}}{2^{1/8}\,3^{D}\,R_0^{1/12}\,r_0^{1/24}} \cdot \left(\frac{L_*}{\dot{M}c}\right)^{1/4} R^{-1/4}, \tag{1}$$

where M_*, L_* are the mass and luminosity of the central collapsed object, $\dot{M}$ is the mass loss rate in the wind with terminal velocity V_∞, r_0 is the initial radius of the bubbles, and R_0 the radius at which they are formed. The terminal velocity is given by

$$V_\infty = \left(\frac{4GM_*\dot{M}cr_0}{3L_*R_0^2}\right) \cdot \left[1 - \frac{\pi^{5/12}R_0^{5/3}}{2^{17/8}(GM_*)^{5/8}\,r_0^{25/24}} \cdot \left(\frac{L_*}{\dot{M}c}\right)^{5/4}\right]. \tag{2}$$

Thus the wind slows as $R^{-1/4}$ towards a constant outflow velocity. This implies that there exists a radius R_c within which $V<V_{esc}$ and outside which the wind is unbound. The outer part of the flow can be character- ised by a 'coasting' solution with a wind density varying as R^{-2}.

We have solved for the conditions necessary to ensure the survival of the bubbles in this region. Both matter and radiation bubbles will sur- vive if the radiation diffusion timescale, τ_{diff}, from them is longer than the dynamical timescale, τ_{dyn}. However, matter bubbles can survive even if $\tau_{diff}<\tau_{dyn}$ provided that the cooling timescale of the matter within them, τ_{cool}, is greater than τ_{dyn}. If we introduce dimensionless variables

$$\mathcal{M} = \sqrt{3}\,U/\bar{c}; \qquad \eta = r/R; \qquad \xi = 2R_c/9R, \tag{3}$$

where $0<\mathcal{M},\eta,\xi<1$, U is the bubble velocity due to buoyancy, $\bar{c}$ the sound- speed in the bubble, and $R_c = 2GM_*/V_\infty^2$ is the critical radius at which $V_\infty=V_{esc}$. With these variables the condition $\tau_{diff}>\tau_{dyn}$ becomes

$$\mathcal{M}^2\eta^2\xi > 16\pi GM_*c/9\kappa_{es}\dot{M}V_\infty^2, \tag{4}$$

where κ_{es} is the electron scattering opacity. Also, if the cooling function of the bubble is $\Lambda(T_b)$ erg cm^{3}s^{-1}, then the condition $\tau_{cool}>\tau_{dyn}$ may be rewritten

$$\eta\mathcal{M}^{-4} > 3\dot{M}\Lambda(T_b)/16\pi\,(\mu m_H)^2\,GM_*V_\infty. \tag{5}$$

Thus all bubbles will survive in a region $\eta>\eta_{crit}$ where $\eta_{crit}\propto\mathcal{M}^2\propto T_b\propto n_b^{-1}$ (T_b and n_b being the temperature and density within the bubble, although in the case of radiation filled bubbles these parameters more properly refer to the shocked dense layer around the bubble).

For bubbles filled with low density gas, characterised by the cooling function $\Lambda(T_b)$, which we write in the form $\Lambda(T_b)= T_b^n$, equation (5) implies survival in a region $\eta>\eta_{crit}$ where $\eta_{crit}\propto\mathcal{M}^4\Lambda(T_b)\propto T_b^{n-2}$. Thus for n<2, large, hot, low density bubbles will survive best. This condition of n<2 is satisfied by, for example, Bremsstrahlung losses.

Equations (4) and (5) suggest that hot matter bubbles probably dominate in the coasting zone, and that these bubbles have the attributes of classical plasmons. Such bubbles will eventually coalesce and burst. Depending on whether the flow region outside the burst points is optically thick to electron scattering in the hot plasma or not, we will see either an electron scattering photosphere in the hot plasma or else down to the radius at which the bubbles burst. The ionising spectrum is determined by these photospheric conditions.

The denser material becomes confined by this hot plasma into small cloudlets which emerge through the photosphere to form the broad line/ coronasphere region. Thus a bubble driven wind will show the first three attributes of radio-quiet quasars listed at the start of this section. The fourth attribute, the absence of relativistic jets, will be satisfied provided that the instabilities involved in the formation of the bubbles are sufficiently developed to advect away the luminosity and plasma generated in the central source and so choke off any nozzles. Radio-loud quasars are therefore objects in which instabilities are relatively less efficient and for which the classical Blandford & Rees (1974) twin-exhaust model applies, and we associate all the non-thermal emission with processes occurring in these jets (which do not necessarily dominate in the ionisation of the broad line region).

CONCLUSIONS

We conclude that a wind model, in which the wind is driven by supersonic hot thermal bubbles generated close to the central collapsed object, promises to be able to describe all of the observed properties of the radio-quiet quasars. Such a system is very attractive to model since many of the free parameters that plague photoionisation models are eliminated. The shape of the ionising spectrum is determined solely by the photospheric conditions, which in turn depend on the conditions in the 'coasting' region of the flow, but are essentially independent of the details of the processes operating near the central source. Indeed, the only free parameters in the 'coasting' zone appear to be M_*, $\dot{M}$ and V_∞. The terminal velocity and possibly the mass loss rate can be determined from observations, whilst the contributions of the various physical processes to the cooling function may be deduced from theoretical considerations, and thus the number of free parameters which determine the ionising spectrum is considerably reduced when compared with other photoionisation models, removing many of the *ad hoc* assumptions implicit in the majority of models of AGN which have been suggested to date.

REFERENCES

Abramowicz MA, Calvani M, Nobili L 1980 Astrophys J 242:772-788
Arp H, Sargent WLW, Willis AG, Oosterbaan CE 1979 Astrophys J 230:68-78
Beltrametti M 1981 Astrophys J 250:18-30
Blandford RD, Rees MJ 1974 Mon Not R Astron Soc 169:395-415
Blandford RD, Rees MJ 1978 In: Wolfe AM (ed) Pittsburgh Conference on
    BL Lac Objects 328-347. University of Pittsburgh, Pittsburgh
Blumenthal GR, Keel WC, Miller JS 1982 Astrophys J 257:499-508
Cheng AYS, O'Dell SL 1981 Astrophys J 251:L49-L54
Falle SAEG 1983 Mon Not R Astron Soc 203:245-252
Grindlay JE, Steiner JE, Forman WR, Canizares CR, McClintock JE 1980
    Astrophys J 239:L43-L48
Kippenhahn R 1978 Physica Scripta 17:301-305
Kippenhahn R, Mestel L, Perry JJ 1975 Astron Astrophys 44:123-138

Kriss GA, Canizares CR, Ricker GR 1980 Astrophys J 242:492-501
Krolik JH, McKee CF, Tarter CB 1981 Astrophys J 249:422-442
Ku WH-M, Helfand DJ, Lucy LB 1980 Nature 288:323-328
Miller JS, French HB 1978 In: Wolfe AM (ed) Pittsburgh Conference on BL
    Lac Objects 228-235. University of Pittsburgh, Pittsburgh
Moore RL, Stockman HS 1981 Astrophys J 243:60-75
Netzer H, Wills BJ, Uomoto AK, Rybski, PM, Tull RG 1979 Astrophys J 232:
    L155-L159
Rybicki GB, Lightman AP 1979 Radiative Processes in Astrophysics 221-222.
    Wiley-Interscience, New York
Savage A, Bolton JG, Wall JV 1982 Mon Not R Astron Soc 200:1135-1141
Sitko ML, Stein WA, Zhang Y-X, Wisniewski WZ 1982 Astrophys J 259:486-
    494
Smith MD, Smarr L, Norman ML, Wilson JR 1983 Astrophys J 264:432-445
Weymann RJ, Scott JS, Schiano AVR 1982 Astrophys J 262:497-510
Wiita PJ 1982 Astrophys J 256:666-680
Yee HKC 1980 Astrophys J 241:894-902
Zamorani G, Henry JP, Maccacaro T, Tananbaum H, Soltan A, Avni Y,
    Liebert J, Stocke J, Strittmatter PA, Weymann RJ, Smith MG, Condon JJ
    1981 Astrophys J 245:375-374

Theoretical Studies of Emission Line Variability in Type 1 Seyfert
Galaxies

A. Robinson

Department of Astronomy, University of Manchester, Manchester M13 9PL,
England.

SUMMARY

Several Type 1 Seyfert galaxies are known to show variability in the
broad components of their permitted lines. The nature of these variations suggests that they are caused by variations in the ionizing part
of the optical-UV power law continuum. To investigate this connection
in detail, a time dependent photoionization model of the broad line
emitting region has been developed. It is shown that, with the inclusion
of light travel time effects within the BLR, the model can account naturally for the general characteristics of the variations.

1. GENERAL FEATURES OF EMISSION LINE VARIABILITY

Recent observations at optical and ultra-violet (UV) wavelengths have
established that a considerable number of type 1 Seyfert galaxies exhibit
both continuum and emission line variability (Peterson et al (1983);
Ulrich et al (1984); Penston et al (1981); Perola et al (1982); Antonucci
and Cohen (1983) and Barr et al (1983)).
   Although there are, as yet, relatively few detailed studies of individual cases, it is possible to pick out the general features of the variations:
   1) The UV power law continuum is variable on timescales of 10 days or
more. There is a correlation between the spectral index (α) and the
intensity in the sense that the slope of the continuum decreases as the
intensity increases.
   2) The emission lines which arise in the dense, compact broad line
region (BLR) are variable and tend to follow the continuum variations.
In some cases the time lag between continuum and line variations has
been measured. For example, the CIV $\lambda1549$ line in NGC 4151 lags by at
least 13 days (Ulrich et al (1984)).
   3) A trend in timescales with ionization state is apparent, the higher
ionization lines (eg. CIV) having shorter variability and lag times than
low and intermediate ionization lines (MGII, CIII], Balmer lines).

2. MECHANISM FOR LINE VARIABILITY

There are good reasons (eg. Davidson and Netzer 1979) for supposing
that the BLR gas is ionized by a continuation of the observed UV power
law to ionizing energies. If so, then a logical explanation for the line
variations is that variations in the continuum cause changes in the
physical conditions in the gas which, in turn, lead to changes in the
emissivity of the lines.

A time dependent photoionization model is being investigated in which the response of the BLR gas to a variable ionizing continuum is calculated. The main assumptions are, firstly, that the gas is photoionized by an extension of the observed UV power law to beyond the Lyman limit (ν_H) and, for simplicity, that the gas is optically thin to ionizing radiation.

The input spectrum is given the form

$$L\nu(t) = L_H(\nu/\nu_H)^{-\alpha(t)}.g(t) \qquad (1)$$

where L_H is the luminosity at ν_H; $g(t)$ is an arbitrary time dependence function and the spectral index $\alpha(t)$ is chosen to reproduce the luminosity slope correlation. The calculation is carried out in two reference frames, the source frame, which is local to the BLR and where time (t_s) is measured from the start of the continuum event, and the observer frame in which a distant observer measures time (t_o) from his first detection of the continuum event.

The source frame: The gas is considered to occupy a large number ($10^3 - 10^4$) of small cloudlets distributed within an overall BLR geometry (eg. spherical shell, thin or thick disk etc.). The diameter of an individual cloudlet is assumed to be insignificant compared with its distance r from the central ionizing source so that each cloudlet can be specified by a single value of the atomic number density n(r) and the ionization parameter, U_H defined by

$$U_H(r,t) = \frac{L_H}{4\pi r^2 cn(r)h}.g(t) \qquad (2)$$

where c is the speed of light and h is Planck's constant.

For the appropriate initial values, the time dependent ionization and energy equations are solved simultaneously to generate the time variation of the ionization ratios, electron density and temperature at each step in r. The line emission coefficients J_λ (erg s^{-1} cm^{-3}) can then be calculated as functions of source time. A cloud at distance r from the source will respond to the continuum emitted at the earlier time $t_s - r/c$. Hence for a line of wavelength λ

$$J_\lambda = J_\lambda (r, t_s - r/c) \qquad (3)$$

The observer frame: Differences in light travel times from the ionizing source S to points (r, θ) in the gas (Fig.1) and then to the observer O cause emission from different clouds to reach O at different times. The result is that, at an observer time t_o after measuring a variation from S the observer detects the resulting effects on the surrounding gas only within the boundary (or light front) defined by

$$ct_o = r_{LF}(1-\cos \theta) \qquad (4)$$

(Morrison and Sartori 1969), where r_{LF} and θ are the polar co-ordinates of the light front (Fig.1).

Therefore, at a given observer time, radiation from clouds at points within the light front would have been emitted at a time t_s in the source frame where

$$t_s = t_o + (r/c)\cos \theta \quad (\text{for } r < r_{LF}) \qquad (5)$$

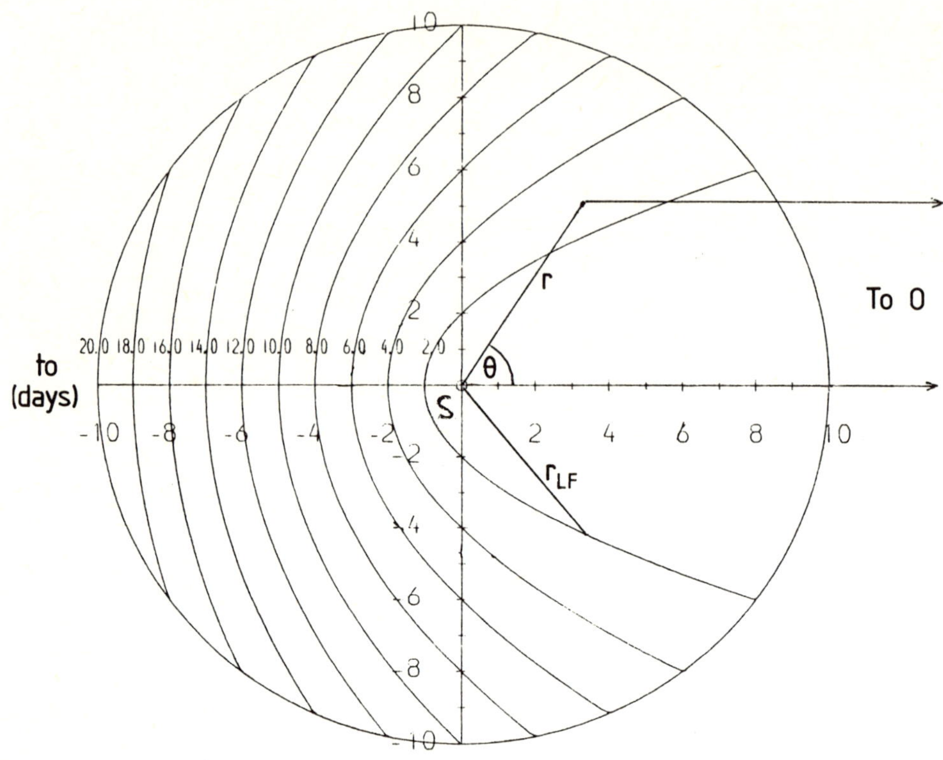

FIG. 1. Line variation light fronts (equation 4) as seen from the
reference frame of a distant observer at increasing observer times
t_o (axes in light days).

Clouds outside the light front will appear unaffected and so in the
source frame, $t_s = 0$.
Thus the "variability status" of each cloud as determined by its
associated source time and its distance from S can be obtained and the
corresponding emission coefficients summed to give the total emission
in each line at any observer time.

3. MODEL CALCULATIONS AND RESULTS

A set of illustrative models has been calculated for various time
dependence functions and different BLR geometries. In all models the
BLR light crossing time (t_L) was 20 days and the hydrogen number density
(n_H) was taken to be constant at 10^9 cm^{-3}. Some examples of the calculation results are shown in Figs. 2 and 3. These results can be summarized
as follows:
Source Frame: (1) Emission lines from a given cloudlet follow variations in the local ionizing continuum effectively instantaneously. This
is to be expected because at BLR densities the recombination time (less
than 1 hour at $n_H = 10^9$ cm^{-3}) is much smaller than continuum variation
timescales.
(2) Changes in the spectral index have a large effect, the variation
of line emissivities being much less dramatic if α is held constant.

(3) Transfer between ionization stages (for example $C^{+2} \rightleftharpoons C^{+3}$) is sometimes important in determining the overall line variability (see below).

Observer Frame: (1) It is clear that the observed delay between continuum and line variations can be accounted for by light travel time effects.

(2) For a given BLR geometry, the magnitude of the time lag is determined by the following factors:

Firstly, for all lines, lag times depend on the form of the continuum time dependence and the scale-size of the BLR, in particular on the relative values of the continuum timescale (t_{max}) and the light crossing time (t_L). If $t_{max} < t_L$ then lag times are longer than for $t_{max} > t_L$. For a periodic time dependence with $t_{max} \ll t_L$ two components are evident in the line variations (Fig.3a), an oscillating component with a timescale t_{max} is superimposed on a slowly increasing component which flattens off after a time t_L.

Secondly, different lines have different lag times. High ionization lines (OVI, NV, CIV) are mainly emitted closer to the source where light travel time delays are smaller and so have shorter response times than low ionization lines which arise further out and Lyman alpha which is more evenly distributed (Table 1).

In some cases, low ionization lines (eg. CIII]λ1909) may be anti-correlated with the continuum in the inner regions where the ion is destroyed by photoionization as the photon flux increases. This results in the line luminosity reaching a minimum at t_{max} before increasing to its maximum. Thus, in these cases the "lag time" is partly due to the anticorrelation (this also explains the comparatively small range in variation of CIII]).

TABLE 1. The trend in lag times with ionization and continuum timescale for a spherical shell BLR

Line	Ioniz. energy (eV)	Model 1	Model 2
OVI1034	114.0	1	1
NV1240	77.6	3	1.5
CIV1549	47.9	7	3.5
NIV]1402	47.5	8	4.5
OIII]1663*	35.2	16	12.5
CIII]1909*	24.4	15.5	12
HI1215†	13.6	11	6

(Lag times measured in days)
Model 1: $g(t) = 1 + t.\exp(-t/t_{max})$, $t_{max} = 5.0$ days
Model 2: $g(t) = 1 + |\sin(\omega.t)|$, $t_{max} = 20.0$ days ($\omega.t_{max} = \pi/2$)
*Anticorrelated with continuum in inner regions
†Collisionally excited component

Changing the BLR geometry has a comparatively minor effect on the line response because the light fronts (Eq.4) are always symmetrical about the line of sight. However, a comparison of spherical and disk models shows that lag times increase with the angle of inclination of the disk axis to the line of sight. This effect is smaller still for the high ionization lines, again because they arise closer to the ionizing source.

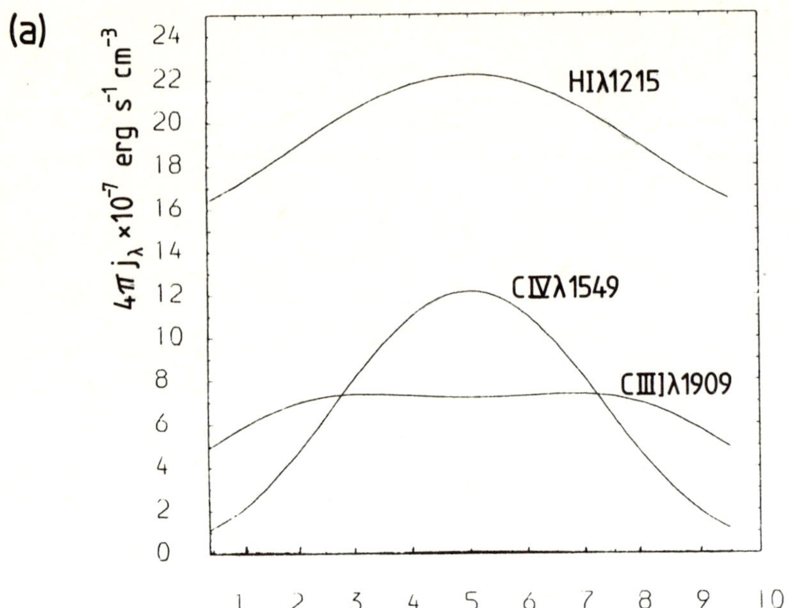

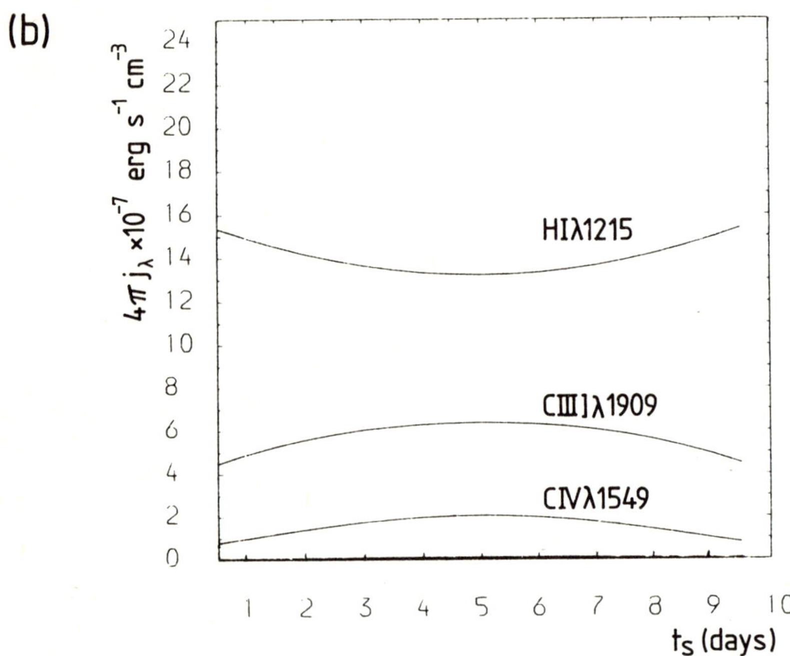

FIG. 2. Volume emission coefficients of strong collisional lines in the source frame at $U_H(t=0) = 2.6 \times 10^4$. $g(t) = 1 + |\sin(\omega.t)|$, $t_{max} = 5.0$ days. a) $\alpha(t) \sim 1/g(t)$; b) $\alpha = 2.0$.

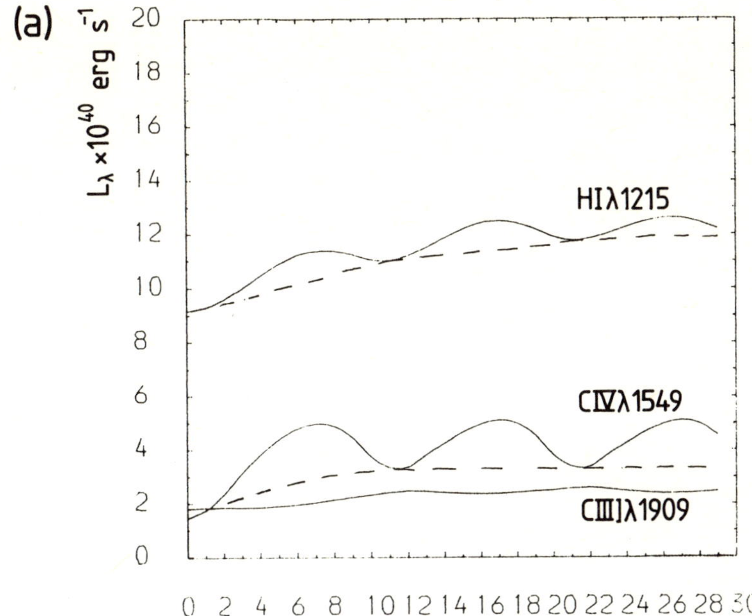

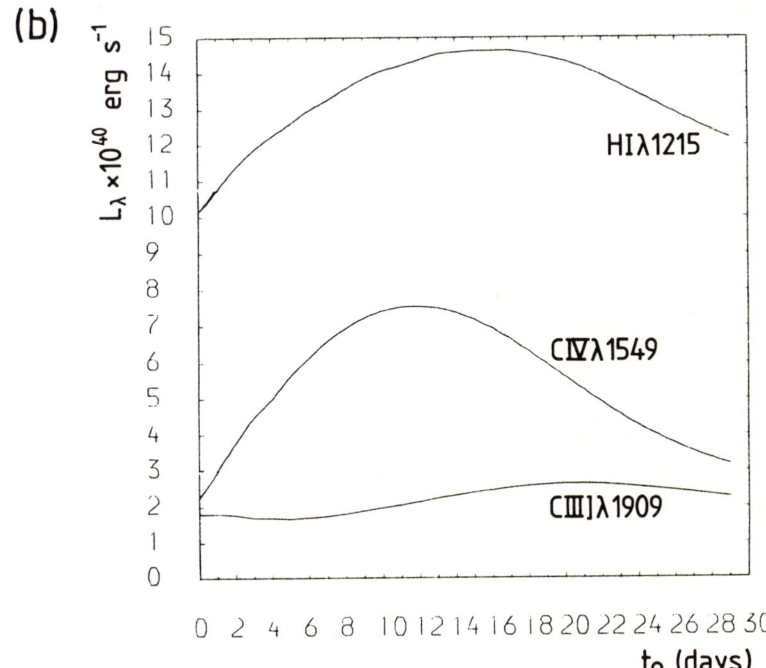

FIG. 3. Integrated line emission from a spherical shell BLR (t_L = 20.0 days) as measured in the observer frame. (a) $g(t) = 1 + |\sin \omega.t|$, t_{max} = 5.0 days; (b), (c), (d) $g(t) = 1 + t_{exp}(-t/t_{max})$, t_{max} = 5.0 days.

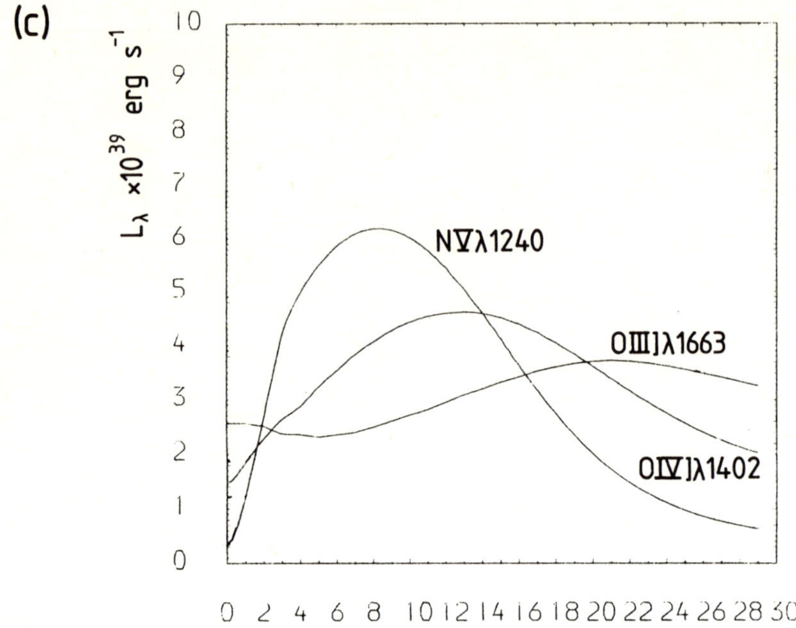
(c)
L_λ ×10^39 erg s^-1
NⅤλ1240
OⅢ]λ1663
OⅣ]λ1402
0 2 4 6 8 10 12 14 16 18 20 22 24 26 28 30

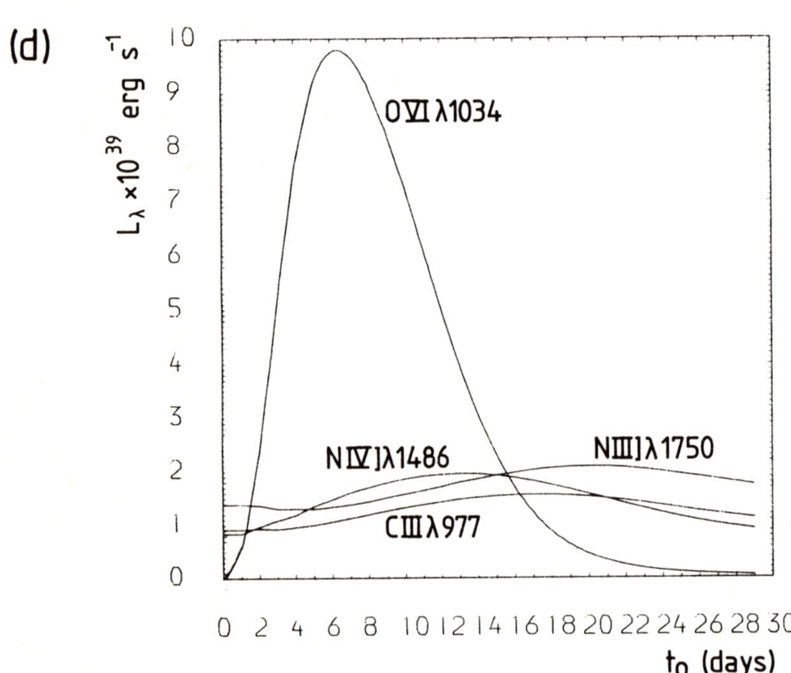
(d)
L_λ ×10^39 erg s^-1
OⅥ λ1034
NⅣ]λ1486
NⅢ]λ1750
CⅢλ977
0 2 4 6 8 10 12 14 16 18 20 22 24 26 28 30
t_0 (days)

4. CONCLUSIONS

The general pattern of the observed emission line variations (the correlation with the continuum, the lag times and the trend in response times with ionization) can be explained in a natural way by a time dependent photoionization model combined with light travel time effects in an extended BLR. The importance of this is that, since the response of any particular cloud depends on its position, detailed model fitting to observations may set constraints on the size and geometry of the BLR. Furthermore, by including appropriate velocity laws in the calculation and generating the time variation of the line profiles it may be possible to deduce the dynamical state of the gas.

Aside from light travel time effects, two other important points emerge: Firstly, changes in the spectral index have a large effect on the degree of variability and secondly, at large values of initial ionization parameter $(U_H(t = 0))$ some lines are likely to be anticorrelated with the continuum, effectively increasing their response times.

An obvious drawback of the current model is that it is optically thin and therefore cannot predict variability in the Balmer lines or in other lines arising in the HI zone such as $MgII\lambda2800$. Future work will involve the generalization of the model to the optically thick case.

Finally, the success of the time dependent photoionization model in explaining line variations is, in itself, strong evidence that photoionization is indeed the main excitation mechanism of the BLR gas.

REFERENCES

Antonucci RJ, Cohen RD 1983 Astrophys. J. 271:564
Barr P, Willis AJ, Wilson R 1983 Mon. Not. R. astr. Soc. 203:201
Davidson K, Netzer H 1979 Rev. Mod. Phys. 51:715
Morrison P, Sartori L 1969 Astrophys. J. 158:541
Penston MV, Boksenberg A, Bromage GE, Clavel J, Elvius A, Gondhalekar PM, Jordan C, Lind J, Lindegren L, Perola GC, Pettini M, Snijders MAJ, Tanzi EG, Tarenghi M, Ulrich MH 1981 Mon. Not. R. astr. Soc. 196:857
Perola GC, Boksenberg A, Bromage GE, Clavel J, Elvis M, Elvius A, Gondhalekar PM, Lind J, Lloyd C, Penston MV, Pettini M, Snijders MAJ, Tanzi EG, Tarenghi M, Ulrich MH, Warwick RS 1982 Mon. Not. R. astr. Soc. 200:293
Peterson BM, Byard PL, Foltz CB, Wagner RM 1982 Astrophys. J. Suppl. Ser. 49:469
Ulrich MH, Boksenberg A, Bromage GE, Clavel J, Elvius A, Penston MV, Perola GC, Pettini M, Snijders MAJ, Tanzi EG, Tarenghi M 1984 Mon. Not. R. astr. Soc. 206:221

Jet Like Features and Filaments in Blue Compact Galaxies.

T.P. Ray,

Dept. of Physics, University College Dublin.

P. Grimley,

Dept. of App. Maths. and Astronomy, University College, Cardiff.

Summary

Examination of UK Schmidt and PSS plates has shown the existence of jet like features and filaments in at least 1/3 of the Haro (1956) sample of blue compact galaxies. Several of these "jets" are associated with dwarfs and might therefore represent a new class of optical jet. The usual explanation for the blue nature of these galaxies is that they are undergoing intense star formation (a starburst). Such a starburst would tend to hide the underlying morphology of the parent galaxy.

Introduction

Haro (1956) showed that emission like galaxies could be found from three exposures (U, B and V) on a single Schmidt plate. This technique is effective because of a strong correlation between emission line strength and U-B magnitudes (Huchra, 1977). The morphological characteristic of these galaxies is usually a strong compact image in the UV and similar objects exist in the Zwicky (1971) and Markarian lists (Markarian, 1967; 1969a,b; Markarian & Lipovetskii 1971, 1972, 1973, 1974, 1976a,b; Markarian et al., 1977a,b). While the blue compacts in these various catalogues do not form a homogeneous group, they do have similar morphological and spectroscopic characteristics. A majority of them are dwarf systems but some are comparable in size to normal spirals. Most of these galaxies are isolated. The presence of emission lines and UV continuum is usually interpreted as indicating the presence of large numbers of early type stars. Unlike their nearest relatives, the Irl galaxies, the HII regions in these systems are concentrated towards the nucleus. Metallicity values are low (typically 0.5 to 0.02 times that found in the solar neighbourhood, see Tully et al., 1979).
Recent HI surveys (Gordon & Gottesman, 1979 and 1981) have shown that blue compacts are rich in neutral hydrogen and that the dimensions of their HI cloud/disk is much larger than their visible structure. It is possible that many possess large HI halos (Gordon & Gottesman, 1981).
One likely scenario is that blue compacts are galaxies whose normal appearance has been temporarily overshadowed by a recent starburst. Thus since most irregular galaxies are dwarfs, it is probable that blue compact dwarfs are irregulars undergoing intense star formation. Similarly, the larger compacts might be spirals with starbursts and

therefore similar to radio loud spirals (Condon et al., 1982).

Examination by us of UK Schmidt and PSS plates has shown the existence of jet like features and filaments in at least 1/3 of Haro's original sample (Haro, 1956). A similar examination by Kiang (1967) of PSS plates has yielded a comparable fraction. It is usually found that these "jets" are more obvious on the blue (O) plate than the red (E) plate and that, in almost all cases, they point towards the nucleus. Some examples are shown in fig. 1 (H3) and fig. 2 (H36). While it is possible that some of these "jets" might turn out, upon deeper imaging, to be bright sections of spiral arms or even bars, this explanation would probably not suffice for the dwarf compacts like H3, H22, etc. Furthermore, tidal interactions (see, e.g., Bothun et al., 1981) seem to be an unlikely explanation since most of these systems are isolated.

Radio observations (6cm) by Sramek & Tovmassian (1975) of blue compact galaxies and optical studies by Kinman & Hintzen (1981) have produced no evidence for these galaxies emitting nonthermal radiation. Thus if we are dealing with "jets", in the true sense of the word, it is unlikely that they are of the synchrotron variety. Instead, they are more likely to resemble the jets in NGC1097 (Wolstencroft, 1981; Wolstencroft et al., 1984), NGC5128 (Cannon, 1981), NGC3310 (Bertola & Sharp, 1983) and NGC4552 (Malin & Carter, 1983) where thermal emission may be dominant.

A link between starbursts in blue compacts and their HI halos/disks has been suggested by Gordon & Gottesman (1981). They propose that infall of HI might fuel/trigger a starburst and that mixing of this HI with recycled gas could explain the low metallicity of these galaxies. Furthermore, Gerola et al., (1980) have successfully applied a stochastic self propagating star formation model to blue dwarfs. They find that, for a radius of approximately 2kpc, bursts of star formation are interrupted by periods of virtual inactivity.

An extension of the above theory is to assume that not only does gas infall fuel/trigger a starburst but it may also be responsible for fuelling one or more jet producing "monsters". Such "monsters" may themselves have formed as a result of the intense star formation. In any event, the passage of a jet through a HI rich environment would be conducive to star formation along its length and this might explain the observed features. A similar model has been applied by Bertola & Sharp (1983) to NGC3310.

Certainly, the number of known optical jets is small (see Sol, 1983) while our ignorance of jets is large! Even amongst the so called "peculiar galaxies", jets are rare (Arp, 1966; Arp & Madore, 1977). It follows that a detailed study of the jet like features in Haro galaxies could be useful as an aid in understanding jets.

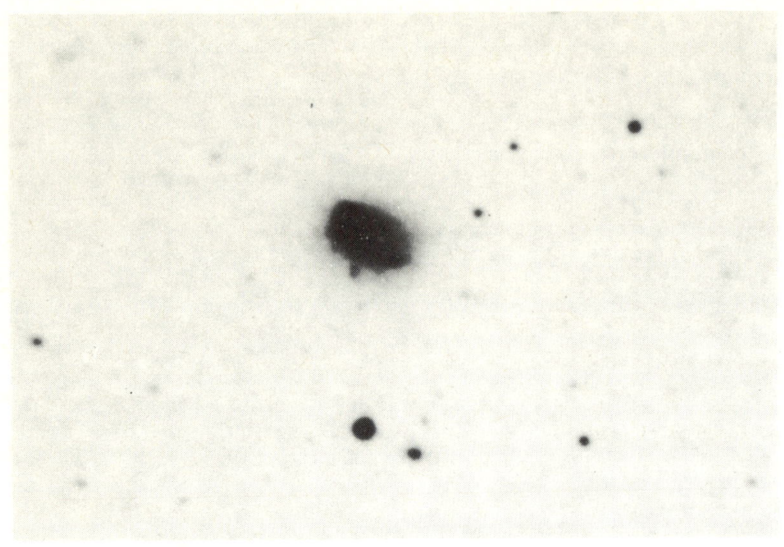

Fig. 1.   H3 (PSS 0 Print).   Jet is pointing downwards.

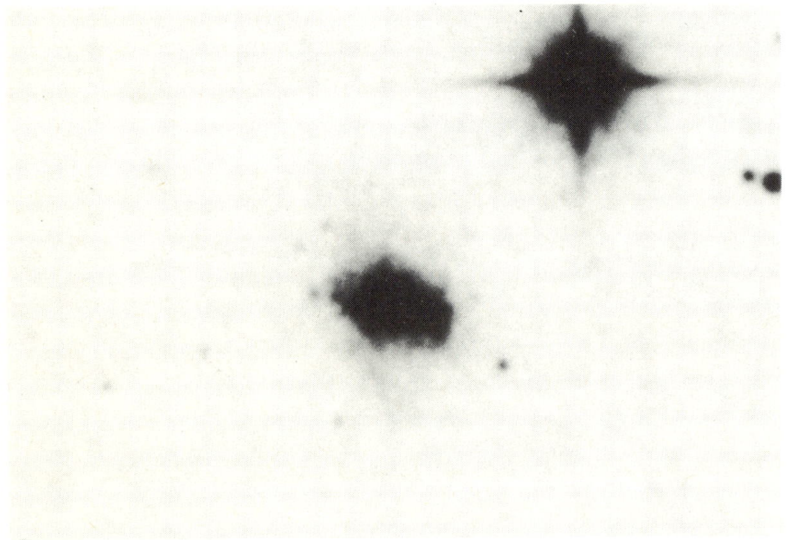

Fig. 2.   H36 (PSS 0 Print).   Jet is pointing towards
bottom right hand corner.

References

Arp, H., 1966, Atlas of Peculiar Galaxies.

Arp, H., Madore, B.F., 1977, Q.J.R. Astr. Soc., 18, 234.

Bertola, F., Sharp, N.A., 1983, in "Astrophysical Jets", 143, editors A. Ferrari and A.G. Pacholczyk, Reidel.

Bothun, G.D., Stauffer, J.R., Schommer, R.A., 1981, Ap. J., 247, 42.

Cannon, R.D., 1981, ESO/ESA Workshop on "Optical Jets in Galaxies", 45.

Condon, J.J., Condon, M.A., Gisler, G., Puschell, J.J., 1982, Ap. J., 252, 102.

Gerola, H., Seiden, P.E., Schulman, L.S., 1980, Ap. J., 242, 517.

Gordon, D., Gottesman, S.T., 1979, in "Photometry, Kinematics and Dynamics of Galaxies", editor D.S. Evans, University of Texas, Austin.

Gordon, D., Gottesman, S.T., 1981, Astron. J., 86, 161.

Haro, G., 1956, Bol. Obs. Tonantzintla y Tacubaya (No.14), 8.

Huchra, J.P., 1977, Ap. J. Suppl., 35, 171.

Kiang, T., 1967, Ap. J., (letters), L31.

Kinman, T.D., Hintzen, P., 1981, Pub. A.S.P., 8, 93, 405.

Malin, D.F., Carter, D., 1983, Ap. J., 274, 534.

Markarian, B.E., 1967, Astrofiz., 3, 55.

Markarian, B.E., 1969a, Astrofiz., 5, 443.

Markarian, B.E., 1969b, Astrofiz., 5, 581.

Markarian, B.E., Lipovetskii, V.A., 1971, Astrofiz., 7, 511.

Markarian, B.E., Lipovetskii, V.A., 1972, Astrofiz., 8, 155.

Markarian, B.E., Lipovetskii, V.A., 1973, Astrofiz., 9, 487.

Markarian, B.E., Lipovetskii, V.A., 1974, Astrofiz., 10, 307.

Markarian, B.E., Lipovetskii, V.A., 1976a, Astrofiz., 12, 389.

Markarian, B.E., Lipovetskii, V.A., 1976b, Astrofiz., 12, 657.

Markarian, B.E., Lipovetskii, V.A., Stepanyan, D.A., 1977a, Astrofiz., 13, 225.

Markarian, B.E., Lipovetskii, V.A., Stepanyan, D.A., 1977b, Astrofiz., 13, 397.

Sol, H., 1983 in "Astrophysical Jets", 135, editors A. Ferrari and A.G. Pacholczyk, Reidel.

Sramek, R., Tovmassian, H., 1975, Ap. J., 196, 339.

Tully, R.B., Boesgaard, A.M., Schempp, W.V., 1979, in "Photometry, Kinematics and Dynamics of Galaxies", editor D.S. Evans, University of Texas, Austin.

Wolstencroft, R.D., 1981, ESO/ESA Workshop on "Optical Jets in Galaxies", 49.

Wolstencroft, R.D., Tully, R., Perley, R.A., 1984, Mon. Not. R. Astr. Soc., 207, 889.

Zwicky, F., 1971, Catalogue of Selected Compact Galaxies and of Post-Eruptive Galaxies (F. Zwicky, Guemligen).

<u>Internal Shocks in Jets</u>.

S.A.E.G.Falle & M.J.Wilson.
Department of Applied Mathematics,
University of Leeds,
Leeds LS2 9JT,
U.K.

SUMMARY

We consider the conditions under which shocks can be induced in supersonic fluid jets propagating through a medium with a pressure gradient. We find that, for plausible galactic atmospheres strong shocks can occur in jets with moderate Mach numbers. These shocks are identified with the radio and optical knots seen in such sources as M87.

<u>Introduction</u>.
Extensive laboratory experiments have shown that shocks are formed in supersonic jets which are not in pressure equilibrium with their surroundings. In this paper we want to consider the relevance of this phenomenon to astrophysical jets. We therefore assume that these can be treated as fluid jets and neglect any radiative losses so that the fluid behaves adiabatically with an adiabatic index $\gamma = 5/3$.

Since astrophysical jets have a high degree of collimation even when the ambient pressure is low and are apparently able to travel many jet diameters undisrupted by fluid instabilities, we can conclude that, in the high power sources at least, the jets are highly supersonic. Furthermore we know that the ambient pressure must vary by a large factor over the length of the jet. This leads us to consider the problem of a supersonic jet propagating through a medium in which the pressure varies. In this paper we only consider steady phenomena so that we are interested in parts of the jet away from the working surface terminating the jet where the flow will be unsteady. This problem has been studied by Sanders (1983) who used the numerical method of characteristics to study the shocks induced in a jet by an external pressure gradient.

<u>Laboratory Jets</u>.
Laboratory experiments generally involve a supersonic jet discharging from a nozzle into a large receiver filled with gas at a different pressure. Since we are considering only steady jets it is only the pressure of the gas in the receiver that matters. If the jet is not diverging or converging, then there are two cases:-

$$\frac{P_{nozzle}}{P_{ambient}} \quad < 1 \qquad \text{Overexpanded,}$$

$$> 1 \qquad \text{Underexpanded,}$$

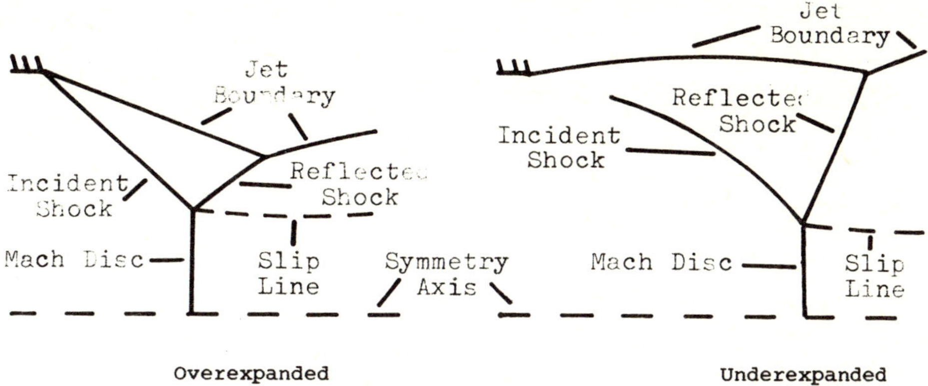

Figure 1

The shock patterns in the two cases are shown schematically in figure 1. An important point is that these patterns can repeat themselves at regular intervals along the jet. We shall argue that the regularly spaced radio and optical knots such as those in M87 can be identified with regularly repeating shock structures induced by pressure imbalance between the jet and the external medium.

For axisymmetric jets it can be shown that the shock always meets the symmetry axis at right angles, i.e. there is always a Mach disc. If the Mach disc is small then one can use a small disturbance approximation together with the Chisnell approximation (Whitham 1974) to show that for an overexpanded jet with infinite Mach number

$$r_m = \left[\frac{2P_{ext}}{(\gamma + 1)\rho_j v_j^2} \right]^\beta ,$$

where $\beta = \beta(\gamma) = 0.11$ for $\gamma = 5/3$. Here r_m is the radius of the Mach disc and ρ_j, v_j are the density and velocity of the jet. Since for such a jet the pressure behind the Mach disc is equal to the ram pressure, it may be much larger than the external pressure. This is in contrast to cases where the radius of the Mach disc is not small compared to the jet radius. In such cases the external pressure must be comparable with the ram pressure of the jet. In our model therefore those radio knots whose pressure is much larger than the ambient pressure must be due to these small Mach discs.

The size of the larger Mach discs is determined by a sonic point condition in the flow downstream of the Mach disc and it is possible to use the numerical method of characteristics together with this condition to determine the shock structure (Ashratov 1966, Chang & Chow 1974, 1975). However these methods are cumbersome and so it is easier to use a finite difference technique to determine the flow. We have used FLIC (Gentry, Martin and Daly 1966) and find that even with only 20 cells across the jet, the size and position of the Mach disc is determined to an accuracy of about 15%. Unfortunately these calculations require a lot of computer time, but we are currently developing a multi grid program which should be several orders of magnitude faster. This will allow us to examine a lot of cases which are of astrophysical interest.

<u>Astrophysical Jets.</u>

In astrophysical jets there is of course no nozzle, but if the external pressure varies on a length scale which is shorter than the scale on which the jet can respond, then the jet behaves as if it has experienced a sudden change in the external pressure. The condition for this is

$$l_p = \frac{P_{ext}}{(dP_{ext}/dx)} < M_j d_j = l_s \ .$$

Here x is the coordinate along the jet, M_j and d_j are the Mach number and diameter of the jet respectively. lp is the external pressure scale length and we shall call l_s the 'shocking scale length' of the jet.

If the jets originate close to the centres of galaxies, then the ambient pressure may drop by a factor $\simeq 10^{10}$ along the length of the jet. Intuitively, one would therefore expect that the jets are analogous to underexpanded jets. To get the equivalent of a laboratory overexpanded jet, the ambient pressure would have to increase along the jet.

However, we now show that in some sense one can get both over and underexpanded jets when the ambient pressure is decreasing monotonically.

First suppose that $l_s < l_p$ so that the jet remains in pressure equilibrium with its surroundings. Then if $r_e(x)$ is the equilibrium radius of the jet, we have

$$r_e(x) \propto \left[\frac{1}{P_{ext}(x)}\right]^{1/2\gamma} \ .$$

If now l_s becomes greater than l_p so that the jet is no longer in pressure equilibrium, then the behaviour depends on the sign of d^2r_e/dx^2. The opening angle of the jet increases if $d^2r_e/dx^2 < 0$ and it is in some sense overexpanded. If $P_{ext} \propto 1/x^n$ then this requires n < 10/3. If $d^2r_e/dx^2 = 0$ then the jet boundary is straight and nothing very interesting can happen even if $l_s > l_p$. For $d^2r_e/dx^2 > 0$ (i.e for n > 10/3) the jet opening angle is increasing. The jet is then underexpanded but no shocks can be produced unless l_s becomes less than l_p at some stage because the rarefaction waves reflecting from the symmetry axis cannot reach the jet boundary before the external pressure has dropped below that in the jet. All that happens is that the opening angle of the jet increases towards its asymptotic value $\theta_0 + \sin^{-1}(1/M_0)$. Here θ_0 is the initial divergence and M_0 is the initial Mach number. The important point is that shocks can only form if l_s/l_p has a maximum > 1. These conclusions are confirmed by numerical experiments with steady jets in a varying external pressure.

This phenomenon will not necessarily be observationally significant in all cases. There is a trade off between the sensitivity of the jet to variations in external pressure, which depends on the ratio l_s/l_p, and the fraction of the jet energy dissipated in the induced shocks. If the Mach number is high then l_s/l_p is large and the jet is sensitive but the dissipation is small (basically because the Mach disc is small and the other shocks are close to the Mach angle and therefore weak). If the jet has low Mach number, then l_s/l_p cannot be so large and so the jet is less likely to contain shocks, although if shocks do form in such cases the Mach disc will be large and a significant fraction of the jet's energy will be dissipated. These effects are therefore most likely to be seen in jets with a moderate Mach number.

Isothermal Galactic Atmosphere.

As an example of a plausible galactic atmosphere we consider an isothermal gas with sound speed c_0 in hydrostatic equilibrium in the gravitational field of a stellar mass distribution of the form

$$\rho_*(R) = \frac{\rho_*(0)}{(1 + R^2/R_c^2)} \, .$$

Here R_c is a core radius which is typically a kiloparsec or so. The gas density ρ_g is then

$$\rho_g(x) = \rho_g(0)\exp\alpha_i\left[1 - \frac{\tan^{-1}x}{x} - \frac{1}{2}\log(1 + x^2)\right] \, ,$$

where

$$\alpha_i = \frac{4\pi G\rho_*(0)R_c^2}{c_0^2} \, .$$

and

$$x = \frac{R}{R_c} \, .$$

For reasonable galaxies $\alpha \simeq 1$. Note that for $\alpha < 2$ l_s/l_p has a maximum at about the core radius. So we would expect a jet propagating through such an atmosphere to form shocks (and therefore radio knots) near the core radius. This seems to be the case in M87 and possibly in other sources as well.

References.
Ashratov,E.A.,1966. Mekhanika Zhidkosti i Gaza $\underline{1}$,158.
Chang,I.S. & Chow,W.L.,1974. AIAA Journal $\underline{12}$,1079.
                1975. AiAA Journal $\underline{13}$,763.
Gentry,R.A.,Martin,R.E. & Daly,B.J.,1966. J.Comp.Phys.,$\underline{1}$,87.
Sanders,R.H.,1983. Astrophys.J. $\underline{266}$,69.
Whitham,G.B.,1974. Linear and Nonlinear Waves p.271. Wiley-Interscience.

Some Exactly Solvable Models of Thick Disks and Radio Jets Near The Black Hole

Sandip K. Chakrabarti

Department of Physics and The Enrico Fermi Institute,
University of Chicago, Chicago, USA

ABSTRACT

The exact solution of the relativistic Euler's equation in the black hole geometry is obtained by identifying surfaces of constant angular momentum with the surfaces of constant angular velocity for the azimuthal component. We show that the solution gives rise to thick disks and cosmic jets for suitable choices of the physical parameters.

Introduction

Various models of thick disks and the cosmic jets have been recently discussed in the literature(For a summary on disk models, see Paczynski (1982) and for a discussion on jets, see Rees (1982) and references therein.). In most of these works the black hole geometry has been replaced by Newtonian or pseudo-Newtonian geometry. But it is well known that the structure of thick disk or the collimation of the jet heavily depends upon the angular momentum distribution assigned near the black hole where the Newtonian approximation clearly fails. On the other hand, uncertainties over the nature of dissipative mechanisms and other physical processes near the hole are so high that an accurate treatment of geometry may seem to be unnecessary. Still one likes to investigate the exact nature of the disks and jets and thereby hopes to eliminate the uncertainties arising out of crude geometric models.

In the present work we have solved *full general relativistic* Euler's equation for ideal fluids with barotropic equation of state. This work actually summerizes the works of Chakrabarti, 1984a,b (paper I and II). Only the azimuthal component of the momentum has been kept in the derivation. We have shown that same solution with different choices of parameters give rise to the thick disk and jet near the black hole. Our treatment does not depend upon the background geometry as long as they are stationary. However, we shall concentrate on the solution in Kerr geometry and neglect the effect of the mass of the disk upon the geometry.

The Analytic Solution

In what follows we shall use the geometrical units, i.e., $G = c = 1$. We shall use the mass of the black hole to be 1 since it is just a scale factor. The signature of the metric is $+ - - -$. We start with the stress-energy tensor of the ideal fluid, namely,

$$T^{\mu\nu} = (p + \epsilon)u^\mu u^\nu - pg^{\mu\nu}, \tag{1}$$

where p and ϵ are the pressure and energy density respectively and u^μ is the component of the four velocity of the fluid element, i.e.,

$$u^\mu = (u^t, 0, 0, u^\phi). \tag{2}$$

One derives the equation of motion for such fluid as,

$$\frac{dp}{p+\epsilon} = -\,d\,(\ln u_t) + \frac{\Omega dl}{1-\Omega l};\tag{3}$$

where,

$$\Omega = \frac{u^\phi}{u^t}\,;\quad l = -\frac{u_\phi}{u_t}.\tag{4}$$

Ω is the "transport" velocity and l is specific angular momentum. We are considering the barotropic equation of state so that $p = p(\epsilon)$ and thus we can define the total potential by,

$$W = -\int \frac{dp}{p+\epsilon}.\tag{5a}$$

W becomes the total potential (gravitational plus centrifugal) in the Newtonian limit. We also define a parameter λ associated with the azimuthal component by,

$$\lambda = \sqrt{\frac{l}{\Omega}}.\tag{5b}$$

It can be argued that there must exist a relationship among Ω and l which would make the second term on the right hand side of equation 3 integrable exactly. Below we write down one such relation, namely,

$$\Omega = k\,l^m,\tag{6a}$$

where, k and m are some constants. This distribution can be re-written as

$$l = C\,\lambda^n,\tag{6b}$$

where $n = \dfrac{2}{1-m}$, and $C = k^{\frac{n}{2}}$. We notice from this distribution that the surfaces of angular momentum are also the surfaces of angular velocities. Using distribution given by equation 6 and definitions of W and λ provided by equation 5, we integrate equation 3 to obtain,

$$\frac{e^W \lambda^{\frac{c^2}{k}}}{u_t} = constant\tag{7a}$$

for $n = 1$, and

$$\frac{e^W}{u_t|1-C^2\,\lambda^{2n-2}|^\alpha} = constant\tag{7b}$$

otherwise. k is given by,

$$k = 1 - C^2,$$

and α is given by,

$$\alpha = \frac{n}{2n-2}.$$

The binding energy u_t can be obtained from,

$$u_t = \left[\frac{-(g_{t\phi}^2 - g_{tt}g_{\phi\phi})}{(g_{\phi\phi} + lg_{t\phi})(1 - C^2\lambda^{2n-2})}\right]^{\frac{1}{2}}.\tag{8}$$

Expressions equivalent to equations 7a and 7b in terms of enthalpy density, h, entropy density, s, and temperature T can be found in paper I. Equations 7a and 7b connect the properties of different parts of the materials distributed around black hole and therefore are of extreme importance. The constant on right hand side can be determined by some boundary values.

We now discuss some restrictions upon the parameters C and n. In the case of thick disks (paper I) the radial and the theta components of the momentum are very small compared to the azimuthal component. Therefore, $u^{r,\theta} \approx 0$. Thus keeping only the azimuthal component is not very restrictive. Stability of the structure requires that $n \geq 0$. For accretion to be possible one has $n \leq 2$. By fixing the inner edge and the center of the disk at r_{in} and r_c one gets the c and the value of n from the following equations, namely,

$$C\lambda_{in}^{n} = l_k(r_{in}),\tag{9a}$$

and

$$C\lambda_{c}^{n} = l_k(r_c),\tag{9b}$$

because it is believed that at the inner-edge and at the center the motion is Keplerian.

In the case of jets (paper II), however, all the components can be non-zero, but we believe that the theta component will be extremely small unless the disk from which the materials are coming is very steep. Also if the jet is pressure confined then the azimuthal component of the angular momentum is going to be important. In the above derivation we kept only this component. But one can add the radial component also. (See paper II for details.) To make the pressure, density, temperature etc. finite on the symmetry axis, i.e., for $\theta = 0$, n must be greater than 1. The other extreme case is the rigidly rotating jet, in which case, maximum value of n is 2.

Results

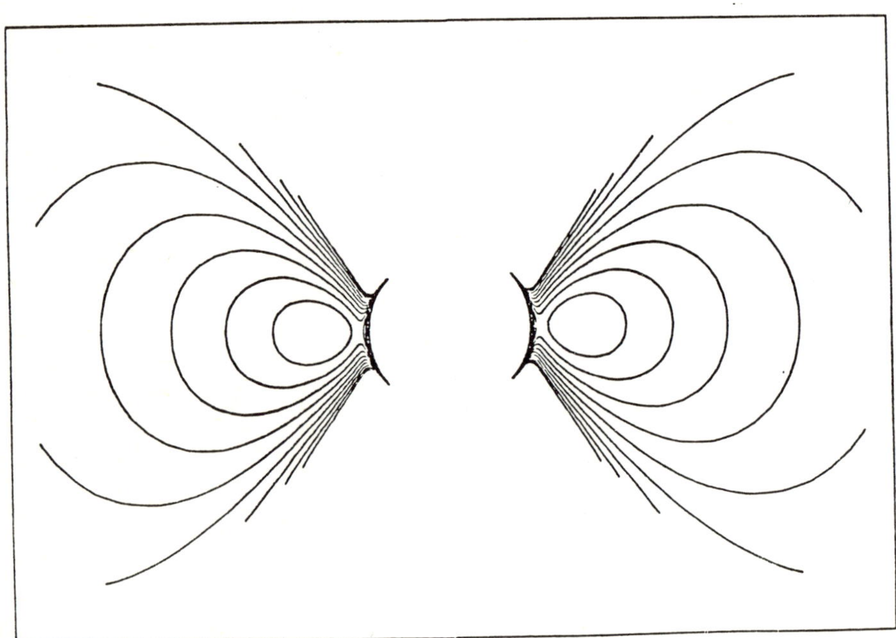

Fig.1:Equipotential contours for thick disks in Kerr hole ($a = 0.99$) geometry. The inner-edge is at $r = 1.3$ and cusp is at $r = 1.8$.

We shall now discuss a typical result in the case of Kerr geometry with Kerr parameter $a = 0.99$. In this case, r_{mb}, the radius of marginally bound orbit is 1.2100, and r_{ms}, the radius of marginally stable orbit is 1.4545. Letting r_{in}, the inner-edge of the disk to be at 1.30, a number somewhere between r_{mb} and r_{ms} inclusive, and r_c, the center of the disk at 1.8, a number beyond r_{ms}, we obtain C = 2.0669 and $n = 0.0472$. The equipotential surface corresponding to this model is shown in figure 1. Notice the existence of cusp at inner-edge, r_{in}. Because the equipotential surfaces are open near the cusp, a small instability will cause the materials to overflow and fall into the hole. Thus, strictly speaking, there should be some non-zero radial velocity of the disk material near the cusp.

Figure 2 shows the behavior of equipotential surfaces near the rotation axis of the Kerr hole for various values of C $(2a - 2c)$. We have chosen $n = 1.5$. In our distribution (equation 6), the azimuthal component of angular momentum increases outward from the axis far away from the hole but almost downward near the hole. This causes the pressure to increase qualitatively in the same way. (See paper II for arguments.) Our jet is thus pressure confined and angular momentum supported. The pressure on the outermost contour corresponds to the inter-galactic medium pressure (p_0) chosen to be constant throughout. So far, the constant on the right hand side of equation 7 or p_0 have been chosen arbitrarily. Any consistent choice would have the features shown in the figure. Especially, if p_0 is chosen to increase as one goes towards the hole, the jet is found to get collimated even at a lower height on the rotation axis.

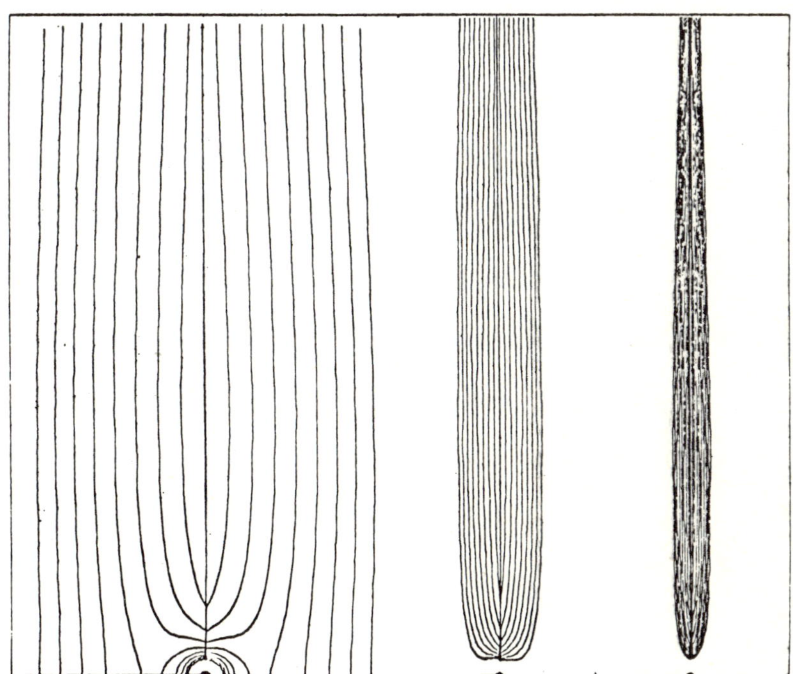

Fig.2:Equipotential contours for cosmic radio jets with $C =$ (a) 0.05, (b) 0.1 and (c) 0.15 respectively.

Conclusions

We have shown that suitable choices of physical parameters do indeed give rise to disk and jet like features near the black hole. We have chosen one of the exact solutions of the relativistic Euler's equation. (We have tried some other solution but no interesting feature came out.) We have chosen very ideal conditions which are obviously not valid especially near the hole because of various dissipative mechanisms which are present. It is possible that the parameters C and n depend upon such viscous mechanisms (Paper II). The merit of the present work is that it provides closed analytic expression for the physical quantities both in the disk and the jet. We have not addressed the question what causes the materials to accelerate radially in the jet. Presumably, the acceleration is due to radiation pressure (Davidson et. al., 1980) or, some kind of electrodynamic processes (Lovelace, 1976; Blandford, 1976). Thus a suitable combination of our result with the acceleration mechanisms of the early workers may provide more complete picture of the jet.

I am indebted to Dr. W. D. Arnett for some stimulating discussions. This work has been supported by NSF grant no. AST80-22876 and NASA grant no. NAG-W123.

References

Blandford, R. D. 1976, *M.N.R.A.S.*, **176** , 465.
Chakrabarti, S. K. 1985, *Ap.J.*, (in press), EFI-5/84 (paper I)
Chakrabarti, S. K. 1985, *Ap.J.*, (in press), EFI-9/84 (paper II)
Davidson, K. and McCray, R. 1980, *Ap. J.*, **241** ,1082.
Lovelace, R. V. E. 1976, *Nature*, **262** , 649.
Paczynski, B. 1982, *Astronomische Gesellschaft Mitteilungen*, **57** , 27.
Rees, M. J. 1982, in *Extra-galactic Radio Sources*, Eds. D. S. Heeschen and C. M. Wade (D. Reidel Publishing Co. Boston), 211.

Amplification caused by Gravitational Bending of Light

Peter Schneider

Max-Planck-Institut für Physik und Astrophysik, Institut für Astrophysik
Karl-Schwarzschild-Str. 1, D-8046 Garching b. München, FRG

SUMMARY

Gravitational bending of light may not only lead to multiple imaging
(gravitational lens effect), but also affects the apparent luminosity of
a source. It is shown here that a mass distribution near the line-of-
sight to any source always increases the observable flux relative to the
case in which the deflector is absent.

Introduction

Our universe certainly is not filled with a homogeneous, isotropic
perfect fluid, but rather matter is concentrated in clumps (stars,
galaxies, black holes(?),...); nevertheless, for the large scale description the Robertson-Walker metric (see, e.g., Weinberg, 1972) may be
appropriate. On the other hand, the propagation of a light bundle depends
on the amount of matter in the beam. As matter is not distributed homo-
geneously in the universe, the mass density in the beam differs from the
average mass density $\rho = \rho_0(1+z)^3$, which in turn affects the luminosity-
redshift relation. Dyer and Roeder (1972, 1973) derived a differential
equation for the angular-diameter distance, D, as a function of redshift, z, for those sources, whose lines-of-sight are well away from
all clumps. Introducing a 'smoothness parameter' $\tilde{\alpha}$ (= ratio of smoothed-
out mass to total mass), D becomes a function $D = D(H_0,\Omega,\tilde{\alpha};z)$, where
H_0 = Hubble constant, $\Omega = 2q_0 = \rho_0/\rho_c$ and ρ_c = 'critical' density, and
the (bolometric) flux of such a source with luminosity L and redshift z
is given by

$$S^* = L/[4\pi D^2(H_0,\Omega,\tilde{\alpha};z)\ (1+z)^4]. \tag{1}$$

In general, the observed flux S will differ from S*, as gravitational
bending of light changes the area of a ray bundle (see, e.g., Misner et
al., 1973, §22). One defines the amplification factor $I \equiv S/S^*$, which
depends on the mass distribution along the line-of-sight to the source
under consideration.

Theorem: Light deflection never leads to deamplification, or $I > 1$!

To prove this, we first remark that any light bending situation,
characterized by a three-dimensional matter distribution, can be trans-
formed into an equivalent single surface mass distribution $\Sigma(\underline{r})$ on a
plane ('lens plane') between source and observer, see Fig.1 (Schneider,
1984b), where $\Sigma > 0$. The 'lens' equation, which relates the position of
the observer, $\underline{x}$, to the point $\underline{r}$ in the lens plane where an image of
the source is seen, is given by

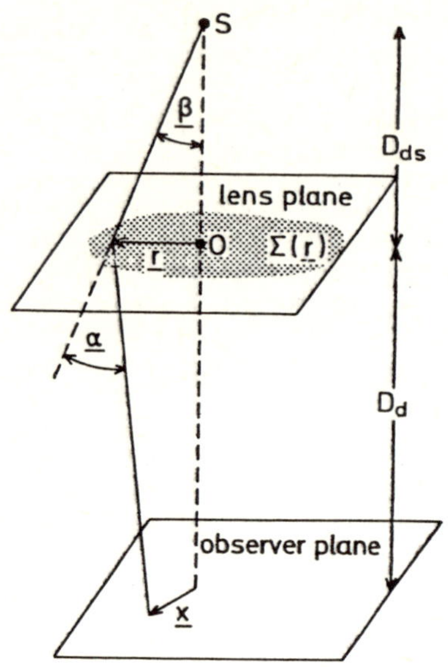

Fig. 1. The geometry of gravitational bending of light

$$\underline{x} = (D_s/D_{ds})\underline{r} - D_d\,\underline{\alpha}\,(\underline{r})\ ,$$

and the deflection angle is

$$\underline{\alpha}(\underline{r}) = 4G/c^2 \int\Sigma(\underline{r}')\,(\underline{r}-\underline{r}')/|\underline{r}-\underline{r}'|^2\,d^2r'\ ;$$

after scaling, $\underline{x} \rightarrow (D_{ds}/D_s)\underline{x}$, $\Sigma \rightarrow (4G/c^2)(D_dD_{ds}/D_s)\Sigma$, this becomes

$$\underline{x} = \underline{r} - \underline{\alpha}(\underline{r}) \tag{2}$$

$$\underline{\alpha}(\underline{r}) = \int\Sigma(\underline{r}')(\underline{r}-\underline{r}')/|\underline{r}-\underline{r}'|^2\,d^2r'\ . \tag{3}$$

The amplification factor I is the area distortion of the mapping (2) and given by $I = \det(\partial\underline{x}/\partial\underline{r})^{-1}$.

An observer at $\underline{x}$ will see an image at $\underline{r}$, if the vector field $\underline{y}(\underline{r}) \equiv \underline{r}-\underline{\alpha}(\underline{r})-\underline{x}$ vanishes at $\underline{r}$; note that $\underline{y} = \mathrm{grad}\ \theta$, where

$$\theta(\underline{r}) = |\underline{r}|^2/2 - \psi(\underline{r})-\underline{r}\cdot\underline{x}\ , \tag{4}$$

and ψ is the Coulomb potential in two dimensions, $\psi(\underline{r}) = \int\Sigma(\underline{r}')\ln|\underline{r}-\underline{r}'|d^2r'$. In terms of $\theta_{ij} \equiv \partial^2\theta/\partial r_i\partial r_j$, the amplification factor is given by

$$I^{-1} = \det\theta_{ij} = (1-\pi\Sigma)^2 - \sigma^2\ , \tag{5}$$

where the first term describes focussing due to matter in the beam

(Ricci focussing), and $\sigma^2 \equiv \psi_{12}^2 + 1/4(\psi_{11} - \psi_{22})^2$ is the shear applied to the beam at $\underline{r}$. The trace of θ_{ij} simply is tr $\theta_{ij} = 2(1-\pi\Sigma)$.

The scalar function $\theta(\underline{r})$ increases as $|\underline{r}|^2/2$ for large $|\underline{r}|$, thus it must have a minimum. Let $\underline{r}_0$ be a minimum; then, grad$\theta(\underline{r}_0) = 0$, hence $\underline{r}_0$ satisfies the lens equation (2), or $\underline{y}(\underline{r}_0) = 0$. Secondly, at a minimum, (i) tr $\theta_{ij} > 0$ and (ii) det $\theta_{ij} > 0$. The first condition implies $(1-\pi\Sigma) \in [0,1]$, therefore (5), det $\theta_{ij} < 1$, and together with (ii) we have det $\theta_{ij}(\underline{r}_0) \in [0,1]$ or $I(\underline{r}_0) > 1$, which completes the proof. Details can be found in Schneider (1984a).

Discussion

We have shown above that for every mass distribution and every observer position at least one image of the source is amplified ($I > 1$) relative to the situation in which no matter concentration is near to the line-of-sight. As can be easily seen, the case $I = 1$ for not identically vanishing surface mass density requires very special conditions – the shear and mass density must vanish at the minimum $\underline{r}_0$ – hence, in general, $I > 1$ for $\Sigma \neq 0$.

Therefore, the gravitational bending of light rays influences the results of source counts. If S_{th} is the threshold flux for a flux limited sample, a source with $S^* < S_{th}$ but $I > S_{th}/S^*$ will be included in the sample. Let $N(S^*(L),z)dS\,dz$ be the inherent luminosity function of a source population, and let $P(I,z)$ be the probability that a source at z is amplified by a factor I, then the observed luminosity function is the convolution of N with P:

$$N_{obs}(S,z)dSdz = \int dI\, P(I,z)N(S^*/I,z)I^{-1}dSdz \tag{6}$$

The function $P(I,z)$ depends on (i) cosmological parameters Ω, $\tilde{\alpha}$, (ii) lens population and (iii) surface brightness distribution of the sources. Investigations of (6) have been carried out (among others) by Canizares (1982) for point mass deflectors, Peacock (1982) for galaxies and Turner et al. (1984) for galaxies and clusters. Because of the uncertainty with which the lens population is known, the function $P(I,z)$ is only poorly determined; therefore it is of valuable interest to obtain model independent relations for $P(I,z)$.

Following an argument of Weinberg (1976) which says that the averaged flux of any source is the one that would be observed in a completely smooth ($\tilde{\alpha} = 1$) universe, we conclude

$$<I(z)> = \int I\, P(I,z)dI = \left[\frac{D(H_0,\Omega,\tilde{\alpha};z)}{D(H_0,\Omega,1;z)}\right]^2 \geq 1, \tag{7}$$

thus the mean of $P(I,z)$ is determined solely by $\tilde{\alpha}$ (note that $\partial D/\partial\tilde{\alpha} < 0$).

The second general relation for $P(I,z)$ is given by the above theorem: $P(I,z) = 0$ for $I < 1$. Furthermore, the large I-tail of $P(I,z)$ can be obtained from a consideration of the 'critical lines' (these points where det$\theta_{ij} = 0$, or where the amplification formally diverges),

$$P(I,z) \propto I^{-3}\exp(-I/I_{max}) \,,$$

where I_{max} is a cut-off which depends on the deflectors as well as on the emitting area of the sources. As long as I_{max} is not too small, one can see from (6) that the amplification caused by gravitational light bending is of great importance for intrinsic luminosity distributions steeper than L^{-3}.

The function $P(I,z)$ for I near 1 is, however, very poorly known, mainly due to the lack of knowledge on the lens population and on $\tilde{\alpha}$;

especially, for (I-1) << 1 multiple light deflection as well as individual stars in galaxies (Chang and Refsdal, 1979, 1984; Young, 1981)
become important which up to now has not been included properly in model
calculations of $P(I,z)$.

Finally I would like to point out that the gravitational lens
theory can be reformulated in terms of the scalar function θ defined
above, thereby simplifying the resulting equations considerably[*];
especially, the calculation of the time-delay between two images (Cooke
and Kantowski, 1975, Kayser and Refsdal, 1983) becomes a simple difference of θ values at the considered image points, rather than a line
integral as in the above mentioned papers. The formulation of gravitational lens theory with the potential θ is given by Schneider (1984b).

[*](I like to thank R. Blandford for pointing this out to me)

REFERENCES

Canizares, CR 1982 ApJ 263:508
Chang K, Refsdal S 1979 Nature 282:561
Chang K, Refsdal S 1984 Astron.Astrophys. 132:168
Cooke JH, Kantowski R 1975 ApJ 195:L11
Dyer CC, Roeder RC 1972 ApJ 174:L115
Dyer CC, Roeder RC 1973 ApJ 180:L31
Kayser R, Refsdal S 1983 Astron.Astrophys. 128:156
Misner CW, Thorne KS, Wheeler JA 1973 Gravitation, Freeman,
    San Francisco
Peacock JA 1982 Month. Not. Roy. Astron. Soc. 199:987
Schneider P 1984 a Astron.Astrophys. : in press
Schneider P 1984 b in preparation
Turner EL, Ostriker JP, Gott JR 1984 (preprint)
Weinberg S 1972 Gravitation and Cosmology, Freeman, New York
Weinberg S 1976 ApJ 208:L1
Young P 1981 ApJ 244:756

Particle acceleration in active galactic nuclei-Coexistence of monoenergetic, power law and Maxwell-Boltzmann electron energy spectra

R. Schlickeiser

Max-Planck-Institut für Radioastronomie, Auf dem Hügel 69,
5300 Bonn 1, West Germany

SUMMARY

The temporal and spectral evolution of relativistic electron momentum spectra in active galactic nuclei is discussed. It is shown that
combination of first and second order Fermi acceleration with synchrotron and inverse Compton radiation losses explains the coexistence of
monoenergetic, Maxwellian and power law with cutoff, particle spectra
in these objects. In particular, a viable mechanism has been found to
establish relativistic thermal particle distribution functions.

INTRODUCTION

It is commonly assumed that active galactic nuclei (quasars, Seyfert
galaxies, BL Lac objects) are powered by the same type of central engine (Rees 1978). Smaller versions of this power house may be responsible for particle acceleration in compact galactic objects (Rees 1981).
Whereas in Seyfert galaxies and "thermal" quasars, surrounding gas
screens our view of processes near the central power house, we believe
that in BL Lac objects and red quasars of moderate redshift we directly
can observe particle acceleration processes in active galactic nuclei.
Existing observations of nonthermal radiation from radio to X-ray
wavelength of these objects do not indicate the presence of a single
unique relativistic electron energy distribution. They rather indicate
the occurence of monoenergetic, power law and relativistic Maxwellian
spectra (Owen et al. 1980). Multifrequency observations of bursting and
flaring objects are particularly interesting. During the violent outburst of the BL Lac object AOO235+164 (Rieke et al. 1976, 1982) the
underlying relativistic electron spectrum developed from a monoenergetic
distribution $N(E) \propto \delta(E-E_O)$ during the early outburst phases to an $E^{-1.8}$-
power law spectrum with sharp cutoff at E_O at later stages. Such power
laws with sharp cutoffs, which resemble themselves in infrared optical
synchrotron emission, are seen in a variety of BL Lac and quasar spectra
(Rieke et al. 1982, Bregman et al. 1981, Ennis et al. 1982). The flare
of 3C273 (Robson et al. 1983) can be interpreted as an evolution of an
initial monoenergetic electron distribution towards a Maxwellian at late
times.
In this work we would like to emphasize that such a diversity of
electron spectra follows as a natural consequence from a simple model
(Schlickeiser 1984) of the evolution of electron particle spectra which
combines Fermi acceleration processes and synchrotron and inverse
Compton losses.

BASIC EQUATIONS

Parker and Tidman (1958) have pointed out that Fermi acceleration is an intrinsic property of any sufficiently agitated magnetized plasma of energetic particles. The momentum spectrum of particles undergoing Fermi acceleration due to irregularly moving magnetized fluid elements can be studied in terms of a diffusion equation in momentum space (Tverskoi 1967)

$$\frac{\partial f}{\partial t} = \frac{1}{p^2} \frac{\partial}{\partial p} \left(p^2 D(p) \frac{\partial f}{\partial p} \right) + Q(p,t) \tag{1}$$

where p is the particle momentum, $4\pi p^2 f(p,t)$ the number of particles per unit momentum and unit volume, $D(p) = a_2 p^2$ is the momentum diffusion coefficient with $a_2 = V_A^2/(9K_{11})$ in the case of scattering off Alfven waves (Skilling 1975) (V_A: Alfven velocity, K_{11}: spatial diffusion coefficient of plasma). $Q(p,t)$ represents sources and sinks of particles. In the presence of ambient magnetic, photon and matter fields, the particles also undergo other types of interactions (synchrotron radiation, inverse Compton scattering and/or ionization), and they may escape from the acceleration region. These effects can be incorporated into equation (2) by adding the momentum change term, $\dot{p}$, for a single particle, and an escape term f/T, where T is the escape lifetime,

$$\frac{\partial f}{\partial t} = \frac{1}{p^2} \frac{\partial}{\partial p} \left(p^2 D(p) \frac{\partial f}{\partial p} \right) - \frac{1}{p^2} \frac{\partial}{\partial p} (p^2 \dot{p} f) - \frac{f}{T} + Q(p,t) \tag{2}.$$

Special versions of equation (2) have been applied to many astrophysical sites, including solar flares (Parker 1957, Ramaty 1979) and supernova remnants (Chevalier et al. 1978, Melrose 1980, Cowsik 1979).

Recently, another aspect has been added to the theory. Near compact objects, due to mass motions at speeds up to 0.03c, we expect the creation of strong shock waves, so that, besides second order Fermi acceleration off magnetized fluid elements particles may gain energy by first-order Fermi acceleration off these strong shocks (Bell 1978, Axford 1980, Drury 1983),

$$\dot{p} = a_1 p \tag{3}$$

with $a_1 \tilde{=} \varkappa V_s^2/(4K_{11})$, where V_s is the shock speed, and $\varkappa$ the volume filling factor of shock waves. Combining this gain mechanism with synchrotron and inverse Compton losses, $\dot{p} = -\beta p^2$, where $\beta \propto (W_H + W_{Ph})$ depends on the ambient photon and magnetic field energy densities, yields for

$$\frac{\partial f}{\partial t} = \frac{a_2}{p^2} \frac{\partial}{\partial p} \left(p^4 \frac{\partial f}{\partial p} \right) - \frac{1}{p^2} \frac{\partial}{\partial p} (p^3 [a_1 - \beta p] f) - \frac{f}{T} + Q(p,t) \tag{4}$$

Complete analytical solutions of both, the steady-state and time-dependent equation (4), are obtained in terms of confluent hypergeometric function (Schlickeiser 1984). Consider each case in turn.

STEADY-STATE SOLUTION

In the steady-state "acceleration phase", with a_1, a_2, β and T different from zero and positive, and $Q(p,t)$ approximated as steady mono-momentum source, $Q(p,t) = q_0 \delta(p-p_0)$, the steady-state solution of (4) can be expressed as

$$N(p)=4\pi p^2 f(p) = \begin{cases} c_1 p^{\mu+((1+a)/2)} & \text{for } p \le p_o \ll p_c \\[2em] c_2 \exp\left(-\dfrac{p}{p_c}\right) p^{\mu+\frac{1-a}{2}} \dfrac{\Gamma(\mu-\frac{a+3}{2})}{\Gamma(1+2\mu)} \, U\left(\mu-\dfrac{a+3}{2}, \, 1+2\mu, \, \dfrac{p}{p_c}\right) & \\[1em] & \text{for } p \ge p_o \end{cases} \tag{5}$$

with $a=a_1/a_2 = 9\varkappa V_S^2/(4\cdot V_A^2) \simeq \varkappa M^2$ (M: shock wave's Mach number), $U(A,B,z)$ denotes Kummer's function,

$$\mu = [(a_2 T)^{-1} + ((a+3)/2)^2]^{1/2} \tag{6},$$

$$p_c = a_2/\beta = 1.8 \ 10^3 \frac{MeV}{C}(V_A/100 \text{ km s}^{-1})^2 (K_{11}/10^{24} \text{cm}^2 \text{s}^{-1})^{-1} \left(\frac{W_H + W_{Ph}}{10^{-7}\text{erg cm}^{-3}}\right)^{-1} \tag{7}.$$

Solution (5) holds for the case where particles are injected at momentum p_o much less than p_c. The particle spectrum (5) is a power law distribution with a Wien bump at momentum $p_M=(2+a)p_c$ (see Figure 1) which is that value of the momentum for which the time scale for first-order Fermi acceleration ($\tau=a_1^{-1}$) equals that for radiation losses ($\tau=\beta^{-1} p^{-1}$), and which depends on the Mach number of the shock waves, the shock waves volume filling factor, the Alfven speed, the spatial diffusion coefficient, the magnetic field and background photon energy densities in the plasma.

The spectral index of the power law particle distribution between $p_o \le p \ll p_c$ below the Wien bump is determined by the ratio of the acceleration time scale to the particle escape lifetime from the

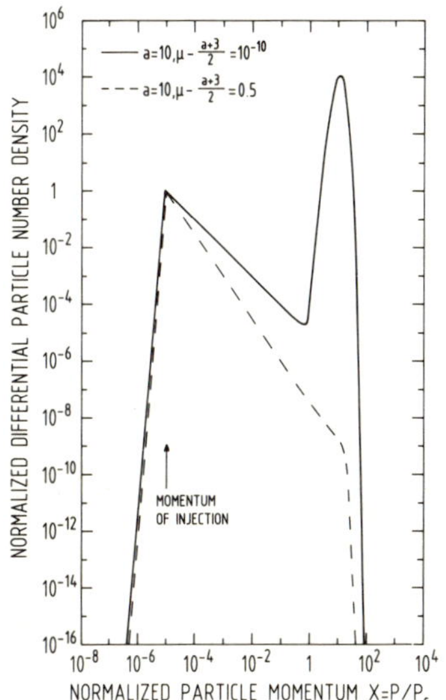

FIG. 1. Resulting particle momentum spectra in the steady-state acceleration phase for very long ($\mu-\frac{a+3}{2} = 10^{-10}$) and moderately long ($\mu-\frac{a+3}{2} = 0.5$) escape lifetimes. Particles are steadily injected with momentum $x_o = p_o/p_c = 10^{-5}$, and a value of $a = 10$ has been assumed.

plasma blob, T. In case of a very long ($T \gg a_2^{-1}$) escape lifetime, this spectral index approaches the value -1, the Wien bump piles up steeply at p_M, and cuts off exponentially beyond p_M (see Figure 1), so that the resulting particle momentum spectrum is nearly a monomomentum distribution. We suggest that this spectrum has been observed during the initial outburst phase of AO 0235+164 (Rieke et al. 1976, 1982) and 3C273 (Robson et al. 1983).

In case of moderately long ($T \approx a_2^{-1}$) escape lifetimes, the spectral index of the power law may obtain any value smaller than -1, depending on the exact ratio of escape lifetime to acceleration time scale, and the spectrum cuts off exponentially beyond p_M without a pile up (see Figure 1). We suggest that such a spectrum has been observed in a variety of BL Lac and quasar spectra (Rieke et al. 1982, Ennis et al. 1982), in particular in QSO 1413+135 (Bregman et al. 1981) and the late phases of the outburst of AO 0235+164 (Rieke et al. 1976, 1982).

TIME-DEPENDENT SOLUTION

The time-dependent equation (4) in the acceleration phase, with a_1, a_2, β and T different from zero and positive, and $Q(p,t)$ taken as impulsive source $Q(p,t) = N(p,t=0)/(4\pi p^2)$ is solved by expanding its solution in eigenfunctions of the Hermitean momentum operator. Substituting $x=p/p_c$ (see (7)), $a=a_1/a_2$, $f(x,t) = F(x,t)\exp(-t/T)$, $y=a_2 t$ yields

$$\frac{\partial F}{\partial y} = x^{-2} \frac{\partial}{\partial x} [x^4 (\frac{\partial F}{\partial x} + F) - ax^3 F] \tag{8}.$$

Developing

$$F(x,y) = \sum_v c_v(y) \chi_v(x) \tag{9}$$

in eigenfunctions $\chi_v(x)$ of the self-adjoint momentum operator

$$\frac{d}{dx} [x^{4-a} e^x \frac{d\chi_v}{dx}] + (4x-3a+v) x^{2-a} e^x \chi_v = 0 \tag{10}$$

with eigenvalues v, gives

$$c_v(y) = c_v(y=0) e^{-vy} = e^{-vy} \int_0^\infty d\xi F(\xi,t=o)\chi_v(\xi) \tag{11}.$$

Evidently, the behaviour at late times ($y \gg 1$) is determined by the smallest eigenvalue v (fundamental mode). Substituting $\chi_v(x)=x^K \exp(-x) U_v(x)$ in (10) yields

$$x \frac{d^2 U_v}{dx^2} + (4+2K-a-x) \frac{dU_v}{dx} - (K-a) U_v = 0 \tag{12}$$

where K depends upon v and is given by the solution of

$$K^2 + (3-a) K + v - 3a = 0 \tag{13}.$$

Equation (13) defines two values of K whose sum is equal to (a-3) for any value of v. Without loss of generality we restrict K such that

$$\text{ReK} \geq \frac{a-3}{2} \tag{14}.$$

Equation (12) is solved by the confluent hypergeometric functions M(K-a, 2K+4-a,x) and U(K-a, 2K+4-a,x) (Abramowitz and Stegun 1964), except if K-a=-m (m: positive integer), in which case equation (12) has to be treated separately. The boundary conditions of the problem (see e.g. Pomraning 1968) restrict the allowable values of K to

$$K = \frac{a-3}{2} + i\sqrt{v - \frac{(3+a)^2}{4}} \tag{15}$$

with v real and $v \geq \frac{(3+a)^2}{4}$. The allowable values of v form a continuum on the real axis with $v \geq \frac{(3+a)^2}{4}$ and corresponding eigenfunctions

$$\chi_v(x) = B(v) \, x^{\frac{a-3}{2} + i\Omega} \, e^{-x} \, U(-\frac{a+3}{2} + i\Omega, 1+2i\Omega, x)$$

$$\Omega = [\, v - \frac{(3+a)^2}{4} \,]^{\frac{1}{2}}, \quad v \geq \frac{(3+a)^2}{4} \tag{16}$$

where B(v) is the normalization constant.

As noted earlier, the cases K-a=-m (m: positive integer) require special treatment. For a given value of a, the restriction (14) limits the allowable values of m to all integers smaller or equal (a+3)/2; e.g. for a=0 one finds m=0,1, whereas for a=4.1 one finds m=0,1,2,3. Equation (13) with K=a-m yields

$$v = -m\,(m-a-3) \tag{17}$$

which is zero for m=0 and positive for m integer and 0<m<(a+3)/2. So the smallest eigenvalue v=0 is obtained for m=0 with the corresponding normalized eigenfunction

$$\chi_0(x) = [\Gamma(3+a)]^{-1/2} \, x^a \, e^{-x} \tag{18},$$

which is the fundamental mode of the problem. Using (18) in (11) and (9) shows that at late times y>>1 ($t>>a_2^{-1}$) $N(p,t) = 4\pi p^2 f(p,t)$ approaches

$$N(p,t) \xrightarrow[t>>a_2^{-1}]{} \frac{1}{\Gamma(3+a)} \, (\frac{p}{pc})^{2+a} \, \exp[-\frac{p}{pc} - \frac{t}{T}] \int_0^{\infty} d\,\xi \xi^{a-2} e^{-\xi} N(p_c \xi, t = 0) \tag{19},$$

a modified (a≠0) ultrarelativistic Maxwellian distribution of mean momentum $<p>= (3+a)p_c$ whose absolute height decays as exp(-t/T). Under the action of synchrotron und inverse Compton radiation losses competing with Fermi acceleration off magnetohydrodynamic waves as well as shock waves, any initial particle distribution at t=0 develops at late times into a modified Maxwell-Boltzmann distribution. In case of no Fermi acceleration at shock waves (a=0), spectrum (19) reduces to a pure ultrarelativistic Maxwellian; the Rayleigh-Jeans part of the spectrum determines the ratio of Fermi acceleration at shock waves to Fermi acceleration off magnetohydrodynamic waves.

The mean temperature of the modified Maxwellian (19) is (mc^2 = 511 KeV)

$$\theta = (m^2 c^4 + (3+a)^2 \, p_c^2 \, c^2)^{1/2} - mc^2 \tag{20},$$

which for large enough p_c, see (7), can be ultrarelativistic. This point is very important for the equilibrium behaviour of relativistic electron-positron plasmas. By comparing "Maxwellianisation" time scales by elastic Møller and Bhabha scattering with bremsstrahlung and synchrotron radiation loss time scales (Rees 1981, Gould 1982, Stepney 1983, Araki and Lightman 1983) it was concluded that a relativistic Maxwellian cannot be maintained. This was a very unsatisfactory result since a number of interpretations of gamma-ray burst sources (e.g.

Cavello and Rees 1978, Liang 1982) and active galactic nuclei (Jones
and Hardee 1979) require the existence of thermal relativistic $e^{\pm}$-
plasmas. Here, we have emphasized that Fermi interaction with MHD
plasma waves and shock waves in connection with radiation losses establish a Maxwellian distribution which for appropriate parameter values
may be of ultrarelativistic temperature. We also suggest that this
thermal distribution has been seen at late times in the recent mm/in-
frared flare of 3C273 (Robson 1983).

In conclusion, consideration of first and second order Fermi acceleration in combination with synchrotron and inverse Compton radiation
losses explains the coexistence of monoenergetic, Maxwellian, and power
law with cutoff, particle spectra in compact objects. In particular, a
viable mechanism has been found to establish relativistic thermal particle distribution functions.

REFERENCES

Abramowitz M, Stegun IA 1964 Handbook of Mathematical Functions,
    National Bureau of Standards, Washington
Araki S, Lightman AP 1983 Ap J 269:49
Axford WI 1980 Proc. IAU/IUPAP Symp No 94, Dordrecht, Reidel, p. 339
Bell AR 1978 MNRAS 182: 147
Bregman JN et al. 1981 Nature 293: 714
Cavallo G, Rees MJ 1978 MNRAS 183: 359
Chevalier RA, Oegerle WR, Scott JS 1978 Ap J 222: 527
Cowsik R 1979 Ap J 227: 856
Drury LOC 1983 Rept. Progr. Phys. 46: 973
Ennis DJ, Neugebauer G, Werner M 1982 Ap J 262: 460
Gould RJ 1982 Ap J 254: 755
Jones TW, Hardee PE 1979 Ap J 228: 268
Liang EPT 1982 Nature 299: 321
Melrose DB 1980 Plasma Astrophysics II., Gordon & Breach, New York
Owen FN, Spangler SR, Cotton WD 1980 Astr. J 85: 351
Parker EN 1957 Phys. Rev. 107: 830
Parker EN, Tidman DA 1958 Phys. Rev. 111: 1206
Pomraning GC 1968 Ap J 152: 809
Ramaty R 1979 In: Particle Acceleration Mechanism in Astrophysics,
    J Arons, C McKee, C Max (eds) Institute of Physics, New York, p. 135
Rees MJ 1978 Physica Scripta 17: 193
Rees MJ 1981 Space Sci Rev 30: 87
Rieke GH, Grasdalen GL, Kinmaw TD, Hintzen P, Wills BJ, Wills D 1976
    Nature 260: 754
Rieke GH, Lebofsky MJ, Wisniewski WZ 1982 Ap J 263: 73
Robson EI et al. 1983 Nature 305: 194
Schlickeiser R 1984 Astr. Ap. in press
Skilling J 1975 MNRAS 172: 557
Stepney S 1983 MNRAS 202: 467
Tverskoi BA 1967 Sov Phys JETP 25: 317

Electron beams in compact sources

P. F. Browne

Department of Pure and Applied Physics, University of Manchester
Institute of Science and Technology, Manchester M60 1QD, England.

SUMMARY

Electrons having ultrarelativistic drift velocities can be
produced in the final stages of magnetic pinch in a collisionless
plasma, Lorentz factors being as high as 10^6 for reasonable
values of axial current J , electron density n_e , and current
sheath radius R . In the current sheath electrons move in an
azimuthal magnetic field B_φ and a radial electrostatic field E_r
such that $E_r \simeq B_\varphi$. Their synchrotron emission whilst in this
region will be anisotropic. On entering the surrounding region
where $B_\varphi \propto r^{-1}$ the emission becomes isotropic. The intensity of
the emission and the spectrum are as observed. Granted such an
electron beam, an explanation is available for the following
phenomena: (i) rotation of polarization position angle through
$> 360°$ either smoothly or in sudden jumps; (ii) a large intrinsic
component of red-shift (due to small angle electron scattering of
comoving photons); (iii) 50 kev absorption in γ-ray burst sources
due to cyclotron resonance absorption strongly blue-shifted; (iv)
properties of broad emission lines, based on a Cerenkov excitation
mechanism.

1. INTRODUCTION

Elsewhere (Browne, 1984) I have proposed a mechanism by which
astrophysical magnetic fields can be generated and also dissipated.
Differential rotation of a plasma associated with a hierarchical
structure of vortices is central to the generation. In no other
way can one reconcile the large scale on which fields are ordered
with the small scale on which induction permits transfer of energy
to or from the field. The fields arising from differential
rotation due to a "viscous battery" (Browne, 1968) are proportional
to vorticity, at least in the early stages of development. In
association with a vortex tube one expects initially an azimuthal
current density j_φ with axial magnetic field B_z . But twisting
of field lines causes (j_φ , B_z) to evolve toward (j_z , B_φ).
Whereas the former field configuration expands due to magnetic
pressure $B_z^2/8\pi$, the latter contracts due to B_φ pinch. In the
final stages of B_φ pinch in a collisionless plasma inductive
maintenance of the current requires that electrons acquire
ultrarelativistic drift velocities in a common direction. Also
operative is a Fermi acceleration process due to repeated reflect-
ions of electrons and protons by the contracting B_φ field. For
long, acceleration of charges has been associated with field line
reconnection, particularly in solar flares. Since the sheet
current between opposite approaching magnetic fields splits into

current filaments as a result of tearing mode instability (Matthaeus
and Montgomery, 1981), and the current filaments then undergo B_φ
pinch, acceleration during field line reconnection is again the
same phenomenon.

The theoretical model is outlined in sections 2 and 3, being
more fully treated elsewhere (Browne, 1984). The parameters for
a pinched current channel which yields Lorentz factors of order
10^6 are determined. The synchrotron flux density and spectrum
from such a current channel then is derived, and shown to be in
agreement with observation for the same parameters.

In sections 4 - 7 phenomena which can be explained by an
approaching ultrarelativistic electron beam (and not satisfactorily
by other means) are considered.

2. PINCH OF A CURRENT CHANNEL

The classical vortex has constant vorticity ω_0 out to $r = R$
and zero vorticity beyond R . The associated velocity field then
is $u_\varphi = \omega_0 r/2$ for $r < R$ and $u_\varphi = \omega_0 R^2/2r$ for $r > R$.
Thus angular momentum per unit mass is constant outside R. The
"viscous battery" provides a magnetic field $\underline{B}$ proportional to
vorticity $\underline{\omega}$, and without dissipation $\underline{B}$ and $\underline{\omega}$ satisfy identical
equations so that their subsequent evolution is the same.

As a first step, let the lines of B_z accumulate at $r \simeq R$,
so that we have a vortex tube. If radial expansion is nonuniform
along the tube, field lines will be twisted into helices, thus
introducing an azimuthal field component B_φ. Associated with
B_φ is an axial component of current density j_z . Thus twisting
causes $(j_z, B_z) \rightarrow (j_z, B_\varphi)$. If the former configuration dominates
the field expands. If the latter dominates the field contracts
(B_φ pinch). There is an intermediate force-free configuration
for which $\underline{j} \times \underline{B} = 0$.

Vortices occur on vastly different scales, and indeed are
manifest by radio jets on vastly different scales. One envisages
that energy in large scale differential rotation is transferred
via a hierarchy of vortices to small scale differential rotation
where the inductive time scale is such that it can be converted
into magnetic field energy. An interesting feature of an array
of vortex tubes each with dominant (j_z, B_φ) fields is that their
mutual attraction forces together opposite B_φ fields in the
inter-tube regions. Thus a field line reconnection situation
develops throughout the volume of the array. As explained in the
introduction the return current will be concentrated into an array
of current filaments. Thus both the outgoing and the return current
are concentrated into channels of small cross sectional area.

In situations in which an inductively maintained current is being
carried by a diminishing number of charges, such charges as remain
acquire increased drift velocity which ultimately becomes ultrarelativistic. Synchrotron dissipation then replaces Ohmic loss.

Let ΔR be the thickness of the cylindrical current sheath.
Initially ΔR may be characteristic of the vortex tube, but as
pinch progresses into a field-free core ΔR will tend to the
depth to which a magnetic field penetrates a perfect conductor.
On this basis Rosenbluth (1957) adopts

$$\Delta R = (4\pi n_e a)^{-\frac{1}{2}} \tag{1}$$

where $a = e^2/mc^2$, the electron radius. If $n_e = 10^{-2} cm^{-3}$ then

$\Delta R = 5.3 \times 10^{6}$ cm.

The axial current J implies that electrons have axial drift velocity βc , where

$$J = j_z 2\pi R \Delta R \qquad ; \qquad j_z = e\beta c \gamma n_e \qquad (2)$$

Here n_e is the density of electrons in their rest frame, so that γn_e is the density in the Earth frame due to Lorentz contraction of volume element. For distances r such that the return current can be neglected the magnetic field is given by

$$B_\varphi = 2J/rc \qquad (r > R) \qquad (3)$$
$$B_\varphi = 0 \qquad (r < R - \Delta R)$$

By combining (1), (2) and (3) one obtains

$$B_\varphi^2(R)/4\pi = \gamma^2 n_e mc^2 \qquad (4)$$

Bearing in mind that the electron density is γn_e, (4) expresses that energy is equipartitioned between field and particles in the current channel. Again adopting $n_e = 10^{-2}$ cm^{-3}, a Lorentz factor of $\gamma = 10^6$ is obtained from (4) if $B_\varphi(R) = 320$ G. The time scale for compact source variability at the highest frequencies suggests $R \simeq 10^{13}$ cm. Then, according to (3) $B_\varphi(1 \text{ pc}) \simeq 10^{-3}$ G. But a considerable smaller value for R may be envisaged.

Charges trapped within the core region will be reflected by the contracting B_φ field, which essentially defines a contracting cylindrical wall. If the radial velocity of a charge is reversed in the rest frame of the contracting wall (velocity $\dot{R}$) then in the Earth frame the velocity of the charge changes from $+ \beta_r c$ to $- \beta_r c - 2(1 - \beta_r^2)\dot{R}$ on neglect of terms of order $\dot{R}/c$ and higher. Then the energy gained by a proton of mass m_i at each reflection will be $- 2\beta_r \gamma_r R m_i c$. Multiplying by $\beta_r c/2R$, one finds the gain per unit time, and equating this to $m_i c^2 (d\gamma_r/dt)$ one obtains a differential equation for β_r whose solution is

$$\beta_r \gamma_r = KR^{-1} \qquad (5)$$

where K is a constant. Thus pinching of the current channel will accelerate the protons trapped within the core by the Fermi process.

The $\underline{j} \times \underline{B}$ force causing pinch acts on electrons, whereas the force opposing the pinch is the excess pressure exerted by positive ions trapped within the core. The separation of electrons and positive ions is limited by a radial electrostatic field E_r due to polarization of the plasma. For constant $\dot{R}$ one has

$$2\beta_r^2 \gamma_r^2 n_i m_i c^2 = B_\varphi^2(R)/8\pi \simeq E_r^2/8\pi \qquad (6)$$

From (4) and (6) it follows that

$$\beta\gamma = 4(m_i/m)^{\frac{1}{2}} \beta_r \gamma_r = 171 \beta_r \gamma_r \qquad (7)$$

Thus electron drift velocity is more relativistic than proton (or electron) radial oscillatory velocity due to the Fermi mechanism. One notes from (6) that $E_r \simeq B_\varphi$, implying that electrons have an electric drift with velocity close to c . Protons do not have this motion because "wall pressure" produces a counter-drift.

The radial extent Δr_E of the field E_r is such as to bring to rest a proton of velocity $\beta_r c$. If b_e is the gyroradius of an electron of velocity βc in the field $B_\varphi(R)$, and b_i is that of a proton of velocity $\beta_r c$ in the field $B_\varphi(R)$, then it is readily shown that

$$b_e \simeq \Delta R \quad ; \quad b_i \simeq \Delta r_E = (m_i^{\frac{1}{2}}/4m^{\frac{1}{2}})b_e \qquad (8)$$

Thus b_e is smaller than Δr_E by a factor of 11 whereas b_i is approximately equal to Δr_E.

3. SYNCHROTRON FLUX AND SPECTRUM

The flux and spectrum of synchrotron emission expected from a pinched current channel of the above type is of interest. Only electrons moving in the thin current sheath where both fields E_r and B_φ are present will have collimated motions, and hence will radiate anisotropically. Electrons outside this sheath will gyrate in the nonuniform magnetic field (3), generating isotropic emission. Assuming that all electrons have much the same energy γmc^2 (ie neglecting radiative losses, and assuming that electrons have been injected into the field $B_\varphi \propto r^{-1}$ with the γ value appropriate for the smallest value of R), one may estimate the flux and spectrum as follows.

Writing $\nu_c = eB_\varphi(r)/2\pi mc$, one notes that synchrotron emission from electrons with Lorentz factor γ occurs predominantly at

$$\nu \sim \gamma^2 \nu_c \propto \gamma^2 B_\varphi \propto B_\varphi \propto r^{-1} \qquad (9)$$

Thus emission in frequency band $\nu \rightarrow \nu + \delta\nu$ comes from electrons in the shell $r \rightarrow r + \delta r$. Let the flux density at Earth in this band be $\mathcal{F}_\nu \delta\nu$. Then for a source at distance d radiating isotropically one has

$$\mathcal{F}_\nu \delta\nu 4\pi d^2 = P_s 2\pi r \delta r L \gamma n_e \qquad (10)$$

where P_s is the power radiated per electron and L is the length of the cylindrical source. Using

$$P_s = 2a^2 c \gamma^2 B_\varphi^2/3 \quad ; \quad |\delta\nu/\nu| = \delta r/r \qquad (11)$$

one finds from (10) that

$$\mathcal{F}_\nu = \frac{4a^2 L n_e J^2 \gamma^3}{3cd^2 \nu} \qquad (12)$$

where B_φ has been eliminated by use of (3).

In section 2 the following values were obtained for the B_φ pinch: $n_e = 10^{-2}$ cm^{-3}, $\gamma = 10^6$, and $B_\varphi(R) = 320$ G. Hence $J/c = \frac{1}{2}RB_\varphi(R) = 1.6 \times 10^{15}$ emu, again taking $R = 10^{15}$ cm. The values $L = 3 \times 10^{14}$ cm and $d = 10^{28}$ cm (for Hubble red-shift of order unity) then imply that $\nu \mathcal{F}_\nu = 2 \times 10^{-11}$ erg s^{-1} cm^{-2}. This agrees with observation for compact sources. The model predicts a spectrum $\mathcal{F}_\nu \propto \nu$ which also agrees with observation, after subtraction of the so-called "blue-uv bump" (Malkan and Sargent, 1982; Cruz-Gonzalez and Huchra, 1984). However, it should be noted that a return current will cause B_φ to fall off more rapidly than $B_\varphi \propto r^{-1}$.

4. POLARIZATION POSITION ANGLE

In the case of end-on viewing of an ultrarelativistic electron beam, the approaching electrons move in a magnetic field B_φ which appears circular or circumferential. The acceleration of the electrons can occur in any radial direction transverse to the line of sight. For an electron path coincident with the axis there would be no preference. For a helical electron path the direction of acceleration (and hence polarization) will precess about the axis, having at any instant the direction specified by the φ coordinate of the electron.

Large rotations of polarization position angle (χ) are a striking characteristic of blazars, being seen at radio frequencies (Ledden and Aller, 1979; Aller et al., 1981), at infrared frequencies (Holmes et al., 1984), and at optical frequencies (Angel and Moore, 1980). Rotations exceeding 360° are common. Sometimes χ increases with time at a constant rate, and on other occasions sudden jumps are observed. This behaviour is explained by the B_φ pinch model. Presumably the electrons emerge from the pinched current channel in bunches, the φ coordinate of the bunch specifying χ. The large values for $\Delta\chi$ are to be expected.

For on-axis viewing there will be no preferred direction of polarization. For slightly off-axis viewing electrons deflected in the direction toward the line of sight will contribute most strongly to the observed flux and their polarization will therefore dominate. This distinction between on-axis and slightly off-axis viewing accounts very well for the observation of two classes of source. For about half the sources χ has no preferred value, $\Delta\chi$ is very large, and the apparent luminosity is high. For the remaining sources χ has much the same value from outburst to outburst, $\Delta\chi$ is relatively small, and apparent luminosity is low (Angel and Stockman, 1980; Wardle et al., 1984).

5. LARGE INTRINSIC RED-SHIFT

The ultrarelativistic motion of the approaching electrons increases their column density by the factor γ, so that the condition for scattering of the nearly comoving photons is

$$\gamma n_e L \sigma_T > 1 \tag{13}$$

where $\sigma_T = 8\pi a^2/3 = 6.65 \times 10^{-25} \mathrm{cm}^2$, the Thomson cross section. If $n_c = 10^{-2} \mathrm{cm}^{-3}$, $\gamma = 10^6$, and $L > 30$ pc this condition is met. Column densities in the absorption line regions of QSOs can reach $10^{20} - 10^{21} \mathrm{cm}^{-2}$ for hydrogen, which is still a factor of 10^{-3} less than σ_T^{-1}. But of course most H atoms may be ionized (eg. Sargent et al., 1979). Welter et al. (1984) find Faraday rotation measures of typically 55 rad m^{-2} for these sources. With the threshold column density for electron scattering such a rotation measure implies that $\langle B_z \rangle = 10^{-2}$ G , and of course this field would be larger for smaller column densities.

Consider Compton scattering by the approaching ultrarelativistic electrons, of photons which are nearly comoving. Let a photon of energy $h\nu$ be scattered from direction θ to direction θ' with respect to the velocity vector βc of the electron, the energy after scattering being $h\nu'$. Then, defining $z_i = \nu/\nu' - 1$, conservation of momentum and energy suffices to show that the red-shift z_i is given by

$$z_i = \frac{\beta(\cos\theta - \cos\theta') + (h\nu/\gamma mc^2)\left[1 - \cos(\theta - \theta')\right]}{1 - \beta\cos\theta} \qquad (14)$$

Since $h\nu \ll \gamma mc^2$ the second term in the numerator can be neglected. Due to relativistic aberration the scattered radiation is confined to a cone of angle $\simeq \gamma^{-1}$ about the direction of β. Hence θ' will be small. In order to obtain a red-shift ($z_i > 0$) one requires that $\theta' > \theta$, so that θ also must be small. Thus (14) reduces to

$$z_i \simeq \frac{\theta'^2 - \theta^2}{\gamma^{-2} + \theta^2} \qquad (15)$$

Suppose that $\theta = 0$, implying that a point-like source of radiation is more distant than the column of scattering electrons. Then $z_i \simeq \gamma^2 \theta'^2$. Relativistic aberration then constrains z_i to be of order unity, which is a large red-shift. The radiation one receives all has been scattered by electrons moving nearly along the line of sight, and if $\theta = 0$ the incident photons comove with these electrons. Motion of the source causes flux density in direction θ relative to the motion to be boosted by a factor $f^3(\theta)$, where $f(\theta) = \gamma^{-1}(1 - \beta\cos\theta)^{-1}$ for a flat spectrum. Due to this effect sources having $z_i > 1$ will have apparent luminosity reduced by a factor $(1 + z_i)^{-3}$. Arp (1983) has noted such an effect.

Evidence for associations of sources having quite different red-shifts has been accumulating (Burbidge, 1979; Arp, 1983; 1984). Of particular note are two chains of QSOs, each spanning about 40 arcmin and each having six members. In one case the sextet spans a galaxy, NGC 3384 (Arp et al., 1979). In the other case the sextet splits into two triplets whose well ordered properties are remarkably similar (Arp and Hazard, 1980). I wish to point out that a rationale exists for chains of QSOs, and indeed other sources.

Chains of galaxies are quite common. Clusters of galaxies also appear to be linked in a chain-like fashion in superclusters (Oort, 1983). On a smaller scale Kronberg et al. (1984) have found chains of knots of 6 cm emission in the nucleus of M82. On a still smaller scale Herbig-Haro sources which are embedded in bipolar nebulae surrounding protostars show remarkable linear alignments (Schwartz, 1983). Elsewhere (Browne, 1984) I have argued that the underlying cause for all of these alignments is a vortex.

6. γ-RAY BURST SOURCES

The B_ϕ pinch of a current channel is essentially a transient phenomenon in which magnetic field energy is dissipated into ultrarelativistic particle energy and synchrotron radiation. The duration will be proportional to J^2 if the rate of pinch $\dot{R}$ is constant. At one extreme QSOs and blazars could have lifetimes of perhaps 10^6 y. At the other extreme one notes hard X-ray spikes during solar flares, the durations being 50 ms - 1 s (Kiplinger et al. 1983). Current filaments produced by tearing mode instability of a current sheet during field line reconnection may be involved in the latter case. Intermediate between these extremes are γ-ray bursts with durations 0.1 - 100 s (Norris et al., 1982). Usually no permanent detectable source can be found, but in one case the source has been identified with the supernova remnant N49 in the LMC at distance 55 kpc (Cline et al., 1982).

Of particular interest is the absorption line at $\simeq 50$ kev in
the spectra of γ-ray bursts (Mazets et al., 1981). It has been
attributed to cyclotron resonance in a magnetic field as strong as
$\simeq 10^{13}$ G, believed to occur in a neutron star. A more plausible
explanation is cyclotron resonance in a beam of approaching
ultrarelativistic electrons. One envisages that the electrons
follow helical paths whilst moving along lines of a magnetic field
B_z parallel to the beam and to the line of sight. If βc is the
velocity of the guiding centres, the line should be seen at

$$\nu = 2\gamma\nu_c = \gamma eB_z/\pi mc \qquad (16)$$

For $\nu = 1.2 \times 10^{19}$ Hz (corresponding to 50 kev) one requires that
$\gamma B_z = 2 \times 10^{12}$ G . It is possible that γ is as large as say 10^9,
in which case one requires merely $B_z \simeq 2$ kG. It should be noted
that the optical depth of the electron column for cyclotron
absorption is likely to exceed unity, being given by $4\pi acn_e Lk^{-1}\zeta$,
where $k = 1 + 4\pi^2(\nu_c - \nu)^2\zeta^2$, ζ being the time between
phase interuptions due to any cause (Robinson, 1973).

A broad emission line also is seen at 430 ± 50 kev in most
burst spectra. This has been attributed to two-photon annihilation
of an electron and a positron at 511 kev in the rest frame. The
red-shift of this line is expected, because the pairs will have
been produced by interaction of the radiant flux with charges
travelling in the opposite direction (initially protons and
eventually pairs).

7. CERENKOV EXCITATIONS

A consequence of a beam of electrons with Lorentz factors as
large as $\gamma \simeq 10^6$ is strong Cerenkov emission at frequencies for
which the threshold condition can be satisfied. The phase velocity
of radiation c/μ (where μ is refractive index) exceeds particle
velocity βc if $\mu\beta > 1$ or $\gamma^2(\mu^2 - 1) > 1$. If ν_{ij} is the
frequency of a resonance line of oscillator strength f_{ij} and
broadened linewidth Γ , then for $\nu \simeq \nu_{ij}$ the threshold condition
for Cerenkov emission becomes (Browne, 1983)

$$N_i\gamma^2 > \Gamma m\nu_{ij}/e^2 f_{ij} \qquad (17)$$

N_i being the density of atoms in state i Doppler tuned within Γ.
Because Cerenkov excitation takes the form of very intense short
duration pulses, broadening of the transition due to AC Stark effect
should be considered (Browne, 1983). In the case of $L\alpha$ radiation
with Γ taken to be the natural line width the threshold condition
is satisfied by $N_{ij} > 5 \times 10^{16}$ cm^{-3} , and if $\gamma \simeq 10^6$ then one
requires only $N_i > 5 \times 10^4$ cm^{-3} .

Strong resonance radiation can excite line emissions in a single
two-photon transition with probability comparable to that for
successive single-photon transitions at frequencies displaced from
resonance by a slight amount. Competition between the two processes
can be seen in the experiments of Wiggins et al. (1978). The more
intense the excitation the broader will be the emission line. In
addition the profile will peak toward the red wing due to the AC
Stark effect. Normally this effect splits the line into two peaks,
but with Cerenkov excitation the blue-shifted peak is weak or
absent due to anomalous dispersion at the line frequency.

Properties of the broad line emissions are most easily explained

by the Cerenkov excitation mechanism. One notes particularly the
following:

(i) Because of the threshold condition condition (17) the appearance of the broad lines is somewhat erratic and unpredictable. Thus
BL Lacs are distinguished from QSOs by the absence of lines. FeII
lines, requiring higher γ due to smaller N_i, tend to be strong in
Seyfert I nuclei, and weak or absent in broad-line radio galaxies
and most QSOs (Osterbrock, 1978).

(ii) Lines have broad often assymmetrical profiles which may
vary from line to line in the same source and even from line to
line of the same element (eg Shuder, 1984).

(iii) A two-photon single transition can photoionize atoms, thus
producing conditions that can explain the "3000 Å bump" which is
superimposed on the ν^{-1} continuum (Malkan and Sargent, 1982).

(iv) The steep Balmer decrements are explained. Lβ excites Hα
and Lγ excites both Hβ and Pα. Excited state populations may
become large enough to render clouds optically thick for Hα .

(v) The weak ($< 2\%$) polarization of most QSOs is directly
explained by off-axis viewing because the dipole moments of atoms
will be directed toward the path of the exciting charge.

REFERENCES

Aller HD, Hodge PE, Aller MF 1981 Astrophys J 248: L5
Angel JRP, Moore RL 1980 Ann NY Acad Sci 336: 55
Angel JRP, Stockman HS 1980 Ann Rev Astron Astrophys 18: 338
Arp H 1983 Astrophys J 271: 479
Arp H 1984 Publ Astron Soc Pac 96 148
Arp H, Sulentic JW, di Tullio G 1979 Astrophys J 229: 489
Arp H, Hazard C 1980 Astrophys J 240: 726
Browne PF 1968 Astrophys Lett 2: 217
Browne PF 1983 Phys Lett 99A: 196
Browne PF 1984 Astron Astrophys: in press
Burbidge G 1979 Nature 282: 451
Cline TL et al 1982 Astrophys J 255: L45
Cruz-Gonzalez I, Huchra JP 1984 Astron J 89: 441
Holmes PA, Brand PWJL, Impey CD, Williams PM 1984 In: Dyson J (ed)
    Active Galactic Nuclei. Manchester University Press, Manchester
Kiplinger AL, Dennis BR, Emslie AG, Frost KJ, Orwig LE 1983
            Astrophys J 265: L99
Kronberg PP, Biermann P, Schwab F 1984 in preparation
Ledden JE, Aller HD 1979 Astrophys J 229: L1
Malkan MA, Sargent WLW 1982 Astrophys J 254: 22
Matthaeus WH, Montgomery D 1981 J Plasma Phys 25: 11
Mazets P, Golenetskii SV, Aptekar RL, Guryan YuA, Il'inskii VN
            1981 Nature 290: 378
Norris JP, Cline TL, Desai UD, Teegarden BJ 1984 Nature 308: 434
Oort JH 1983 Ann Rev Astron Astrophys 21: 373
Osterbrock DE 1978 Phys Scripta 17: 137
Robinson LC 1973 Physical Principles of Far-Infrared Radiation:
            p239. Academic Press, New York.
Rosenbluth M 1957 In: Landshoff RKM (ed) Magnetohydrodynamics:
            p57. Stanford University Press, Stanford
Sargent WLW et al 1979 Astrophys J 230: 49
Schwartz RD 1983 Ann Rev Astron Astrophys 21: 209
Shuder JM 1984 Astrophys J 280: 491
Wardle JFC, Moore RL, Angel JRP 1984 Astrophys J 279: 93
Welter GL, Perry JJ, Kronberg PP 1984 Astrophys J 279: 19
Wiggins J, Drozdowicz Z., Temkin RJ 1978 IEEE J Quant Electr 14:23

VIOLENT EVENTS IN EXTRAGALACTIC OBJECTS

Geoffrey Burbidge
Center for Astrophysics & Space Sciences,
University of California, San Diego
and
Kitt Peak National Observatory*
Tucson, Arizona

We have heard much discussion of new results in the continuing
study of violent, nonthermal events in extragalactic objects. All
observers have interpreted their observations — when they felt so moved
— in terms of the canonical explanation in which black holes, accretion
disks and relativistic jets are invoked.

My role is to attempt to inject some element of reality into the
discussion by describing both what we know and what we don't know.

First a comment about nomenclature. When certain characteristics
of a morphological kind or of a spectroscopic kind are used to define a
class of objects and a name is given to the class, i.e. Seyfert
galaxies, the name should not have the connotation that we know what is
going on. A particularly bad example of that kind of thing is the term
"Star-Burst Nuclei". Here we are talking about the nuclei of galaxies
with strong narrow emission lines and a continuum which some observers
have speculated may be due to recent outbursts of star formation. To
use the term "Star-Burst Nuclei" suggests that we <u>know</u> rather than only
speculate on the origin of the spectrum. Frequent use of this term
will transform yet another speculation into an apparently hard fact
quoted in the textbooks.

Let us turn now to the more general problem associated with the
results of observations of galactic nuclei and QSOs.

*Operated by the Association of Universities for Research in Astronomy
Inc., under contract from the National Science Foundation

Twenty years ago or more we had reached the situation where we understood that the energy released in the nuclear regions of galaxies or QSOs, manifested in a variety of non-thermal forms or in high velocity clouds, is either gravitational energy or energy released in some creation process (cf. the work of Hoyle and Narlikar, and of Ambartsumian). This latter proposal has been largely ignored on the ground that "new physics" should not be invoked until all of the possibilities associated with conventional physics have been exhausted. What has then been done is to consider models which allow the release of large amounts of gravitational energy. In my view, little real progress in understanding has been made since the work of Fowler and Hoyle, and Hoyle, Fowler, Burbidge, and Burbidge which was published in 1963 and 1964. The preferred scenario at present is a massive black hole whose origin is never really worked out, and an accretion disk. It is assumed that matter originally in the form of stars or gas and dust is accreted, and ultimately falls into the black hole with the release of the maximum possible gravitational energy, about 6% of the rest mass.

Now what we see coming from the so-called central machine is a strong continuum in the optical region of the spectrum, sometimes extending into the radio and x-ray spectral regions, and a broad emission line region. Also in many compact objects and in all radio sources, there is evidence that relativistic particles are present in great abundance, and that ejection in preferred directions over great distances up to megaparsecs can occur. Not only this, but in many cases, variability on time-scales of days, months and years is seen in optical wavelengths and at high radio frequencies, and even variability at low radio frequencies now appears to be commonplace.

It was shown in the 1960s' that the variability posed many severe difficulties for any conceivable models for energy generation. The so-called Compton paradox, first discussed by Hoyle, Burbidge and Sargent in 1966, led to the conclusion that either large bulk relativistic motions were present, or that the objects showing these effects, nearly all of which were QSOs, were much closer than was suggested by their redshifts.

VLBI measurements have now shown quite conclusively that a number of the brightest radio QSOs show changes in angular size which indicate highly relativistic bulk motion (so-called superluminal motion) unless the objects are much closer than their redshifts would suggest. Without exception, radio astronomers have chosen to accept the cosmological interpretation of the redshifts, and thus a small industry of superluminal modelmaking has developed.

The ideas are not unrelated to the attempts to make models of the extended radio structures where the preferred scheme is that originally due to Blandford and Rees in 1974. It is argued that a large amount of very hot relativistic plasma is generated (from gravitational energy) and that this is surrounded by a thick disk of cooler gas. The hot gas inflates a cavity and then blows holes out of which it steadily escapes along the rotation axis. The steady state solutions lead to the idea that a pair of Laval nozzles convert hot relativistic plasma into a relativistic but thermally "cool" outflow. It is then argued, without proof, that these streams travel outward very great distances until they finally interact strongly enough with the intergalactic medium to dissipate their energy, which is then used to accelerate the electrons which give rise to the radio emission in the lobes. It is assumed that this energy conversion takes place with high efficiency.

While the relativistic jet models are very popular, evidence is growing that they cannot be correct. In a recent review, deYoung (*Science*, **225**, 677 (1984)) has argued that the observational evidence favors non-relativistic jets with velocities $\sim$ 1000 km sec^{-1}. Apart from this evidence, I have always found it difficult, if not impossible, to understand how relativistic jets, well collimated over hundreds of kiloparsecs, could be stable against various kinds of instabilities. It may be argued that the beam is electrically neutral, and this may increase its stability. But the contrast between what is expected of nature in this situation compared to what is required in fact for confinement and stability in a man-made linear accelerator is so great as to be, for me, quite unbelievable. Confinement by hot or cold thermal gas without instabilities developing also seems highly unlikely.

Problems also arise when one takes seriously the idea of superluminal motions. It has now been shown that one sided radio jets are often seen in QSOs and the nuclei of radio galaxies. Thus both in those cases, and in the classical objects like 3C237 and 3C345 where the changes in the radio structure are attributed to superluminal motions, it is argued that we are looking at relativistic beaming effects, i.e.,

the jet that we see has a large component of relativistic motion directed towards the observer (blue shifted), while the redshifted component is not seen. Now it is known that some compact jets are strongly curved or bent, and this is extremely difficult to understand in terms of projection or effects of precession. A further difficulty is associated with the fact that the highly relativistic bulk motions required must occur, and be maintained, in a region of high density where dissipation is very high, a difficulty originally pointed out by Jones and Burbidge in 1973. In summary, it seems to me that the idea that a black hole or an accretion disk actually gives rise to what we see is fraught with serious physical problems which have not been solved, but have simply been ignored.

Can a case be made for looking at more radical explanations? My answer is yes, because there is a great deal of evidence that the redshifts are not good indicators of distance. This leads me to a few comments about the evidence for non-cosmological redshifts. As is usually the case, at this meeting the evidence that the redshifts are of cosmological origin has been stressed. The strongest evidence of this kind is that some low redshifts QSOs lie adjacent to galaxies with similar redshifts, and that many low redshift QSOs are surrounded by fuzz which is likely to be the underlying galaxies. I do not want to criticize this evidence except to say that in looking for fuzz, astronomers still will not look at a sample of QSOs independent of redshift, but look preferentially at QSOs with redshifts small enough so that if their preferred hypothesis is correct, they will see galaxies, which they claim to do. One QSO with a redshift of 3 with fuzz around it would be enough to stop this bandwagon in its tracks, and that is probably why it is not looked for!

But what I want to stress is that there is by now a great deal of evidence provided by several of us, mostly in modern times by Arp, which shows that there are many QSOs with large redshifts close to bright galaxies not more than 50 Mpc away, and however you do the statistical analysis, the conclusion is that these cannot be accidental configurations, given that the surface density of QSOs is not more than about 1 per square degree at 18^m or about 10 at 20^m. A large number of the QSOs associated with nearby galaxies are bright (18^m). In some cases, several QSOs are found very close to a single bright galaxy. The most striking case is NGC 1073, where there are three high-redshift QSOs lying within 2' of its nucleus.

I analyzed much of this evidence in <u>Nature</u> about two years ago. However, the astronomical community appears determined to ignore such results.

Recently there has been much discussion of close pairs of QSOs with the same redshifts which are considered to be gravitational lenses. For some of the cases described this may be the case. However, completely ignored in the discussion are the many close pairs of QSOs with very different redshift. J.V. Narlikar, A. Hewitt and I have recently carried out a study of all of those pairs with very different redshifts and separations less than 2' and have concluded that they cannot be due to chance.

The existence of the phenomena described above means first that many QSOs are nearby. Thus the energetic mechanisms and the arguments for the existence of relativistic motions in superluminal sources must be re-examined. Second, we have to consider alternative explanations for the redshifts as Narlikar and others have been trying to do. Third, we must admit openly that there is a great deal concerning violent events in galaxies and QSOs that we simply do not understand, and approach the observational evidence with much more open minds. New physics or even a new approach to cosmology is indicated.

Summary

B E J Pagel

Royal Greenwich Observatory, Herstmonceux Castle, Hailsham,
East Sussex BN27 1RP, England

We have come to the end of a very enjoyable week of high thinking and plain living (not too plain, fortunately) at the University where some of the important developments that led to the discovery of quasars were initiated about 30 years ago and which it is gratifying to see still in the forefront of extragalactic radio astronomy at the present time thanks to the magic of MERLIN. It was therefore very appropriate that we should meet here in Manchester to discuss the topic of Active Galactic Nuclei, the last such conference that I attended having been at Cambridge 7 years ago (Hazard and Mitton 1978).

What have we learned this time? To some extent I have the impression that I have unlearned more than I have learned, and that I come away knowing less than I thought I knew before I arrived. Perhaps this is healthy. Anyway, the only thing to do is take a deep breath and carry on.

A question that has been raised several times is: What is an active galactic nucleus? To answer this question, I would remind you that, apart from being the 21st birthday of the observation of 3C273, which was the Rosetta Stone of the quasars, 1984 is also the year of George Orwell who supplied an appropriate slogan for the occasion: <u>All galactic nuclei are active but some are more active than others.</u>

I was especially impressed with the very fascinating Introductory Review by Cyril Hazard who if he is not the onlie begetter is nevertheless and has been for over 20 years the midwife or <u>sagefemme</u> of so many new developments arising from the surveys and subsequent investigations of quasars and other possibly equally interesting objects such as compact HII galaxies. His historical review merged imperceptibly into an account of where we now stand on some very basic questions:-

To begin with, how many QSOs are there? The survey in selected regions is apparently complete and free from observational selection effects, at least for certain types of objects, between redshifts of 1.8 and 3.3 and to a magnitude somewhere between 19 and 19.5. Subject to possible effects of clustering, the number of QSOs found is very large, 60 positive identifications and 80 more candidates in an area of 4 square degrees with a distinct peak in the distribution at z = 2. This invites comparison with the frequently quoted figure of 10 QSOs per square degree down to 20^m. Radio and X-ray surveys in the same area are turning up yet more objects. From this wealth of material he showed us a great variety of spectra and derived a number of conclusions and suggestions, eg.

1. Broad absorption line spectra are common for z > 3, suggesting that this is the usual early phase in their development.

2. Quasars are not passive objects but stir up their environment

Talk given at Workshop on <u>Active Galactic Nuclei</u>,
Manchester University, 2-6 April 1984

something fierce and eject heavy elements.

3.   Quasars come in many different varieties, just as stars do, and may require the same sort of efforts at classification.

One wonders whether one should consider setting up an HD-type catalogue using criteria such as line width, ratio of NV/Lyα, absorption characteristics etc. The imagination boggles at such a task, and in any case the classification system is not ready yet.

4.   He suggested that quasars actually make heavy elements, replacing in this role the Population III objects that have sometimes been postulated. I am by no means convinced that Population III is needed for any purpose, but it is certainly impressive that we look back now to $z = 3.8$ and still find heavy elements present in proportions that cannot be distinguished from normal, ie solar. The implication on the dull old conventional view that my prejudices incline me to support is that galaxies of stars must have been formed at a still higher red shift, and that they developed active nuclei afterwards. Models of this kind were developed by Larson (1974) about 10 years ago.

Apart from these considerations and Geoff Burbidge's talk just now, cosmology has been but barely touched on at this meeting, though one did feel it hovering in the background, that is, literally, in the X-ray and γ-ray background. One has the feeling that we are going to need a lot more data, especially at the highest energies, before we can definitely say the background is entirely due to such-and-such objects, or alternatively that such-and-such objects violate the constraints imposed by the background to an embarrassing extent. It is an interesting corner to watch, however. We also heard from Schneider some general considerations on the effect of gravitational lensing on source counts.

The main thrust of the conference was devoted to the nature of active nuclei in themselves: what have we learned from observations at various different wavelengths, what are the detailed processes underlying the phenomena, and what is the nature of the basic engine that powers all these processes?

We had some problems in defining what we mean by the various classes of active nuclei and indeed it is difficult to place them in any coherent order with respect to their properties and their environment. Not only do we have a tremendous zoo with the various different beasts, but we don't even know how many kinds of beasts there are, or what sort of cages they are kept in. What are the host galaxies of quasars like? There has been quite a lot of work on this recently, but apart from Mike Perryman's talk the question came up here mostly in connection with the Blazars. It is well known that Seyfert galaxies are spirals, but broad-line radio galaxies which are, of course, ellipticals, have optical properties much like Seyfert 1's and narrow line radio galaxies much like Seyfert 2's or Liners. Anyway, I feel that any summary talk should try to make some sort of presentation of the beasts in the zoo, which I shall do in much the same way as Roger Blandford did yesterday.

Table 1 is a sketch list, without prejudice to any interrelation, merging or splitting that might eventually transpire between the classes. Most groups show definite evidence for non-thermal phenomena, either as radio emission, hard X-rays and γ-rays or the quasi-nonthermal phenomenon of a power-law continuum in the optical. Until a week or two ago I would have asserted that the line spectra of all except the starburst nuclei themselves indicated a power law continuum, but as was mentioned by Mrs Díaz, there is some very new work by Terlevich and Melnick (1984) which may provide an alternative explanation of the high-excitation line spectra of Seyfert galaxies and Liners, for which neither conventional photo-ionization by O stars as in HII regions, nor

<u>Table 1: Beasts in the zoo</u>

Beast		**Host galaxies**	**Characteristics**
QSOs	Radio quiet	S?	UVOIR, BAL, NAL, BLR, NLR (X-r)
	Radio loud	E?	UVOIR, BAL, NAL, BLR, NLR, X-r γ-r, double RS, jets, superlum
BLAZARS	OVV	E?	γ-r, double RS, jets, superlum Pol. var.
	BL	E	UVOIR, NAL, X-r, Pol., var. jets (end-on?)
BLRG(N-gal)		E, Ep	UVOIR, dust, double RS, jets, BLR
Sy 1, 1.5, 1.8		Sa, b	UVOIR, BLR, NLR, X-r, γ-r
(WARMERS?)	NLRG	E, Ep	RS, jets, NLR (often less exc.)
	Sy 2	Sa, b	UVOIR, (RS), NLR, stars, HII
	LINERS	Sa,b,E	Stars (nucl. RS), BLR, NLR (less exc.) HII (arms near nucl.), (X-r)
	STARBURST	Sc, I	Stars, HII, dust, IR, SN, X-r
GC (quiescent BH?)		Sb	(RS), winds, shocks, IR, stars, HII

shock-excitation, provide a satisfactory explanation. The idea is that very massive stars with high metallicity, such as may perhaps be expected to be formed in the nuclei of early-type spirals and giant ellipticals, lose mass to such an extent as to reveal carbon-oxygen cores with effective temperatures up to 200,000 K which they call 'Warmers', and the resulting radiation field produces much the same sort of line spectrum as does the conventional power-law continuum. An attractive feature of this approach is that it explains why Seyferts and Liners appear in the nuclei of early-type spirals, because of the high metallicity (obviously an argument specially designed to appeal to me!), but an unattractive feature is that it offers no explanation of the featureless power-law type of continuum that is actually seen in Seyferts of both types in the optical region. One should remember, of course, that other non-thermal phenomena, like radio and X-ray emission, can come just as well from a starburst with its associated supernovae (of which we heard a graphic account in the case of M82 from Unger and from Kronberg) and binary X-ray sources which as Andy Fabian reminded us have many features in common with the most active of active galactic nuclei.

In any case, I believe that there is an association between non-thermal activity and starbursts, such as we see in typical Liners and Seyfert galaxies. NGC 1365 and NGC 1097, which Malcolm Smith discussed, are cases in point. In each case we see a nucleus, the first with an intermediate Seyfert spectrum complete with broad Balmer lines, high ionisation and a power-law continuum, surrounded by dusty supergiant HII regions (Edmunds and Pagel 1982), and in the second case we see a Liner nucleus surrounded by inner spiral arms consisting of dusty HII regions (Meaburn et al. 1981, Phillips et al. 1984). Similar effects are seen in NGC 1068, with extended high-excitation emission-line regions and giant HII regions rubbing shoulders. NGC 1097 is also a member of an interacting system with mysterious optical jets that have been illustrated by Wolstencroft et al. (1975), Arp (1976) and Lorre (1978) and were referred to in Tom Ray's paper on the Haro galaxies. Several times at this meeting we have been reminded by Rod Davies and others that interacting systems are prone to showing signs of activity, sometimes in the form of starbursts and sometimes in the forms of radio emission and Seyfert-like activity, eg NGC 2992. This reminds me that, at a still lower level, Kormendy and Norman (1979) some years ago also pointed out that the best spiral structures are associated either with the presence of companions or with the presence of a bar; 1365 and 1097 are in fact both barred spirals, as is NGC 5643 which was described to us by Simon Morris.

This all suggests that the different types of activity are closely related, presumably to some sort of perturbation that sends down material to feed the monster lying in wait in the nucleus. We could be seeing various stages in some sequence of events involving nuclear activity which in turn gives rise to star formation, or vice-versa as suggested by Malcolm Smith. Perhaps as David Axon suggested to me in private conversation, in a starburst system like M82, we are seeing a lot of activity after the villain of the piece has gone into hibernation! Furthermore, the increase in virulence of activity of all kinds with redshift could be attributed to all galaxies having been closer together at that time.

That is enough of riding my personal hobby horses. What other considerations can we take away from the results that have been presented at this meeting?

We have heard quite a lot about the radio observations using the new facilities. Porcas told us that all quasars have a very compact radio core, often with jets, and there have been more results on superluminal expansion: the core retains its position and in one case there is marginal evidence for superluminal contraction. Theoreticians I have spoken to agree that this is not difficult to explain, but they are not convinced that the result is significant. Rod Davies and Alan Pedlar reported interesting radio studies of optically selected samples of Sbc galaxies and Seyfert galaxies respectively. In this connection I should like to recommend the presentation used by Osterbrock (1984) in a recent article in Q.J.R.A.S. where he forms a radio-optical colour index P-RF and contrasts the results for different kinds of Seyfert galaxies. A bolometric index would be even better. I think this is a useful way to decide whether a galaxy is active in the radio sense, whether it be a Seyfert galaxy or not. Osterbrock's diagram shows clearly that the narrow-line region is the one most closely associated with radio emission from Seyfert galaxies and from Alan Pedlar, following up earlier studies by Andrew Wilson and others, we learned that this may be a region where plasma dribbles out from the nucleus and attains pressure equilibrium with the optically emitting gas.

Optical, u.v. and infra-red spectroscopy were reviewed for us by Bob Carswell, Mike Penston and Malcolm Smith. Bob Carswell explained how difficult it is to understand the excessively smooth and similar profiles of broad emission lines in QSOs, but managed to find some asymmetries and excitation-dependent line shifts that indicate expansion combined with obscuration. Mike Penston was rather more optimistic. He took us entertainingly through the Fe II and Lyα/Hβ problems, cheerfully demolished the 2200 Å dereddening technique and updated by a factor of 10 the careful calculation by the EEC of the mass of the black hole in NGC 4151 (assuming the broad-line region to be gravitationally bound), which now comes out at precisely 5×10^8 solar masses if I understand correctly. The response of the emission lines to changes in the exciting spectrum has been modelled in somewhat more detail by Robinson. However, the EEC method of fitting the continuum to a power law plus other things is different from that used by Oke et al. and by Wamsteker and Gilmozzi in fitting other objects that are not intrinsically different. Penston also pointed out a slight problem with the gravitationally lensed double quasar, where the optical and ultra-violet observations disagree as to which component is varying. Malcolm Smith presented us with a veritable explosion of interesting infra-red results which I hardly feel capable of summarising beyond the ideas that I have pinched from him already. Important points emerging from his talk, and from that of Andy Lawrence of the Che Guevara cap, are that broad-line radio galaxies are dusty with a power law in the infra-red, whereas narrow-line radio galaxies and Liners just show a normal stellar population; Seyfert 1s likewise show a power law with a featureless spectrum (normally), whereas Seyfert 2s are not a well-defined class, being either like Seyfert 1s or a mixture of old stars with HII-region or equivalently starburst type spectrum with recombination, free-free and hot dust.

Optical polarisation was reviewed for us by P G Martin in a very clear fashion, contrasting the highly polarised BL Lac objects and somewhat similar OVV quasars (known collectively as Blazars), where we may see a relativistic jet end on, as was strongly suggested by Perez-Fournon and Antonucci but not proven, with low polarisation quasars in which there is just some sort of scattering phenomenon that also occurs with differences in detail among Seyfert galaxies. A high optical polarisation of 12.5% in the end of the jet of 3C273 was found by Roeser and we wait on tenterhooks to see whether the electric vector is parallel or perpendicular to the jet. We heard more about Blazars in a study of polarisation variations by Peter Brand and in a very elegant presentation of sub-mm wave observations by Gear, in which the remarkably flat spectrum does not necessarily require there to have been a cosmic conspiracy. In fact, according to Schlickeiser it does not even require self-absorption.

Andy Fabian gave an exciting account of X-ray emission and the interpretation of the constant spectral index of 0.65 as a consequence of pair production, and Dean extended this to γ-ray energies where additional data are eagerly awaited. These are important regions because in some objects at least the power per octave has a peak at 1 Mev where one can expect to look pretty deeply into the basic engine; on the other hand, the idea of a truly naked engine would seem to be limited by the pair production process.

Roger Blandford, apart from anticipating substantial chunks of the present summary talk, gave us a lucid explanation of how various types of black-hole accretion disks can be tailored to fulfil the requirements of different sorts of thermal and non-thermal phenomena, the thermal model bearing a close resemblance to the supermassive objects first postulated by Hoyle and Fowler 20 years ago or more. This raises the question of how fundamental is the distinction between radio-noisy and radio-quiet objects which in other respects appear remarkably similar. One thinks of the Sun, for instance, which is not so very different during radio bursts from what it is when it is radio quiet. If there is a fundamental distinction, then the zoo table suggests that it may be related to environment, in particular whether this is a spiral or elliptical galaxy, and according to Roger Blandford's interpretation this comes out in having totally different black-hole masses and mass accretion rates in the various systems. From Chakrabarty's paper it seems that the production of collimated relativistic jets from thick accretion disks around rotating black holes is no problem.

Ian Evans, Leon Mestel and Judith Perry discussed theoretical aspects of the formation and acceleration of the broad-line region by radiation-filled bubbles, radiation-driven winds and shocks resulting from obstacles encountered on the way. It seems possible that such models can account for some features of the broad-line profiles, but this seems to be a difficult area in which to come to any definite conclusions. If these models are relevant, then presumably one cannot come to conclusions about the mass of any central compact object on the basis of line widths. We were also treated to a display of the superb CFHT picture by Wlérick and his colleagues of the jet in M 87 and to an interpretation by Wilson in terms of Mach disks and other laboratory shock phenomena.

Conclusion

To conclude this summary I should like to address briefly the question: Where do we go from here? What are the likely future developments over the next few years?

This is very much the kind of subject that is driven by observation rather than theory. The black-hole-accretion-disk model supplies a plausible explanatory framework that can be adapted, as Roger Blandford explained to us, to account for the various funny things that people observe, but it does not make very specific predictions and has not yet been subjected to crucial observational tests, as Geoff Burbidge pointed out. The observations in turn have been driven by technical developments: the heroes of this conference have been the VLA, MERLIN, VLBI, UKIRT, the large optical telescopes and their associated instrumentation, IUE, the Einstein Observatory, HEAOA, balloons and γ-ray satellites, with IRAS data in the pipeline and further detailed results to be awaited from EXOSAT.

Where, then, will progress come from in the next few years? One healthy trend is the convergence of different astronomical techniques which has been going on for some time: nowadays it is a matter of routine to follow up optical observations with radio observations, X-ray observations with infra-red observations etc, so that we are studying well-defined objects and looking at their spectra over the entire frequency range, rather than mere "sources". The new objects coming out of the surveys will keep all our major telescopes occupied for a long time to come. As to the next really major advances in our field, it is pretty obvious, and Roger Blandford has already pointed out, that these are likely to come from the Space Telescope, especially as a consequence of the high spatial resolution of the order of 0.1 arcsec and more particularly by the use of this in conjunction with the long-slit facility on the Faint Object Camera. Of course there are lots of things that people want to do with the Space Telescope in general and with the Faint Object Camera in particular, and I should just like to focus on a particular application that has a bearing on problems raised at this conference, was itself pointed out by Roger Blandford yesterday and was indeed the main reason why the long slit was put in.

One specific prediction of the black-hole model, made first by Lynden-Bell (1969), I believe, is that every major galaxy should have a black hole in the centre. It is not clear what the masses of such black holes should be. In M87, Sargent et al. (1978) claim to have detected a compact mass of several times 10^9 solar masses, although this has been disputed on observational grounds by Dressler (1980) and on theoretical grounds by Binney and Mamon (1982). At this conference we have heard various estimates of masses of black holes in active galaxies, ranging from 10^6 to several times 10^8 solar masses. A mass of several times 10^6 has been suggested in the centre of our own Galaxy. For galaxies within a few Mpc, the Space Telescope resolution of 0.1 arcsec corresponds to a distance of the order of 1 pc at which distance from a compact object with 2.10^6 solar masses we have a circular velocity of 100 km s^{-1}, which we might expect to show up in nearby galaxies either as a rotational velocity or as an enhanced velocity dispersion. Thus if there are reasonably massive black holes in the nearer galaxies, ranging in activity from, say, NGC 4151 to humble "Liners" like M 81, it should be possible to detect these, essentially by spatially resolving the broad-line region, assuming it to be gravitationally bound.

That is just one idea and all of you undoubtedly have various ideas of your own which will cause a still richer and perhaps more coherent picture of active galactic nuclei to emerge in a few years' time.

It remains for me to thank John Dyson and his colleagues for organising this enjoyable meeting and looking after all our needs with a maximum of efficiency and a minimum of fuss. Thank you very much.

References

Arp HC 1976 Astrophys. J. $\underline{207}$, L147

Binney J and Mamon GA 1982 Mon. Not. R. astr. Soc. $\underline{200}$ 361

Dressler A 1980 Astrophys. J. $\underline{240}$ L11

Edmunds MG and Pagel BEJ 1982 Mon. Not. R. astr. Soc $\underline{198}$ 1089

Hazard C and Mitton S 1978 <u>Active Galactic Nuclei</u> Cambridge University
    Press

Kormendy J and Norman CA 1979 Astrophys. J. $\underline{233}$ 539

Larson RB 1974 Mon. Not. R. astr. Soc $\underline{166}$ 585

Lorre JJ 1978 Astrophys. J $\underline{222}$ L99

Lynden-Bell D 1969 Nature $\underline{223}$ 690

Meaburn J Terrett DL, Theokas A and Walsh JR 1981 Mon. Not. R. astr.
    Soc. $\underline{195}$ 39

Osterbrock DE 1984, Q.J.R. astr. soc. $\underline{25}$, 1

Phillips MM, Pagel BEJ, Edmunds MG and Díaz A 1984 Mon. Not. R. astr.
    Soc. in press

Sargent WLW, Young PJ, Boksenberg A, Shortridge K, Lynds CR and Hartwick
    FDA 1978 Astrophys. J. $\underline{221}$ 731

Terlevich R and Melnick J 1984 in preparation

Wolstencroft RD and Zealey WJ 1975 Mon. Not. R. astr. Soc. $\underline{173}$ 51P